MODERN
Materia Medica and
THERAPEUTICS

MAVEN BOOKS

Chennai   Trichy   Tirunelveli   New Delhi

MODERN
Materia Medica and
THERAPEUTICS

A. A. STEVENS, A.M., M.D.

Lecturer on Physical Diagnosis in the University of Pennsylvania;
Physician to the Episcopal Hospital and to St. Agnes's Hospital;
Fellow of the College of Physicians of Philadelphia, etc.

Chennai   Trichy   Tirunelveli   New Delhi

MAVEN BOOKS

An Imprint of **MJP Publishers**

ISBN 978-93-87867-72-7 **MAVEN Books**

All rights reserved No. 44, Nallathambi Street,
Printed and bound in India Triplicane, Chennai 600 005

MJP   773 © Publishers, 2020

Publisher : **C. Janarthanan**

Publisher's Note

The legacy of a country is in its varied cultural heritage, historical literature, developments in the field of economy and science. The top nations in the world are competing in the field of science, economy and literature. This vast legacy has to be conserved and documented so that it can be bestowed to the future generation. The knowledge of this legacy is slowly getting perished in the present generation due to lack of documentation.

Keeping this in mind, the concern with retrospective acquiring of rare books has been accented recently by the burgeoning reprint industry. MAVEN Books is gratified to retrieve the rare collections with a view to bring back those books that were landmarks in their time.

In this effort, a series of rare books would be republished under the banner, "MAVEN Books". The books in the reprint series have been carefully selected for their contemporary usefulness as well as their historical importance within the intellectual. We reconstruct the book with slight enhancements made for better presentation, without affecting the contents of the original edition.

Most of the works selected for republishing covers a huge range of subjects, from history to anthropology. We believe this reprint edition will be a service to the numerous researchers and practitioners active in this fascinating field. We allow readers to experience the wonder of peering into a scholarly work of the highest order and seminal significance.

MAVEN Books

MODERN

MATERIA MEDICA

AND

THERAPEUTICS

PREFACE TO THE THIRD EDITION.

SINCE the appearance of the last edition of this book such rapid advances have been made in materia medica and therapeutics that the author has felt it imperative to rewrite the work entirely. Instead of considering the drugs in alphabetic order, as in previous editions, he has thought it best in the present revision to classify them according to their pharmacologic action. Although in the present unsettled state of pharmacology such a classification must necessarily be a very imperfect one, it has seemed to the author to possess, notwithstanding its imperfections, certain advantages in that it aids the student to correlate established facts and to apply them more readily to the treatment of disease.

An earnest effort has been made to bring the articles on the old and generally approved remedies thoroughly up to date. Of the many new drugs, only those have been considered which have been shown by competent observers to possess real merit and to be worthy of a more extended trial at the hands of the profession.

In order to economize space, references to authorities have been omitted; but the author wishes to fully acknowledge his indebtedness to the numerous writers of works and monographs on general and special therapy which he has freely consulted.

314 S. SIXTEENTH ST., PHILADELPHIA.

CONTENTS.

MATERIA MEDICA.

APPLIED THERAPEUTICS.

MODERN

MATERIA MEDICA

AND

THERAPEUTICS

MODERN

MATERIA MEDICA AND THERAPEUTICS.

GENERAL CONSIDERATIONS.

Materia medica is that branch of medical science which treats of the remedies used in medicine. It deals with their sources, physical characters, composition, preparations, and doses.

Pharmacy is the art of preparing, compounding, and dispensing medicines.

Pharmacodynamics is the study of the action of drugs upon living organisms. Our knowledge of the action of drugs upon human beings is obtained in two ways : First, by clinical experience, and secondly, by comparative study of their action on the lower animals.

Therapeutics is that branch of medicine which deals with the application of remedies to disease. In its broadest sense, it has to do not only with drugs, but with all other agents which are of service in restoring health, prolonging life, or affording comfort to the sick.

Treatment based solely on clinical experience, and instituted without reference to the physiologic action of the remedies employed, constitutes *empirical therapeutics.* Thus, the giving of opium in diabetes and of salicylic acid in rheumatism, simply because experience has taught that these drugs often do good, are illustrations of empirical therapeutics.

The treatment of disease by drugs which are expected, from a knowledge of their physiologic actions, to antagonize certain known pathologic conditions, constitutes *rational therapeutics.* The employment of chloral in the convulsions of tetanus may be cited as an illustration of rational therapeutics, since it is expected that chloral, through its sedative effect, will counteract the spinal irritation which is manifesting itself in convulsions.

COMPOSITION OF DRUGS.

The composition of the inorganic drugs is indicated by their names and formulas. The composition of the organic drugs is often exceedingly complex. They may contain fixed oils, volatile or essential oils, resins, oleoresins, balsams, gums, gum-resins, alkaloids, glucosids, neutral principles, and organic acids.

Fixed oils are oils obtained by simple pressure, and are not readily volatilized. Croton oil, cod-liver oil, castor oil, and olive oil are examples of fixed oils.

Volatile or essential oils are obtained chiefly by distillation. Oils of cinnamon, turpentine, mint, eucalyptus, juniper, cloves, copaiba, and lemon are volatile oils.

Resins are exudations allied to and probably derived from volatile oils. They are oxidized hydrocarbons, amorphous, brittle, insoluble in water, but freely soluble in alcohol. They unite with alkalies to form soaps. They melt at a low heat, and solidify again on cooling. The chief official resins are those of copaiba, jalap, podophyllum, scammony, and guaiacum-wood.

Oleoresins are stable mixtures of a volatile oil and a resin. Their two constituents can be separated by distillation. The oleoresins used in medicine are those of aspidium, capsicum, cubeb, lupulin, ginger, and black pepper.

Balsams are resins or oleoresins containing benzoic acid or cinnamic acid. The chief balsams are those of peru, tolu, and storax.

Gums are desiccated exudations obtained by incising the limbs and branches of certain plants. They contain two principles: arabin, which is soluble in water, and bassorin, which is not soluble in water, but swells up in it. The most important gums are those of acacia and tragacanth.

Gum-resins are mixtures of gum with resins or oleoresins. Asafetida, ammoniac, myrrh, gamboge, and scammony are examples of gum-resins.

Alkaloids are nitrogenous compounds occurring in plants, resembling ammonia in being basic and capable of forming salts with acids. In many instances, as in the case of atropin, morphin, aconitin, and quinin, the alkaloid represents to a great extent the active properties of the drug.

Glucosids are compounds which, treated with mineral acids, alkalies, or certain ferments, are resolved into a sugar, an acid, and sometimes another organic principle. They may be either neutral or acid in reaction, and nearly all are soluble in alcohol. The chief glucosids are arbutin, amygdalin, glycyrrhizin, convallamarin, adonidin, digitalin, digitoxin, and strophanthin.

Neutral principles are neutral or feebly acid constituents of plants, differing from glucosids in not being resolvable into glucose, and from alkaloids in not being precipitated by tannic acid or mercuric potassium iodid. Aloin, elaterin, santonin, picrotoxin, salicin, and chrysarobin belong to this class of compounds.

Organic Acids.—Many vegetable drugs contain organic acids, either free or combined with alkaloids or inorganic bases. The organic vegetable acids of most value in medicine are gallic, tannic, salicylic, benzoic, citric, acetic, and tartaric acids.

PREPARATIONS OF DRUGS.

Vinegars (aceta) are liquid preparations made by treating vegetable drugs with diluted acetic acid. Only two are official:

Acetum Opii (vinegar of opium).

Acetum Scillæ (vinegar of squill).

Waters (aquæ) are aqueous solutions of volatile substances. The waters most commonly used are:

Aqua Ammoniæ (water of ammonia).

Aqua Ammoniæ Fortior (stronger water of ammonia).

Aqua Anisi (anise-water).

Aqua Aurantii Florum (orange-flower water).

Aqua Camphoræ (camphor-water).

Aqua Chlori (chlorin-water).

Aqua Chloroformi (chloroform-water).

Aqua Cinnamomi (cinnamon-water).

Aqua Fœniculi (fennel-water).

Aqua Hydrogenii Dioxidi (solution of hydrogen dioxid).

Aqua Menthæ Piperitæ (peppermint-water).

Aqua Rosæ (rose-water).

Solutions (liquores) are solutions of non-volatile substances in water. The official solutions contain only inorganic salts. Among the most important of this class of preparations are:

Liquor Acidi Arsenósi (solution of arsenous acid).

Liquor Ammonii Acetatis (solution of ammonium acetate, spirit of Mindererus).

Liquor Arseni et Hydrargyri Iodidi (solution of arsenic and mercuric iodid, Donovan's solution).

Liquor Calcis (solution of lime, lime-water).

Liquor Ferri et Ammonii Acetatis (solution of iron and ammonium acetate, Basham's mixture).

Liquor Ferri Subsulphatis (solution of ferric subsulphate, Monsel's solution).

Liquor Iodi Compositus (compound solution of iodin, Lugol's solution).

Liquor Magnesii Citratis (solution of magnesium citrate).

Liquor Plumbi Subacetatis Dilutus (diluted solution of lead subacetate, lead-water).

Liquor Potassæ (solution of potassa).

Liquor Potassii Arsenitis (solution of potassium arsenite, Fowler's solution).

Liquor Potassii Citratis (solution of potassium citrate, neutral mixture).

Liquor Sodæ (solution of soda).

Decoctions (decocta) are liquid preparations made by boiling vegetable drugs for fifteen minutes in a closely covered vessel, allowing to cool, and then straining. They should be freshly prepared, since they readily decompose. Only two decoctions are official :

Decoctum Cetrariæ (decoction of cetraria).

Decoctum Sarsaparillæ Compositum (compound decoction of sarsaparilla).

Infusions (infusa) are liquid preparations made by adding to vegetable substances hot or cold water, allowing the mixture to stand for a certain period, and then straining it. The large dose and the tendency to decompose are disadvantages. The official infusions are :

Infusum Cinchonæ (infusion of cinchona).

Infusum Digitalis (infusion of digitalis).

Infusum Pruni Virginianæ (infusion of wild cherry).

Infusum Sennæ Compositum (compound infusion of senna, black draught).

Mixtures (misturæ) are liquid preparations holding in suspension medicinal substances. In some mixtures mucilage or syrup is used to prevent rapid precipitation of the insoluble substance. Mixtures should be well shaken before being administered. The official mixtures are :

Mistura Cretæ (chalk mixture).

Mistura Ferri Composita (compound iron mixture, Griffith's mixture).

Mistura Glycyrrhizæ Composita (compound mixture of glycyrrhiza, brown mixture).

Mistura Rhei et Sodæ (mixture of rhubarb and soda).

Mucilages (mucilagines) are aqueous solutions of gums or other mucilaginous substances. They are used as emollients, as excipients for pills, and for suspending insoluble substances in liquids.   The official mucilages are:

Mucilago Acaciæ (mucilage of acacia).

Mucilago Sassafras Medullæ (mucilage of sassafras pith).

Mucilago Tragacanthæ (mucilage of tragacanth).

Mucilago Ulmi (mucilage of elm).

Syrups (syrupi) are concentrated aqueous solutions of sugar containing medicinal or flavoring agents.   The official syrups in most common use are :

Syrupus Acaciæ (syrup of acacia).

Syrupus Acidi Citrici (syrup of citric acid).

Syrupus Acidi Hydriodici (syrup of hydriodic acid).

Syrupus Aurantii (syrup of orange).

Syrupus Calcii Lactophosphatis (syrup of calcium lacto-phosphate).

Syrupus Ferri Iodidi (syrup of ferrous iodid).

Syrupus Ferri, Quininæ, et Strychninæ Phosphatum (syrup of the phosphates of iron, quinin, and strychnin).

Syrupus Hypophosphitum (syrup of hypophosphites).

Syrupus Ipecacuanhæ (syrup of ipecac).

Syrupus Lactucarii (syrup of lactucarium).

Syrupus Pruni Virginianæ (syrup of wild cherry).

Syrupus Rhei (syrup of rhubarb).

Syrupus Rhei Aromaticus (aromatic syrup of rhubarb).

Syrupus Sarsaparillæ Compositus (compound syrup of sar-saparilla).

Syrupus Scillæ (syrup of squill).

Syrupus Scillæ Compositus (compound syrup of squill, Coxe's hive syrup).

Syrupus Senegæ (syrup of senega).

Syrupus Tolutanus (syrup of tolu).

Syrupus Zingiberis (syrup of ginger).

Simple syrup (Syrupus) consists of water and 85 per cent. (by volume) of sugar.

Elixirs (elixiria) are sweetened aromatic alcoholic preparations.   The official elixirs are :

Elixir Aromaticum (aromatic elixir).

Elixir Phosphori (elixir of phosphorus).

Spirits (spiritus) are alcoholic solutions of volatile substances.   The most important spirits are :

Spiritus Ætheris Compositus (compound spirit of ether, Hoffmann's anodyne).

Spiritus Ætheris Nitrosi (spirit of nitrous ether).

Spiritus Ammoniæ (spirit of ammonia).

Spiritus Ammoniæ Aromaticus (aromatic spirit of ammonia).

Spiritus Camphoræ (spirit of camphor).
Spiritus Chloroformi (spirit of chloroform).
Spiritus Cinnamomi (spirit of cinnamon).
Spiritus Frumenti (whiskey).
Spiritus Glonoini (spirit of glonoin, spirit of nitroglycerin).
Spiritus Juniperi Compositus (compound spirit of juniper).
Spiritus Lavandulæ (spirit of lavender).
Spiritus Limonis (spirit of lemon).
Spiritus Menthæ Piperitæ (spirit of peppermint).
Spiritus Vini Gallici (brandy).

Tinctures are alcoholic or hydro-alcoholic solutions of non-volatile (except iodin) substances. They are made by simple solution, maceration, percolation, or maceration and percolation. They are not so strong as the fluid extracts, and, unlike the latter, they are not all of the same definite strength. Most of the tinctures are made with water and alcohol, but some are made with undiluted alcohol. The following are among the most important tinctures that are made with alcohol and water:

Tinctura Aconiti (tincture of aconite).
Tinctura Aurantii Amari (tincture of bitter orange peel).
Tinctura Belladonnæ Foliorum (tincture of belladonna leaves).
Tinctura Calumbæ (tincture of calumba).
Tinctura Capsici (tincture of capsicum).
Tinctura Cardamomi Composita (compound tincture of cardamom).
Tinctura Catechu Composita (compound tincture of catechu).
Tinctura Cinchonæ (tincture of cinchona).
Tinctura Cinchonæ Composita (compound tincture of cinchona, Huxham's tincture).
Tinctura Colchici Seminis (tincture of colchicum seed).
Tinctura Digitalis (tincture of digitalis).
Tinctura Gelsemii (Tincture of gelsemium).
Tinctura Gentianæ Composita (compound tincture of gentian).
Tinctura Guaiaci Ammoniata (ammoniated tincture of guaiac).
Tinctura Hyoscyami (tincture of hyoscyamus).
Tinctura Kino (tincture of kino).
Tinctura Krameriæ (tincture of krameria).
Tinctura Lavandulæ Composita (compound tincture of lavender).
Tinctura Lobeliæ (tincture of lobelia).
Tinctura Nucis Vomicæ (tincture of nux vomica).

Tinctura Opii (tincture of opium, laudanum).

Tinctura Opii Camphorata (camphorated tincture of opium, paregoric).

Tinctura Opii Deodorati (tincture of deodorized opium).

Tinctura Quassiæ (tincture of quassia).

Tinctura Rhei (tincture of rhubarb).

Tinctura Stramonii Seminis (tincture of stramonium seed).

Tinctura Strophanthi (tincture of strophanthus).

Tinctura Sumbul (tincture of sumbul).

Tinctura Valerianæ Ammoniata (ammoniated tincture of valerian).

The following official tinctures are made with undiluted alcohol:

Tinctura Asafœtidæ (tincture of asafetida).

Tinctura Aurantii Dulcis (tincture of sweet orange-peel).

Tinctura Benzoini (tincture of benzoin).

Tinctura Benzoini Composita (compound tincture of benzoin).

Tinctura Cannabis Indicæ (tincture of Indian cannabis).

Tinctura Cantharidis (tincture of cantharides).

Tinctura Cimicifugæ (tincture of cimicifuga).

Tinctura Cubebæ (tincture of cubeb).

Tinctura Guaiaci (tincture of guaiac).

Tinctura Iodi (tincture of iodin).

Tinctura Physostigmatis (tincture of physostigma).

Tinctura Veratri Viridis (tincture of veratrum viride).

Fluid Extracts (extracta fluida) are liquid preparations (alcoholic or hydro-alcoholic) of organic drugs, so made that one cubic centimeter represents the active properties of one gram of the crude drug. The most important fluid extracts are:

Extractum Belladonnæ Radicis Fluidum (fluid extract of belladonna root).

Extractum Buchu Fluidum (fluid extract of buchu).

Extractum Cannabis Indicæ Fluidum (fluid extract of cannabis indica).

Extractum Cimicifugæ Fluidum (fluid extract of cimicifuga).

Extractum Cinchonæ Fluidum (fluid extract of cinchona).

Extractum Cocæ Fluidum (fluid extract of coca).

Extractum Conii Fluidum (fluid extract of conium).

Extractum Ergotæ Fluidum (fluid extract of ergot).

Extractum Gelsemii Fluidum (fluid extract of gelsemium).

Extractum Glycyrrhizæ Fluidum (fluid extract of glycyr-rhiza).

Extractum Hydrastis Fluidum (fluid extract of hydrastis).

Extractum Hyoscyami Fluidum (fluid extract of hyoscyamus).

Extractum Nucis Vomicæ Fluidum (fluid extract of nux vomica).

Extractum Pilocarpi Fluidum (fluid extract of pilocarpus).

Extractum Rhamni Purshianæ Fluidum (fluid extract of rhamnus purshiana).

Extractum Rhois Glabræ Fluidum (fluid extract of rhus glabra).

Extractum Sarsaparillæ Fluidum (fluid extract of sarsaparilla).

Extractum Sarsaparillæ Fluidum Compositum (compound fluid extract of sarsaparilla).

Extractum Scoparii Fluidum (fluid extract of scoparin).

Extractum Sennæ Fluidum (fluid extract of senna).

Extractum Spigeliæ Fluidum (fluid extract of spigelia).

Extractum Uvæ Ursi Fluidum (fluid extract of uva ursi).

Extractum Veratri Viridis Fluidum (fluid extract of veratrum viride).

Wines (vina) are alcoholic liquids made by fermenting the juice of fresh grapes. *Medicated wines* are alcoholic preparations in which white wine is used as a menstruum. The official wines are:

Vinum Album (white wine).

Vinum Rubrum (red wine).

Vinum Antimonii (wine of antimony).

Vinum Colchici Radicis (wine of colchicum root)

Vinum Colchici Seminis (wine of colchicum seed)

Vinum Ergotæ (wine of ergot).

Vinum Ferri Amarum (bitter wine of iron).

Vinum Ferri Citratis (wine of ferric citrate).

Vinum Ipecacuanhæ (wine of ipecac).

Vinum Opii (wine of opium).

Emulsions (emulsa) are aqueous preparations in which, by the aid of some mucilaginous material, insoluble, oily, or resinous substances are suspended in the form of minute globules or particles. The chief excipients used for making emulsions are acacia, tragacanth, and yelk of egg. Alcoholic preparations in large quantity are incompatible with emulsions, since they precipitate the excipient, the gum, or egg. The official emulsions are:

Emulsum Ammoniaci (emulsion of ammoniac).

Emulsum Amygdalæ (emulsion of almond).

Emulsum Asafœtidæ (emulsion of asafetida).

Emulsum Chloroformi (emulsion of chloroform).

Honeys (mellita) are liquid preparations in which honey is used as a menstruum. The official honeys are:

Mel (honey).

Mel Despumatum (clarified honey).

Mel Rosæ (honey of rose).

Liniments (linimenta) are liquid preparations containing oleaginous substances, and intended for external application. With the exception of belladonna liniment and lime liniment, which are used as sedative applications, all the official liniments are of a stimulating character, and are to be applied with friction. The official liniments are:

Linimentum Ammoniæ (ammonia liniment).

Linimentum Belladonnæ (belladonna liniment).

Linimentum Calcis (lime liniment, carron oil).

Linimentum Camphoræ (camphor liniment).

Linimentum Chloroformi (chloroform liniment).

Linimentum Saponis (soap liniment).

Linimentum Saponis Mollis (liniment of soft soap, tincture of green soap).

Linimentum Sinapis Compositum (compound liniment of mustard).

Linimentum Terebinthinæ (turpentine liniment).

Lotions (lotiones) are weak medicated solutions or mixtures for external use. When intended for the eyes, they are known as *collyria*. Collyria are usually applied by means of a pipet with a rubber bulb. There are no official lotions, but the following are in common use:

Lotio Plumbi et Opii (lead-water and laudanum: lead acetate, 120 gr. (8.0 gm.); tincture of opium, ½ fl. oz. (15 c.c.); water, 16 fl. oz. (475 c.c.)).

Lotio Hydrargyri Flava (yellow wash: corrosive sublimate, 24 gr. (1.5 gm.); lime-water, 16 fl. oz. (475 c.c.)).

Lotio Hydrargyri Nigra (black wash: calomel, 64 gr. (4.0 gm.); lime-water, 16 fl. oz. (475 c.c.)).

Collodions (collodia) are liquid preparations having for a menstruum a solution of pyroxylin (gun-cotton) in ether and alcohol. They are used externally, and are applied by means of a brush. With the exception of cantharidal collodion, which is used for blistering the skin, they are chiefly employed as protectants. The official collodia are:

Collodium (collodion).

Collodium Cantharidatum (cantharidal collodion, blistering collodion).

Collodium Flexile (flexile collodion).

Collodium Stypticum (styptic collodion).

Glycerites (glycerita) are solutions of drugs in glycerin. The following are official:

Glyceritum Acidi Carbolici (glycerite of carbolic acid).
Glyceritum Acidi Tannici (glycerite of tannic acid).
Glyceritum Amyli (glycerite of starch).
Glyceritum Boroglycerini (glycerite of boroglycerin).
Glyceritum Hydrastis (glycerite of hydrastis).
Glyceritum Vitelli (glycerite of yelk of egg).

Enemas or Clysters (enemata) are liquid preparations intended for injection into the rectum. There are no official enemas.

Extracts (extracta) are solid or semisolid preparations made by evaporating solutions of vegetable drugs. The menstrua are alcohol, diluted alcohol, water, water and ammonia-water, diluted acetic acid, or acetic acid and diluted alcohol. Being concentrated and solid or semisolid, they are well adapted for administering in the form of pills. The most important extracts are:

Extractum Aconiti (extract of aconite).
Extractum Aloes (extract of aloes).
Extractum Belladonnæ Foliorum Alcoholicum (alcoholic extract of belladonna leaves).
Extractum Cannabis Indicæ (extract of cannabis indica).
Extractum Cimicifugæ (extract of cimicifuga).
Extractum Cinchonæ (extract of cinchona).
Extractum Colchici Radicis (extract of colchicum root).
Extractum Colocynthidis (extract of colocynth).
Extractum Colocynthidis Compositum (compound extract of colocynth).
Extractum Conii (extract of conium).
Extractum Digitalis (extract of digitalis).
Extractum Ergotæ (extract of ergot).
Extractum Euonymi (extract of euonymus).
Extractum Gentianæ (extract of gentian).
Extractum Glycyrrhizæ (extract of glycyrrhiza).
Extractum Hæmatoxyli (extract of hematoxylon).
Extractum Jalapæ (extract of jalap).
Extractum Nucis Vomicæ (extract of nux vomica).
Extractum Opii (extract of opium).
Extractum Physostigmatis (extract of physostigma).
Extractum Podophylli (extract of podophyllum).
Extractum Quassiæ (extract of quassia).
Extractum Rhei (extract of rhubarb).
Extractum Stramonii Seminis (extract of stramonium seed).
Extractum Uvæ Ursi (extract of uva ursi).

Masses (massæ) are soft-solid preparations preserved in bulk and intended for forming into pills. The official masses are:

Massa Copaibæ (mass of copaiba, solidified copaiba).

Massa Ferri Carbonatis (mass of ferrous carbonate, Vallet's mass).

Massa Hydrargyri (mass of mercury, blue mass, blue pill).

Pills (pilulæ) are small rounded masses containing one or more medicinal substances held together by some adhesive material, known as an excipient. The usual excipients are water, glycerin, tragacanth, acacia, honey, bread-crumbs, soap, confection of rose, extract of glycyrrhiza, and extract of gentian. Ordinarily, pills should not exceed in weight 5 grains, unless made of very heavy drugs of small bulk. The most important official pills are:

Pilulæ Aloes (pills of aloes).

Pilulæ Aloes et Asafœtidæ (pills of aloes and asafetida).

Pilulæ Aloes et Mastiches (pills of aloes and mastich).

Pilulæ Aloes et Myrrhæ (pills of aloes and myrrh).

Pilulæ Asafœtidæ (pills of asafetida).

Pilulæ Catharticæ Compositæ (compound cathartic pills).

Pilulæ Cartharticæ Vegetabiles (vegetable cathartic pills).

Pilulæ Ferri Carbonatis (pills of carbonate of iron ; Blaud's pills).

Pilulæ Ferri Iodidi (pills of ferrous iodid).

Pilulæ Opii (pills of opium).

Pilulæ Phosphori (pills of phosphorus).

Pilulæ Rhei (pills of rhubarb).

Pilulæ Rhei Compositæ (compound pills of rhubarb).

Confections (confectiones) are soft masses made by mixing drugs with sugar and water or honey. Only two are official:

Confectio Rosæ (confection of rose).

Confectio Sennæ (confection of senna).

Powders (pulveres) are drugs in the form of fine, loose, uncompacted particles. They are commonly dispensed in small paper packages, known as *chartulæ*. But the latter are not suitable for highly volatile, hygroscopic, efflorescent, or deliquescent substances, or drugs with a disagreeable taste. When not suitable for papers, powders may be dispensed in capsules or cachets. *Capsules* are small gelatin cases or envelopes (hard or soft) in which liquid or solid medicines are enclosed to be swallowed. *Cachets* are round concave wafers made of flour and water. The drug is contained in the cavity formed by bringing together the concave surfaces of two wafers. Their edges are fastened by being moistened and then pressed together. They are well adapted for enclosing powders having a disagreeable taste. They should be taken floated on a small quantity of water.

Triturates (triturationes) are fine powders of medicinal substances intimately mixed with sugar of milk. There is only one official triturate:

Trituratio Elaterini (triturate of elaterin).

Tablets (tabellæ) are small disks containing medicinal substances mixed with sugar, and pressed into shape by metallic moulds. They are a convenient form in which to administer concentrated remedies, such as calomel, morphin, and strychnin. There are no official tablets.

Troches or **lozenges** (trochisci) are small, cylindrical or flat, solid masses, containing medicinal agents mixed with sugar and mucilage. Most of them are intended to influence the mucous membrane of the throat. The official troches are:

Trochisci Acidi Tannici (troches of tannic acid).

Trochisci Ammonii Chloridi (troches of ammonium chlorid).

Trochisci Cretæ (troches of chalk).

Trochisci Catechu (troches of catechu).

Trochisci Cubebæ (troches of cubeb).

Trochisci Ferri (troches of iron).

Trochisci Glycyrrhizæ et Opii (troches of glycyrrhiza and opium).

Trochisci Ipecacuanhæ (troches of ipecac).

Trochisci Krameriæ (troches of krameria).

Trochisci Menthæ Piperitæ (troches of peppermint).

Trochisci Morphinæ et Ipecacuanhæ (troches of morphin and ipecac).

Trochisci Potassii Chloratis (troches of potassium chlorate).

Trochisci Santonini (troches of santonin).

Trochisci Sodii Bicarbonatis (troches of sodium bicarbonate).

Trochisci Zingiberis (troches of ginger).

Suppositories (suppositoria) are solid bodies, of various shapes, intended to be inserted into the rectum, vagina, urethra, or nostril. The usual base is oil of theobroma. They are poured while in liquid form into suitable moulds and are solidified by cooling on ice. The only official suppositories are:

Suppositoria Glycerini (suppositories of glycerin).

Ointments (unguenta) are soft fatty preparations, melting at the temperature of the body, and containing medicinal substances incorporated with a base of lard, a fixed oil, petrolatum, wax, or spermaceti. The official ointments are:

Unguentum Acidi Carbolici (ointment of carbolic acid).

Unguentum Acidi Tannici (ointment of tannic acid).

Unguentum Aquæ Rosæ (ointment of rose-water, cold cream).

Unguentum Belladonnæ (belladonna ointment).

Unguentum Chrysarobini (chrysarobin ointment).

Unguentum Diachylon (diachylon ointment).

Unguentum Gallæ (nutgall ointment).

Unguentum Hydrargyri (mercurial ointment, blue ointment).

Unguentum Hydrargyri Ammoniati (ointment of ammoniated mercury).

Unguentum Hydrargyri Nitratis (ointment of mercuric nitrate, citrine ointment).

Unguentum Hydrargyri Oxidi Flavi (ointment of yellow mercuric oxid).

Unguentum Hydrargyri Oxidi Rubri (ointment of red mercuric oxid).

Unguentum Iodi (iodin ointment).

Unguentum Iodoformi (iodoform ointment).

Unguentum Picis Liquidæ (tar ointment).

Unguentum Plumbi Carbonatis (ointment of lead carbonate).

Unguentum Plumbi Iodidi (ointment of lead iodid).

Unguentum Potassii Iodidi (ointment of potassium iodid).

Unguentum Stramonii (stramonium ointment).

Unguentum Sulphuris (sulphur ointment).

Unguentum Veratrinæ (veratrin ointment).

Unguentum Zinci Oxidi (ointment of zinc oxid).

Oleates (oleata) are solutions of metallic oxids or alkaloids in oleic acid. There are three official oleates:

Oleatum Hydrargyri (oleate of mercury).

Oleatum Veratrinæ (oleate of veratrin).

Oleatum Zinci (oleate of zinc).

Cerates (cerata) are solid fatty preparations resembling ointments, but made firmer by the addition of wax. They are of such consistence that they soften but do not melt at the body-temperature. The official cerates are:

Ceratum (cerate).

Ceratum Camphoræ (camphor cerate).

Ceratum Cantharidis (cantharides cerate).

Ceratum Cetacei (spermaceti cerate).

Ceratum Plumbi Subacetatis (cerate of lead subacetate, Goulard's cerate).

Ceratum Resinæ (resin cerate, basilicon ointment).

Plasters (emplastra) are solids having a fatty or resinous base, and of such consistence that they do not melt but become adhesive when applied to the surface of the body. The official plasters are:

Emplastrum Ammoniaci cum Hydrargyro (ammoniac plaster with mercury).

Emplastrum Arnicæ (arnica plaster).

Emplastrum Belladonnæ (belladonna plaster).

Emplastrum Capsici (capsicum plaster).

Emplastrum Ferri (iron plaster, strengthening plaster).

Emplastrum Hydrargyri (mercurial plaster).

Emplastrum Ichthyocollæ (isinglass plaster, court-plaster).

Emplastrum Opii (opium plaster).

Emplastrum Picis Burgundicæ (Burgundy-pitch plaster).

Emplastrum Picis Cantharidatum (cantharidal pitch plaster, warming plaster).

Emplastrum Plumbi (lead plaster, diachylon plaster).

Emplastrum Resinæ (resin plaster, adhesive plaster).

Emplastrum Saponis (soap plaster).

Papers (chartæ) are strips of paper impregnated or coated with some medicinal substances. They are intended to be moistened and applied as a plaster, or else to be burned and the fumes inhaled. Two papers are official:

Charta Potassii Nitratis (potassium nitrate paper).

Charta Sinapis (mustard paper).

Poultices (cataplasmata) are soft semiliquid preparations made of some cohesive substance, mixed with water, to be used for applying heat and moisture to the tissues, or for securing a local stimulant effect. They are usually made of flaxseed meal, slippery elm, or bread and milk. Charcoal may be added as an absorbent, mustard as a stimulaht, and laudanum as an anodyne.

INCOMPATIBILITY IN PRESCRIPTIONS.[1]

Incompatibility in a prescription has been defined as that condition in which there exists either "a chemical decomposition, a pharmaceutic dissociation, or a therapeutic opposition" of its constituents. The term is thus susceptible of three meanings. A prescription is chemically incompatible where chemical change results; it is pharmaceutically incompatible where there is violation of correct pharmaceutic procedure; and there is therapeutic incompatibility where there is antagonism in physiologic action. Now, accepting these definitions, a prescription may be chemically incompatible, and yet be just what the physician wants. It may be pharmaceutically incompatible, and yet be desirable for the same reason. But it is never desirable when a change of chemical composition or

---

[1] This article has been written by Joseph W. England, Ph.G.

pharmaceutic character results in the formation of new prodncts having totally different therapeutic effects than those obviously intended. And this view—the intended therapeutic action of the prescription—the pharmacist should ever bear in mind.

Every new prescription is largely a law unto itself until tried. Expertness in pharmaceutic manipulation, of which prescription-work is the highest type, is a matter of individual ability, which can be acquired only in the largest and best measures by personal experience. The subject of incompatibility is not a formidable one if there primarily exists a clear knowledge of the chemical or pharmaceutic properties of the substances used, so that any deviation from the right standard may be detected; but here is the puzzling question, How are we to know but that, in the event of some chemical or pharmaceutic change, the physician does not mean just such a change, and nothing else?

At first glance it seems strange, but there are some most successful physicians who, every now and then, write pharmaceutically and chemically, the most incompatible prescriptions. Yet they have success; and their happy results can only be due to the certain formation of new products or an alteration in pharmaceutic character of old ones. It does not follow that all prescriptions thus written are of the highest therapeutic value. Far from it. The tendency of the times is steadily in the direction of greater simplicity in prescription-writing.

It is to be regretted that the physician often depends in large measure upon the pharmacist for detecting any chemical or pharmaceutic incompatibility, and that the pharmacist depends solely and alone upon the physician for recognizing any therapeutic incompatibility. A physician, with his many duties, cannot be expected to have at his command the vast detail of pharmaceutic facts, nor can the pharmacist be considered negligent in not possessing an extended acquaintance with the application of drugs in medicine; but it is clear that some elementary knowledge as to how drugs act and for what purposes they may be employed would be of great practical value to the pharmacist in affording him a clear idea of the therapeutic intent of the prescriber, and the ability to detect any deviation through a chemical or pharmaceutic error. An argument for therapeutic knowledge is not a step in the direction of counter-prescribing. It is only a plea for broader education—for elementary therapeutics on distinctly pharmaceutic lines. With therapeutics pure and simple the pharmacist has nothing whatever to do. That is solely the province of the physician. Medicine and pharmacy are making rapid scientific progress—not in the same way, but upon certain definite lines of work and study, yearly becoming more distinct and widely separated, rendering each the more dependent on the other.

To render a knowledge of the solubilities and insolubilities of inorganic compounds readily accessible, the following table is presented, based almost wholly upon Professor Attfield's "Statement of the Solubilities and Insolubilities of Salts," which expresses, directly or by inference, nearly five hundred soluble and insoluble compounds of the following inorganic basylous radicals: aluminum, ammonium, antimony, barium, bismuth, cadmium, calcium, chromium, cobalt, copper, ferric, ferrous, gold, lead, lithium, magnesium, manganese, mercuric, mercurous, nickel, potassium, silver, sodium, stannic, stannous, strontium, and zinc.

In using this table it is only needful to remember the well-known chemical law that when a solution of a compound is brought in contact with a solution of another compound, and, by an interchange of radicals, an insoluble compound is rendered possible, that compound will be precipitated.

Acetates are soluble.

Arsenates are insoluble, except those of the alkali metals.

Arsenites are insoluble, except those of the alkali metals.

Bromids are soluble, except mercurous and silver; those of antimony and bismuth are decomposed by water to form oxysalts.

Carbonates are insoluble, except those of the alkali metals.

Chlorids are soluble, except those of lead (s),[1] mercurous, and silver.

Citrates are soluble, except those of manganese, mercurous, silver, and strontium, aluminum (s), barium (s), bismuth (s), cadmium (s), calcium (s), lead (s), zinc (s).

Cyanids are insoluble, except the mercuric and those of the alkaline metals and earths.

Hydrates are insoluble, except those of barium, strontium, calcium (s), lead (s), and the alkali metals.

Iodids are soluble, except those of antimony, bismuth, gold, lead (s), mercuric, mercurous, platinum (s), and silver.

Nitrates are soluble.

Oxalates are insoluble, except those of antimony (s), chromium, ferric (s), ferrous (s), stannic, and the alkali metals.

Oxids are insoluble, except those of barium, strontium, calcium (s), and the alkaline metals.

---

[1] (s) means sparingly soluble.

Phosphates (ortho) are insoluble, except those of the alkali metals.

Sulphates are soluble, except those of barium, strontium, calcium (*s*), antimony, lead, mercurous (*s*), and silver (*s*).

Sulphids are insoluble, except those of barium, calcium (*s*), strontium, and the alkali metals.

Sulphites are soluble, except those of aluminum, antimony, barium, bismuth, calcium (*s*), cobalt (*s*), copper, ferrous (*s*), lead, manganese (*s*), nickel (*s*), silver, stannous, strontium, and zinc (*s*).

Tartrates are soluble, except those of antimony, barium, bismuth, cadmium (*s*), calcium (*s*), copper (*s*), ferrous (*s*), lead, manganese (*s*), mercuric, mercurous, nickel (*s*), silver, strontium (*s*), and zinc (*s*).

Acids decompose hydrates, carbonates, and acid carbonates to form salts; the stronger acids, which are largely inorganic, set free the weaker acids, which are largely organic; or, brought in contact with alcohol or alcoholic solutions, form ethers; alkaline hydrates, carbonates, and acid carbonates neutralize free acids, decompose some glucosids, and precipitate all alkaloids, some of which precipitates are soluble in excess of the precipitant, or in alcohol, if that liquid be present in sufficient amount to dissolve them.

Oxidizing agents, such as nitric, hydrochloric, nitrohydrochloric, picric, and chromic acids, and potassium bichromate and permanganate, with readily oxidizable substances, such as carbohydrates, alcohols, ethers, sulphur, phosphorus, sulphids, and organic matter in general, form explosive compounds. Potassium permanganate, if ordered in pill form, can best be made with cacao butter and cosmolin in very small quantity, and enclosed in gelatin capsules. Silver nitrate is reduced by organic matter to oxid, with the exception, it is said, of opium and extract of hyoscyamus. A very good way of making pills of it is with cacao butter and cosmolin, etc., as mentioned above under potassium permanganate; syrup of ferrous iodid and potassium chlorate form a poisonous compound, and potassium iodid and potassium chlorate form a mixture which yields the poisonous iodate on being taken internally.[1]

Iodin and iodids yield precipitates with the alkaloids; bromids precipitate morphin and strychnin salts on standing, but a few drops of dilute hydrochloric acid added, after the addition of the alkaloid, prevents the change. Sodium biborate precipitates morphin and cocain salts, but on the addition of a small quantity of boric acid, or with boric acid alone, precipi-

[1] *Am. Jour. Phar.*, p. 277, 1876.

tation does not take place. Mercuric chlorid with acidulated solutions of the alkaloids forms crystalline double salts ; potassium-mercuric iodid precipitates alkaloidal solutions. Solutions of quinin salts with those of the alkaline acetates, or with Basham's mixture, precipitate the sparingly soluble quinin acetate. Morphin solutions give the phenol reaction if mixed with tincture of ferric chlorid.

Glucosids are decomposed by free acids and precipitated by tannin ; tannic and gallic acids precipitate alkaloids, albumin, gelatin, and the majority of metallic salts, and yield inks with iron solutions.

Resinous tinctures and fluid extracts prescribed with aqueous solutions should always be emulsified with acacia ; tinctures and fluid extracts made of stronger alcohol, mixed with those made of diluted alcohol, become turbid and precipitate, since the special solvent power of alcohol or of water for a substance diminishes in proportion to the quantity of the other liquid present. A "shake" label should always be used.

When for internal use, fixed and volatile oils and oleoresins and aqueous solutions should always be emulsified, whether ordered or not, and to better emulsify the volatile oils they should have mixed with them, prior to emulsification, an equal volume of olive, almond, or cotton-seed oil.

Tincture of ferric chlorid gelatinizes mucilage of acacia ; free acids separate insoluble carminic acids from compound tincture of cardamom ; free acids precipitate glycyrrhizin from fluid extract of licorice.

Commercial spirit of nitrous ether liberates iodin from solutions of iodids, decomposes antipyrin solutions to form a green nitroderivative, and precipitates mucilage of acacia, but if it be well diluted with water it can usually be added last without precipitating. Tincture of guaiac and spirit of nitrous ether are stated to be pharmaceutically incompatible by Potter, although they are often prescribed together ; likewise infusion of wild cherry with compound infusion of gentian, infusion of cinchona with compound infusion of gentian, and infusions with metallic salts generally.

Sodium salicylate in solution precipitates the sparingly soluble salicylic acid if mixed with acids, and yields, if dispensed in powders with potassium acetate, the very deliquescent potassium salicylate. Sodium salicylate in strong solution is decomposed by tincture of ferric chlorid, but if well diluted first changes only into ferric salicylate. Sodium-benzoate solution is decomposed by acids to yield the sparingly soluble benzoic acid.

Mercuric chlorid is decomposed by solution of potassium arsenite, but if the alkaline solution has first added to it, in slight excess, diluted hydrochloric acid, no precipitation will take place on the addition of the mercurial salt. Pyrophosphate and phosphate of iron solutions precipitate with dilute phosphoric acid. The National Formulary recommends the usage of dilute metaphosphoric acid in place of the official "ortho" variety, as yielding a permanently clear solution.

In conclusion, the writer would say that in this brief article he has endeavored to present, not an exhaustive list of special incompatibles, but simply a general expression of those liable to occur in the every-day routine of prescription-work.

METHODS OF ADMINISTERING DRUGS.

By the Mouth.—This is the most common method. Drugs may be administered by the mouth for their local action on the mouth, throat, stomach, or intestines, or for the purpose of being absorbed. When it is desired to influence the mouth or throat, the remedy is generally given in the form of a lozenge, and the patient advised not to take food or water for at least an hour afterward. Remedies intended to act directly on the mucous membrane of the stomach should be given when the stomach is empty, that is, half an hour or an hour before meals. Drugs intended to exert a direct influence on the intestine should be given two or three hours. after meals, and preferably in firm pills. Sometimes such pills. are coated with a substance, like keratin, that will resist the action of the gastric juice.

Absorption may take place from any part of the alimentary canal. Powerful remedies, as nitroglycerin and tincture of aconite, are readily absorbed from the tongue. Absorption from the stomach is effected most quickly, as a rule, when the drug is given in solution on an empty stomach. However, in the case of remedies that have a local irritant action, it is far better to give them immediately after meals, so that by becoming mixed with the food their local effect is more or less avoided.

A few remedies, such as calomel and salol, are digested almost entirely in the intestine.

Hypodermic Method.—This consists in injecting medicinal solutions into the subcutaneous tissue by means of the hypodermic needle and syringe. The advantages of this method are the great rapidity with which absorption is effected, and the certainty of securing the action of the entire dose, since partial destruction of the remedy by the digestive organs is avoided. The disadvantages of this method are the pain inflicted, the liability of causing abscess, and the risk of throwing the solution directly into a vein. The alkaloidal salts, on account of their small bulk, are especially adapted for hypodermic use. Some remedies, such as toxins and antitoxins, are effective only when administered subcutaneously. The best places to select for the injections are the extensor surface of the arm, the calf of the leg, the abdominal wall, and the buttock. Both the needle and the solution should be sterile. All the air should be expelled from the syringe before the injection is given, and great care should be exercised to avoid entering a vein. The dose of a drug administered hypodermically is ordinarily about one-half of that given by the mouth.

Intravenous Injection.—This method is employed only in great emergency. Saline solution may be given intravenously in copious hemorrhage, uremia, diabetic coma, and in the algid stage of cholera. In the collapse of acute poisoning, diffusible stimulants—ammonia-water, ether, whiskey—may be introduced into the system through the veins.

By the Rectum.—Absorption through the rectum is not effected near so rapidly as through the stomach. To accomplish the same result it is generally necessary to give twice the dose by the rectum that would be required by the mouth. This method is convenient when the stomach is unretentive. It also affords a means of acting directly on the bowel with drugs, and of removing hardened feces, flatus, and parasites. When the enema is to be retained in the bowel for its local effect, the fluid should be warm, the quantity should not exceed two ounces, and the injection should be made very slowly. When the enema is intended to bring about defecation, one or two pints of fluid may be employed. When flatus is to be removed, an enema consisting of from 4 to 6 ounces of the emulsion of asafetida is effective. Both local and constitutional effects may also be secured by means of suppositories containing medicinal agents.

Inunction.—Many medicinal substances, if dissolved in fats, are fairly well absorbed through the unbroken skin, especially if friction be used in their application. When this method of administration is employed systematically to secure a constitutional effect, different surfaces should be selected each day. The best surfaces for inunction are those having a thin skin, such as the axillæ, popliteal spaces, the groins, and the inner sides of the thighs. The drugs that are most frequently introduced by inunction are mercury and iodin. A convenient method of treating syphilis in infants is by smearing mercurial ointment on a flannel binder worn around the abdomen.

Endermic Method.—This method consists in raising a blister, removing the epidermis, and sprinkling on the raw derm the drug that is to be absorbed. It is very painful, and it is far less satisfactory than the hypodermic method.

Fumigation.—This method is sometimes employed in syphilis instead of inunction. A mercurial salt, preferably calomel, is volatized from a tin plate suspended over a spirit-lamp. The latter is placed under a cane-seated chair, upon which is seated the patient, disrobed and surrounded by a blanket fastened to the neck.

Inhalation.—Volatile drugs are rapidly absorbed from the respiratory tract, and this method is especially employed for the administration of drugs which induce general anesthesia, such as ether, chloroform, and nitrous oxid. The effect of other volatile substances, like amyl nitrite, is also secured by inhalation. Medicated vapors and sprays are employed to influence directly the mucous membrane of the respiratory tract.

CIRCUMSTANCES MODIFYING THE EFFECT OF DRUGS.

The effect of drugs is modified by many circumstances, chief of which are the age of the patient, sex, race, body-weight, temperament, disease, temperature, habit, idiosyncrasy, method of administration, time of administration, preparation of the drug, and dose.

Habit.—With many drugs continuous use induces toleration. This is particularly true of opium. Increased elimination and diminished absorption may play a part in the development of this condition, but it is probably due very largely to a change in the tissues themselves, whereby they are rendered less susceptible to the influence of the drug.

Idiosyncrasy.—This is a peculiar susceptibility or insusceptibility of one or more of the tissues to the influence of certain drugs. Less than ¼ of a grain of calomel has been known to induce salivation. The smallest dose of quinin will in some individuals cause a diffuse erythematous rash.

Cumulative Action.—This is the property which some drugs have, after they have been given for a certain length of time, of producing a severe and more or less sudden effect, owing to their accumulation in the body. Cumulative action of a drug may be due to its remaining for a long time in the intestines without being absorbed; to the very strong affinity which it possesses for the tissues; or, as Brunton suggests, to the stopping of its own excretion by its action upon the kidney.

DOSAGE.

The circumstances which modify the effect of drugs must also necessarily influence dosage. The most important conditions to be considered are the age, sex, weight, and disease.

Age.—Children require smaller doses than adults. Two rules are in common use for determining the proper dose for children of different ages:

Young's rule is as follows: Add 12 to the age and divide by the age, and the quotient will be the denominator of a fraction the numerator of which is 1. Thus, for a child of four years, $\dfrac{4 + 12}{4} = 4$, and the dose is ¼ that for an adult.

Cowling's rule is to divide the age of the child at its next birthday by 24. Thus, for a child 3 years old, the dose would be $\frac{4}{24} = \frac{1}{6}$ of the adult dose.

These rules, of course, are only approximately correct, and each drug must be considered by itself in reference to dose. Thus, children are very susceptible to opium, and, therefore, the dose of this drug must be smaller than the age would apparently indicate. On the other hand, arsenic and belladonna are well borne by children, and, therefore, relatively larger doses of these drugs can be prescribed.

Sex.—Women generally require smaller doses than men. The dose for a woman is about four-fifths of that for a man.

Weight.—Weight has less influence on dosage than either age or sex, but a very heavy muscular individual will generally require a larger dose than one of slight frame.

DRUGS.

CIRCULATORY STIMULANTS.

THE following drugs stimulate the heart-muscle or its contained ganglia:

Ammonia.	Strychnin.
Alcohol.	Spartein.
Ether.	Convallaria.
Nitrites.	Adonidin.
Digitalis.	Cactus grandiflorus.
Strophanthus.	Belladonna.
Caffein.	Camphor.
Adrenalin.	Cocain.

The Effect of Circulatory Stimulants upon the Arterial Pressure.—A drug that stimulates the heart, as a rule raises the blood-pressure, but not necessarily so. Thus, the *nitrites* in small doses stimulate the heart, but they more than counterbalance the effect of this stimulation by dilating the blood-vessels, and so they induce a marked fall of the arterial pressure.

Many heart-stimulants serve to maintain the arterial pressure also by constricting the peripheral blood-vessels. Vaso-constriction may be brought about by direct action of the drug on the muscular coats of the arteries, or by its indirect action on the vasomotor center.

Digitalis and probably *cocain* raise the arterial pressure·not only by stimulating the heart, but also the vasomotor center and the muscular coat of the arteries.

Strychnin, caffein, ether, and *belladonna* raise the arterial pressure by stimulating the heart and the vasomotor center.

Strophanthus and *adrenalin* raise the arterial pressure by stimulating the heart and the muscular coat of the arteries.

A heart-stimulant with a well-marked constricting effect upon the arteries, like digitalis, is often very valuable in valvular disease of the heart with simple dilatation, especially when there is dropsy. For a certain amount of tension in the arteries is necessary to equalize the two circulations, to secure for the heart-muscle the proper amount of nourishment and to promote diuresis. When, however, the myocardium is the seat of advanced degenerative changes, digitalis must be used with considerable caution, since its constricting influence on the blood-vessels may be more pronounced than its stimulant effect on the heart. In such cases a vasodilator, like one of the nitrites, should be associated with the digitalis, or a heart-stimulant should be selected that will cause less contraction of the muscular fibers of the vessels.

The Effect of Circulatory Stimulants upon the Pulse-rate.—A circulatory stimulant may increase the pulse-rate: (1) By stimulating the heart alone; (2) By stimulating the accelerator nerves; (3) By paralyzing the inhibitory nerves (vagi) centrally or peripherally; and (4) By dilating the peripheral arteries, and so diminishing the resistance to the flow of blood.

Alcohol, ether, adrenalin, and *caffein* apparently increase the pulse-rate by stimulating the heart alone.

Ammonia quickens the circulation by stimulating the heart and the accelerator nerves.

The *nitrites* act chiefly by paralyzing the inhibitory nerves and by dilating the peripheral arteries.

Belladonna acts chiefly by paralyzing the inhibitory nerves and by stimulating the accelerator nerves.

The circulatory stimulants that slow the pulse usually do so, either by stimulating the inhibitory nerves or by constricting the peripheral arteries. Increased tension in the vessels serves to slow the pulse directly by offering resistance to the flow of blood, and indirectly by influencing the inhibitory center in the medulla.

Digitalis and probably *adonidin* slow the pulse by stimulating the inhibitory nerves, and partly also by constricting the peripheral arteries.

AMMONIA.

(NH$_3$.)

Ammonia is a colorless gas having an intensely pungent odor and acrid taste, and obtained as a by-product in the manufacture of coal-gas. It acts as a base in uniting with acids to form salts.

PREPARATIONS.	DOSE.

Aqua Ammoniæ, U. S. P. (10 per cent. of
the gas) . . . . 10–30 min. (0.6–2.0 c.c.).
Aqua Ammoniæ Fortior, U. S. P. (28 per
cent. of the gas).
Spiritus Ammoniæ, U. S. P. (10 per cent. of
the gas) . . . . 20–60 min. (1.0–4.0 c.c.)
Spiritus Ammoniæ Aromaticus, U. S. P.
(ammonium carbonate, 34; ammonia-
water, 90; aromatic oils; water to
make 1000 parts) . . . . 20–60 min. (1.0–4.0 c.c.).
Linimentum Ammoniæ, U. S. P.

Physiologic Action.—Circulatory System.—Moderate doses of ammonia increase both the strength and rapidity of the pulse, and this effect is produced by a direct stimulation of the heart and of its accelerator nerves. When injected directly into the jugular vein, the blood-pressure falls from cardiac paralysis.

Nervous System.—In large doses ammonia stimulates the motor centers in the spinal cord and increases reflex activity. When injected into a vein in a toxic dose it induces violent convulsions of spinal origin. Upon the cerebrum and sensory nervous mechanism, the drug is without appreciable effect.

Respiratory System.—Moderate doses increase the depth and rapidity of the respirations by stimulating the respiratory center.

Alimentary Canal.—In the stomach ammonia, in weak solution, like other alkalies, stimulates the flow of gastric juice. Toxic doses produce intense gastro-enteritis.

Local Action.—Locally, it acts as a decided irritant, and in concentrated solution, it speedily produces vesication.

Elimination.—As ammonia undergoes rapid decomposition in the system, its action is very evanescent. One of the products of its oxidation being nitric acid, it rather increases than diminishes the acidity of the urine.

Toxicology.—Ammonia-poisoning is characterized by severe burning pain in the fauces, esophagus, and stomach, dyspnea, persistent vomiting and purging, a rapid, thready pulse, and the symptoms of collapse. The mind may remain clear until the end, but coma often precedes death. The intense irritation of the throat sometimes causes edema of the larynx, in consequence of which death may result in a few minutes from asphyxia.

Treatment.—The ammonia should be neutralized by the administration of some weak acid, such as vinegar. Subsequently the inflammation should be allayed by demulcents and opium. Asphyxia from laryngeal edema will demand tracheotomy.

Therapeutics.—Ammonia is employed chiefly as a quickly acting cardiac and respiratory stimulant, as an antacid in gastric acidity, and as a counter-irritant.

In *syncope, collapse,* and other forms of *sudden heart-failure,* ammonia is an invaluable stimulant. In *gastralgia* and *heartburn* dependent upon unnatural acidity of the stomach-contents, it is a useful antacid. As an antacid it is employed also externally to neutralize the poison in the *bites of certain insects.* As a rubefacient, ammonia liniment is a useful remedy in *chilblains, chronic rheumatism,* and *myalgia.*

Administration.—The spirits of ammonia are usually selected for internal administration. They should be given well diluted. For subcutaneous or intravenous injection ammonia-water is the most suitable preparation. The latter is also used in various rubefacient liniments. Stronger ammonia-water, applied on a pledget of cotton, covered with a watch-crystal, has been used as a vesicant, but the process is very painful and the blister is slow in healing.

Incompatibles.—Ammonia is incompatible with mineral and vegetable acids, chloral, and alkaloids. With solutions of corrosive sublimate it forms the insoluble ammoniated mercury. With solutions of ferric salts it forms the sesquioxid of iron, which is the antidote to arsenic. With tincture of iodin it yields a black precipitate of nitrogen iodid, which when dry is very explosive. With solutions of formaldehyd it forms urotropin.

AMMONII CARBONAS, U. S. P.
(Ammonium Carbonate, $NH_4HCO_3.NH_4NH_2CO_2.$)

Carbonate of ammonium occurs in the form of white, translucent, crystalline masses having an extremely pungent odor and acrid taste. When exposed to the air it breaks up into a white powder—bicarbonate of ammonium. It is soluble in 4 parts of water. Its dose is 5–10 gr. (0.3–0.6 gm.).

Physiologic Action.—It is a cardiac and respiratory stimulant, and a stimulating expectorant. Its action is prompt, but evanescent.

Therapeutics.—It is used for the same purposes as the solutions of ammonia. It is particularly efficacious in *acute pulmonary diseases* associated with cardiac and respiratory weakness, such as severe bronchitis, croupous pneumonia, and bronchopneumonia.

Administration.—To guard against its irritant effect on

the stomach, it is usually prescribed in mucilage or syrup of acacia, as in the following formula :

```
R  Ammonii carbonatis,                        ℥ij (8.0 gm.) ;
   Pulveris acaciæ et sacchari,               āā q. s ;
   Tincturæ lavandulæ compositæ,              f℥ij (8.0 c.c.) ;
   Aquæ,                        q. s. ad f℥ iij (90.0 c.c.).—M.
      Sig.  A teaspoonful in water every two or three hours.
```

Incompatibles.—The same as the preparations of ammonia. Ammonium carbonate cannot be prescribed with syrups of squill, citric acid, or garlic on account of the free acid which they contain.

ALCOHOL, U. S. P.
(Ethyl Alcohol, C_2H_5HO.)

Ethyl alcohol is obtained from the distillation of fermented saccharine material. The official preparation contains 91 per cent. by weight of absolute alcohol, and appears as a colorless, inflammable liquid, having a pungent odor and a burning taste.

PREPARATIONS.

Alcohol Absolutum, U. S. P. (99 per cent. alcohol)
Alcohol Deodoratum, U. S. P. (92.5 per cent. alcohol).
Alcohol Dilutum, U. S. P. (41 per cent. absolute alcohol).

Spirits :
Spiritus Frumenti, U. S. P., or whiskey (44 to 50 per cent. alcohol), distilled from fermented grain.
Spiritus Juniperi Compositus, U. S. P., equivalent to gin (about 60 per cent. alcohol).
Spiritus Vini Gallici, U. S. P., or brandy (39 to 47 per cent. alcohol), distilled from fermented grapes.

Malt Liquors :
Ale, Beer, Porter, } obtained from the fermentation of malted grain, and contain from 3 to 6 per cent. alcohol.

Wines :
Vinum Album, U. S. P., or white wine (10 to 14 per cent. alcohol).
Vinum Rubrum, U. S. P., or red wine (10 to 14 per cent. alcohol).
Vinum Portense, or port wine (30 to 40 per cent. alcohol).
Vinum Xericum, or sherry wine (20 to 35 per cent. alcohol).
The official wines are made by fermenting the unmodified juice of the grape.

Physiologic Action.—**Circulatory System.**—In moderate doses alcohol increases the force and rapidity of the pulse, probably by direct action on the heart. Large doses paralyze the heart and the vasomotor mechanism.

Nervous System.—It first stimulates and then paralyzes all parts of the nervous system.

Alimentary Canal.—Small doses of alcohol sharpen the appetite and favor digestion. In the mouth it stimulates reflexly the flow of saliva. Its presence in the stomach tends to retard digestion; but this effect is more than counterbalanced by the greater flow of gastric juice, and the more energetic movements of the viscus which it calls forth. Large doses produce severe gastritis.

Effect on Metabolism.—Moderate amounts of alcohol undergo combustion in the body, protect fats and carbohydrates (and probably proteids) from combustion, impart energy to the system, and serve to maintain the body-weight when, for any reason, the diet is insufficient. To this extent alcohol may be regarded as a food.

Temperature.—Small doses impart a sense of warmth to the body by stimulating circulatory activity. Large doses cause a distinct fall of temperature by favoring radiation of heat, by increasing perspiration, and by retarding metabolism.

Elimination.—When taken in excess, it is eliminated unchanged through all the emunctories; small quantities, however, are completely oxidized in the body.

Local Action.—When applied to the skin and allowed to evaporate, it constricts the vessels and causes a sensation of coldness. If evaporation be prevented it acts as a rubefacient.

Action on Lower Organisms.—Alcohol possesses moderate germicidal properties. According to the researches of Harrington and others the best results are obtained with alcohol of 60 per cent. strength, its germicidal power diminishing with departures in both directions from this dilution.

Toxicology.—The ingestion of large quantities of alcohol produces the following symptoms: Flushing of the face, mental excitement, quickening of the pulse and respiration; then incoherent speech, delirium, dilated pupils, loss of co-ordination, subnormal temperature, vomiting, and, finally, stupor and coma. Not infrequently the coma is interrupted by convulsive seizures. In most cases, if the dose has not been too large, recovery follows in a day or two

Care must be taken to distinguish acute alcoholism from uremia, opium-poisoning, and apoplexy. The urinous odor of the breath, the accentuation of second aortic sound, the small pupils, and the presence of albumin in the urine will serve to distinguish *uremia*. The small pupils, slow respiration, and slow pulse will indicate *opium-poisoning*. The unequal pupils, hemiplegia, and elevated temperature will serve to separate *apoplexy* from alcoholism.

Treatment.—The stomach should be emptied by the stomach-pump, a stimulating emetic, or the hypodermic injection of apomorphin (gr. $\frac{1}{10}$—0.007 gm.). If the pulse weakens, stimulants like strychnin, ammonia, or digitalis should be administered hypodermically.

Chronic alcoholism is characterized by disturbed sleep, fine tremors, mental impairment, injection of the eyes, redness of the nose, and the symptoms of gastro-intestinal catarrh. When the habit is long-continued, degenerative and cirrhotic changes in the heart, blood-vessels, liver, and kidneys are apt to develop.

A common complication of chronic alcoholism is *delirium tremens,* which is generally excited by temporary excess, an injury, or some acute intercurrent disease, especially pneumonia. It is manifested by great mental excitement, insomnia, incoherent speech, tremors, disordered intellect, and terrifying hallucinations of sight or hearing. The pulse is rapid and feeble, the appetite is lost, the bowels are constipated, and the temperature slightly elevated. In favorable cases convalescence follows in a few days, but not infrequently typhoid symptoms develop and the attack ends in death.

Among other sequels of dipsomania may be mentioned pneumonia, chronic meningitis, multiple neuritis, amblyopia, epilepsy, and dementia.

Treatment of Delirium Tremens.—As there has usually been a complete abstinence from food during the debauch leading to the delirium, nutritious foods are always necessary, and the best are milk with lime-water and highly seasoned beef-tea. Sleep must be secured by chloral (gr. xx—1.3 gm.), hyoscin (gr. $\frac{1}{100}$—.00065 gm.), potassium bromid (ʒj—4.0 gm.), or paraldehyd (f℥ss-f℥j—2.0-4.0 c.c.). Active catharsis sbould be encouraged. When the pulse is weak strychnin and digitalis will be found useful stimulants. In many cases physical restraint will be required; it may be secured by strapping the patient to the bed with sheets. Should profound stupor develop, the application of a blister to the back of the neck or a few light touches of the actual cautery will often serve to arouse the patient.

Therapeutics.—Alcohol is employed internally as a diffusible circulatory stimulant, a stomachic, a food, and a chemical antidote. Externally it is used as an antiseptic, a stimulant, a hemostatic, and an antihydrotic.

In all forms of sudden heart-failure, as in *syncope, shock, snake-bite,* and *acute febrile disease,* alcohol is an invaluable stimulant. In the continued fevers, like *typhoid,* it fills a triple rôle: it serves as a food, as a general stimulant, and as a promoter of digestion. In these cases, however, it is well to withhold it until the pulse flags or the heart-sounds indicate a weakening of the circulation. Brunton's rule for administering alcohol is a good one: " Sit by the side of your patient for a while and watch him after the administration of a dose of alcohol, and if you find that the alcohol brings back the various functions nearer to the normal, then it is doing good; if the functions of the organs diverge further from the normal after the administration of alcohol, then it is doing harm."

In chronic diseases, like *phthisis, obstinate atonic dyspepsia*, and *valvular affections of the heart*, it often does good, but it should be administered tentatively, with due regard for the danger of producing chronic alcoholism.

The power which alcohol possesses of counteracting the escharotic action of carbolic acid was first pointed out by Phelps and Powell in 1899. Since then sufficient evidence has accumulated to warrant the belief that alcohol, when administered promptly, is a valuable *antidote in carbolic-acid poisoning*. From 1 to 3 ounces of alcohol should be poured into the stomach through a tube, and lavage immediately practised with water containing a soluble sulphate.

External Uses.—Alcohol is a useful adjuvant to other drugs in disinfecting the skin. A mixture of alcohol and water makes a useful evaporating lotion in the treatment of contusions with ecchymosis. Applied to wounded surfaces it is effective in controlling capillary oozing. Sponging the body with alcohol and water is often serviceable in checking the night-sweats of phthisis.

DIGITALIS, U. S. P.

(Foxglove.)

Digitalis is the dried leaves of *Digitalis purpurea*, a biennial herb growing in Central and Southern Europe. It contains several principles, all of which are probably of the nature of glucosids, the most important being *digitoxin, digitalin, digitalein, digitonin*, and *digitin*. The first three have a stimulating action on the heart and represent more or less imperfectly the virtues of the crude drug. Digitonin is antagonistic to the other principles in that it depresses the heart, and digitin is inert. Digitoxin and digitonin are decided irritants, and are probably responsible for much of the gastric disturbance which digitalis often occasions. Digitalin is a stable compound, and represents very fully the actions of the crude drug, but for general use it is not so reliable as the tincture or the fluid extract. Pure digitoxin, on account of its slow absorption and irritant properties, should not be employed.

Since digitalin and digitoxin are freely soluble in alcohol, and are almost insoluble in water, it follows that the most potent principles are contained in the tincture and fluid extract, and the least potent in the infusion.

Preparations.	Dose.
Tinctura Digitalis, U. S. P.	5–20 min. (0.3–1.2 c.c.).
Extractum Digitalis Fluidum, U. S. P. . .	1–2 min. (0.06–0.12 c.c.).
Infusum Digitalis, U. S. P. . . .	1–4 fl. dr. (4.0–15.0 c.c.).
Extractum Digitalis, U. S. P. . .	$\frac{1}{6}$–$\frac{1}{4}$ gr. (0.010–0.016 gm.).
Pulvis Digitalis, . .	$\frac{1}{2}$–2 gr. (0.032–0.13 gm.).
Digitalinum (Merck)	$\frac{1}{20}$–$\frac{1}{4}$ gr. (0.003–0.016 gm.).

The digitalinum of Nativelle is an impure digitoxin, its dose is $\frac{1}{200}$–$\frac{1}{60}$ (0.0003–0.001 gm.).

Physiologic Action.—Circulatory System.—The dominant action of digitalis is on the circulation. In therapeutic doses it slows the pulse and raises the blood-pressure. The slowing of the pulse results from a prolongation of the diastole, and this in turn is due to stimulation of the vagi. The increased blood-pressure is due to a powerful stimulant effect on the heart, and to a constriction of the arterioles, resulting indirectly from the stimulation of the vasomotor center, and directly from the action of the drug on the vessel-walls.

Toxic doses of digitalis make the pulse rapid, weak, and irregular. The rapid pulse is due to paralysis of the vagi, and the fall of blood-pressure and irregularity to an undetermined action of the drug on the heart itself. In man the heart is finally arrested in diastole.

Nervous System.—Therapeutic doses have no effect on the nervous system, but toxic doses lessen reflex activity, first by stimulating the inhibitory centers in the medulla, and later by depressing the spinal cord.

Respiratory System.—Only in poisoning are the respirations affected, and then they are somewhat slowed.

Alimentary Canal.—Large doses frequently excite nausea, vomiting, and diarrhea.

Kidneys.—In health digitalis has but little diuretic action, but when the urine is scanty from low arterial tension it induces free diuresis. It may, therefore, be assumed that the drug has little or no direct influence on the renal epithelium, but that it indirectly increases the quantity of urine through its action on the heart and blood-vessels. The elimination of digitalis is effected very slowly.

Toxic doses, by causing a very high blood-pressure in the

kidneys, diminish the flow of urine. According to Brunton, the cumulative effects are probably due to its power of arresting the action of the kidneys, and thus stopping its own excretion.

Toxicology.—Digitalis-poisoning is characterized by obstinate vomiting, diarrhea, headache, disordered vision, and a very slow full pulse. The pulse, however, often becomes suddenly rapid and feeble when the patient sits up. Later, even in recumbency, the pulse becomes rapid, irregular, and thready, the urine scanty or suppressed, the surface cold, and the mind cloudy. Intelligence, however, is not lost until shortly before death. Occasionally convulsions develop during the last stage.

The uninterrupted use of digitalis in full doses is sometimes followed by the sudden appearance of toxic symptoms; this untoward effect seems to be due to the *cumulative action* of the drug, and is especially liable to occur when there is no diuretic effect, and after the removal of serous effusions by paracentesis. In the first instance elimination is interfered with, and in the second, absorption is effected with undue rapidity. The earliest manifestations of this accident are usually irregularity of pulse, precordial distress, and the appearance of the heart-beats in couples, the first of which only can be detected at the wrist. When a prolonged use of the remedy is required, it should be suspended at intervals for a definite period, so as to allow of complete elimination.

Treatment of Poisoning.—The patient must be kept in the recumbent position. Emetics or purges may be necessary for the removal of any part of the poison that is still unabsorbed. Tannic acid should be administered as a chemical antidote. It is important to maintain the body-temperature by the application of external heat. Alcoholic stimulants may prove useful. Aconite has been recommended on theoretic grounds as a physiologic antidote, but it should be used with considerable caution.

Therapeutics.—Digitalis fills three important offices: It serves as a powerful cardiac stimulant; it slows the heart and regulates its rhythm, and it acts as a diuretic.

As a Cardiac Stimulant.—Digitalis is especially indicated when there is simple dilatation of the heart. By prolonging diastole it rests the heart and allows the ventricles to become more completely filled; and by strengthening the systole it aids in the more thorough emptying of these chambers, this combined action tending to readjust the inequality in the arterial and venous circulations. Moreover, by forcing more blood into the coronary arteries, and by stimulating the vagi, which are probably trophic as well as inhibitory nerves, digitalis may permanently influence for good the tone of the heart-muscle.

Hypodermic injections of the tincture or of digitalin are often useful in *syncope*, in *collapse* from various causes, and in *poisoning* by cardiac depressants.

In *chronic valvular disease with symptoms of failing compensation* digitalis is often of the utmost value, but it must not be used indiscriminately. The chief indications for its use are dropsy, deficient urination, and a, rapid, weak, and irregular pulse. When cardiac hypertrophy overbalances dilatation and the symptoms of arterial hyperemia exist, its administration will be productive of harm.

The best results are seen in *mitral regurgitation.* In this disease digitalis secures a more perfect closure of the mitral valves by causing a more vigorous contraction of the muscles surrounding the orifice; moreover, it supports the right ventricle, which is subjected to a severe strain on account of the increased resistance in the pulmonary circuit. In some cases of mitral regurgitation, digitalis, although apparently indicated, fails to afford relief because of the existence of pericardial adhesions or of advanced degeneration of the myocardium. When the right ventricle is much distended and the evidences of venous stasis are marked, the drug may fail to act until the tension in the veins has been lessened by the abstraction of a few ounces of blood.

In *mitral stenosis* digitalis is far less serviceable than it is in mitral regurgitation. It should be used only when, as the result of a failing right ventricle, dropsy and anuria are present. Theoretically, by strengthening the contractions of the right ventricle, and by lengthening diastole, so allowing a longer time for blood to enter the left ventricle through the constricted orifice, it should be effective in lessening venous stasis, but unfortunately, in many cases of mitral stenosis, there is extensive disease of the myocardium, and in the presence of the latter—which in all forms of endocarditis is more important than the mere mechanical defect in the valve— digitalis is either useless or harmful.

In *aortic stenosis* digitalis is rarely indicated, and used injudiciously it may provoke serious injury. When, however, there are symptoms of back-pressure in the lungs, with dropsy, it may be employed with advantage.

The existence of *aortic regurgitation* is not necessarily a contraindication to the employment of digitalis. It is true that by prolonging diastole regurgitation is favored, but under certain conditions the drug more than compensates for this drawback by strengthening the contractions of the ventricle, and by improving the tone of the myocardium. It should be withheld when the pulse is slow, full, and regular, but when the pulse becomes rapid, weak, and irregular, and symptoms of venous stasis appear, it may be administered tentatively.

In *tricuspid regurgitation* not dependent upon disease of the valves, but due to simple dilatation of the right ventricle, digitalis is sometimes effective, but when used too freely it may excite, as Potain has observed, pulmonary hemorrhage.

In *advanced fatty degeneration* digitalis often entirely fails or acts unfavorably, but in the early stages of the disorder the exhibition of minute doses, as recommended by Balfour, is often followed by gratifying results. In these cases it is nearly always necessary to counteract the constricting effect of the drug on the arterioles by giving with it some vasodilator, like nitroglycerin, sodium nitrite, or potassium iodid. When symptoms of *angina pectoris* are present, digitalis must be used with extreme caution, but even here small doses are sometimes of value in the intervals between the attacks, and especially after severe attacks, when symptoms of cardiac failure present themselves.

In *pneumonia*, when there is failure of the right ventricle in its efforts to drive blood into the partially consolidated lung, and in consequence the pulse becomes rapid and feeble, digitalis often causes a decided improvement in the symptoms. In such cases it has seemed to us to be more effective when given hypodermically.

As a Vagus Stimulant.—In Graves's disease digitalis sometimes quiets the heart and lessens the pulse-rate, but generally its action is disappointing, and in not a few cases it is badly borne. In the *irritable heart* resulting from overwork, and in *arhythmia* dependent upon simple dilatation, it is often of great service, but in *nervous palpitation* and *paroxysmal tachycardia* it rarely affords relief.

As a Diuretic.—We regard digitalis as the most valuable remedy we possess in *cardiac* and *renal dropsy*. The best results are obtained when the pulse is soft and rapid. In *hepatic dropsy*, in common with other diuretics, it often proves useless. In *pleural effusion* and in other inflammatory transudations it is rarely of value.

Contraindications.—Digitalis is contraindicated in aneurysm, advanced degeneration of the myocardium, and well-developed atheroma.

Administration.—For general use the tincture is the most reliable preparation; as a diuretic, however, some prefer the infusion. When digitalis is not well borne by the stomach, or when a prompt action is desired, the tincture or digitalin may be given hypodermically. When it is desirable to give the drug in pill form the powdered leaves or the extract will be found convenient. In *cardiac* or *hepatic dropsy* it may be advantageously combined with mercury, as in the well-known Niemeyer's pill:

```
R  Pulveris digitalis,
   Pulveris scillæ,
   Massæ hydrargyri,        āā gr. xx (1.3 gm.).—M.
      Fiant pilulæ, No. xx.
   Sig.  One thrice daily after meals.
```

In renal dropsy the following combination of digitalis with juniper and potassium acetate is well thought of:

```
R  Junip. contus.,              ℥x (40.0 gm.);
   Pulv. scillæ,                ℥j (30.0 gm.);
   Pulv. digitalis,             ℥j (4.0 gm.);
   Vin. xerici,                 Oj (½ liter).
Macerate for four days and add:
   Potass. acetatis,            ℥iij (12.0 gm.).—M.
      Express and filter.
   Sig.  Tablespoonful three times a day.
```

Patients taking digitalis should be seen at frequent intervals, on account of the suddenness with which untoward symptoms may develop during its administration. When large doses are being used it is essential that the patient should be confined to bed. While some patients can take small doses of digitalis with advantage for long periods without interrupting the treatment, it nevertheless is advisable in most cases to suspend the administration every six or seven days for a like period, employing in the interval some alternate, such as caffein, strophanthus, or convallaria.

Incompatibles.—Digitalis is incompatible with preparations containing tannic acid. With iron salts it forms an inky mixture.

STROPHANTHUS, U. S. P.

Strophanthus is the seed of *Strophanthus hispidus,* a native of Africa. It contains a glucosid, *strophanthin,* which represents, in a measure, the therapeutic properties of the drug. This principle appears as a white, amorphous, or crystalline powder, of an intensely bitter taste, and soluble in water and alcohol.

PREPARATIONS.	DOSE.
Tinctura Strophanthi, U. S. P.	3–10 min. (0.18–0.65 c.c.).
Strophanthin	$\frac{1}{100}$–$\frac{1}{50}$ gr. (0.0006–0.0013 gm.).

Physiologic Action.—Strophanthus resembles digitalis in its physiologic action; like the latter, it stimulates the heart, slows the pulse, contracts the peripheral vessels, and increases the flow of urine. There are, however, certain points of difference in the action of the two drugs; thus, strophanthus slows the pulse by acting on the heart itself, and not by stimulating the vagi; constriction of the peripheral arteries by strophanthus is much less pronounced than that caused by digitalis, and appears to be due solely to the effect of the drug on the vessel-walls, and not to stimulation of the vasomotor center. Strophanthus is less efficient as a diuretic than digitalis, but what power it possesses of increasing the quantity of urine results from its direct action on the renal epithelium as well as from increased arterial tension; in moderate doses strophanthus is less likely to disturb the stomach than digitalis, but in large amounts it causes epigastric distress, nausea, vomiting, and diarrhea.

In concentrated form strophanthus acts as a muscle-poison, causing paralysis and tonic contraction of the fibers. As it is both absorbed and eliminated rapidly it has little or no cumulative effect.

Therapeutics.—Strophanthus is a valuable cardiac tonic, and may be employed in the class of cases in which digitalis is indicated; it is more prompt in its action, but, as a rule, it is less reliable than the latter. It may be recommended whenever the older remedy fails, or is not well borne, or when an alternate is desired.

Contraindications.—The contraindications to the use of strophanthus are the same as those referred to in connection with digitalis.

CAFFEINA, U. S. P.

(Caffein.)

Caffein is an approximate principle obtained from the leaves of *Thea sinensis*, or from the seeds of *Coffea arabica*. It appears in the form of colorless, fine, silky crystals, having a bitter taste and a neutral reaction, and soluble in about 80 parts of cold water. The drug is freely soluble in solution of sodium benzoate and sodium salicylate.

PREPARATIONS.	DOSE.
Caffeina Citrata, U. S. P.	2–5 gr. (0.13–0.3 gm.).
Caffeina Citrata Effervescens, U. S. P.	2–4 dr. (8.0–15.0 gm.).
Caffeinæ Sodio-benzoas (45 per cent. of caffein),	5–10 gr. (0.3–0.6 gm.).

Physiologic Action.—**Circulatory System.**—Moderate doses of caffein raise the blood-pressure and increase the frequency of the pulse; these effects being due partly to the stimulant action of the drug on the vasomotor center and partly to its direct influence on the heart.

Nervous System.—Caffein is a powerful cerebrospinal stimulant. The brain is especially susceptible to its influence; even small doses sharpen the intellect and induce wakefulness. This action, however, is only temporary, since sooner or later the brain fails to respond to usual doses. Unlike opium, it does not affect the imaginative so much as the reasoning powers.

Large doses cause extreme nervous excitement, exaggeration of the reflexes, and even convulsions of an epileptiform or á tetanic character.

Muscles.—In concentrated form it acts as a muscle-poison and excites tonic contraction of the fibers.

Respiratory System.—In medicinal doses caffein is a powerful stimulant to the respiratory center.

Kidneys.—In ordinary doses it acts as a diuretic, partly by raising the blood-pressure but chiefly by stimulating the secreting structure of the kidneys.

Tissue Metabolism.—It is generally believed that caffein lessens to some extent tissue-waste.

Elimination.—It is largely converted into xanthin bodies in the tissues, and as such eliminated by the kidneys. As its elimination and decomposition are both quickly effected there is no tendency to cumulative action.

Toxicology.—The characteristic symptoms of caffein-poisoning are mental excitement followed by delirium, rapid pulse, rapid breathing, frequent urination, intestinal colic, numbness of the extremities, tremors, convulsions, and, finally, collapse.

Therapeutics.—As a stimulant in *heart-disease* caffein is distinctly less useful than digitalis. When for any reason the administration of the latter must be suspended, however, caffein will be found an efficient *alternate.* A combination of caffein with digitalis often acts more favorably than either drug alone. It is preferable to digitalis when the *myocardium is seriously affected.* It is also more useful than digitalis in combating the cardiac insufficiency of *pneumonia* and other *infective processes.*

Being both a direct and indirect diuretic, it often induces diuresis in *chronic parenchymatous nephritis with dropsy* when digitalis fails. In *chronic interstitial nephritis,* when the left ventricle dilates, and, in consequence, the urine decreases, dyspnea develops and dropsy appears, it may prove an excellent substitute for digitalis. In *acute pulmonary edema* no drug is more useful than caffein when administered hypodermically.

As a respiratory stimulant it is valuable in *poisoning by opium* and other drugs which depress the respiratory center. In combination with phenacetin or one of the bromids it will often afford relief in *migraine* and *headache from nervous strain.*

Contraindications.—It is contraindicated in all conditions associated with troublesome insomnia. In some persons even small doses, when administered late in the day, cause wakefulness at night.

Administration.—Since caffein and citrated caffein are readily decomposed in the presence of water, they are generally prescribed for internal use in pills, powders, or capsules. When a prompt action is desired the drug should be given hypodermically, and for this purpose the sodiobenzoate of caffein, as it is freely soluble and quite stable in solution, will be found a suitable preparation. The following formulæ will illustrate the method of prescribing caffein:

> ℞ Caffeinæ citratæ,   gr. xxiv (1.6 gm.);
>   Phenacetini   ʒj (4.0 gm.).—M.
> Fiant chartulæ, No. xii.
> Sig. One every two hours. (In *nervous headache.*)

> ℞ Caffeinæ,   gr. xlviii (3.0 gm.);
>   Sodii benzoatis, ʒij (8.0 gm.).—M.
> Fiant chartulæ, No. xii.
> Sig. One every four hours. (In *chronic nephritis with dropsy.*)

> ℞ Caffeinæ sodio-benzoatis, gr. xij (o 8 gm.);
>   Aquæ destillatæ, ʒj (4 c.c.).—M.
> Sig. 20 minims = 4 grains, or 1 c.c. = 0.2 gm.

SPARTEIN.

Spartein is a liquid alkaloid obtained from the tops of *Cytisus Scoparius,* or common broom. It is official in the form of *Sparteinæ sulphas,* a white, crystalline, or granular powder, of a somewhat bitter taste, and freely soluble in water and alcohol. The dose of spartein sulphate is ¼ gr. (0.015 gm.) cautiously increased to 1 gr. (0.06 gm.).

Physiologic Action.—In sufficient dose spartein slows the pulse, increases its volume, and causes a rise of the arterial pressure. Toxic doses accelerate the pulse and induce a fall of arterial pressure. The primary slowing of the pulse is probably due to stimulation of the vagi, and the later quickening of the circulation to paralysis of inhibition. The rise of arterial pressure is due in part to direct stimulation of the heart, and in part, according to some authorities, to stimulation of the vasomotor center. The final fall of arterial pressure is due to cardiac depression.

Therapeutic doses of spartein have little if any effect on the nervous system. Toxic doses at first appear to stimulate the spinal cord, since they exaggerate reflex action and induce tonic convulsions. Later, however, they depress the spinal cord and cause paralysis. According to some observers spartein is a distinct depressant to the peripheral motor nerves.

The marked diuretic action of scoparius is probably dependent upon *scoparin* and not upon spartein, since the latter does not materially increase the amount of urine.

Spartein-poisoning is characterized by a weak, rapid pulse, dyspnea, tremors, inco-ordination, convulsions, muscular weakness, and collapse. Death results either from paralysis of the respiratory center or from paralysis of the peripheral motor nerves.

Therapeutics.—While, as a rule, spartein is decidedly inferior to digitalis in valvular disease with failing compensation, there are cases in which it proves to be the more effective remedy of the two. Its action is more often favorable, however, in cases in which the rapidity and irregularity of the pulse are of functional origin. Clarke, Pawinski, and Wood have found it useful in Graves's disease.

Incompatibles.—Alkalies, tannic acid, potassium iodid.

CONVALLARIA, U. S. P.
(Lily of the Valley.)

Convallaria is the rhizome and roots of *Convallaria majalis.* It contains two glucosids, *convallarin* and *convallamarin.*

Convallarin in full doses acts simply as a drastic cathartic. so that the value of the drug as a cardiac stimulant is chiefly due to the convallamarin.

Preparations.	Dose.
Extractum Convallariæ Fluidum, U. S. P.	5–20 min. (0.3–1.2 c.c.).
Convallamarin	½–1 gr. (0.03–0.065 gm.).

Physiologic Action and Therapeutics.—The results obtained from the study of the physiologic action of convallaria are somewhat conflicting, but clinical experience indicates that it acts much like digitalis in stimulating the heart and increasing the flow of urine. It has the advantages of not having any cumulative effect, and of not disturbing the stomach.   Most authorities, however, agree that, as a rule, its action is inferior to that of digitalis, but that it sometimes gives excellent results in cases in which the latter remedy has failed.   Its best effects are seen in the *dropsy of heart-disease.*

Administration.—Convallaria should be prescribed only in the form of the fluid extract, since the convallamarin of commerce is a variable product.

ADONIDIN.

Adonidin is a glucosid obtained from the *Adonis vernalis,* a perennial plant indigenous to Central Europe and Asia.   It is an amorphous, yellow or brown, odorless powder, having an intensely bitter taste.   It is soluble in water and alcohol.   The dose is from $\frac{1}{12}$–$\frac{1}{4}$ gr. (0.005–0.01 gm.).

Physiologic Action.—Adonidin affects the circulation somewhat like digitalis.   It slows the pulse, probably by stimulating the vagi, and increases the arterial pressure probably by stimulating both the heart and the vasomotor center.   Its action is more prompt than that of digitalis, but not so powerful.   It does not seem to have any cumulative action.   When the arterial pressure is low, adonidin causes some increase in the quantity of urine, but the drug apparently exerts very little influence on the kidney itself.   Overdoses may cause nausea, vomiting, and diarrhea.

Therapeutics.—Adonidin is used in about the same class of cases as digitalis.   While generally less reliable than digitalis, it may be employed as a substitute when the older remedy is badly borne or is ineffective.   Oliver speaks favorably of its action in *aortic regurgitation.*

CACTUS GRANDIFLORUS.

This plant, known as *Cereus grandiflora,* or *night-blooming cereus,* though unofficial, has been employed medicinally for a number of years on account of its stimulating effect on the heart.   The tincture and fluid extract are available preparations, the dose of the former being 20–30 min. (1.0–2.0 c.c.); of the latter, 5–10 min. (0.3–0.6 c.c.).

Physiologic Action and Therapeutics.—Like digitalis, cactus increases the arterial pressure, probably by stimulating the heart itself and the vasomotor center.   It does not disturb the stomach, and it appears to be free from cumulative action. It is sometimes employed with advantage in *functional affections of the heart,* in *simple dilatation,* and in *valvular disease* with failing compensation when digitalis is not well borne.

OTHER CIRCULATORY STIMULANTS.[1]

✝ **Ether** (see p. 105).—In ordinary doses ether is a quickly acting heart-stimulant. It raises arterial pressure by a direct action on the heart and by stimulation of the vasomotor center. On account of the promptness of its action it is a valuable remedy in *sudden heart-failure*, as in *shock, poisoning*, collapse of *cholera*, and *low fevers*. In these cases it may be given hypodermically in doses of from 15–30 min. (1.0–2.0 c.c.).

Nitrites—nitroglycerin, amyl nitrite, sodium nitrite—(see p. 133) affect the circulation in a similar manner, but differ in the rapidity and duration of their action, nitrite of amyl being the most rapid, but the least permanent, and nitrite of sodium being the least rapid, but the most permanent. They primarily stimulate the heart, but this action is very evanescent and is soon followed by one of depression. By paralyzing the vagi they increase the rate of the pulse, and by depressing the muscular coats of the arteries they cause a marked fall of the arterial pressure.

The value of the nitrites in diseases of the circulatory system depends not so much on their influence upon the heart itself as on their power to lesson arterial tension. They are especially useful in *valvular disease*, when there is precordial distress and increased pulse-tension. When there is *myocardial degeneration* they may be combined advantageously with small doses of digitalis, in order to counteract the constricting effect of the latter on the blood-vessels. The dyspnea, headache, and precordial discomfort attendant upon *chronic nephritis* with arteriosclerosis is often relieved by a nitrite. Inhalations of amyl nitrite are often exceedingly efficacious in the paroxysms of *angina pectoris*. Its action, however, is uncertain, and the state of the pulse is no guide to its use. It succeeds quite as often when the pulse-tension is low as when it is high.

When a very prompt action is desired, nitrite of amyl (3–5 drops) may be given by inhalation. Nitroglycerin may be given in the form of the official 1 per cent. alcoholic solution, the dose of which is 1–2 min. (0.06–0.1 c.c.), cautiously increased, or in the form of tablets containing a dose of from $\frac{1}{200}$–$\frac{1}{50}$ gr. (0.0003–0.001 gm.).

On account of the marked variation in individual susceptibility small doses should be administered at first, and these gradually increased until the limit of tolerance is reached.

---

[1] The properties of these drugs are more fully considered under other headings.

Strychnin (see p. 126) in full doses raises arterial pressure by stimulating the heart and the vasomotor center. This action, together with its tonic influence upon the stomach and nervous system, renders it one of the most valuable remedies that we possess in combating circulatory depression. In *simple dilatation of the heart,* with or without valvular disease, it is often of the greatest service when combined with digitalis, and occasionally succeeds alone when the latter fails. In *pneumonia* and other *infective processes* it is more valuable than digitalis as an adjuvant to alcohol. In the *weak heart of the old,* also, it is generally a better remedy than digitalis. Administered hypodermically it is especially valuable in *surgical shock.* In *chronic bronchitis* and *emphysema* it is indispensable in strengthening the right ventricle.

Belladonna (see p. 70) quickens the pulse by depressing the inhibitory nerves and stimulating the accelerators, and raises the arterial pressure by stimulating the vasomotor center and the heart itself As a circulatory stimulant it is especially useful in *shock* and *collapse,* in which conditions it is best given hypodermically in the form of atropin sulphate. Pushed to the limits of tolerance it is sometimes efficacious in *tachycardia.*

Adrenalin (see p. 338).—When injected directly into a vein, adrenalin causes a marked increase in both the force and rate of the pulse. The increased arterial pressure is due to the direct action of the drug on the heart itself, and in part, also, to its powerful constricting influence on the peripheral blood-vessels. The increased rapidity of the pulse is probably caused by stimulation of the heart itself. These effects of adrenalin on the circulation are prompt but evanescent, and are only pronounced when the drug is given intravenously or in large doses subcutaneously. Administered in one of these ways, adrenalin chlorid (1 : 1000), in doses of from 1–2 drams (4.0–8.0 c.c.), diluted fifty times by normal salt solution, has been distinctly serviceable in maintaining the arterial pressure in *shock* and *collapse* the result of severe injury or operation.

Camphor (see p. 115).—Camphor is a useful diffusible cardiac stimulant in *adynamic fevers.* It not only serves to strengthen the pulse, but it also acts favorably in lessening the restlessness and delirium. It is best given hypodermically in doses of ½–2 gr. (0.03–0.13 gm.), dissolved in sterilized olive oil.

CIRCULATORY DEPRESSANTS.

The following drugs decrease the force and rapidity of the
pulse by depressing the heart:

Aconite.	Hydrocyanic acid.
Veratrum viride.	Veratrin.
Antimony (tartar emetic).	Pulsatilla.

Cold applied to the region of the heart also acts as a circulatory depressant

A circulatory depressant may lower arterial pressure by its
action on the heart itself or its contained ganglia, or by dilating the blood-paths.

Aconite acts on the heart or its contained ganglia ; *veratrum
viride* acts by depressing the heart and, also, through its alkaloid, *jervine,* by depressing the vasomotor center. The action
of *tartar emetic* is somewhat similar to that of veratrum viride,
but it is undetermined whether the widening of the blood-paths
is due to a centric or a peripheral influence.

The *nitrites,* although they primarily stimulate the heart,
cause a marked fall of arterial pressure by dilating the peripheral arteries.

**The Effect of Circulatory Sedatives on the Pulse-
rate.**—All of the heart sedatives in small doses slow the pulse.
This they may do, as in the case of *tartar emetic*, by depressing
the heart itself, or, as in the case of *aconite* and *veratrum viride*,
by stimulating the vagi as well as by depressing the heart.

As a class, circulatory sedatives are indicated when the
action of the heart is unduly forcible. Thus, they may be of
service in sthenic fevers, in acute local inflammations, in valvular disease when there is excessive hypertrophy of the heart,
and in aneurism when the tension in the arteries is high.

ACONITUM, U. S. P.
(Monkshood.)

Official aconite is derived from the root of the *Aconitum
Napellus,* growing in Europe, Asia, and America. The root is
conical in shape, 2 or 3 inches long, blackish in color, and
closely resembles horse-radish ; from this, however, it may be
distinguished by its lack of pungent odor when scraped.
When slowly chewed it produces in the mouth a sense of
warmth and tingling, soon followed by numbness.

Active Principle.—The most important active principle
of aconite is *aconitin,* a variable alkaloid occurring in both an
amorphous and a crystalline form. It is feebly soluble in water,

but freely so in alcohol and ether. When rubbed on the skin it causes tingling, and, finally, more or less anesthesia. The dose of aconitin is $\frac{1}{200}$–$\frac{1}{150}$ gr. (0.0003–0.0004 gm.).

PREPARATIONS.	DOSE.
Tinctura Aconiti, U. S. P.	1–5 min. (0.06–0.3 c.c.).
Extractum Aconiti Fluidum, U. S. P.	1–2 min. (0.06–0.12 c.c.).
Extractum Aconiti, U. S. P.	$\frac{1}{8}$–$\frac{1}{4}$ gr. (0.008–0.016 gm.).
Linimentum Aconiti.	

Physiologic Action.—**Circulatory System.**—The dominant action of aconite is on the circulation. It slows the pulse by stimulating the peripheral vagi, and lowers the blood-pressure by directly depressing the heart or its motor ganglia. Toxic doses cause an arrest of the heart in diastole.

Nervous System.—Medicinal doses do not affect the brain, but toxic doses sometimes induce clonic convulsions by interfering with the cerebral circulation. Large doses depress the sensory neurons of the spinal cord and the sensory filaments of the peripheral nerves. When applied locally it first irritates and then paralyzes the peripheral sensory nerve-fibers.

Respiration.—Medicinal doses slow the respiration by depressing the respiratory center.

Stomach.—To the stomach aconite acts as a sedative; even toxic doses may fail to induce vomiting.

Temperature.—In the febrile state moderate doses of aconite cause a fall of temperature, which is chiefly due to an increase of the heat-dissipation.

Toxicology.—Aconite-poisoning is characterized by a very slow, weak pulse, which later may become rapid and irregular; slow and feeble respirations; a sensation of tingling about the lips, tongue, and fingers, and occasionally throughout the body; a subnormal temperature; and, ultimately, by all the phenomena of profound collapse.

Treatment.—The patient should be kept in the recumbent position, with the feet somewhat higher than the head. While in this position the stomach should be thoroughly emptied with the stomach-pump. The temperature should be kept up by the external application of heat, and such stimulants as ammonia, atropin, strychnin, and digitalis should be given hypodermically for their effect upon the heart and respiration.

Therapeutics.—In the early stages of acute inflammations, such as *tonsillitis, pharyngitis, laryngitis,* and *bronchitis,* and in the beginning of the *acute infectious diseases of childhood,* aconite is often very useful in quieting the circulation and lowering the temperature. In *excessive hypertrophy of the heart* it sometimes acts favorably. In *acute cerebral conges-*

tion with high arterial tension, it may be given in full doses. In tachycardia and palpitation it is of little value unless these symptoms be associated with hypertrophy of the heart. Small doses ($\frac{1}{2}$ drop), at frequent intervals, are sometimes efficacious in *hyperemesis.* An ointment of aconitin (5–10 gr. to the ounce—0.3–0.6 to 30.0 gm.) was formerly much used as a local remedy in *neuralgia.* The liniment of aconite, official in the British Pharmacopeia, is often beneficial in *muscular rheumatism.*

Administration.—The tincture is the preparation most commonly selected for internal use; small doses should be given in water, frequently repeated, and the effect watched. Aconitin, on account of its variableness and pronounced toxicity, is rarely given internally. If used externally, care should be taken to avoid the eye, as the drug is highly irritating. On cutaneous surfaces, provided there are no abrasions, it may be applied with safety. The following time-honored combination is efficient in the mild febrile diseases of childhood:

R Tincturæ aconiti, ⋅ ⋅ ⋅ ⋅ ⋅ ⋅ ⋅ ⋅ ⋅ ⋅ ⋅ ⋅ ⋅ ⋅ ⋅ ⋅ ⋅ ⋅ ⋅ mv (0.3 c.c.);
    Spiritus ætheris nitrosi, ⋅ ⋅ ⋅ ⋅ ⋅ ⋅ ⋅ ⋅ ⋅ ⋅ ⋅ fȝvj (23.0 c.c.);
    Liquoris ammonii acetatis,   q. s. ad fȝiij (90.0 c.c.).—M.
Sig. A teaspoonful every two or three hours for a child of two years.

VERATRUM VIRIDE, U. S. P.
(American Hellebore.)

The rhizome and roots of *Veratrum viride,* a native of North America. It contains several alkaloids, but the most important are *jervin* and *veratroidin.*

PREPARATIONS.	DOSE.
Tinctura Veratri Viridis, U. S. P. ⋅ ⋅ ⋅ ⋅ ⋅	2–6 min. (0.12–0.4 c.c.).
Extractum Veratri Viridis Fluidum, U. S. P.	1–3 min. (0.06–0.2 c.c.).

Physiologic Action.—The dominant action of veratrum viride is on the circulatory system : it lessens in a marked degree both the force and the rate of the cardiac pulsations. The lowered arterial tension results from depression of the vasomotor center and of the heart itself; the slowing of the pulse, from stimulation of the inhibitory nerves of the heart and from weakening of the cardiac muscle. The drug is irritating to the stomach, and even moderate doses often cause nausea and vomiting. Large doses depress the respiratory center and spinal cord.

Action of the Alkaloids.—The action of veratrum viride represents the sum of the actions of the alkaloids. According to Wood, veratroidin causes the slowing of the pulse by stimulating the vagi ; jervin depresses the heart and vasomotor center ; and both, in large doses, depress the respiratory center and spinal cord.

Toxicology.—Veratrum-viride-poisoning is characterized by nausea, vomiting, a slow weak pulse, great muscular relaxation, vertigo, impaired vision, partial unconsciousness, slow shallow breathing, and death from asphyxia or syncope.

On account of the vomiting which follows the ingestion of large doses, veratrum viride, although a powerful drug, is not so dangerous a poison as aconite.

Treatment.—This consists in keeping the patient in the recumbent posture, in favoring emesis by copious draughts of warm water, in keeping up the temperature by the external application of heat, and in the administration hypodermically of full doses of diffusible cardiac and respiratory stimulants, such as ammonia, alcohol, strychnin, and digitalis.

Veratrum Viride and Aconite Compared.— Aconite differs in its action from veratrum viride in not influencing, except in toxic doses, the vasomotor center ; in allaying gastric irritability, instead of exciting nausea and vomiting ; and in depressing the sensory neurons, thereby causing numbness and tingling.

Therapeutics.—In excessive *hypertrophy of the heart*, with or without valvular lesions, veratrum viride is a useful circulatory sedative. In *aneurism*, when there is much pain and the pulse is full and strong, veratrum viride, in conjunction with potassium iodid and rest, sometimes affords relief. In the beginning of *acute inflammatory diseases* it may replace aconite as a cardiac sedative and febrifuge.

In the first stage of *croupous pneumonia* of a sthenic type it may affect favorably the pulse and temperature, but it has no power of aborting the disease. In *puerperal eclampsia*, by diverting blood from the brain and by depressing the motor neurons of the spinal cord, its administration has often proved a valuable addition to other treatment.

Contraindications.—Cardiac and systemic weakness and irritability of the stomach are the chief contraindications to its use.

Administration.—The tincture is the best preparation ; it should be given at frequent intervals, the dose being cautiously increased until the physiologic effect is produced.

ANTIMONIUM.
(Antimony, Sb.)

The following preparations of antimony are official : Tartar emetic (*antimonii et potassii tartras*) ; sulphid of antimony (*antimonii sulphidum*) ; purified sulphid of antimony (*antimonii sulphidum purificatum*) ; sulphurated antimony (*antimonium sulphuratum*) ; oxid of antimony (*antimonii oxidum*) ; compound pills of antimony, or Plummer's pills (*pilulæ antimonii compositæ*), each containing $\frac{1}{2}$ gr. (0.03 gm.) of calomel, $\frac{1}{2}$ gr. (0.03 gm.) of sulphurated antimony, and 1 gr. (0.06 gm.) of guaiac; and antimonial or James's powder (*pulvis antimonialis*), which is composed of about one-third oxid of antimony and two-thirds phosphate of calcium. Of these preparations, tartar emetic is the only one in common use.

ANTIMONII ET POTASSII TARTRAS, U. S. P.
(Tartar emetic, $2K(SbO)C_4H_4O_6.H_2O.$)

Tartar emetic occurs in the form of colorless transparent crystals, or as a white granular powder having a sweetish metallic taste. It is soluble in 17 parts of cold water, and almost insoluble in alcohol. Its dose as an emetic is $\frac{1}{2}$–1 gr. (0.03–0.06 gm.); as an expectorant, $\frac{1}{16}$–$\frac{1}{12}$ gr. (0.004–0.005 gm.).

PREPARATIONS.	DOSE.
Vinum Antimonii, U. S. P. 2 grs. of tartar emetic to 1 oz. (0.4 gm.–100 c.c.).	
As an expectorant . . . . 10–30 min. (0.6–1.8 c.c.).	
As an emetic . . . . $\frac{1}{2}$–1 fl. oz. (15.0–30.0 c.c.).	
Syrupus Scillæ Compositus, U. S. P.	
Tartar emetic, 2; squill, senega, and water, to	
1000 . . . . 5–60 min. (0.3–4.0 c.c.).	

Physiologic Action. — Circulatory System. — Tartar emetic lessens the rate and force of the pulse by directly depressing the heart. The fall of blood-pressure is probably also in part due to the action of the drug on the vasomotor nerves.

Nervous System.—Large doses paralyze the spinal cord, especially the sensory neurons.

Respiratory System.—Toxic doses greatly embarrass the respiration and produce extreme irregularity of rhythm. This effect is due to the depressant action of the drug on the respiratory center, and indirectly to the pulmonary congestion incident to the heart-failure.

Alimentary Canal.—Small doses excite a sensation of warmth in the stomach; large doses cause nausea and vomiting; and lethal doses produce gastro-enteritis with serous discharges from the stomach and intestines. The emesis is partly due to the irritant action of the drug on the stomach, and partly to its action on the vomiting center in the medulla oblongata.

Local Action.—Applied to the skin, tartar emetic acts as an irritant, and its prolonged application in concentrated form is followed by the appearance of papules, vesicles, pustules, and even sloughs.

Elimination.—Its elimination is effected through all the mucous membranes, but especially through the stomach and intestines.

Toxicology.—In toxic doses tartar emetic causes intense burning in the esophagus and stomach, followed by persistent vomiting and purging of serous or " rice-water " material, painful cramps, not only in the stomach, but in the muscles of the extremities, and all the symptoms of collapse—thready pulse, suppressed voice, shallow respiration, subnormal temperature, pinched features, cold sweats, and scanty urine. The mind may be clear until the close, but frequently death is preceded by delirium, convulsions, and coma.

Postmortem examination reveals gastro-enteritis, engorgement of the venous system, and widespread fatty degeneration of the tissues.

From the close resemblance of antimony-poisoning to Asiatic cholera and other choleraic diseases, considerable difficulty may be experienced in arriving at a correct diagnosis. The history of the case and the recovery of antimony from the discharges are the decisive factors.

Treatment.—Tannic acid should be administered as the chemical antidote, and the stomach should be emptied by means of the pump. The temperature should be kept up by the external application of heat, and stimulants like ether, alcohol, and strychnin should be given hypodermically to combat the heart-failure. To check the vomiting and to relieve the irritation, morphin may be given hypodermically and demulcents by the mouth.

Therapeutics.—As a remedy, tartar emetic is fast falling into disuse, its place having been filled largely by other drugs, which are more prompt in their action and are less depressing. It may be employed, however, as an emetic, a diaphoretic, cardiac sedative, sedative expectorant, and counter-irritant.

Its slowness of action and depressant effect render it less valuable as an emetic than ipecac, alum, zinc sulphate, or apomorphin. As a cardiac sedative and diaphoretic it was formerly much used in the beginning of acute sthenic inflammations—*pneumonia, pleurisy, pericarditis,* and *acute rheumatism,* but aconite and veratrum are preferable, since their action is less permanent. As a sedative expectorant it is of considerable value in *acute bronchitis,* when there is hard cough, little expectoration, and the chest is full of dry râles. The ointment of tartar emetic has been employed to produce a persistent counter-irritant effect in deep-seated inflammation, such as *chronic arthritis* and *chronic meningitis.*

Administration.—The wine of antimony ($\frac{1}{2}$–1 tablespoonful), compound syrup of squill ($\frac{1}{2}$–1 teaspoonful for a child), or tartar emetic ($\frac{1}{2}$–1 gr.—0.03–0.06 gm.) may be employed as an emetic. In acute bronchitis the wine of antimony or the compound syrup of squill may be combined with other sedative expectorants and with opium, thus:

   ℞ Vini antimonii,   f℥iiss (9.0 c.c.);
     Tincturæ opii camphoratæ, f℥j (30.0 c.c.);
     Liquor ammonii acetatis, q. s. ad f℥iv (120.0 c.c.).—M.
  Sig. A dessertspoonful in water every two or three hours.

   ℞ Tincturæ aconiti,   ♏vj (0.04 c.c.);
     Syrupi scillæ compositi,
     Tincturæ opii camphoratæ, aa f℥ij (8.0 c.c.);
     Syrupi tolutani,   f℥j (30.0 c.c.);
     Liquoris potassii citratis, q. s. f℥iij (90.0 c.c.).—M.
  Sig. Dessertspoonful every three hours for a child of five years.

Incompatibles.—Tartar emetic is incompatible with tannic acid and drugs containing it, other acids, and alkalies.

OTHER CIRCULATORY DEPRESSANTS.

Veratrin.—Veratrin is an alkaloidal principle obtained from the seed of *Asagræa officinalis.* It is a whitish powder, having an acrid taste, odorless, but when inhaled causing intense irritation and persistent sneezing. It is official as Unguentum veratrinæ (4 per cent.) and Oleatum veratrinæ (2 per cent.).

Locally, veratrin is a powerful irritant. Internally, after large doses, the cardiac inhibitory center and the vasomotor center are primarily stimulated, in consequence of which the pulse is slowed and the blood-pressure is raised; later, however, the medullary centers, and probably the heart also, are depressed, so that the pulse becomes rapid and weak. On account of its irritant action on mucous membranes large doses occasion vomiting and purging. Toxic doses first excite the voluntary muscles and induce tetanic seizures, and then cause muscular paralysis. In fatal poisoning death results from asphyxia.

Veratrin is never used internally. Externally it has been employed chiefly in *neuralgia* and *muscular rheumatism*. As absorption can be effected through the skin, considerable caution should be exercised in its application.

Pulsatilla.—This drug is the herb of *Anemone Pulsatilla* and *Anemone prætensis.* It contains a colorless, crystalline, acrid principle, known as *anemonin.* The dose of the latter is from $\frac{1}{8}$–$\frac{1}{2}$ gr. (0.008–0.03 gm.). The drug may be used also in the form of the tincture, the dose of which is 10–20 min. (0.6–1.2 c.c.).

Locally, pulsatilla acts as an irritant. Internally, large doses may cause vomiting and even purging. On the circulation its action resembles that of aconite. Toxic doses depress also the respiration and spinal cord.

Pulsatilla has been used as a circulatory sedative in acute inflammatory diseases, as a respiratory sedative in asthma, and as an emmenagogue, but as a therapeutic agent it is of little value.

Hydrocyanic Acid (see p. 68).—Large toxic doses of hydrocyanic acid kill instantly, the respiratory center being paralyzed and the heart arrested in diastole almost simultaneously. Smaller doses first slow the pulse and then quicken it by stimulating and subsequently paralyzing the vagi. There is also a primary rise of the arterial pressure from stimulation of the vasomotor center, and then a fall, from paralysis. Hydrocyanic acid is never used as a circulatory sedative.

RESPIRATORY STIMULANTS.

The following drugs stimulate the respiratory center in the medulla oblongata:

Ammonia.	Cocain.
Strychnin.	Atropin.
Caffein.	

In addition to these drugs, *external heat* increases the depth of the respirations. The expiratory and inspiratory fibers of the vagus are stimulated by the rhythmic expansion and collapse of the lung, which takes place in the practice of artificial respiration by *Sylvester's method.* In pursuing this method the patient is placed in a supine position; his arms are drawn upward and outward so as to fully expand the chest; after a pause of a few seconds the arms are brought down forcibly to the sides of the chest and pressed against them for a second. These movements are repeated about sixteen or eighteen times in a minute. The respiratory center may be stimulated also reflexly by irritating certain peripheral sensory nerves. Thus, in *Laborde's method* of treating asphyxia by rhythmic traction of the tongue, the center is affected through the sensory nerves of the tongue. This method consists in drawing the tongue outward and upward by means of forceps for from twelve to fourteen times a minute.

The *vapor of ammonia* when inhaled stimulates the nasal branches of the fifth nerve, and thus reflexly excites the respiratory center. The use of the cold douche, the pouring of ether on the bare chest, and the application of an electric current to the skin are also means of arousing the respiratory center through peripheral irritation.

Drugs which stimulate the respiratory center are indicated in pulmonary affections associated with marked dyspnea, such as *pneumonia, severe bronchitis*, and *emphysema;* and in poisoning by drugs which cause asphyxia, such as *opium* and *ether*.

The various measures by which the respiratory center is stimulated reflexly are especially indicated in poisoning.

RESPIRATORY DEPRESSANTS.

The following drugs depress the respiratory center:

Hydrocyanic acid.
Opium and its chief alkaloids.
Hyoscin.
Nitrites.
Bromids.
Chloral. } Spinal cord depressants in large doses.
Physostigma.
Gelsemium.

In addition to these drugs there are many others, such as aconite, veratrum viride, alcohol, ether, and belladonna, which in *toxic doses* kill by asphyxia.

Certain of the respiratory depressants, notably hydrocyanic acid and opium and its alkaloids, are employed in pulmonary diseases to *allay cough.*

ACIDUM HYDROCYANICUM DILUTUM, U. S. P.

(Diluted Hydrocyanic or Prussic Acid, HCN.)

Diluted hydrocyanic acid is a 2 per cent. aqueous solution of pure hydrocyanic acid. It appears as a colorless liquid, of feeble acid reaction, having an odor and taste of bitter almonds. The dose is from 1–3 min. (0.05–0.2 c.c.).

Physiologic Action.—**Circulatory System.**—Moderate doses slow the pulse by stimulating the inhibitory nerves of the heart or by directly acting on the cardiac ganglia. The drug first causes a distinct rise of the arterial pressure by stimulating the vasomotor center, and then a fall from paralysis. Large toxic doses kill instantly, the respiratory center being paralyzed and the heart arrested in diastole almost simultaneously.

Blood.—For a short time after death from prussic-acid poisoning the blood may have a uniformly bright red color. This phenomenon appears to be due to the formation of cyano-hemoglobin. It is not probable, however, that this compound is produced during life.

Nervous System.—In lethal doses it acts as a paralyzant to all parts of the nervous system. Clinical experience indicates that in medicinal doses it has a distinct sedative influence on the sensory nerve-filaments.

Respiratory System.—It is a powerful depressant to the respiratory center. In poisoning, unless death be instantaneous, the respiration usually fails before the circulation.

Elimination.—Being highly volatile, it is absorbed and eliminated with great rapidity; its action, therefore, is very fugacious.

Toxicology.—Hydrocyanic acid is one of the most quickly acting of poisons, death often resulting immediately or after the lapse of a few minutes. When death is not instantaneous, the characteristic phenomena are slow respiration, cyanosis, coma, protruding eyeballs, dilated pupils, frothing at the mouth, convulsive movements of the muscles, involuntary discharge of urine and feces, and the odor of bitter almonds about the patient.

Therapeutics.—Hydrocyanic acid is used to allay irritability of the respiratory center and of the peripheral nerves, especially the gastric and cutaneous nerves. It is an excellent remedy for allaying the troublesome cough of *phthisis* and *bronchitis*. It is one of the best sedatives we possess for relieving *gastralgia* and for controlling the pain and vomiting of *gastric ulcer* and *catarrh*. A lotion of hydrocyanic acid may be employed to relieve itching in *eczema* and *pruritus*, provided there is no abrasion of the skin.

Administration.—As solutions of hydrocyanic acid tend to deteriorate with age, the preparations on the market will be found to vary considerably in their effectiveness. The following formulæ will illustrate the manner in which the drug may be prescribed:

℞ Codeinæ hydrochloratis, gr. iv (0.25 gm.);
 Acidi hydrocyanici diluti, ♏xxxij (2.0 c.c.);
 Syrupi tolutani, q. s. ad f℥ij (60.0 c.c.).—M.
Sig. A teaspoonful every four hours. (*Cough of phthisis.*)

℞ Bismuthi subnitratis, ℥ss (15.0 gm.);
 Acidi hydrocyanici diluti, ♏xl (2.5 c.c.);
 Aquæ, q. s. ad f℥iv (120.0 c.c.).—M.
Sig. Shake well. A dessertspoonful before meals. (*Gastric ulcer or catarrh.*)

℞ Acidi hydrocyanici diluti, f℈ss (2.0 c.c.);
 Glycerini, f℈j (4.0 c.c.);
 Aquæ, q. s. ad f℥ij (60.0 c.c.).—M.
Sig. Apply as directed. (*Pruritus.*)

Potassii Cyanidum (*Potassium Cyanid*, KCN).—This is an intensely poisonous salt of potassium, appearing as a white powder or as white, opaque, amorphous pieces, having an odor of bitter almonds and a sharp alkaline taste. It is soluble in 2 parts of water. Dose, $\frac{1}{12}$–$\frac{1}{10}$ gr. (0.005–0.006 gm.).

Therapeutics.—It is converted by the acids of the stomach into hydrocyanic acid, and may be used as a substitute for the latter. It has the same toxic properties as the acid from which it is derived, but is somewhat slower in its action.

CEREBRAL EXCITANTS.

Our knowledge concerning the action of drugs on the brain is very imperfect. We know, however, that some of them stimulate the motor centers, and in large doses cause epileptiform convulsions, while others affect chiefly the psychic centers. Among the latter there is probably only one that increases the reasoning power, and that is *caffein;* the others merely excite the imagination and emotion, and in large doses induce delirium.

The following drugs in toxic doses cause epileptiform convulsions probably by their direct action on the motor centers of the brain:

Cocain.	Salicylic acid.	Salts of copper.
Camphor.	Tansy.	Salts of silver.
Quinin.	Salts of lead.	

The cardiac sedatives (*aconite, veratrum viride*, and *antimony*) in poisonous doses also produce cerebral convulsions, which, according to H. C. Wood, are not due to their direct influence on the motor centers, but to disturbance of the circulation at the base of the brain.

The following drugs before they depress the brain exalt imagination and emotion:

Alcohol.	Coca.
Opium.	Cannabis indica.

In certain doses the following drugs affect the cerebral cells in such a way as to cause delirium:

Alcohol.	Hyoscyamus.
Belladonna.	Cannabis indica.
Stramonium.	Coca.

Anesthetics (*ether* and *chloroform*) also occasion delirium before complete anesthesia is attained.

A few of the cerebral excitants, particularly atropin, caffein, and cocain, are useful in arousing the cerebral centers when they have been profoundly depressed by poisons, such as opium.

BELLADONNA.

Belladonna is official as the leaves (*belladonna folia*) and as the root (*belladonna radix*) of *Atropa Belladonna*. The active principle is *atropin*, which is official as *Atropina* and *Atropinæ sulphas*. Both the alkaloid and its salt appear in the form of a white crystalline powder, having an acrid, bitter taste. The former is sparingly soluble in water, the latter in 0.4 part of water and 6.5 parts of alcohol. The dose of atropin is $\frac{1}{200}$–$\frac{1}{75}$ gr. (0.00032–0.00086 gm.); of atropin sulphate, $\frac{1}{150}$–$\frac{1}{50}$ gr. (0.00043–0.0013 gm.).

· PREPARATIONS.	DOSE.
Tinctura Belladonnæ Foliorum, U. S. P. . .	5–20 min. (0.3–1.2 c.c.).
Extractum Belladonnæ Foliorum Alcoholicum, U. S. P.	$\frac{1}{8}$–$\frac{1}{4}$ gr. (0.008–0.016 gm.).
Extractum Belladonnæ Radicis Fluidum, U. S. P.	1–2 min. (0.06–0.12 c.c.).
Emplastrum Belladonnæ, U. S. P.	
Unguentum Belladonnæ, U. S. P. (10 per cent.).	

Physiologic Action.—When taken internally in large doses belladonna produces dryness of the throat, dilatation of the pupils, quickening of the pulse, elevation of temperature, and sometimes talkative delirium and an erythematous rash. The rash generally appears on the face and neck and resembles that of scarlatina, but it lacks the characteristic punctations of the latter.

Circulatory System.—Medicinal doses quicken the pulse and raise the arterial pressure. The quickening of the pulse results from depression of the inhibitory nerves and stimulation of the accelerators. The increased blood-pressure is due to stimulation of the vasomotor center and the heart itself. Toxic doses paralyze the heart.

Respiratory System.—Large doses stimulate the respiratory center; toxic doses paralyze it.

Nervous System.—The action of belladonna on the nervous system is a complex one. The active delirium induced by poisonous doses indicates that the drug is a cerebral excitant. Ordinary doses do not affect the spinal cord. Toxic doses first cause paralysis and abolish the reflexes, and later excite tetanic seizures and heighten the reflex functions. The primary effect, which appears promptly and gradually wears away, appears to be due to depression of the motor nerves and the motor paths of the cord, and the secondary effect (often delayed for several hours) is due, according to Ringer and Murrell, to depression of the inhibitory centers of the cord. It depresses both the motor and sensory nerve-fibers, but the former far more than the latter.

Muscles.—It has no action on striated muscles, but large doses paralyze non-striated muscles.

Alimentary Canal.—Small doses increase peristalsis by paralyzing the inhibitory fibers of the splanchnic; very large doses arrest peristalsis. It decreases to some extent the flow of gastric juice.

Secretion.—By an inhibitory influence on the nerves supplying the various secretory glands, belladonna lessens all secretions except those of the kidneys and intestines.

Local Action.—In concentrated form it acts as a depressant to all highly organized tissues.

Eye.—It dilates the pupil by paralyzing the peripheral ends of the oculomotor nerve and by stimulating the peripheral ends of the sympathetic. It destroys the power of accommodation and increases intra-ocular tension.

Elimination.—While it escapes from the body through other channels, it is chiefly eliminated through the kidneys.

Toxicology.—The characteristic phenomena of belladonna-poisoning are dryness of the throat, dilated pupils, rapid pulse, hurried respiration, restlessness, talkative delirium, an erythematous rash, and, finally, stupor, paralysis, and collapse.

The urine of patients suffering from belladonna-poisoning when dropped into the eye of an animal (except birds) causes dilatation of the pupil.

Treatment of Poisoning.—After evacuating the stomach, tannic acid should be administered as a chemical antidote. Respiratory and circulatory stimulants are frequently required. Opium, physostigma, and jaborandi are somewhat antagonistic to belladonna, and, therefore, have been recommended as physiologic antidotes. Considerable caution should be exercised in their employment.

Therapeutics.—Belladonna is employed chiefly to relax spasm, to check excessive secretion, to stimulate the vasomotor center, to dilate the pupil, to allay peripheral irritation, to stimulate intestinal peristalsis, to impress the nervous mechanism of the heart, and to antagonize certain poisons.

To Relax Spasm.—It is of little value in general convulsions of cerebral or spinal origin, but it is of great service in local spasms excited by peripheral irritation. Thus, it is useful in *whooping-cough* and *laryngismus stridulus.* In the various forms of *colic*—renal, biliary, and intestinal—atropin is a valuable adjuvant to morphin, and should be given freely to guard the latter and to relax spasm. In *asthma* it may be given internally, but inhalation of the smoke made by burning the leaves often gives better results. When *incontinence of urine* is due to vesical hyperesthesia and spasm of the sphincter, as it frequently is in children, belladonna is a very reliable remedy. To be effective, however, the remedy must be pushed to the point of intolerance, that is, until it causes dryness of the throat and dilatation of the pupils. In obstinate *torticollis* or *wry-neck* intramuscular injections of atropin are often followed promptly by relief. In *spasm of the anal sphincter* and in *dysmenorrhea* suppositories of belladonna are often efficacious.

To Check Excessive Secretion.—Atropin is one of the best remedies we possess for controlling the *night sweats* of phthisis. A dose of $\frac{1}{200}$–$\frac{1}{100}$ gr. (0.00032–0.00065 gm.) at bedtime is sufficient. Unfortunately, untoward effects—dryness of the throat, thirst, and dimness of vision—often militate against its use. In *salivation* from mercury it is also useful. In *hyperidrosis* of the hands and feet belladonna is used both as an internal remedy and as a local application. When the mother is unable to suckle her child and the breast becomes swollen and tender, the thorough application of an ointment of belladonna will often serve to arrest the secretion of milk and to prevent suppuration, even though the latter seems imminent. In *leucorrhea* with ulceration of the os uteri, Trousseau's suppositories containing 1 gr. (0.06 gm.) of the extract and 8 gr. (0.5 gm.) of tannin will be found beneficial.

In the *bronchopneumonia* of childhood belladonna in large doses (2 drops of the tincture every hour) is often of the utmost value in lessening secretion and in relieving dyspnea. In *pneumonia* with edema it is also useful.

To Stimulate the Vasomotor Center.—In *shock* and *collapse* atropin hypodermically is a valuable remedy, though less effective than strychnin. It has also been recommended as a vasomotor stimulant in the algid stage of *cholera.*

To Dilate the Pupil.—If a drop or two of a solution of atropin, having the strength of 4 gr. (0.26 gm.) to the ounce (30.0 c.c.), be dropped into the conjunctival sac dilatation of the pupil begins in fifteen minutes, and attains its maximum in about half an hour. Accommodation is not affected quite so quickly: paralysis not being complete within an hour and a half. On the other hand, mydriasis persists somewhat longer than the suspension of accommodation. The effect of a 4 grain solution usually lasts from ten days to two weeks.

Atropin is sometimes employed as a mydriatic to *facilitate ophthalmoscopic examination*, but homatropin or cocain may be substituted with less inconvenience to the patient. For the *estimation of refraction* it is necessary to select a mydriatic that will suspend the accommodation, and for this purpose atropin may be used, although there are cycloplegics equally efficient and less persistent in their effects. In *iritis* it is indispensable in preventing and in breaking up adhesions between the iris and the capsule of the lens. In *acute keratitis* it is also very useful in allaying the ciliary irritation.

The sulphate, on account of its ready solubility, is the salt usually selected. As a simple mydriatic ¼ gr. (0.016 gm.) to the ounce (30.0 c.c.) is sufficient, and the mydriasis from this solution will not last more than four or five days. As a cycloplegic in refraction work a strength of 4 gr. (0.26 gm.) to the ounce (30.0 c.c.) should be employed. A like solution is generally used in iritis. In keratitis a solution containing 1–2 gr. (0.065–0.13 gm.) to the ounce (30.0 c.c.) will be effective.

To Allay Peripheral Irritation.—In the form of an ointment or plaster belladonna makes a very useful sedative application in *muscular rheumatism* (pleurodynia, wry-neck, lumbago), in *inflammatory affections of the joints, orchitis,* and *chronic heart-disease.*

Suppositories containing the extract are much used in *pain-*

ful hemorrhoids, and for the relief of *painful defecation* due to inflammation of the pelvic organs.

To Stimulate Intestinal Peristalsis.—Belladonna is a useful adjuvant to other drugs in *chronic constipation.* By allaying spasm of the intestinal muscles it also prevents griping.

To Impress the Nervous Mechanism of the Heart.—Although the manner of its action is obscure, belladonna is often efficacious in *nervous palpitation* and *tachycardia.* It may be applied over the region of the heart in the form of a large plaster, or given internally, alone or in combination with sodium bromid. In some cases of *exophthalmic goiter,* when given in doses large enough to cause some dryness of the throat, it is more useful than any other drug.

To Antagonize Certain Poisons.—It is a useful antidote in *poisoning by opium, physostigma, muscarin* (poisonous principle of toadstools), *hydrocyanic acid, aconite, jaborandi,* and *chloroform.*

Administration.—The root and leaves are not used internally. The tincture is the most reliable of the liquid preparations. For subcutaneous administration the sulphate of atropin should be employed, since atropin is only sparingly soluble in water. Children are relatively less susceptible to belladonna than adults, and, therefore, often tolerate much larger doses than their age would naturally indicate. The following formulæ will serve to illustrate the manner of prescribing the drug:

```
R  Atropinæ sulphatis,              gr. j (0.065 gm.);
   Acidi borici,                    gr. x (0.65 gm.);
   Aquæ destillatæ,                 fℨj (30.0 c.c.).—M.
```
Sig. A drop or two to be instilled into the conjunctivæ twice a day. (*Acute keratitis.*)

```
R  Tincturæ belladonnæ,            fℨiss (5.5 c.c.);
   Glycerini,                      fℨij (7.4 c.c.);
   Aquæ menthæ piperitæ,  q. s. ad fℨij (60.0 c.c.).—M.
```
Sig. Thirty minims three times a day for a child of three years. The dose to be gradually increased until slight flushing of the face is induced. (*Whooping-cough.*)

```
R  Extracti belladonnæ,            gr. x (0.65 gm.);
   Acidi tannici,                  gr. vj (0.4 gm.);
   Hydrargyri chloridi mitis,      gr. xxx (2 0 gm.);
   Cocainæ hydrochloratis,         gr. vj (0.4 gm.);
   Unguenti petrolati,             ℨj (31.0 gm.).—M.
```
Sig. Apply night and morning after washing the parts. (*Hemorrhoids.*)

Incompatibles.—Atropin should not be prescribed in solutions containing caustic alkalies or tannic acid.

STRAMONIUM.
(Jamestown Weed; Thorn-apple.)

The drug is official as the leaves (*stramonii folia*) and the seed (*stramonii semen*) of *Datura Stramonium*, a weed growing in waste places in Europe, Asia, and America. The active principle is an alkaloid, *daturin*, which is identical in its physiologic action with atropin. The dose of daturin is from $\frac{1}{150}$–$\frac{1}{60}$ gr. (0.0004–0.001 gm.).

PREPARATIONS.	DOSE.
Tinctura Stramonii Seminis, U. S. P.	10–20 min. (0.6–1.0 c.c.).
Extractum Stramonii Seminis Fludium, U. S. P. .	1–3 min. (0.06–0.2 c.c.).
Extractum Stramonii Seminis, U. S. P.	$\frac{1}{4}$–$\frac{1}{2}$ gr. (0.015–0.03 gm.).
Unguentum Stramonii, U. S. P., contains 10 per cent. of the extract.	

Therapeutics.—Stramonium is a therapeutic equivalent of belladonna, and may be employed in the same class of diseases in which the latter is prescribed. It has a special reputation in *asthma*, in which disease it may be used in the form of cigarettes made of the dried leaves. Ointment of stramonium with an equal amount of ointment of galls makes a useful application for *hemorrhoids*.

HYOSCYAMUS, U. S. P.
(Henbane.)

The leaves and flowering tops of *Hyoscyamus niger*, a plant growing in Europe, Asia, and North America. Two alkaloids are present in the leaves, *hyoscyamin* and *hyoscin*. Hyoscyamin is official in the form of the sulphate (*Hyoscyaminæ sulphas*) and the hydrobromate (*Hyoscyaminæ hydrobromas*), both of which are freely soluble in water. Hyoscin is official in the form of the hydrobromate (*Hyoscinæ hydrobromas*), which is soluble in about 2 parts of water and in 13 parts of alcohol. The dose of the salts of hyoscyamin is from $\frac{1}{100}$–$\frac{1}{50}$ gr. (0.0006–0.0013 gm.). The dose of hyoscin hydrobromate is from $\frac{1}{150}$–$\frac{1}{80}$ gr. (0.0004–0.00085 gm.).

PREPARATIONS.	DOSE.
Tinctura Hyoscyami, U. S. P.	20–60 min. (1.2–3.7 c.c.).
Extractum Hyoscyami Fluidum, U. S. P. . .	5–20 min. (0.3–1.2 c.c.).
Extractum Hyoscyami, U. S. P.	$\frac{1}{2}$–3 gr. (0.03–0.2 gm.).

Physiologic Action.—In full doses hyoscyamus acts like belladonna, causing dilatation of the pupils, dryness of the throat, and quickening of the pulse and respiration. On the brain its action is somewhat more sedative than that of belladonna.

Therapeutics.—Like belladonna, it is used to relax *spasm* and to allay *peripheral irritation*. It may be employed in *whooping-cough*, in *bronchitis* with harassing cough, when opium is contraindicated, and in *incontinence of urine* due to vesical irritation. It is a useful vesical sedative in *acute cystitis*.

HYOSCYAMINA.

Hyoscyamin is almost equivalent as a therapeutic agent to atropin. Like the latter, it dilates the pupil, but its action is less prolonged. In painful *facial spasm* and *torticollis* it is sometimes very useful. Dana has recommended it in the *chorea of childhood*. It may afford relief in the tremors of *paralysis agitans*, *disseminated sclerosis*, and *chronic mercurial poisoning*.

HYOSCINA.

In the main, the action of hyoscin resembles that of atropin, but, unlike the latter, it quiets the brain, acting as a powerful hypnotic, and rather depresses than stimulates the respiratory center. To the spinal cord also it seems to be a more powerful depressant than atropin. In full doses it has a tendency to produce paresis of the throat-muscles, and for this reason is contraindicated in affections involving the throat, like diphtheria and scarlet fever. Toxic doses of hyoscin induce dryness of the throat, dilatation of the pupils, muttering delirium, stupor, slowing of the respirations and pulse, free perspiration, and muscular relaxation. In animals death results from asphyxia.

On account of the very pronounced susceptibility of certain persons to the action of hyoscin, the initial dose should always be small.

Therapeutics.—Hyoscin hydrobromate is used as a hypnotic, a depressomotor, and an anaphrodisiac. It is a very efficient hypnotic in *insomnia*, associated with great mental or emotional excitement; thus it often acts admirably in *mania*, *delirium tremens*, *melancholia agitata*, and *typhoid fever* with active delirium. As it has very little effect upon the circulation it sometimes acts favorably as a somnifacient in *chronic heart-disease*, but usually it is inferior to morphin, chloralamid, and chloretone. Given hypodermically in doses of $\frac{1}{200}$ gr. (0.0003 gm.) twice daily, gradually increased, it is sometimes useful in quieting the violent movements of *acute chorea*. It is the best remedy that we have in diminishing the severity of the tremors and in allaying the restlessness of *paralysis agitans*. A dose of $\frac{1}{150}$ gr. (0.0004 gm.), gradually increased to

$\frac{1}{8.0}$ gr. (0.0008 gm.), may be given in chloroform-water two or three times a day. It is an excellent remedy in *spermator-rhea, erotomania*, and allied sexual disorders.

Administration.—It may be given hypodermically, or by the mouth in pills, granules, or aqueous solution.

CANNABIS INDICA, U. S. P.
(Indian Hemp; Hashish).

The flowering tops of an East Indian plant, *Cannabis sativa*. It contains a resinous principle known as *cannabin*.

PREPARATIONS.	DOSE.
Tinctura Cannabis Indicæ, U. S. P.	10–30 min. (0.6–1.8 c.c.).
Extractum Cannabis Indicæ Fluidum, U. S. P.	5–10 min. (0.3–0.6 c.c.).
Extractum Cannabis Indicæ, U. S. P.	$\frac{1}{8}$–1 gr. (0.008–0.06 gm.).

Physiologic Action.—In large doses it produces a condition of mental exhilaration associated with hallucinations and disordered consciousness of time, locality, and personality. This stage of excitement finally gives way to sleep, which may last several hours. Sensation is perverted and benumbed, and before sleep is induced there is often more or less general anesthesia. Upon the circulatory and respiratory systems the drug has little influence.

Although alarming symptoms sometimes follow overdoses of cannabis indica, no death from its use has been recorded.

Therapeutics.—Cannabis indica is used chiefly as a mild analgesic, sedative, and hypnotic. Although much inferior to opium, it may replace the latter when, for any reason, the more active drug is unavailable. It is often efficient in the persistent cough of *chronic bronchitis* and *phthisis*, in the dyspnea of *asthma*, and in the restlessness and insomnia of *chronic nephritis*. As an analgesic it is very valuable in certain forms of *neuralgia*, especially *migraine*, in which affection it is not only useful between, but also during, the paroxysms. It is occasionally effective in relieving the violent pains of *locomotor ataxia*. It may be employed as an adjuvant to chloral in *tetanus*. In some forms of *dysmenorrhea* and *menorrhagia* it proves serviceable. In combination with the bromid of potassium it is sometimes useful in *mania* and *delirium tremens*.

Sée and Suckling have strongly recommended it in *gastralgia*.

Administration.—As the preparations of cannabis indica vary considerably in strength, and as some individuals are far more susceptible to the action of the drug than others, it is always advisable to begin with small doses and to increase them gradually. A dose of 10 drops of the tincture, repeated in four hours, has caused in an adult intense excitement, dilatation of the pupils, and disordered sensation.

As the addition of water or of an aqueous liquid to the fluid extract or tincture precipitates the resinous principle, an alcoholic menstruum should be selected for these preparations, or, in case water is used, enough mucilage should be added to hold the resin in suspension.

COCA, U. S. P.

The leaves of *Erythroxylon Coca*, a native of Peru and other South American States. It contains several alkaloids, the most important being *cocain*, which is official in the form of cocain hydrochlorate (*Cocainæ hydrochloras*). The latter appears as colorless crystals, which have a slightly bitter taste, and are freely soluble in water and alcohol. Its dose is from $\frac{1}{4}-\frac{1}{2}$ gr. (0.015–0.03 gm.).

Preparation.	Dose.
Extractum Cocæ Fluidum, U. S. P.	1–2 fl. dr. (4.0–8.0 c.c.).

Physiologic Action.—In full doses cocain raises the arterial pressure and increases the rate of the pulse. The rise of arterial pressure is due to its stimulant action on the vasomotor center and the heart, and also, perhaps, to its direct constricting influence on the peripheral vessels. The increased pulse-rate, according to Reichert, is due to depression of the cardio-inhibitory centers or the cardio-inhibitory ganglion. After very large doses there is a pronounced fall of the arterial pressure from vasomotor paralysis and a decrease of the pulse-rate from depression of the cardio-inhibitory centers.

Locally, it is a powerful vasoconstrictor; the ischemia, however, is finally replaced by congestion.

Nervous System.—Cocain is a cerebral excitant, in moderate doses inducing a state of mental exhilaration. Lethal doses, however, cause stupor and convulsive seizures, which are probably of cerebral origin. Toxic doses also depress the spinal cord and the peripheral nerves, especially the sensory columns and sensory nerve-filaments. When locally applied its depressant action on the sensory filaments is sufficiently pronounced to induce complete analgesia and anesthesia.

Respiratory System.—In small doses it increases both the depth and the rate of the respirations by stimulating the respiratory center. Toxic doses render the respiratory movements irregular in rhythm and shallow, and finally kill through asphyxia.

Muscles.—It appears first to stimulate and then to paralyze striated muscles.

Kidneys.—Cocain has very little influence on the excretion of urine. The nitrogenous elimination is probably slightly decreased.

Alimentary Canal.—Small doses increase and large doses diminish peristaltic movement.

Temperature.—Therapeutic doses are without effect, but toxic doses cause a rise of temperature, probably by increasing heat-production.

Local Action.—An important physiologic property of cocain is its power, when applied directly to a mucous membrane, or when injected under the skin, of inducing anesthesia of the part by paralyzing the sensory nerve-filaments. The application of cocain solutions directly to cutaneous surfaces is without effect, since absorption through the epidermis is very slight. In addition to anesthesia, it causes a blanching of the part, which is subsequently followed by congestion.

Eye.—The instillation into the eye of a few drops of a 4 per cent. solution of cocain causes in from ten to fifteen minutes, in addition to anesthesia, marked dilatation of the pupil. The mydriasis attains its maximum in about an hour, and persists for from twelve to twenty-four hours. It is of peripheral origin, and is probably caused by stimulation of the sympathetic nerve. Cocain has but little effect on accommodation. Strong solutions have an injurious effect upon the corneal epithelium.

Toxicology.—The symptoms of *acute cocain-poisoning* are somewhat variable ; the usual phenomena, however, are nervous excitement, followed by delirium, and ultimately by stupor ; nausea, vomiting, a very rapid pulse, hurried respirations, elevation of temperature, dilatation of the pupils, and convulsive seizures.

The **treatment** is purely symptomatic. When there are symptoms of circulatory and respiratory failure, ammonia, atropin, and strychnin are indicated. When cerebral excitement and convulsions are prominent features, inhalations of chloroform may be employed, and bromids and chloral may be administered tentatively.

Untoward Effects of Medicinal Doses.—There is a very wide variation in individual susceptibility to cocain. In some subjects the local application of the drug to the nose or throat, even in solutions of moderate strength, is followed by nausea, vomiting, blindness, syncope, or convulsions. The application of a 10 per cent. solution to the mouth has been followed by convulsions lasting five or ten minutes.

Chronic Poisoning (*Cocainism*).—A potent cause of the cocain-habit has been the frequent use of the drug in diseases of the nose and throat. Tyson states that he has known three successive chiefs of clinic in nose and throat dispensary service to acquire the habit. The habit is also common among those addicted to other narcotics, such as opium, alcohol, and chloral. The chief symptoms are emotional excitement, physical unrest, mental impairment, moral turpitude, hallucinations, mild epileptiform attacks, dilatation of the pupils, a rapid, feeble pulse, severe gastric disturbance, wasting, and anemia. When cocainism is uncomplicated the prognosis is guardedly favorable. The drug should be withdrawn rapidly but not suddenly. Treatment in a sanatorium is always advisable. Stimulants like strychnin are often useful. Hygienic and dietetic measures calculated to improve general nutrition are indicated.

Therapeutics.—Cocain is employed as a local anesthetic, mydriatic, and respiratory and circulatory stimulant.

As a Local Anesthetic.—As a local anesthetic cocain has a very wide range of usefulness. In the various operations on the eye, nose, throat, urethra, and rectum it is indispensable. In minor surgical operations—like amputation of the fingers, removal of small neoplasms, opening of small abscesses—it may also be employed with advantage. In these cases care must be taken to prevent undue absorption of the drug, and this is best accomplished by the application of a tight rubber band to the proximal side of the part to be operated on, thus controlling the circulation. A 2 per cent. solution may be injected to the extent of a dram with safety. The capillary oozing should not be checked immediately, as it serves to carry off the cocain. The same quantity of cocain in weak solution is distinctly less toxic than it is when in strong solution (Custer).

In operations in which the circulation cannot be controlled, and in which much of the solution must be allowed to remain in the tissues, the method employed by Schleich, known as *infiltration anesthesia,* is of very great value. Schleich employs solutions of three strengths. The one of medium strength consisting of—

Cocain hydrochlorate,	1 gr. (0.065 gm.);
Morphin hydrochlorate,	⅓ gr. (0.013 gm.);
Sodium chlorid,	2 gr. (0.13 gm.);
Sterilized distilled water,	2 fl. oz. (60.0 c.c.).

The stronger solution contains 2 gr. (0.13 gm.) of cocain, and the weaker solution, $\frac{1}{10}$ gr. (0.0065 gm.).

The important factors in the induction of anesthesia by this method are the ischemia of the tissues caused by the pressure of the injected fluid, the compression of the terminal nerve-filaments from the same cause, and the direct paralysis of the sensory nerves by the cocain. The effect of the morphin is not local, but general. Physiologic salt-solution is employed as the vehicle, because it is unirritating to the tissues.

The field of operation may be infiltrated *en masse*, or it may be surrounded by an anesthetic edematous zone. In addition, the nerves leading to the part may also be cocainized either some distance from the prospective incision, or directly as they are exposed in the wound.

An ordinary antitoxin syringe capable of holding at least 10 c.c. may be employed. The derm should be infiltrated before the deeper tissues are injected. The needle having been introduced obliquely, a few drops of the solution are forced out; the needle is then thrust deeper and deeper, while more of the fluid is injected in various directions until the field is thoroughly edematized. The subsequent application of an ice-bag to the part will intensify the analgesic effect of the injection. At least five minutes should elapse after the infiltration before the incision is made. The period of analgesia generally lasts from twenty to thirty minutes. The strength of the solution and the amount employed will depend upon the sensitiveness of the part, the extent of the operation, and the amount of cocain that is likely to be retained in the tissues. It is not safe to allow more than 1 grain of cocain to remain in the part. Of the strong solution, 1 ounce (30.0 c.c.); of the solution of medium strength, 2 ounces (60.0 c.c.); and of the weak solution, 10 ounces (300.0 c.c.) or more may be employed without risk.

Medullary Cocainization.—The employment of spinal sub-arachnoid injections for the purpose of inducing analgesia was suggested by Corning in 1885, and made practical by the researches of Bier, published in 1899. The technic of medullary cocainization is as follows: The entire lumbar and sacral region is prepared with antiseptic care as for operation; the patient is placed in a sitting posture, and the injection is made at the level of the crest of the ilium, that is, just above or below the fourth lumbar vertebra, and 1 cm. ($\frac{1}{3}$ in.) external to its spinous process. The solution is injected through a fine iridoplatinum needle, about 9 cm. (3–4 in.) long, attached to an ordinary hypodermic syringe

6

capable of holding 2 c.c. (30 min.). As the injection should not be made before a flow of cerebrospinal fluid is observed, the needle should be introduced into the subarachnoid space before the syringe is attached. The injection is made very slowly, and to insure the retention of the solution the needle is held *in situ* for about a minute, then carefully withdrawn, and the puncture sealed with collodion. A 2 per cent. solution (9 gr.–1 oz.—0.6 gm.–30.0 c.c.) of cocain is usually employed; it should be freshly prepared and sterile. From 15–20 min. (1.0–1.2 c.c.) are injected. The amount of cocain should not exceed $\frac{2}{5}$ gr. (0.025 gm.). Analgesia extending downward from about the level of the diaphragm follows in from five to ten minutes, and usually persists about an hour and a half.

The relative value of this method of inducing analgesia has not yet been determined; it is certainly not altogether free from danger. Unpleasant after-effects—fever, headache, vertigo, nausea, and vomiting—are of frequent occurrence. Cocain analgesia is not suitable for prolonged, complicated operations, especially when they are intra-abdominal; and since it does not induce unconsciousness, it should not be employed in young children nor in insane or hysteric adults.

In *inflammations* and *ulcerations of the nose, pharynx,* and *larynx* cocain may be employed alone or in combination with antiseptic sprays or powders. In *acute coryza* and *hay-fever* it gives great relief by lessening the sensibility and constricting the turgid tissues.

In *laryngeal tuberculosis* it is invaluable for the relief of the intense pain and dysphagia. It may be given in insufflation in combination with iodoform, iodol, or morphin. As a local anesthetic it is also used internally to relieve the pain of *gastric ulcer* and to check *excessive vomiting.*

The application of a 10 per cent. solution of cocain on a cotton pledget has been found useful in *rigidity of the os uteri.*

As a Mydriatic.—As a mydriatic, *cocain* is not so generally useful as atropin. Since, however, it only slightly impairs accommodation, and its effects pass off in a few hours, it is a convenient mydriatic for *retinal examinations.* In iritis atropin is distinctly preferable on account of its forcible action and lasting effect. The mydriatic effect of cocain is readily neutralized by the instillation into the eye of a few drops of a $\frac{1}{2}$ per cent. solution of eserin.

As a Respiratory Stimulant.—Cocain is one of the most powerful of the respiratory stimulants, and for this reason, and also on account of its general excitant influence, it is an excellent physiologic antidote in *opium-poisoning.*

As a Circulatory Stimulant.—In *adynamic pneumonia* cocain, administered hypodermically, is a valuable adjuvant to strychnin.

Administration.—As a local anesthetic to mucous membranes it is employed in solutions varying in strength from 2 to 20 per cent. The mucous membrane of the larynx is less susceptible than that of the nose or throat. On the latter, from 4 to 6 per cent. solutions are usually sufficient. A solution stronger than 4 per cent. should not be used in the eye on account of the danger of inducing degenerative changes in the corneal epithelium.

The drug is not well borne by children and women of a neurotic temperament. The danger of causing the habit from the frequent use of cocain in chronic diseases of the nose and throat must never be lost sight of, and under no circumstances should the remedy be placed in the patient's own hands. In tuberculous laryngitis with dysphagia small tablets containing from $\frac{1}{20}$–$\frac{1}{12}$ gr. (0.003–0.005 gm.) of cocain may be allowed to dissolve slowly in the mouth, or the drug may be used in the form of a powder or spray:

℞ Cocainæ hydrocbloratis,	gr. xx (1.3 gm.);
Resorcini,	gr. x (0.65 gm.);
Aquæ,	f℥j (30 0 c.c.).—M.

Sig. To be used as a spray. (*Tuberculous laryngitis.*)

Incompatibles.—Cocain is incompatible with tannic acid, alkaline carbonates, iodids, borax, and zinc sulphate. It cannot be added to Dobell's solution, since the latter contains borax.

CEREBRAL DEPRESSANTS.

The cerebral depressants may be classed as somnifacients, general anesthetics, general analgesics, and anticonvulsants.

Somnifacients or *hypnotics* are remedies which induce sleep. The factors favoring natural sleep are the withdrawal of afferent stimuli, decreased responsiveness of the brain-cells as the result of fatigue, the accumulation in the blood of products evolved by the cells in the exercise of their functions, and, finally, lessening of the blood-supply to the brain. Pure somnifacients inhibit the functional activity of the brain-cells, but they do not completely suspend the power of the cells to recuperate (the main object of sleep) nor to react to stimuli which reach them from without.

General anesthetics are drugs which induce unconsciousness and insensibility. Their effect upon the brain-cells is much more profound than that of the hypnotics, in that they temporarily inhibit the responsiveness of the cells to external stimuli, and suspend more or less completely recuperative changes.

General analgesics or *anodynes* are remedies which relieve pain. They may act either by directly influencing the receptive centers in the brain or by preventing the transmission of painful impressions through the sensory nerves and spinal cord to the brain.

Anticonvulsants are remedies which check excessive motor activity. In the case of cerebral convulsions they may act by directly depressing the cells in the cortex of the brain, by lessening the power of the sensory tract in the peripheral nerves and spinal cord to transmit to the brain-cells stimuli from without, or by lessening the power of the motor tract in the spinal cord and peripheral nerves to transmit to the muscles the excessive discharges emanating from the brain-cells.

SOMNIFACIENTS.

The most important element in the treatment of insomnia is the removal of the cause. A potent factor in many cases of persistent insomnia is the circulation in the blood of the products of faulty metabolism, as seen in dyspeptic and gouty subjects. In such affections a careful regulation of the diet is of the utmost importance. The evening meal especially should be simple and light. Systematic exercise, coupled with the use of mild mercurial or saline aperients, often proves more effective in promoting sleep than the exhibition of hypnotics. In the aged insomnia is very often associated with lowered vascular tone, and when such is the case, small doses of strychnin are beneficial. In chronic valvular disease stimulants like digitalis, strophanthus, and nux vomica are indicated. In chronic Bright's disease insomnia is a common symptom, and it may depend upon the retention in the blood of excrementitious materials or it may be due to high arterial tension. In the former case a milk-diet and the use of eliminants will be most helpful, and in the latter case the administration of some vasodilator, such as nitroglycerin or erythrol tetranitrate, will often afford marked relief. Tea, coffee, and tobacco are sometimes preventives of sleep, and if so they must be used sparingly or perhaps altogether avoided. When anemia is a causal factor, iron and arsenic will be needed. Insomnia from overwork, mental anxiety, or grief will demand mental rest, diversion, or change of scene.

When the habit of sleeplessness has become well established it may require more than the removal of the cause to effect a complete cure. Simple measures, however, should always be tried first before resorting to hypnotics. The bedroom should be quiet, well ventilated, and of moderate temperature. When the feet are cold the blood should be diverted from the head by hot foot-baths, which may be made more effective by the addition of mustard. Vigorous rubbing of the limbs is also useful. A glass of hot milk, a cup of bouillon, or a toddy just before retiring often acts favorably as a derivative. In some persons light reading for half an hour, in others a brisk walk, will invite sleep. In neurasthenia, massage and the systematic use of the wet-pack are invaluable aids.

Hypnotic drugs must be used with caution. Their continued exhibition predisposes to invalidism and favors the development of vicious habits. In simple habitual insomnia they should not be prescribed until general measures have been tried and found wanting.

The chief somnifacients are:

Opium.	Paraldehyd.
Chloral.	Hyoscin.
Sulphonal.	Bromids.
Trional.	Chloralose.
Tetronal.	Urethan.
Chloralamid.	Duboisin.
Chloretone.	Pellotin.

OPIUM, U. S. P.

Opium is the inspissated juice obtained by incising the unripe capsules of *Papaver somniferum*, a poppy indigenous in Western Asia. To be up to the official standard, it should contain not less than 9 per cent. of crystallized morphin. It appears in the form of irregular lumps, having a dark-brown color, a gummy consistence, a peculiar narcotic odor, and a bitter taste. In addition to *mcconic acid* and a neutral principle, *meconin*, it contains a number of alkaloids, the most important of which are *morphin, codein, thebain, papaverin, narcotin,* and *narcein.*

PREPARATIONS.	DOSE.
Opium, U. S. P. (9 per cent. of morphin)	$\frac{1}{2}$–2 gr. (0.03–0.13 gm.).
Opii Pulvis, U. S. P. (13–15 per cent. of morphin)	$\frac{1}{2}$–2 gr. (0.03–0.13 gm.).

PREPARATIONS.	DOSE.
Opium Deodoratum, U. S. P. (denarcotized opium; 13–15 per cent. of morphin)	½–2 gr. (0.03–0.13 gm.).
Pilulæ Opii, U. S. P. (1 gr.—0.06 gm. of powdered opium in each)	1–2 pills.
Extractum Opii, U. S. P. (18 per cent. of morphin)	¼–1 gr. (0.016–0.65 gm.).
Tinctura Opii, U. S. P. (laudanum; 10 per cent. of powdered opium)	5–20 min. (0.3–1.2 c.c.).
Tinctura Opii Deodorati, U. S. P. (10 per cent. of powdered opium)	5–20 min. (0.3–1.2 c.c.).
Tinctura Opii Camphorata, U. S. P. (paregoric; contains camphor, benzoic acid, oil of anise, and about 2 gr.—0.12 gm. of powdered opium per ounce)	½–4 fl. dr. (1.8–15.0 c.c.).
Tinctura Ipecacuanhæ et Opii, U. S. P.	5–15 min. (0.3–1.0 c.c.).
Acetum Opii, U. S. P. (black drop; 10 per cent. of powdered opium)	5–20 min. (0.3–1.2 c.c.).
Vinum Opii, U. S. P. (10 per cent. of powdered opium)	5–20 min. (0.3–1.2 c.c.).
Pulvis Ipecacuanhæ et Opii, U. S. P. (Dover's powder; contains 10 per cent. of powdered opium; 10 per cent. of ipecac; and 80 per cent. of sugar of milk)	5–10 gr. (0.3–0.6 gm.).
Trochisci Glycyrrhizæ et Opii, U. S. P. (each contains $\frac{1}{12}$ gr.—0.005 gm. of powdered opium).	
Emplastrum Opii, U. S. P. (6 per cent. of extract of opium).	

Morphin is the chief alkaloid of opium and represents its physiologic activity. It occurs in white prismatic crystals, or fine needles, of a bitter taste, and very slightly soluble in water. The salts of morphin, being much more soluble than the alkaloid itself, are generally prescribed.

The following alkaloids of opium and their preparations are used medicinally :

PREPARATIONS.	DOSE.
Morphina, U. S. P.	⅛–⅓ gr. (0.008–0.032 gm.).
Morphinæ Acetas, U. S. P.	⅛–⅓ gr. (0.008–0.032 gm.).
Morphinæ Hydrochloras, U. S. P.	⅛–⅓ gr. (0.008–0.032 gm.).
Morphinæ Sulphas, U. S. P.	⅛–⅓ gr. (0.008–0.032 gm.).
Pulvis Morphinæ Compositus, U. S. P. (Tully's powder; contains licorice, camphor, calcium carbonate, and morphin sulphate)	5–10 gr. (0.3–0.6 gm.).
Trochisci Morphinæ et Ipecacuanhæ, U. S. P. ($\frac{1}{40}$ gr.—0.0015 gm. of morphin sulphate in each).	
Codeina, U. S. P.	½–2 gr. (0.03–0.13 gm.).
Codeinæ Sulphas	½–2 gr. (0.03–0.13 gm.).
Codeinæ Phosphas	½–2 gr. (0 03–0.13 gm.).

The following compounds are derived from the alkaloids of opium. From morphin: apomorphin (see p. 265), heroin, dionin, and peronin; from codein: apocodein (see p. 266); and from narcotin: cotarnin hydrochlorate (see p. 376).

Physiologic Action.—Circulatory System.—Moderate doses of opium have little effect upon the circulation; large doses, however, slow the pulse by stimulating the vagi, both centrally and peripherally, and increase to some extent the force of the pulse by stimulating the heart or its contained ganglia. Toxic doses finally paralyze both the heart and vagi, and thereby induce a rapid feeble pulse.

Nervous System.—In man the dominant action of opium is on the brain, which it soon depresses, causing sleep. When the dose has not been large, a stage of excitement or exhilaration often precedes the narcosis.

To what action its power of relieving pain is due is not known, but the drug probably acts by depressing both the perceptive centers in the brain and the afferent paths in the spinal cord. Locally, morphin acts as a direct depressant to the sensory nerve-filaments.

Respiratory System.—In large doses opium is a powerful respiratory depressant, and in fatal cases of poisoning by the drug death usually results from paralysis of the respiratory center.

Alimentary Canal.—Opium checks the intestinal secretions, and in moderate doses arrests peristalsis by stimulating the inhibitory nervous apparatus; on account of this double action it has a decided tendency to induce constipation. Toxic doses paralyze the inhibitory nerves and stimulate peristalsis.

Secretions.—All the secretions, except the sweat, are diminished by opium.

Eye.—Large doses contract the pupils by stimulating the oculomotor centers. In poisoning, the pupils sometimes dilate just before death, probably from paralysis of the oculomotor centers.

Elimination.—Morphin is eliminated, in part as morphin, by the various emunctories, especially the stomach and kidneys; much, however, of that which is ingested is oxidized in the body.

Conditions Modifying the Action of Opium.— Certain symptoms and diseases counteract the narcotic effect of opium, and when these are present the drug may be used in much larger doses than are generally prescribed; thus, patients suffering from diabetes or affections attended with severe pain, such as acute peritonitis, renal or biliary colic, can frequently take with advantage doses that under other circumstances would produce profound narcosis. On the other hand, when there are evidences of nephritis the drug must be used with great caution, since small doses not infrequently produce lethal effects.

Age and sex modify its action. Children, on account of the sensitiveness of their nervous systems, are peculiarly susceptible to opium, so that it must be given in smaller doses than the age would naturally indicate. A drop of laudanum has produced a fatal result in a child less than six months old. Its action is more pronounced on women than on men, and in the former it is more apt to cause disagreeable after-effects.

Patients rapidly acquire a tolerance of the drug through frequent repetition of the dose, so that habitués can often take enormous amounts without experiencing the usual effects.

The Action of Opium and Morphin Compared.—Apart from being more powerful than opium, morphin differs from the latter in having a less constipating, less nauseating, and less diaphoretic effect.

Toxicology.—Acute Opium-poisoning.—Unless the dose is very large, there may be at first a stage of excitement, in which the imagination is stimulated and the feelings are exalted. This stage is soon followed by one of depression, in which the patient becomes listless and drowsy, and finally falls into a deep sleep. The pulse is slow and full, the pupils are contracted to a pin-point size, the respirations are slow and labored, the muscles are relaxed, and the face is pale. In this stage it is still possible to arouse the patient by loud noises, flagellation, or shaking. In the third stage coma is absolute, the pulse is rapid and irregular, the skin is clammy, and the breathing is shallow and irregular. Finally, death results from paralytic asphyxia.

Treatment of Acute Opium-poisoning.—The stomach should be emptied by means of the stomach-pump or a stimulating emetic, such as zinc sulphate or mustard flour. Since morphin is eliminated by the stomach and then reabsorbed, lavage should be repeated at short intervals. Black coffee may be given by the mouth or by the rectum ; it promotes wakefulness and also stimulates respiration. The best chemical antidote is potassium permanganate ; 3 to 5 grains (0.2–0.3 gm.) of this drug, dissolved in a glassful of water, should be given at once and repeated in thirty minutes. Tannic acid is also recommended as a chemical antidote, but the sulphate and hydrochlorate of morphin are not precipitated by it. The best physiologic antidotes are the powerful respiratory stimulants, such as cocain, strychnin, and atropin; one or all of these may be given hypodermically in full doses.

Artificial respiration is of the greatest value, and should be maintained so long as the beating of the heart continues. It is necessary that the patient should be aroused and kept awake, so that he himself can aid in keeping up respiration. Flicking with a wet towel, the use alternately of hot and cold water, and the application of the faradic current are among the measures that may be employed in staving off sleep. In carrying out the treatment it is important to avoid chilling or exhausting the patient.

Chronic Poisoning (*Morphinomania, Morphinism*).—The symptoms resulting from the habitual use of opium are an irresistable craving for the drug, loss of flesh and strength, tremors, anemia, a peculiar sallow complexion, anorexia, deranged digestion, a tendency to diarrhea, disturbed sleep, mental depression, irritability, and a characteristic propensity for lying and deceiving.

Treatment of Chronic Opium-poisoning.—Isolation in a special institution or asylum is almost imperative. As a rule, the drug should be withdrawn rapidly, but in aggravated cases not too abruptly for fear of collapse. The diet should consist of nutritious, easily digested food. Strychnin, while it is without specific action, is often extremely valuable for its stimulating effect. Bromids and cannabis indica are sometimes useful in ameliorating the distress which follows the withdrawal of opium. Sulphonal, paraldehyd, and chloretone are the best hypnotics. Massage, graduated exercise, and the Turkish bath are useful roborant measures in the convalescent stage.

Therapeutics.—Opium is used to relieve pain and distress; to induce sleep; to allay peripheral irritation; to check excessive secretion; to promote diaphoresis; to control convulsions; and to check hemorrhage.

To Relieve Pain and Distress.—Opium is by far the best analgesic that we possess. In allaying the severe pain of gross lesions—*fractures, malignant growths*, and *acute inflammation* of serous membranes—it has no rival. In *neuralgia* and other forms of recurrent pain it should be used only after all other measures have failed, and then with extreme caution, since the danger of forming the opium-habit in these cases is very great. In the painful crises of *locomotor ataxia* its use may become imperative. Osler regards morphin, hypodermically, as the most useful drug in those attacks of *angina pectoris* in which amyl nitrite proves ineffective.

In various forms of *colic*—renal, biliary, and intestinal—it is well to combine atropin with morphin, since the former aids the latter in relaxing the spasm.

Acute inflammation of the brain and its membranes cannot be considered, as was formerly the case, a contraindication to the use of opium ; indeed, in *cerebrospinal fever* and other forms of *acute meningitis* it may be the only remedy that will afford relief from the intense suffering. In most cases of *acute appendicitis* morphin should be withheld, since it masks the symptoms and thus interferes with an accurate study of the local conditions.

In *acute melancholia* and other psychoses associated with profound mental anxiety, opium often exerts an excellent calmative effect. In many cases of *chronic heart-disease* it is invaluable in relieving the general discomfort and in promoting sleep. To obtain the best results, however, it should be given hypodermically.

To Induce Sleep.—While opium will relieve *insomnia* from almost any cause, it is especially suitable when the cause of the sleeplessness is *pain.* It is also useful as a somnifacient in *continued fevers,* like typhoid fever, in *delirium tremens,* and in *chorea,* when the movements prevent sleep

Occasionally, owing to some peculiar idiosyncrasy, opium causes excitement and wakefulness instead of sleep. In states of *extreme nervous excitement* chloral is generally superior to opium, but not infrequently better results are obtained from a combination of opium with chloral than from either drug singly.

To Allay Peripheral Irritation.—No remedy is so useful as opium or one of its derivatives in relieving the *irritative cough* of bronchitis and phthisis. It should not be employed, however, in pulmonary affections when the expectoration is copious, and in *pulmonary edema* it is a dangerous drug. In *asthma* morphin hypodermically, with or without atropin, is a reliable remedy. In some cases of *acute vomiting* injections of the drug act happily, but in the *pernicious vomiting of pregnancy* its effects are not good. In *chordee* and in *threatened abortion* it may be employed in the form of suppositories.

To Check Excessive Secretion.—In *acute inflammatory affections of the bowel,* after irritating matters have been removed, opium is often indispensable in checking the excessive secretion. In *cholera infantum* hypodermic injections of morphin ($\frac{1}{100}$ gr.—0.00065 gm.) are sometimes of benefit.

In *diabetes mellitus* it does more good than any other remedy, and as the disease establishes a tolerance for the drug, it should be given in ascending doses until it induces favorable or untoward results. The manner in which it acts in this disease is unknown. Codein is preferred by some practitioners to morphin and opium, since it causes less disagreeable after-effects. In *diabetes insipidus* the drug is of far less value.

To Promote Diaphoresis.—In the beginning of " colds," and in so-called *muscular rheumatism*, it is useful for its sudorific effect. In these cases Dover's powder is the best preparation, since the ipecac which it contains assists the opium in promoting diaphoresis.

To Control Convulsions.—Morphin hypodermically is of service in controlling the painful spasmodic seizures of *tetanus*. Mackenzie, Loomis, and Osler speak favorably of its action in the convulsions and other nervous manifestations of *uremia*, and unquestionably it may be given with great benefit in many cases, but it should always be employed with considerable caution, especially when there are evidences of chronic interstitial nephritis.

Flechsig's method of using opium in the treatment of epilepsy is occasionally of service in old and obstinate cases. This method consists in giving opium in large doses for a period of six weeks (beginning with 1 grain of the extract the dose is rapidly increased until 15 grains a day are taken at the end of the first week, this amount being kept up for six weeks). The opium is abruptly withdrawn and potassium bromid is substituted. Large doses of the latter drug (30 grains three times a day) are given for about two months, and are then gradually reduced until less than 40 grains a day are being taken.

To Check Hemorrhage.—It is a common practice to combine opium with hemostatics in the various hemorrhages, such as *hemoptysis, hematemesis,* and *enterorrhagia.* The good which it accomplishes is probably the result of its calmative influence on the nervous system.

External Use.—In the form of a lotion it makes an efficient sedative application in *sprains, bruises, articular rheumatism,* and *erysipelas.* In these cases, notwithstanding the chemical incompatibility of the two drugs, it is generally combined with lead subacetate, as in the following formula:

    ℞ Tincturæ opii,
        Liquoris plumbi subacetatis,   aa ℥ij;
        Aquæ,   q. s. ad Ōj.—M.

Untoward Effects.—In many persons the use of opium, even in moderate doses, is followed by headache, depression, anorexia, nausea, and vomiting. The gastric disturbance is probably due to the excretion and subsequent reabsorption of some of the products of the decomposition of morphin, such as oxydimorphin, and possibly apomorphin. In some persons, on account of a peculiar idiosyncrasy, morphin causes excitement and wakefulness instead of sleep. In rare instances it occasions general pruritus or an erythematous rash.

Contraindications.—Opium should not be used in pulmonary affections associated with embarrassed respiration, such as edema, and pneumonia during the stage of consolidation. For reasons already mentioned it is best to withhold it in appendicitis. When chronic nephritis exists it must be used with the utmost circumspection.

Administration.—When a prompt analgesic effect is desired, one of the salts of morphin should be selected in preference to opium, and should be administered hypodermically. When large doses are required it is well to add a small dose of atropin ($\frac{1}{150}$ gr.—0.0004 gm.) to each injection. Pulverized opium, deodorized opium, and the extract are suitable preparations for use in pills. Dover's powder is the best preparation for promoting diaphoresis. The most agreeable liquid preparations are the deodorized tincture and paregoric. Paregoric, on account of its weakness and pleasant taste, is especially suitable for children. The disagreeable after-effects of opium are often prevented by combining with it potassium bromid.

Children are extremely intolerant to the drug in all forms, and, in consequence, the dose must be considerably less than the age would apparently justify.

Incompatibles.—Ammonia, alkaline carbonates, preparations containing tannic acid, or salts of the metals should not be prescribed in solution with morphin. Chlorids, bromids, iodids, and borates, when present in large amount, precipitate morphin from solutions. Even dilute hydrocyanic acid may slowly precipitate from solutions of morphin an insoluble cyanid of the alkaloid.

CODEINA, U. S. P.

Among the alkaloids of opium codein ranks next in importance to morphin. It occurs in the form of colorless crystals, of a bitter taste, and soluble in 80 parts of water and in 3 parts of alcohol. The sulphate and phosphate, although unofficial, are to be preferred to the alkaloid itself, on account of their greater solubility. The phosphate, being soluble in 4 parts of water, is adapted to hypodermic use.

Codein resembles morphin in its action, but it is far less powerful as a hypnotic and analgesic. It is less liable, however, than morphin to induce nausea and constipation. It is an excellent remedy for allaying irritative cough in *bronchitis* and *phthisis*. It is preferred by some clinicians to opium in the treatment of *diabetes*, but, as a rule, the crude drug or morphin will be found more efficacious.

HEROIN.

Heroin is the diacetic ester of morphin. It appears as a white crystalline powder, of slightly bitter taste. It is insoluble in water, but freely soluble in acidulated solutions. According to Dreser, who first studied its action, it is a respiratory stimulant rather than a depressant; it is without special influence on the circulation; and it is a more powerful sedative than codein, although it is distinctly less toxic than that drug. Later observations have not confirmed Dreser's conclusions as to the relative toxicity of heroin; Harnack, Winternitz, and Fränkel believe that it is more poisonous than codein. It has been employed chiefly as a sedative for the alleviation of cough and dyspnea. It has feeble analgesic and hypnotic properties. The dose is from $\frac{1}{12}$–$\frac{1}{8}$ gr. (0.005—0.008 gm.). Large doses are frequently followed by nausea, headache, dryness of the throat, vertigo, diplopia, and numbness of the limbs.

Heroin is useful in allaying the cough of *bronchitis* and *phthisis*, although it is not so generally efficacious as codein. In *emphysema* and *asthma* it is an excellent sedative. It may be given in pill, powder, or in water to which a little acid has been added. The hydrochlorid of heroin is freely soluble, and may be given in aqueous solution by the mouth or hypodermically. Heroin is incompatible with apomorphin.

DIONIN.

Dionin is the mono-ethyl-ester of morphin hydrochlorid. It is freely soluble in water. According to Winternitz, it does not depress the respiration, and it has no unpleasant effect upon the stomach. As an analgesic and a hypnotic it is far less powerful than morphin; as a sedative it stands probably midway in efficiency between codein and morphin. It may be prescribed in the same dose as codein.

PERONIN.

Peronin is the hydrochlorid of the benzyl ester of morphin. It was first studied by von Mering, who found it less active than morphin, but freer from disagreeable after-effects. Later researches indicate that, while it is a somewhat more powerful sedative than codein, it is also more toxic. A sufficient number of clinical observations have not yet been made to determine the exact rank of peronin as a therapeutic agent, but, in view of its slight solubility, unpleasant taste, and relatively high toxicity, it would seem to have few, if any, advantages over codein. The dose is from $\frac{1}{3}$–1 gr. (0.02–0.06 gm.).

CHLORAL, U. S. P.
(Chloral Hydrate, $CCL_3.CH(OH)_2$.)

Chloral or, more correctly, chloral hydrate, is obtained from the union of a molecule of water with trichloraldehyd (chloral), the latter being a product of the action of chlorin on alcohol. It occurs in the form of colorless, transparent crystals, having a bitterish, caustic taste, a pungent odor, and a neutral reaction. It is freely soluble in water, alcohol, ether, and chloroform. The dose is from 10–20 gr. (0.65–1.3 gm.).

Physiologic Action.—Therapeutic doses of chloral do not affect the circulation or respiration, but induce quiet sleep of a natural character.

Circulatory System.—In large doses it lowers arterial pressure by depressing the heart, and, perhaps, also the vasomotor centers.

Nervous System.—The dominant action of chloral is on the nervous system. It induces sleep by directly influencing the cerebral cells and lessens reflex activity by depressing the motor neurons of the spinal cord. It seems probable, however, that before abolishing reflex activity it temporarily exaggerates it by stimulating the spinal ganglia. Very large doses induce anesthesia. It has very little influence on the peripheral nerves.

Respiratory System.—Under toxic doses the respirations first become slow, then irregular, rapid and shallow, and, finally, cease from paralysis of the respiratory center.

Temperature.—Toxic doses cause a marked lowering (3°–6° F.) of the body-temperature.

Local Action.—Upon mucous membranes and raw surfaces it is a powerful irritant.

Action on Lower Organisms.—It is destructive to lower organisms and prevents decomposition.

Elimination.—Chloral circulates in the blood as chloral, and is eliminated by the kidneys as urochloralic acid. When taken in excess it escapes in part unchanged. Urine containing urochloralic acid responds to Fehling's test for sugar.

Toxicology.—The characteristic features of acute chloral-poisoning are sleep, deepening into coma, a rapid feeble pulse, slow breathing, followed by rapid shallow breathing, contracted pupils, followed by dilated pupils, muscular relaxation, and collapse. Death may result from either respiratory or cardiac paralysis.

Treatment of Poisoning.—The temperature must be maintained by external heat. Brunton and Stricker found that an animal would stand a very much larger dose of chloral when the body-temperature was kept up artificially than when it was allowed to fall. The patient should be aroused by friction, flagellation, douches, etc., but should not be shaken or forcibly made to walk, as in opium-poisoning, on account of the danger of heart-failure. Cardiac and respiratory stimulants like strychnin, atropin, ammonia, and digitalis should be given freely. Artificial respiration should be resorted to early, before the development of asphyxia.

Chronic Poisoning (*Chloralism*).—The chief symptoms of chronic chloral-poisoning are nervousness, dyspnea, palpitation, insomnia, mental impairment, and a group of phenomena apparently due to vasomotor disturbance, such as temporary lividity, erythematous rashes, ecchymoses, and edema.

Therapeutics.—Chloral is used chiefly to induce sleep and to allay spasm.

Somnifacient.—It is a valuable somnifacient, inducing quiet and refreshing sleep. It is especially indicated in the *insomnia* of overwork, excitement, sthenic fevers, delirium tremens, mania, and chronic nephritis. In sleeplessness resulting from pain it is much inferior to opium.

Antispasmodic.—It may be employed with advantage to arrest *uremic* and *puerperal convulsions*. In conjunction with the bromid of potassium it is perhaps the best sedative we possess in the *convulsions of tetanus* and *strychnin-poisoning*. In grave cases of *epilepsy*, in *whooping-cough* attended with violent paroxysms, and in *chorea insaniens* chloral is often useful, but in the milder manifestations of these diseases it should not be selected, on account of the depressing effect which follows its continued use.

Chloral has been employed in solution as a wash for *foul ulcers*, as a vaginal douche in *cancer of the uterus*, and as a local application in *diphtheria*, but at present it is rarely used as an antiseptic except in preventing the decomposition of urine. It may be employed to preserve urine for microscopic examination and to purify the urinals of paraplegics.

When equal parts of chloral and camphor are rubbed together a clear syrupy liquid is formed, which is termed *chloral camphor.* It is used as a local anesthetic in *neuralgia* and *toothache.* In the form of an ointment, 30–60 gr. to the ounce (2.0–4.0 gm. to 30.0 gm.), it is useful in *pruritus.*

Contraindications.— Marked cardiac and respiratory weakness are contraindications to the use of chloral. In the chronic neuroses it must be given with considerable caution, on account of the danger of inducing the chloral habit.

Administration.—It is best given dissolved in some agreeable vehicle, as in the following formula :

> ℞ Chloralis,   gr. lxxx (5.2 gm.) ;
>   Tincturæ aurantii dulcis,   fʒss (2.0 c.c.) ;
>   Syrupi aurantii,   fℨj (30.0 c.c.) ;
>   Aquæ,   q. s. fℨij (60.0 c.c.).—M.
> Sig. A tablespoonful at bedtime.   (*Insomnia.*)

Morphin and the bromids may be given with chloral to enhance its effect :

> ℞ Chloralis,   gr. xvj (1.0 gm.) ;
>   Sodii bromidi,   gr. xl (2.5 gm.) ;
>   Syrupi aurantii,   f ʒvj (22.5 c.c.) ;
>   Aquæ,   q. s. fℨij (60.0 c.c.).—M.
>   Sig. A dessertspoonful every four hours for a child of three years. (*Whooping-cough.*)

To lessen its irritant effect when given by the bowel, chloral should be mixed with some demulcent, such as milk.

Incompatibles.—Alkalies and strong solutions of anti-pyrin.

BUTYL-CHLORAL HYDRAS.
(Butyl-chloral Hydrate, $C_4H_5Cl_3O_1H_2O$.)

Butyl-chloral is an oily liquid formed by the action of chlorin on acetic aldehyd. It is converted into the solid butyl-chloral hydrate by the addition of water. Butyl-chloral hydrate occurs in the form of white pearly scales having a pungent odor and an acid, disagreeable taste. It is sparingly soluble in water, but freely so in glycerin and alcohol. The dose is from 5–20 gr. (0.3–1.3 gm.).

Therapeutics.—The action of butyl-chloral hydrate is much the same as that of chloral hydrate, except that it is less powerful. It is generally supposed that the fifth nerve is especially sensitive to its sedative influence, and therefore it has been extensively employed in the treatment of *trifacial neuralgia* and *migraine.*

On account of its nauseous taste, it is best given in capsules or in pills or dissolved in some aromatic mixture. Solutions should be freshly made when wanted, as the drug gradually decomposes when dissolved. It is incompatible with alkalies.

SULPHONAL.

$$((CH_3)_2C(SO_2C_2H_5)_2.)$$

Sulphonal is a synthetic compound obtained by oxidizing a mixture of ethyl hydrosulphid and acetone. It occurs as a colorless, crystalline powder, free from odor and taste. It is soluble in 500 parts of cold water, 15 parts of boiling water, and 65 parts of alcohol. The dose is from 15–30 gr. (1.0–2.0 gm.).

Physiologic Action.—In moderate doses sulphonal is a pure hypnotic, inducing sleep by a direct action on the cerebrum. Owing to its insolubility in gastric juice it is very slowly absorbed from the stomach, and when given in the form of a powder its soporific effect may not be manifest for several hours or, after a late dose, not until the following day. In ordinary amounts it does not influence the circulation or respiration, and rarely the digestion. Drowsiness, headache, vertigo, mental confusion, and cutaneous rashes are sometimes noted after its use. When exhibited in single doses, or in short courses not exceeding a week or two, sulphonal may be regarded as a safe hypnotic, rarely producing unpleasant symptoms. It has no analgesic properties.

Toxicology.—Acute sulphonal-poisoning does not commonly prove fatal, probably because much of the drug escapes from the stomach before absorbtion takes place. Neisser reports the case of a person who took a tablespoonful of the drug, and slept in consequence four days and nights, and then rapidly convalesced; and another patient, aged fifteen years, who took nearly 3 ounces and was unconscious for five days, recovering entirely in eight days.

Death, however, has followed the ingestion of comparatively small amounts. Murrell has cited two instances of death from 30 and 40 grains respectively.

The symptoms of acute poisoning are headache, vertigo, tinnitus aurium, marked cyanosis, vomiting, diarrhea, shallow breathing, feeble pulse, dilated pupils, unconsciousness, and collapse. The treatment consists in administering saline cathartics and quickly acting stimulants (strychnin, ammonia), and in maintaining the body-temperature.

Chronic sulphonal-poisoning has frequently followed the

continuous use of the drug for weeks or months. Seventeen out of 20 cases collected by H. C. Wood, Jr., terminated fatally. Probably from the gradual accumulation of the drug in the body the symptoms often appear abruptly, sometimes even after the habit has been suspended for several days. The chief phenomena of chronic poisoning are a peculiar red coloration of the urine, languor, ataxia, progressive weakness, paralysis, paresthesia, nausea, vomiting, colicky pains, serous diarrhea or obstinate constipation, insomnia, and mental confusion. In unfavorable cases the urine becomes scanty and albuminous, and the exhaustion deepens into collapse. In a fatal case reported by Taylor and Sailer well-marked degenerative changes were found in the liver, kidneys, and heart. The discoloration of the urine, which is of considerable diagnostic import, is due to the presence of hematoporphyrin, a compound thought to be derived from hematin.

The treatment of chronic poisoning consists in withdrawing the drug at once, and in freely administering alkaline carbonates (sodium carbonate) with saline cathartics. Hypodermoclysis and enteroclysis are also recommended by Wood.

Therapeutics.—It is an excellent hypnotic in the *insomnia of nervous excitement, overwork*, and *mental disease*. When sleeplessness is due to pain, it is ineffective. In chronic disease of the heart, lungs, or kidneys, sulphonal is not, as a rule, to be chosen.

Administration.—Sulphonal is best administered in some hot liquid, and, as its action is slow, it is well to give it an hour or two before bedtime. Compressed tablets should be avoided. On account of its tendency to accumulate in the body, it should not be given continuously for more than a week. Even after its exhibition for several successive days, it is advisable to order a saline cathartic for the purpose of removing from the bowel any portion of the drug remaining unabsorbed.

TRIONAL.

$$(C_2H_5.CH_3.C(So_2C_2H_5)_2.)$$

Trional, like sulphonal, belongs to the group of "sulphones," *i. e.*, combinations of So_2 with organic radicles. Sulphonal, trional, and tetronal contain respectively two, three, and four molecules of ethyl. It occurs in the form of colorless, flat crystals, free from odor, and almost tasteless. It is soluble in 320 parts of cold water, and freely soluble in hot water, ether, and alcohol. The dose is from 15–30 gr. (1.0–2.0 gm.).

Physiologic Action and Therapeutics.—The action of trional is almost identical with that of sulphonal. Owing to its greater solubility, however, it is more prompt in its effect, and has less tendency to accumulate in the body. When not used too continuously, it is a safe and reliable hypnotic, suitable for the same class of cases in which sulphonal is most effective. It is best given in some warm vehicle—milk, tea, or whisky—an hour or two before retiring. Occasionally it gives rise to headache, drowsiness, and nausea on the day following its administration. When used continuously for several weeks, it causes the following toxic symptoms: headache, vertigo, impairment of memory, disordered speech, tremors, nausea, constipation or diarrhea, colicky pains, hematoporphyrinuria, oliguria, and even strangury. These symptoms, however, do not, as a rule, develop so abruptly as they do after the prolonged use of sulphonal. The treatment does not differ from that of sulphonal-poisoning.

TETRONAL.

$$((C_2H_5)_2 \; C.(So_2C_2H_5)_2.)$$

Tetronal is an analogue of trional and sulphonal, containing four ethyl groups. It is used in the same doses and for the same purposes as trional. It has no advantages over the latter; on the contrary, it is less soluble and much more costly.

CHLORALAMID.

(Chloralformamid, $CCL_3CH(OH).CONH_2$.)

Chloralamid is a synthetic compound formed by the union of chloral with formamid. It occurs in the form of colorless crystals, having a bitter taste, and soluble in 20 parts of water and $1\frac{1}{2}$ parts of alcohol. The dose is from 10–30 gr. (0.65 –2.0 gm.).

Physiologic Action.—The dominant action of chloralamid is upon the cerebrum, which it depresses, causing profound sleep. Unlike chloral, it has little effect on the circulation, and is only a feeble depressant to the spinal cord. The experiments of Wood and Cerna indicate that it rather stimulates than depresses the respiratory center.

Therapeutics.—Chloralamid is a pure hypnotic, less depressing and less irritating to the stomach than chloral, but at the same time less powerful and less prompt in its action. It is not so slow in its action as sulphonal, and it is less liable than that drug to cause languor and drowsiness next day.

Headache, giddiness, exhilaration, nausea, and inco-ordination are occasionally observed after its use. It has no cumulative effect, and patients do not readily become addicted to its use. It is an excellent somnifacient in *chronic heart, lung*, and *kidney disease*, in *hysteria, neurasthenia, delirium tremens, senility*, and *acute fevers.* When wakefulness is due to pain, opium is a better remedy, although chloralamid seems to possess some analgesic properties. It has been highly recommended as a sedative in *cardiac asthma*, and, in combination with potassium bromid, as a superior remedy in *sea-sickness.*

Administration.—Chloralamid should be given about half an hour before bedtime. On account of its bitter taste and imperfect solubility it is better not to prescribe it in the form of powders. It may be conveniently given in some aromatic aqueous vehicle, to which a little alcohol or weak acid has been added to facilitate solution. It is decomposed by hot water.

R Chloralamid, gr. lxxx (5.2 gm.) ;
Alcoholis, f℥ij (8.0 c.c.) ;
Elixiris aromatici, f℥j (30.0 c.c.) ;
Aquæ, q. s. ad f℥ij (60.0 c.c.).—M.
Sig. A tablespoonful in water half an hour before bedtime.

Incompatibles.—Alkalies convert it into chloral.

CHLORALOSE.
(Anhydroglucochloral, $C_8H_{11}Cl_3O_6$.)

Chloralose is a compound prepared by subjecting a mixture of equal parts of anhydrous chloral and glucose to a temperature of 212° F. for one hour. It appears in the form of fine colorless crystals having a bitter, disagreeable taste. It is freely soluble in hot water and in alcohol, and slightly so in cold water. The dose is from 5–8 gr. (0.32–0.5 gm.).

Physiologic Action.—Chloralose, like chloral, induces sleep by directly depressing the brain. During sleep the reflex excitability is not diminished, but rather exaggerated, indicating that in ordinary doses the drug does not depress the spinal cord. Its somnifacient effect cannot be due to chloral set free in the alimentary canal, since chloralose is active in less than half the dose of the older hypnotic. It has little effect on the sensory nervous system, and is, therefore, only a feeble analgesic. Small doses do not generally affect the circulation, but large doses depress the heart.

The great drawback to the general use of chloralose is the irregularity of its action ; it often acts very promptly, inducing a refreshing sleep, lasting from six to eight hours, but not infrequently alarming symptoms follow its exhibition, even in small doses. The untoward effects are retardation of the pulse, dilatation of the pupils, deafness, tremors, extreme prostration, cyanosis, incoherent speech, convulsive seizures, nausea, and incontinence of urine and feces. In neurotic subjects it may induce delirious excitement instead of sleep.

Therapeutics.—Chloralose has been used with some success in *functional insomnia*. Tyson has found it effective in the sleeplessness of *chronic heart-disease* and *Bright's disease*. Balfour also speaks favorably of its action in the insomnia associated with *heart-disease*. The drug should be avoided when there is great debility.

Administration.—Chloralose is best administered ir cachets or capsules, followed by a draught of hot milk or tea

CHLORETONE.

(Aneson, Trichlortertiary Butyl Alcohol, $CCl_3C(CH_3)_2OH$.)

Chloretone is a compound formed by the addition of caustic potash to equal weights of chloroform and acetone. It is a white crystalline powder, with a camphoraceous odor and taste. It is sparingly soluble in cold water, but freely so in ether and strong alcohol. The dose is from 5–20 gr. (0.3–1.3 gm.).

Physiologic Action.—Chloretone has two important actions: it is a somnifacient and a local anesthetic. In animals large doses reduce the temperature and also lower the blood-pressure, probably by depressing the heart and the vasomotor center. Impens considers the drug to be more than twice as toxic as chloral. In man we have not observed any material change in the pulse or temperature from therapeutic doses of chloretone, but several instances of its untoward effects have been reported. Donald cites the case of a morphin habitué who, during a period of but little more than forty-eight hours, took 192 grains of chloretone, and in consequence slept almost continuously for nearly six days, when he awoke none the worse for his experience. It has an advantage over chloral in not being irritating to the stomach. It probably undergoes decomposition in the body, since after its ingestion neither chloretone itself nor its components (chloroform and acetone) are found in the urine. It has pronounced germicidal properties.

Therapeutics.—Chloretone may be regarded as a reasonably safe hypnotic of moderate power. It is prompt in its action, sleep usually following in from half an hour to an hour after its administration. Beyond causing slight drowsiness on the day following its exhibition, it rarely gives rise to unpleasant after-effects. Toleration of the drug is rather rapidly acquired by continuous use. It is especially adapted for use in cases of *insomnia unattended with severe pain, high fever*, or *pronounced nervous excitement.* It has been used with some success as a sedative in *epilepsy.*

As a local anesthetic and antiseptic it has proved serviceable in minor surgery. It may be applied in the form of a dusting-powder to *wounds, burns*, and *painful ulcers.* In the form of a 1 per cent. solution (1–2 fl. dr.—4.0–8.0 c.c.) it has been substituted for cocain in producing infiltration anesthesia. In the eye it does not affect the pupil or accommodation, and does not disturb the nutrition of the corneal epithelium.

Chloretone (5 per cent.) serves as a preservative of solutions of organic compounds (extract of suprarenal gland, homatropin).

Administration.—It may be given in tablets, cachets, or powders, followed by a draught of water.

PARALDEHYDUM, U. S. P.
(Paraldehyd, $C_6H_{12}O_3$.)

Paraldehyd is a product obtained by treating aldehyd with dilute sulphuric or nitric acid. It appears as a colorless liquid, having a strong ethereal odor and a disagreeable, pungent taste. It is soluble in 8.5 parts of water, and is miscible in all proportions with alcohol, ether, chloroform, and oils. The dose is $\frac{1}{2}$–1 fl. dr. (2.0–4.0 c.c.).

Physiologic Action.—The action of paraldehyd somewhat resembles that of chloral. Like the latter, it depresses the brain and spinal cord. Although it causes a sensible fall in the arterial pressure, it is not so depressing to the heart itself as chloral. Large doses are prone to disturb the stomach. It is eliminated by the lungs and kidneys. Toxic doses produce coma, feeble breathing, great muscular relaxation, abolition of reflex activity, and, finally, death from paralysis of the respiratory center. Its long-continued use is sometimes followed by cutaneous eruptions. It has no analgesic properties.

Therapeutics.—Paraldehyd is a safe and reliable hypnotic. The great drawbacks to its use are its unpleasant taste and the disagreeable odor which it imparts to the breath for twenty-four hours after its ingestion. Toleration is somewhat rapidly acquired. It may be employed with advantage in the *insomnia of mania, delirium tremens, hysteria*, and *heart-*

disease. It is sometimes efficient as an antispasmodic in *asthma.* It has also proved serviceable in controlling the convulsions of *tetanus.*

Administration.—It may be prescribed in capsules or in some aromatic vehicle. A few drops of alcohol render it freely miscible with water. The dose should be well diluted.

℞ Paraldehydi,
    Glycerini,
    Spiritus aurantii compositi,   āā f℥iv;
    Aquæ,   q. s. ad f℥ij.—M.
  Sig. A dessertspoonful in water, repeated in an hour if necessary.

URETHANE.

(Ethyl Urethan, Ethyl Carbamate, $NH_2.CO.OC_2H_5$.)

Urethane is made by the interaction of ethyl alcohol and nitrate of urea. It occurs as colorless tabular crystals, odorless, and having a cooling saline taste. It is readily soluble in water, alcohol, ether, and chloroform. The dose is from 15–45 gr. (1.0–3.0 gm.).

Physiologic Action and Therapeutics.—Urethane is a comparatively safe hypnotic, somewhat uncertain in action, and without influence in ordinary doses upon the respiration and circulation. It has no analgesic effect. Toxic doses are followed by coma, a fall of temperature, weakness of the pulse, abolition of reflex activity, and death from respiratory paralysis. It may be successfully employed in *insomnia from nervous excitement* or *insanity.* It is not without value as an anticonvulsant. Dana and Jacobi have spoken very favorably of its action in some cases of *epilepsy.*

Urethan is incompatible with acids and alkalies. Rubbed with chloretone it produces a liquid compound.

DUBOISIN.

Duboisin is an alkaloid obtained from *Duboisia myoporoides.* It occurs in the form of microscopic crystals which are soluble in alcohol, but very slightly so in water. The salts of the alkaloid are soluble in water. The dose is from $\frac{1}{80}$–$\frac{1}{30}$ gr. (0.0008–0.002 gm.).

Physiologic Action and Therapeutics.—Duboisin is used as a hypnotic, depressomotor, and mydriatic. As a hypnotic it resembles hyoscin in its action, although the sleep which follows its use is not so quiet or so refreshing as that induced by hyoscin. Large doses cause mydriasis, dryness of the throat, vertigo, and headache. Tolerance is soon established. Weakness, emaciation, and gastric disturbance have been observed after its prolonged use. It has proved most efficacious as a somnifacient in *insanity* associated with physical excitement.

Duboisin sulphate may be employed in doses of from $\frac{1}{100}$–$\frac{1}{60}$ gr. (0.0006–0.001 gm.) two or three times a day, as a substitute for hyoscin, to lessen the tremors of *paralysis agitans*.

The action of duboisin sulphate on the eye resembles that of atropin, but it is more energetic, more prompt, and of shorter duration. Its chief drawback is its greater tendency to excite constitutional disturbance. Like atropin, it dilates the pupil, paralyzes accommodation, and increases intra-ocular tension. As it is nearly twice as effective as atropin, a solution of the sulphate containing 2 grains to the ounce (0.13 gm. to 30.00 c.c.) is sufficient. In this strength it is perhaps less irritating to the eye than a 4 grain solution of atropin.

PELLOTIN.

Pellotin is a crystalline alkaloid obtained from *Anhalonium Williamsii*, a species of Mexican cactus. The hydrochlorid, on account of its ready solubility in water, is preferred for medicinal purposes. Its physiologic action has not been thoroughly studied. According to Heffter, it produces in the frog slight narcosis, followed by exaggerated reflex irritability and tetanic convulsions. In man the chief effect of an ordinary dose is drowsiness, followed by sleep and some retardation of the pulse. Unpleasant after-effects, such as headache, vertigo, and nausea, are not uncommon. It may be given by the mouth in doses of from $\frac{1}{2}$–1 gr. (0.03–0.06 gm.); or subcutaneously, in doses of from $\frac{1}{4}$–$\frac{2}{3}$ gr. (0.016–0.04 gm.).

OTHER SOMNIFACIENTS.

Hyoscin Hydrobromate (see p. 76).—Hyoscin is an excellent somnifacient when sleeplessness is attended with great mental excitement. Thus it often acts very favorably in *mania, melancholia agitata, delirium tremens*, and *typhoid fever* with active delirium. In *chronic heart-disease*, while it is sometimes efficacious, it is not so generally useful as opium, or chloralamid. The dose is from $\frac{1}{150}$–$\frac{1}{80}$ gr. (0.0004–0.00085 gm.).

Bromids (see p. 137).—Potassium bromid, sodium bromid, and ammonium bromid depress the cerebral cortex and in sufficient dose induce sleep. They are not very energetic hypnotics, but in doses of from 20–30 gr. (1.3–2.0 gm.) they frequently serve a good purpose in the milder forms of *insomnia*, such as occur from *nervous irritability, overstudy*, and *anxiety*. In some cases the combination of a bromid with a small dose of chloral is very effective. The disagreeable after-effects of opium are often prevented by combining it with a bromid. Bromids should be given in solution, well diluted.

GENERAL ANESTHETICS.

General anesthetics are drugs which induce unconsciousness and insensibility. They are employed to abolish or lessen pain and to relax spasm. They are used for both of these purposes in surgical operations, in labor, and in biliary and renal colic. They are of service in subduing the convulsive seizures of tetanus, strychnin-poisoning, and uremia. By suspending muscular irritability they also render valuable aid in the reduction of luxations, fractures, and strangulated hernias, and in the recognition of deep-seated abdominal lesions. The most important general anesthetics are:

Ether.	Nitrous oxid.
Chloroform.	Ethyl bromid.

ÆTHER, U. S. P.

(Ether, Ethyl Oxid, $C_4H_{10}O$.)

Ether is obtained by the action of sulphuric acid on alcohol, and appears as a colorless, volatile, highly inflammable liquid, having a strong characteristic odor. It is soluble in alcohol, chloroform, oils, and water. The official preparation contains about 4 per cent. of alcohol.

PREPARATIONS.	DOSE.
Spiritus Ætheris, U. S. P.	1–3 fl. dr. (3.5–11.0 c.c.).
Spiritus Ætheris Compositus, U. S. P. . . .	½–1 fl. dr. (2 0–3.5 c.c.).

Physiologic Action.—When freely inhaled ether first causes coughing, choking, and a sense of strangulation from its irritant effect on the mucous membrane of the respiratory tract. Soon it induces flushing of the face, injection of the eyes, deepening of the respiration, and an increase in the rate and strength of the pulse. Frequently marked emotional excitement develops, characterized by shouting, crying, laughing, erotic dreams, or pugilistic manifestations. These phenomena gradually subside, and are followed by complete unconsciousness and muscular relaxation. If the anesthetic be still further pushed, the breathing becomes stertorous from paresis of the palatal muscles, the face pale or deeply cyanosed, the muscles become flaccid, the respirations slow and shallow, the pulse grows rapid and feeble, and, finally, death results, usually from respiratory failure.

Circulatory System.—In ordinary amounts ether is a quickly acting heart-stimulant. It raises the blood-pressure by direct action on the heart and by stimulation of the vaso-motor center. Only in very large amounts does it depress the circulation.

Blood.—Etherization slightly increases the number of red cells, but very decidedly lessens their hemoglobin value. It also increases the number of leukocytes.

Respiratory System.—It first stimulates and then depresses the respiratory center. In fatal cases respiration is usually arrested before the cardiac pulsations cease. In the early stage of ether-anesthesia respiration may suddenly cease from reflex irritation of the trigemini.

Temperature.—Etherization materially lowers the temperature of the body, and therefore it is important that the operating-room should be well heated and that the patient should be guarded against exposure.

Nervous System.—The brain is first affected by ether, the anesthesia being due to the direct depressant action of the drug on the cerebrum; subsequently it depresses the sensory neurons of the spinal cord, then the motor neurons of the spinal cord, and, finally, the medulla oblongata.

Local Action.—When applied to the surface of the body it evaporates rapidly, produces intense cold, and thus benumbs the peripheral nerves.

Elimination.—It is eliminated chiefly by the lungs and kidneys; on the latter it has an irritant effect, and, therefore, considerable caution must be exercised in its employment as a general anesthetic in cases in which there is serious nephritis.

Therapeutics.—Ether was first employed as an anesthetic in general surgery by Dr. John C. Warren at the Massachusetts General Hospital in 1846; since that time it has been, in this country, the most extensively used anesthetic in all operations excepting the most trivial.

The Administration of Ether as an Anesthetic.—When administered as an anesthetic the following precautions should be observed: No solid food should be taken for several hours before the operation. If there has been nausea or if the patient is very weak, a little brandy may be given prior to the anesthetic. When given to a female, a third person should always be present. Before its administration the teeth should be examined, and, if false, they should be removed. The throat and waist should be freed from tight clothing, but the patient should not be exposed, as inflammation of the lungs is liable to follow when this precaution is unheeded.

The anesthetizer must guard against placing the head and arms of the patient in positions likely to cause pressure on nerve-trunks. The eyes should be protected from the irritant action of the ether. As the vapor of ether is highly inflammable, and since it is heavier than air, it is necessary when operating by gas-light to have the jet above the operator. In using the actual cautery special care is necessary to prevent ignition. No more ether should be used than is absolutely necessary to induce and maintain the desired degree of anesthesia. The anesthetizer should be continually on the lookout for any irregularities in the respiration or pulse and for changes in the facial appearance.

Ether may be administered from a towel loosely folded in the form of a cone, surrounded by stiff paper, and enclosing at its tip a small sponge, or, better, from an open inhaler made especially for the purpose. At first the inhaler should be held some distance from the nose, to accustom the patient to the irritant effects of the ether, but soon it should be brought close to the nose, so that the anesthetic may be inhaled in a more concentrated form. Insensibility of the conjunctivæ and muscular relaxation are the indications that the patient is prepared for the operation.

Accidents During Anesthesia.—The most common accident during etherization is failure of respiration. When this occurs after the first few inhalations, it is generally the result of a reflex *spasm of the laryngeal muscles* excited by the ether. The admixture of a little more air with the vapor will serve to relax the spasm, when respiration will be resumed. Embarrassed respiration is often due to the *accumulation of mucus in the upper air-passages*. In this event the head should be turned to one side, when much of the fluid will run out, or will collect in the dependent cheek, from which it may be removed by means of a sponge held by forceps. Finally, respiratory failure may result from the *direct action of the anesthetic on the respiratory center*. Should such an accident occur, the method of treatment to pursue is as follows : Withdraw the ether ; push the jaw forward by pressing on its angles ; draw the tongue forward by means of a volsella-forceps or a tenaculum. Rhythmic traction of the tongue is useful. Pour a little ether on the chest or abdomen, as recommended by Hare, to stimulate inspiration by reflex action. Administer strychnin and atropin hypodermically, but not alcohol, since its action closely resembles that of ether. Practise artificial respiration and persist in it for at least half an hour.

Heart-failure is a far less common accident than asphyxia. It is indicated by lividity of the face and feebleness of the pulse. In the event of its occurrence the head should be lowered, strychnin and atropin should be used hypodermically, and artificial respiration should be practised. The last serves to remove the ether from the chest, favors the entrance of fresh air, and stimulates the circulation.

After-effects.—The most common after-effects are *nausea and vomiting.* The administration of oxygen with the anesthetic will often prevent their occurrence. The inhalation of vinegar immediately after the withdrawal of the ether is also efficacious. Should the nausea persist, a hypodermic injection of morphin and the administration of cracked ice with a little brandy will usually suffice to allay it.

Pneumonia is an occasional sequel. It has been ascribed to exposure of the patient, to chilling of the lungs by the rapid evaporation of the anesthetic, to the irritant action of the ether, to the accumulation of secretion in the bronchi, and, finally, to the lowered vitality of the tissues induced by prolonged etherization.

Paralysis.—Central paralysis from hemorrhage or embolism is rare. Peripheral palsies are not infrequent, and are due to compression of nerve-trunks. The anesthetic contributes only indirectly to the latter by rendering the patient oblivious to pain.

Contraindications.—The contraindications to the employment of ether as an anesthetic are advanced arteriosclerosis, acute inflammatory infections of the respiratory tract, severe nephritis, especially when associated with cardiovascular lesions, and anemia when the hemoglobin is less than 30 per cent. The presence of valvular disease is not in itself a contraindication, provided compensation is well maintained. Diabetes, especially when well established and associated with acetonuria, should be considered a contraindication, since Becker has shown that the anesthetic increases the acetone and renders the patient more liable to coma.

Other Uses of Ether.—Ether may be employed in the form of a spray to induce *local anesthesia* preliminary to opening small abscesses or performing paracentesis.

Administered hypodermically (15–30 min.—1.0–2.0 c.c.), it is a valuable heart-stimulant, on account of the promptness of its action. It may be so employed in *shock*, in *poisoning*, and in the collapse of *cholera* and *low fevers.*

It has been given internally as an anthelmintic against tape-worms, and as a sedative in various forms of colic.

To allay its irritant effect on the fauces it is best given in capsules or in ice-water.

CHLOROFORMUM, U. S. P.

(Chloroform, Chloroformum Purificatum, $CHCl_3$.)

Chloroform is a heavy, colorless, non-inflammable liquid, obtained by the action of chlorin on alcohol. It is soluble in two hundred times its volume of water, and in all proportions in alcohol, oils, and ether. The dose is 5–30 min. (0.3–2.0 c.c.).

PREPARATIONS.	DOSE.
Aqua Chloroformi, U. S. P.	½–2 fl. oz. (15.0–60.0 c.c.).
Emulsum Chloroformi, U. S. P. (4 parts of chloroform in 100)	½–1 fl. oz. (15.0–30.0 c.c.).
Linimentum Chloroformi, U. S. P.	
Spiritus Chloroformi, U. S. P.	20–60 min. (1.0–4.0 c.c.).

Physiologic Action.—When freely inhaled, it produces a set of phenomena which may be grouped under three stages: The first stage is characterized by excitement, muscular rigidity, and lessened sensibility; the second, by anesthesia and muscular relaxation; and the third, by stertorous breathing, abolition of reflexes, profound narcosis, and absolute muscular relaxation.

Circulatory System.—Anesthesia induced by chloroform is always accompanied by a distinct lowering in the arterial pressure. This effect is certainly due in part to the depressant action of the drug on the heart, but the researches of Bowditch and Minot, Hill, Hare, and Thornton would indicate that it is partly due also to paralysis of the vasomotor center. When death occurs during chloroformization it is usually caused by arrest of circulation and not by asphyxia.

Nervous System.—It depresses the nervous system in the following order: First, the brain; next, the sensory side of the spinal cord; next, the motor side of the spinal cord; and, finally, the nerve-centers in the medulla. When applied locally it benumbs the part by depressing the peripheral nerves.

Respiratory System.—In large amounts chloroform acts as a respiratory paralyzant.

Alimentary Canal.—In small doses it stimulates the

mucous membrane of the stomach, and co-ordinates the peristaltic movements of the intestines.

Temperature.—During chloroform-anesthesia there is a distinct fall in the body-temperature, though the latter is not quite so pronounced as it is in etherization.

Local Action.—When applied to the skin it evaporates rapidly and induces a sense of coldness. When evaporation is prevented it causes redness of the surface, and even vesication.

Nutrition.—Prolonged chloroformization produces degenerative changes, especially fatty metamorphosis, in the tissues of the vital organs.

Elimination.—It escapes from the body for the most part unchanged through the lungs and kidneys. Chloroform is an irritant to the kidneys, but as the quantity employed to induce anesthesia is comparatively small, its deleterious effects on these organs are less pronounced than those of ether.

Lower Organisms.—Chloroform is destructive to many of the lower forms of animal and vegetable life.

The Actions of Chloroform and Ether Compared.—Chloroform is more agreeable to the patient than ether, less irritating to the respiratory tract, and more rapid in its action. On the other hand, it is more dangerous than ether, the mortality from chloroformization being from four to five times greater than from etherization. Upon the circulation the action of the two drugs is almost diametrically opposite. Ether primarily stimulates the heart and vasomotor center, and chloroform primarily depresses them. Death from ether usually results from asphyxia; from chloroform, it usually results from circulatory failure. The nutrition of specialized tissues is more profoundly affected by chloroform than by ether.

Therapeutics.—Chloroform is employed as a general anesthetic, a nerve-sedative, a carminative, an anthelmintic, and a counterirritant.

General Anesthetic.—The precautions to be observed in administering chloroform as an anesthetic are much the same as those already described in dealing with ether. The vapor, however, must never be concentrated, but well mixed with air. To insure this chloroform is best administered from a folded handkerchief or piece of lint held four or five inches from the patient's nose. Not more than half a dram should be dropped on the inhaler at one time. Insensitiveness of the cornea and muscular relaxation indicate complete anesthesia. An ordinary operation will not usually require more than an ounce or two of chloroform. The anesthetizer should not rely upon any one sign as a warning of danger, but should keep careful watch upon the face, the respiration, and the pulse. Lividity of the face, stertorous or irregular breathing, or feebleness of the pulse call for immediate withdrawal of the anesthetic, lowering of the head, practice of artificial respiration, and the hypodermic administration of cardiac stimulants, such as strychnin, digitalis, and ammonia.

When chloroform is administered in the presence of lighted gas-jets, the room should be well ventilated, since the burnt vapor of the anesthetic yields irritant products (especially carbonyl chlorid) capable of inducing pneumonia when inhaled in concentrated form.

Conditions Favoring the Use of Chloroform as an Anesthetic.—While chloroform is more dangerous than ether, it is to be preferred to the latter when the patient suffers from acute inflammation of the bronchi or lungs, from well-developed nephritis, or advanced diseases of the blood-vessels (atheroma or aneurism). In military practice chloroform is also preferable on account of the rapidity of its action and the small quantity required to cause insensibility.

Nerve-sedative.—A few whiffs of chloroform may be employed to allay motor excitation in *puerperal eclampsia, tetanus,* and *chorea insaniens,* and to allay local spasm in *asthma, whooping-cough,* and severe attacks of *renal* and *biliary colic.*

In attacks of *angina pectoris* of great intensity Balfour and Osler recommend inhalations of chloroform. It may be dropped upon a handkerchief, or, as Balfour recommends, poured on a sponge in a smelling-bottle, and the patient told to breathe it through the nose as deeply as possible. In a minute or two relief is obtained.

In the obstinate cough of *phthisis* the addition of chloroform to the usual cough-mixture is often advantageous.

<pre>
℞   Codeinæ sulphatis,   gr. iv. (0.26 gm.);
     Acidi hydrocyanici diluti,   ℳ xl (2.5 c.c.);
     Spiritus chloroformi,   fʒij (8.0 c.c.);
     Glycerini,   f℥ss (15.0 c.c.);
     Elixiris aurantii,   q. s. ad f℥iij (90.0 c.c.).—M.
  Dose. One teaspoonful.
</pre>

Carminative.—It is a useful carminative and sedative in *gastralgia* and *intestinal colic.* The following combination will be found useful in gastralgia:

R Chloroformi,   fʒiss–ij (5.5–8.0 c.c.);
   Spiritus ammoniæ aromatici,
   Spiritus vini gallici,
   Tincturæ cardamomi compositæ, āā fʒv (18.0 c.c.).—M.
  Sig. A teaspoonful in hot water every fifteen or thirty minutes until relief is obtained.

Anthelmintic.—It has been used in large doses (fʒss–j— 2.0–4.0 c.c.) as an anthelmintic against *tapeworms,* but it is an unreliable remedy.   In nineteen cases reported by Whyte and Leichtenstern it proved successful in only two instances.

Counterirritant.—In the form of a liniment chloroform is extensively used as a stimulating application in *muscular rheumatism, sprains,* and *contusions.*

NITROUS OXID.

(Nitrogen Monoxid, N_2O.)

Nitrous oxid is obtained from the distillation of ammonium nitrate, and appears as a colorless, odorless gas having a somewhat sweetish taste.   By pressure it is converted into a liquid, and in this form is stored in iron cylinders for subsequent use.

Physiologic Action.—When freely inhaled, nitrous oxid causes an increase of blood-pressure, a sense of exhilaration, ringing in the ears, and lividity of the face; in a minute or two these symptoms are followed by complete unconsciousness. Great excitement, laughter, or a pugilistic tendency is frequently induced by inhalation of the gas when the supply of air is not completely cut off.   It is a perfectly safe anesthetic, and quite free from disagreeable after-effects.   According to the researches of Wood and Kemp it has no direct action on the heart or vasomotor center, the increased blood-pressure being due to the asphyxia.   The anesthesia is probably due in part to the displacement of oxygen from the blood, and in part to the direct action of the gas upon the cerebrum.

Therapeutics.—As an anesthetic, nitrous oxid is especially suitable for *minor surgical operations* requiring but a short period of unconsciousness.   The severe symptoms inseparable from asphyxia can be mitigated in a measure by the administration of oxygen as soon as anesthesia has been induced by the nitrous oxid.

ÆTHYL BROMIDUM.

(Hydrobromic Ether, C_2H_5Br.)

Ethyl bromid is obtained by the distillation of a mixture of ethyl alcohol, sulphuric acid, and potassium bromid.   When perfectly pure it forms a colorless, highly volatile, inflammable liquid having a sweetish taste and a chloroformic odor. On exposure to air it liberates bromin and hydrobromic acid and becomes unfit for use. It should be kept in dark amber-colored, well-stoppered bottles. It must not be confounded with the poisonous *ethylene bromid.*

Physiologic Action and Therapeutics.—Ethyl bromid somewhat resembles chloroform in its anesthetic properties, but it acts more quickly, is inhaled with even less difficulty, is less depressing, and its effects are less lasting. Large doses depress the respiration and lower arterial pressure. After its administration there is a garlicky odor on the breath which may persist for a day or two. With this exception it rarely causes disagreeable after-effects, such as headache, nausea, and vomiting Although its use is not entirely devoid of danger, there is good reason for believing that with a pure preparation, cautiously administered, the risk is very little greater than with ether. While not quite so safe as nitrous oxid, it has the advantage of not requiring cumbersome apparatus for its administration.

Ethyl bromid is a useful anesthetic for *short operations.* The dose for a child is from 1–3 drams (4–11 c.c.), and for an adult, from 2–6 drams (8–22 c.c.). The patient should be instructed to breathe deeply; the full dose should be poured at once upon the inhaler (preferably a towel folded into an air-tight cone), and the inhaler should be held firmly over the nose and mouth, and not removed until the anesthesia is complete. Anesthetization is usually accomplished within from thirty to forty seconds, and lasts from two to three minutes. It is not safe to continue the administration for more than two minutes, and under no circumstances should the anesthesia be prolonged by removing the mask and adding more of the drug.

The contraindications are severe lesions of the heart, lungs, or kidneys. ————————

GENERAL ANALGESICS OR ANODYNES.

General analgesics or anodynes are remedies which relieve pain without necessarily inducing unconsciousness or general anesthesia. They may accomplish their object by acting upon the perceptive centers in the brain, the afferent paths in the spinal cord, or the peripheral nerves, through which the painful impression is transmitted. The most powerful analgesic is *opium.* It probably acts by depressing the cerebral centers and the afferent paths in the spinal cord. While it has the power of relieving pain arising from almost any cause, it is especially useful in pain occasioned by gross lesions—inflammation, traumatism, and morbid growths. In the various so-called functional pains, such as headache and neuralgia, the coal-tar derivatives—*antipyrin*, *phenacetin*, and *acetanilid*—are more serviceable analgesics. The manner of their action is not definitely known, but they probably act upon the entire sensory nervous mechanism. *Cannabis indica* somewhat resembles opium in its action as an analgesic, but it is far less powerful. It is especially useful in neuralgia and migraine. The *bromids*, by depressing the cerebral cortex, often afford relief in headache, but in other forms of pain they usually fail. Some drugs seem to have an affinity for certain nerves: thus *gelsemium* and *butyl chloral hydrate* are often distinctly valuable in trifacial neuralgia.

ANTICONVULSANTS.

Certain drugs have the power of controlling cerebral convulsions by directly diminishing the excitability of the motor area in the cortex of the brain. A drug that thus depresses the motor centers, and also lessens the power of the motor columns of the spinal cord to convey discharges from the brain to the muscles, or lessens the power of the sensory columns of the spinal cord to transmit to the brain stimuli from without, has additional value as an anticonvulsant.

Anesthetics (ether and chloroform) profoundly affect the motor centers of the brain, and to a less extent the sensory and motor columns of the spinal cord. They are especially useful in severe convulsions, such as may occur in hysteria, puerperal eclampsia, and the status epilepticus. *Bromids* also depress the brain and the sensory and motor pathways of the spinal cord. While their action is not so powerful as that of the anesthetics, it is much more permanent. They are very useful in epilepsy and in the reflex convulsions of childhood. The value of *chloral* in cerebral convulsions is due chiefly to its depressant action on the motor tract of the spinal cord. Chloral may be employed with advantage in uremic, infantile, and puerperal convulsions. It is too depressing for continuous use in essential epilepsy.

Amyl nitrite checks convulsions also by depressing the motor side of the spinal cord, its effect in therapeutic doses on the motor cortex of the brain being inconsiderable. It is the most rapidly acting, and at the same time the most fugacious, of all the anticonvulsants. It is of great service in aborting threatened convulsions or in breaking up a series of seizures such as occur in the status epilepticus. *Hydrastinin* is a depressant to the motor area of the cerebral cortex, and in a few cases of epilepsy in which it was tried it is said to have given favorable results.

Antipyrin and *Solanum Carolinense* (horse-nettle) are of value in essential epilepsy, but the manner in which they act is not definitely known.

ANTISPASMODICS.

Antispasmodics are remedies used to control minor grades of motor excitation and to lessen states of general nervous irritability. While the manner of their action is by no means clear, it seems likely that many of them owe their sedative power to their stimulating influence on the higher or inhibitory centers of the brain, whereby the lower motor centers are brought into more complete subjection. As a class, antispasmodics are useful in hysteria, nervous excitability, and local spasms. The following drugs are the most important members of this group:

Camphor.	Cimicifuga.
Asafetida.	Lactucarium.
Valerian.	Hops.
Compound spirit of ether.	Sumbul.
Bromoform.	Viburnum prunifolium.
Musk.	Oil of amber.

CAMPHORA, U. S. P.

(Camphor, $C_{10}H_{16}O$.)

Camphor is a whitish, translucent, volatile substance (solid volatile oil or stearopten), having a penetrating odor and a pungent taste, and obtained by distilling the wood of an Oriental tree—*Cinnamomum Camphora*. It dissolves freely in all ordinary menstrua except water, in which it is only sparingly soluble. The dose is from 1–5 gr. (0.06–0.3 gm.).

PREPARATIONS.   DOSE.

Aqua Camphoræ, U. S. P. (0.8 per cent.) . . ½–1 fl. oz, (15.0–30.0 c.c.).
Linimentum Camphoræ, U. S. P. (20 per
   cent. in cotton-seed oil).
Linimentum Saponis, U. S. P. (4.5 per cent.).
Spiritus Camphoræ, U. S. P. (10 per cent.) . . 10–30 min. (0.6–2.0 c.c.).

It also enters into *Tinctura opii camphorata.*

Physiologic Action.—In therapeutic doses camphor stimulates the higher cerebral centers, exerting a calmative influence in the presence of nervous irritability. It has a stimulating effect on the heart, although after large doses the blood-pressure falls, probably from paresis of the vasomotor center. In the stomach it acts as a mild carminative, inducing a sensation of warmth, increasing peristalsis, and expelling flatus. The action of camphor on the respiration has not been fully determined, but the researches of Stockman and Binz indicate that the drug is a stimulant to the respiratory center. Moderate doses exert a feeble anaphrodisiac effect. Applied externally it acts as a stimulant and antipruritic. When taken internally much of it is destroyed in the body, but a part is eliminated in the urine as camphoglycuric acid.

Toxic doses of camphor produce languor, vertigo, headache, confusion of thought, tinnitus aurium, gastric irritation, a rapid, feeble pulse, convulsions, and collapse. The convulsions are of an epileptiform character, and are due to the action of the drug on the cerebral cortex.

Therapeutics.—Camphor is employed as an antispasmodic, a circulatory stimulant, carminative, and rubefacient.

As an antispasmodic, it is useful in *hysterical excitement, hiccough, dysmenorrhea, whooping-cough, chordee,* and the various *climacteric neuroses.*

As a circulatory stimulant, camphor is exceedingly valuable in *adynamic fevers*, particularly in typhoid fever and pneumonia. It not only serves to strengthen the pulse in these affections, but it also acts favorably in lessening the restlessness and delirium.

As a carminative, camphor is serviceable in *tympanites, intestinal colic,* and the various forms of *choleraic diarrhea.*

As a rubefacient, it is much used in the form of a liniment in *sprains, bruises,* and *muscular rheumatism.* Anderson's powder is a useful antipruritic dusting-powder in *simple erythema, intertrigo,* and *urticaria.* It consists of from 1–2 drams (4.0–8.0 gm.) of finely pulverized camphor to ½ ounce (15.0 gm.) each of powdered starch and zinc oxid.

Camphor is a useful remedy in *acute coryza*, particularly when it is given early in the attack. It may be used internally and also by inhalation, a teaspoonful of powdered camphor being added to a tumbler of boiling water, and the fumes inhaled.

Administration.—Camphor may be given internally in the form of the water or spirit, or in substance in pill or capsule. In adynamic fevers it is best given subcutaneously dissolved in sterilized olive oil. From 1–2 gr. (0.06–0.13 gm.) may be dissolved in 15 min. (1.0 c.c.) of the oil, and injected every three or four hours, or even every two hours when the symptoms are very urgent. In coryza the following combination may be employed:

> ℞ Pulveris opii,   gr. iij–vj (0.2–0.4 gm.);
>   Pulveris camphoræ,   gr. xij (0.8 gm.);
>   Ammonii carbonatis,   gr. xxiv–xxxvj (1.5–2.3 gm.).—M.
>   Fiant capsulæ No. xii.
> Sig. One every three hours.

Incompatibles.—Liquefaction results when camphor is triturated with chloral hydrate, menthol, phenol, or thymol. Aqueous solutions precipitate camphor from its alcoholic solution.

CAMPHORA MONOBROMATA, U. S. P.

(Monobromated Camphor, $C_{10}H_{15}BrO$.)

Monobromated camphor is obtained from the union of bromin and camphor in the presence of heat, and occurs as colorless prismatic needles or scales, having a mild camphoraceous odor and taste. It is very sparingly soluble in water and glycerin, but freely soluble in alcohol, ether, and chloroform. The dose is from 2–5 gr. (0.13–0.3 gm.) in pill or capsule.

Physiologic Action.—In small doses it acts as a sedative to the nervous system. Toxic doses are followed by tremors, muscular relaxation, a marked slowing of the pulse, embarrassed respiration, epileptiform convulsions, and coma.

Therapeutics.—Monobromated camphor is employed as an antispasmodic and anaphrodisiac. It has been recommended in *hysteria, delirium tremens, chorea,* and *whooping-cough,* but its use in these affections is only rarely attended with success. In *petit mal* it is sometimes more effective than the ordinary bromids. Its more important use is as an anaphrodisiac. It often proves very efficacious in *abnormal sexual excitement, chordee, spermatorrhea,* etc. Its chief drawback is its tendency to induce heartburn and eructations. In spermatorrhea it may be combined with hyoscin, as in the following formula:

> ℞ Hyoscinæ hydrobromatis,   gr. 1/16 (0.006 gm.);
>   Camphoræ monobromatæ,   ʒj (4.0 gm.).—M.
>   Fiant pilulæ No. xx.
> Sig. One or two at bedtime.

ASAFŒTIDA, U. S. P.
(Asafetida.)

Asafetida is a gum-resin obtained by incising the roots of *Ferula fœtida*, an umbelliferous plant abounding in Persia and Afghanistan. It appears in the form of irregular masses or tears of a yellowish-brown color, and has a persistent garlicky odor and an acrid taste. It consists of gum, resin, and an essential oil (about 4 per cent.) which is rich in sulphur. The oil is the most active ingredient. The dose is from 3–10 gr. (0.2–0.65 gm.).

PREPARATIONS.	DOSE.
Emulsum Asafœtidæ, U. S. P. (4 per cent. watery emulsion)	½–1 fl. oz. (15.0–30.0 c.c.).
Tinctura Asafœtidæ, U. S. P.	½–1 fl. dr. (2.0–4.0 c.c.).
Pilulæ Asafœtidæ, U. S. P. (3 gr.—0.2 gm.)	1–3 pills.
Pilulæ Aloes et Asafœtidæ, U. S. P. (1⅓ gr.—0.085 gm. of each)	1–3 pills.

Physiologic Action.—On the circulatory and nervous systems asafetida acts as a mild stimulant; on the alimentary canal, as a carminative.

Therapeutics.—Asafetida is employed chiefly as an antispasmodic and as a carminative. As an antispasmodic it is used in *nervous excitement, whooping-cough,* and *hysteria.* Suppositories of asafetida are sometimes valuable in *infantile convulsions* and in the *restlessness of the specific fevers.* As a carminative, it is an excellent remedy in *infantile colic* and the *flatulent dyspepsia of the aged.* An enema of asafetida is one of the most efficacious remedies that we possess in *excessive tympanites.* Asafetida has also been employed as a stimulating expectorant in *chronic bronchitis.*

Administration.—For continued use the drug is best administered in the form of pills. To children it may be conveniently administered in the form of suppositories. In tympanites the official emulsion (4–6 ounces—120.0–175.0 c.c.) should be employed as an enema.

VALERIANA, U. S. P.
(Valerian.)

Valerian is the rhizome and roots of *Valeriana officinalis*, a native of Europe. Its active principles are a volatile oil and valerianic acid.

PREPARATIONS.	DOSE.
Tinctura Valerianæ, U. S. P.	1–3 fl. dr. (4.0–11.0 c.c.).
Tinctura Valerianæ Ammoniata, U. S. P. (made by macerating valerian in aromatic spirits of ammonia)	1–3 fl. dr. (4.0–11.0 c.c.).
Extractum Valerianæ Fluidum, U. S. P. . .	½–1 fl. dr. (2.0–4.0 c.c.).

Physiologic Action.—In moderate doses valerian acts as a sedative to the brain and spinal cord. Very large doses produce a sense of warmth in the stomach, quickening of the pulse, vomiting, diarrhea, frequent micturition, headache, vertigo, and hallucinations.

Therapeutics.—Valerian is a valuable remedy for subduing the various nervous phenomena occurring in patients of excitable temperaments. The *spasms, headache, palpitation, globus, sleeplessness,* and *flatulent distention of hysteria* are often remarkably influenced by the drug.

AMMONII VALERIANAS, U. S. P.

(Ammonium Valerianate, $NH_4C_5H_9O_2$.)

Valerianate of ammonium occurs in the form of colorless quadrangular plates having the odor of valerianic acid and a sharp, sweetish taste. It is freely soluble in water and alcohol. Its dose is from 5–10 gr. (0.3–0.6 gm.) in the form of an elixir.

It is employed for the same purposes as valerian.

FERRI VALERIANAS, U. S. P.

(Ferric Valerianate; Iron Valerianate, $Fe_2(C_5H_9O_2)_6$.)

Iron valerianate appears as a dark-red amorphous powder, having the odor of valerianic acid. It is a useful combination of an antispasmodic with a chalybeate, and may be prescribed with advantage in *anemia* associated with *nervous excitement* or *hysteria*. The dose is from 1–3 gr. (0.06–0.2 gm.) in pill form. The following tonic and antispasmodic pill of three valerianates is frequently prescribed:

 ℞ Ferri valerianatis,
  Zinci valerianatis,
  Quininæ valerianatis, aa gr. xx (1.3 gm.).—M.
  Fiant pilulæ No. xx.
  Sig. One after meals.

ZINCI VALERIANAS, U. S. P.

(Zinci Valerianate, $Zn(C_5H_9O_2)_2 + 2H_2O$.)

Zinc valerianate appears in the form of white, pearly scales, having the odor of valerianic acid, and a sweetish, astringent, and metallic taste. It is soluble in 100 parts of water and 40 parts of alcohol. The dose is from $\frac{1}{2}$–3 gr. (0.03–0.2 gm.). It has about the same therapeutic value as valerian.

SPIRITUS ÆTHERIS COMPOSITUS, U. S. P.

(Compound Spirit of Ether ; Hoffmann's Anodyne.)

Hoffmann's anodyne is a colorless, inflammable liquid, having an ethereal odor and taste. It consists of 325 parts by volume of ether, 650 parts of alcohol, and 25 parts of ethereal oil, the last being a product of the action of sulphuric acid on alcohol. The dose is from $\frac{1}{2}$–2 fl. dr. (2.0–8.0 c.c.).

Physiologic Action and Therapeutics.—Hare concludes, from a careful study of this compound, that its calmative effects are largely due to the ether which it contains rather than to the ethereal oil, and that each of its ingredients stimulates the system—the ether the most, and the ethereal oil the least powerfully.

Hoffmann's anodyne is an excellent carminative for expelling gas from the stomach. It affords prompt relief in severe *paroxysmal palpitation* dependent upon gastric flatulence. It is often of service in *angina pectoris*, especially when the attacks are precipitated by flatulency. In *chronic valvular disease*, when compensation is partially lost and the patient suffers from restlessness, dyspnea, and insomnia, the compound spirit of ether often acts most favorably. It is sometimes very efficacious in *hiccough*. A combination like the following will be found useful in *asthma :*

<pre>
R   Ammonii bromidi,   ʒij (8.0 gm.) ;
    Tincturæ belladonnæ,   fʒj (4.9 c.c.) ;
    Tincturæ lobeliæ,   fʒij (8.0 c.c.) ;
    Spiritus ætheris compositi,   fʒj (30.0 c.c.) ;
    Elixiris aromatici,   q. s. ad fʒiij (90.0 c.c.).—M.
  Sig. A dessertspoonful in water every hour or two during the paroxysm.
</pre>

The following mixture is often serviceable in severe gastric flatulency :

<pre>
R   Spiritus ætheris compositi,
    Spiritus camphoræ,   aa ʒss (2.0 c.c.) ;
    Spiritus menthæ piperitæ,   ♏vj (0.4 c.c.).—M.
  Sig. To be taken in water.
</pre>

BROMOFORM.

(Methyl Tribromid ; Tribrommethane, $CHBr_3$.)

Bromoform is made by the action of hypobromite of sodium on acetone, or of bromin on a mixture of methyl alcohol and milk of lime. It appears as a clear, colorless liquid having a pleasant odor and a sweetish taste. It is sparingly soluble in water, but freely so in alcohol and glycerin. The dose is from 1–6 min. (0.06–0.4 c.c.). To be fit for use it should be free from color and acidity.

Therapeutics. — Bromoform has anesthetic properties somewhat analogous to those of chloroform, but as a therapeutic agent it is of interest chiefly as an internal remedy in *whooping-cough.* While it has a definite value in lessening the severity of the paroxysms, it has no power to shorten the duration of the disease.

Toxicology.—Many cases of poisoning by bromoform have been reported since its first introduction as a remedy for whooping-cough. Very often the accident has resulted from a lack of familiarity with the physical properties of the drug. As bromoform is heavier than mucilage, its suspension in the latter can only be temporary; it naturally tends to fall to the bottom of the bottle, and unless this is prevented by frequent and thorough shaking, a much larger dose may be taken on emptying the bottle than was intended. The chief phenomenon of bromoform-poisoning is a kind of intoxication, followed by coma, marked cyanosis, difficult breathing, a rapid, feeble pulse, muscular relaxation, and collapse. In some cases the coma has been interrupted by general convulsions. Death usually results from asphyxia.

Treatment consists in the evacuation of the stomach, the practice of artificial respiration, and the administration of cardiac and respiratory stimulants.

Administration.—It may be administered by simply dropping it on sugar or into a little peppermint-water. On account of its high specific gravity a minim is equal to about 6 drops. If suspended in mucilaginous liquids, the bottle must be well shaken before each administration. As it is soluble in alcohol and glycerin, a mixture of these liquids may be made the vehicle, as in the following formula:

℞	Bromoform,	ℳ xx–xxx (1.2–1.8 c.c.);
	Alcoholis,	f℥ij (8.0 c.c.);
	Spiritus menthæ piperitæ,	ℳ xx (1.3 c.c.);
	Glycerini, .	f℥ij (60.0 c.c.). —M.

. Sig.—A teaspoonful four times a day for a child of two years.

MOSCHUS, U. S. P.

(Musk.)

Musk is the dried secretion from the preputial follicles of *Moschus moschiferus,* a species of deer found in Central Asia.

It appears as reddish-black unctuous grains having a peculiar penetrating odor and a bitter aromatic taste. The dose is from 5–15 gr. (0.3–1.0 gm.), given preferably by the rectum, in suppository or suspended in mucilage.

PREPARATION.	DOSE.
Tinctura Moschi, U. S. P. (5 per cent.)	1–2 fl. dr. (4.0–8.0 c.c.).

Therapeutics.—Musk is used as an antispasmodic and as a general stimulant. As Graves first pointed out, it is a most valuable stimulant in *typhoid fever* and *kindred affections*, when there are profound nervous exhaustion and circulatory failure. It is also of service in *obstinate hiccough*, the *convulsions of childhood*, and *laryngismus stridulus*. The great drawbacks to its use are its costliness and the difficulty in obtaining an unadulterated preparation.

CIMICIFUGA, U. S. P.

(Black Snakeroot, Actæa Racemosa.)

Cimicifuga is the dried rhizome and rootlets of *Cimicifuga racemosa*, a perennial plant growing in the woodlands of North America. It contains a resin, a bitter neutral substance, and a volatile oil.

PREPARATIONS.	DOSE.
Tinctura Cimicifugæ, U. S. P.	½–2 fl. dr. (2.0–8.0 c.c.).
Extractum Cimicifugæ Fluidum, U. S. P.	10–30 min. (0.6–2.0 c.c.).
Extractum Cimicifugæ, U. S. P.	2–8 gr. (0.1–0.5 gm.).

Physiologic Action.—In large doses cimicifuga causes nausea, headache, vertigo, tremors, muscular relaxation, slowing and weakening of the pulse, anesthesia, and, finally, paralysis of respiration. The anesthesia is due, according to Hutchinson, to paralysis of the sensory columns of the spinal cord. The weakness of the pulse is probably due to a direct depressant effect exerted on the heart itself. In full doses it seems also to possess the power of stimulating uterine contractions.

Therapeutics.—Cimicifuga is a valuable remedy in *simple chorea*, ranking next to arsenic in efficiency. It should be given in doses of 10 min. (0.6 c.c.) of the fluid extract three times a day, after meals, and gradually increased to from ½–1 dr. (2.0–4.0 c.c.). It is sometimes of service in chronic rheumatic affections of the nerves and muscles, like *myalgia, pleurodynia, sciatica*, and *lumbago*, but in articular rheumatism it is of little value. It is said to be useful in *atonic amenorrhea, menorrhagia*, and *sudden cessation of the menses* from cold or nervous shock. It has been used in labor to stimulate uterine contractions, its action resembling that of quinin rather than that of ergot. Ringer strongly recommends it in combination with gelsemium for the distressing symptoms attending the menopause; and, according to Simpson, it is highly beneficial in the *mental disturbances* which sometimes follow pregnancy.

It is often of distinct service in relieving the *tinnitus aurium* accompanying chronic catarrh of the middle ear. It has been used as an expectorant in *bronchitis*, with free expectoration, and in *phthisis*.

LACTUCARIUM, U. S. P.

(Wild Lettuce.)

Lactucarium is the concrete milk-juice of *Lactuca virosa*, a biennial herb cultivated chiefly in Central and Southern Europe. It contains lactucin (a bitter principle), lactucic acid, lactucopicrin, and lactucerin. The dose is from $\frac{1}{2}$–1 dr. (2.0–4.0 gm.).

PREPARATIONS.	DOSE.
Tinctura Lactucarii, U. S. P.	$\frac{1}{8}$–1 fl. dr. (2.0–4.0 c.c.).
Syrupus Lactucarii, U. S. P.	1–4 fl. dr. (4.0–15.0 c.c.).
Extractum Lactucarii Fluidum	20–30 min. (1.0–2.0 c.c.).

Therapeutics.—Lactucarium is a feeble hypnotic, an anodyne, and an antispasmodic. Added to cough-mixtures, the syrup is a useful sedative in the *acute bronchitis* of children.

HUMULUS, U. S. P.

(Hops.)

Hops are the dried strobiles of *Humulus Lupulus*, a perennial climber largely cultivated in Europe and the United States. They have an aromatic odor, and a bitter, slightly astringent taste. A glandular powder separated from the dried strobiles is also official as *Lupulinum* (lupulin). Lupulin, of which hops contain from 8 to 15 per cent., appears in the form of minute yellowish-brown, resinous granules, having a bitter aromatic taste. The chief constituents of hops are cholin (a volatile, liquid alkaloid), a volatile oil, lupamaric acid (the bitter principle), tannic acid, and a resin. The dose of lupulin is from 3–5 gr. (0.2–0.3 gm.).

PREPARATIONS.	DOSE.
Tinctura Humuli, U. S. P.	2–4 fl. dr. (8.0–15.0 c.c.).
Extractum Lupulini Fluidum, U. S. P.	$\frac{1}{2}$–1 fl. dr. (2.0–4.0 c.c.).
Oleoresina Lupulini, U. S. P.	5–10 min. (0.3–0.6 c.c.).

Therapeutics.—Hops act as a bitter tonic and as a feeble hypnotic. In the *restlessness of fevers* and in *delirium tremens* they are sometimes useful sedatives. They have been found especially serviceable in *chordee, spermatorrhea,* and *vesical irritation.* The following suppositories, containing lupulin, afford relief in chordee:

R Pulveris opii,	gr. vj (0.4 gm.);
Pulveris camphoræ,	gr. xxlv (1.5 gm.);
Lupulini,	3j (4 0 gm.);
Olei theobromatis,	q. s.—M.

Fiant suppositoria No. vi.
Sig. Introduce one into the bowel at bedtime.

SUMBUL, U. S. P.
(Musk-root.)

Sumbul is the root of *Ferula Sumbul,* a perennial herb growing in Central Asia. Its chief constituents are a resin, to which its musk-like odor is due, a fixed oil, angelic acid (sumbulic acid), and valerianic acid. The dose of the root is from 10–30 gr. (0.6–2.0 gm.).

PREPARATIONS.	DOSE.
Tinctura Sumbul, U. S. P.	$\frac{1}{2}$–2 fl. dr. (2.0–8.0 c.c.).
Extractum Sumbul	2–5 gr. (0.1–0.3 gm.).

Therapeutics.—Sumbul is sometimes useful as an antispasmodic in *hysteria* and allied neuroses. In *neurasthenia* the extract may be given in pill with iron and arsenic. The following combination, suggested by Bradbury, is very effective in relieving the insomnia, mental depression, and flushing heats commonly occurring at the *menopause :*

R Ammonii bromidi,	3ij (8.0 gm.);
Tincturæ sumbul,	
Tincturæ humuli,	āā f3iss (45.0 c.c.);
Spiritus camphoræ,	f3iij (11.0 c.c.);
Elixiris aromatici,	q. s. ad f3vj (175.0 c.c.).—M.

Sig. Tablespoonful in water after meals.

VIBURNUM PRUNIFOLIUM, U. S. P.
(Black Haw.)

Officially viburnum prunifolium is the bark of a tall shrub of the same name, growing in the eastern and southern States of North America. Its chief constituents are viburnin (a bitter principle), valerianic acid, resin, and tannin.

PREPARATION. DOSE.

Extractum Viburni Prunifolii Fluidum, U. S. P. ½–2 fl. dr. (2.0–8.0 c.c.).

Therapeutics.—Large doses of the fluid extract are said to cause in warm-blooded animals drowsiness, a fall of blood-pressure, muscular relaxation, and a diminution of reflex irritability. It is employed chiefly as a uterine and ovarian sedative. It has been found useful in *habitual abortion* occurring in non-syphilitic patients. Combined with opium it is very efficacious in *threatened abortion*. In *non-inflammatory dysmenorrhea* it affords great relief when given in doses of 1 dr. (4.0 c.c.) three times a day, beginning five or six days before menstruation. It is sometimes very successful in subduing the *menstrual irregularities* and *nervous distress* occurring about the time of the menopause.

OLEUM SUCCINI.

(Oil of Amber.)

Oil of amber is a volatile oil obtained by the destructive distillation of a fossil resin which is yielded by submerged fir trees washed ashore along the coast of Prussia. It is a thin, transparent, yellow liquid, having a balsamic odor and a warm, acrid taste. The dose is from 5–10 min. (0.3–0.6 c.c.), in capsules or in an emulsion.

Therapeutics.—Oil of amber, with camphor or turpentine, was at one time held in high repute as a stimulating liniment for the chest in *whooping-cough* and *acute bronchitis*. Internally it is sometimes efficacious in *obstinate hiccough*.

SPINAL CORD EXCITANTS.

Spinal cord excitants, or excitomotors, are drugs which increase the functional activity of the spinal cord. By stimulating the cell-bodies of the lower motor neurons they exaggerate reflex activity, and if given in toxic dose, induce tetanic convulsions. The origin of the convulsions excited by a drug can be determined in animals by dividing the spinal cord between the occiput and the atlas. Convulsions of cerebral origin cease after section, while those of spinal origin do not. The most important spinal cord excitants are:

Strychnin.	Caffein.
Brucin.	Ammonia.
Hydrastin.	Carbolic acid.

With the exception of strychnin and brucin, however, these drugs do not sensibly excite the spinal cord unless given in toxic doses.

The indications for using excitomotors as such cannot be stated very definitely. They often prove efficacious, however, when loss of power is dependent upon depression of the motor neurons, and not upon destructive lesions. When there is evidence of inflammation they are absolutely contraindicated.

NUX VOMICA, U. S. P.

Nux vomica is the seeds of the *Strychnos Nux vomica*, a small tree growing in the East Indies. It contains two alkaloids, *strychnin* and *brucin*. Strychnin fully represents the action of the crude drug. The sulphate of strychnin (Strychninæ sulphas, U. S. P.) appears as white prismatic crystals, odorless, of an intensely bitter taste, and soluble in 50 parts of cool water. Brucin resembles strychnin in its action, but is less powerful. The dose of strychnin sulphate is from $\frac{1}{60}$–$\frac{1}{20}$ gr. (0.001–0.003 gm.).

Preparations.	Dose.
Extractum Nucis Vomicæ, U. S. P.	$\frac{1}{8}$–$\frac{1}{4}$ gr. (0.008–0.016 gm.).
Extractum Nucis Vomicæ Fluidum, U. S. P.	1–5 min. (0.06–0.3 gm.).
Tinctura Nucis Vomicæ, U. S. P.	5–15 min. (0.3–1.0 c.c.).
Pulvis Nucis Vomicæ,	1–5 gr. (0.06–0.3 gm.).

Physiologic Action.—Circulatory System.—Full medicinal doses of strychnin stimulate the heart and vasomotor center, the pulse under its influence becoming stronger and more rapid.

Toxic doses tend to lower the blood-pressure by depressing the vasomotor center; but during the convulsive seizures induced by the drug an enormous increase in the pressure may be observed, resulting from the violent muscular contractions.

Nervous System.—The dominant action of the drug is on the nervous system; it is a powerful stimulant to the spinal cord, especially to the motor centers. In poisoning the reflex excitability of the spinal cord is so exaggerated that the slightest external irritation will call forth violent tetanic convulsions. Ordinary doses have little or no effect on the peripheral nerves, but toxic doses paralyze the efferent or motor fibers.

Eye.—The nervous mechanism of the eye shares in the general nervous stimulation induced by the drug, and vision is rendered more acute.

Respiratory System.—It is a powerful stimulant to the respiratory center, quickening and deepening the respiratory movements.

Alimentary Canal.—In small doses it whets the appetite, increases the secretion of gastric juice and of free acid, strengthens the muscular movements of the stomach, and increases intestinal peristalsis.

Uterus.—Indirectly through its action on the central nervous system strychnin is of some value in overcoming uterine inertia.

Elimination.—Strychnin is eliminated through the kidneys, partly as strychnin and partly as strychnic acid; much of the drug, however, is oxidized in the body.

Toxicology.—The first evidences of the toxic effects of strychnin are restlessness, anxiety, twitching and starting of the muscles, and stiffness of the neck. If the dose has been sufficiently large, spinal convulsions speedily develop, throwing the patient into opisthotonos, so that he rests on his head and his heels. The convulsions are for the most part intermittent, and are repeated under the slightest external irritation. The pupils are dilated, the vision is hyperacute, the mind is unaffected, and the stomach is usually retentive. In fatal cases the convulsions become more rapid and severe, and finally death results from asphyxia due to spasm of the respiratory muscles, from paralysis of the motor nerves, or more rarely from exhaustion.

The history of the case, the absence of a wound, the intermittent character of the convulsions, and the late involvement of the muscles of the jaw (trismus) will serve to distinguish strychnin-poisoning from tetanus.

The history of the case, the emotional excitement, the closed eyes, and the persistent opisthotonos will serve to distinguish hysteria from strychnin-poisoning.

Treatment.—Since the slightest stimulus will provoke convulsions, the patient should be disturbed as little as possible. If convulsions have already begun, no attempt should be made to use a stomach-pump until the reflex excitability has been subdued by inhalations of chloroform or amyl nitrite. Tannic acid in solution may be administered as a chemical antidote. The best physiologic antidotes are potassium bromid (3j–ij—4.0–8.0 gm.) and chloral (20 gr.—1.3 gm.); when the patient is unable to swallow the chloral, 1 dr. (4.0 gm.) may be given in an enema. Osterwald observed very beneficial effects from oxygen inhalations in strychnin-poisoning in animals.

Therapeutics.—Nux vomica or strychnin is used chiefly as a circulatory stimulant, a respiratory stimulant, an excito-motor, and a stomachic.

As a Circulatory Stimulant.—As a heart-stimulant strychnin has a wide range of usefulness. In *simple dilatation of the heart*, with or without valvular disease, it often renders good service when combined with digitalis, but used alone it is distinctly inferior to the latter drug. In *pneumonia* and other *infective processes* it is an excellent adjuvant to alcohol and as such is superior to digitalis. In *the weak heart of the old* also it is generally a better remedy than digitalis.

In *chronic bronchitis* and *emphysema* it is often indispensable in toning up the right ventricle. It is a valuable circulatory stimulant in *surgical shock*.

As a Respiratory Stimulant.—Being both a respiratory and cardiac stimulant, strychnin is indicated in many diseases of the lungs associated with enfeeblement of the respiration, such as *pneumonia, chronic bronchitis, phthisis,* and *asthma*. In emphysema there is no single drug so useful as strychnin. Employed hypodermically in cases of *acute pulmonary edema* it is scarcely less efficient than caffein.

In *poisoning* by drugs which depress the heart or respiratory center strychnin is of value. In poisoning by opium, chloral, aconite, ether, or chloroform it should be given hypodermically in full doses.

As an Excitomotor.—When paralysis is the result of complete destruction of nerve-cells or fibers, strychnin, like all other drugs, will be ineffective; but when the loss of power is the result simply of depression or exhaustion of the nervous apparatus, it may prove useful. It is commonly prescribed in the *hemiplegia* following apoplexy, but it is valueless in this condition except, perhaps, as a general tonic. On the other hand, in the various forms of *peripheral neuritis*, such as plumbic, alcoholic, diphtheritic, and that due to pressure, it is often of considerable service. It should not be used, however, until all acute symptoms of the affection have subsided. In these cases the best results are obtained by injecting the drug directly into the paralyzed muscles.

In *neurasthenia* minute doses of strychnin with arsenic are sometimes efficacious, but more often the drug is useless or actually harmful. In *amaurosis* from tobacco, alcohol, or lead it is invaluable.

It is thought that intramuscular injections of strychnin aid in restoring power to the affected muscles in *acute poliomyelitis*.

This treatment, however, should not be instituted until three or four weeks after the incidence of the disease. Nux vomica often affords relief in *incontinence of urine* from atony of the bladder—a condition sometimes met with in old people. It is likewise an excellent remedy in *constipation* dependent upon atony of the intestinal walls.

As a Stomachic.—Nux vomica is often very useful in *chronic dyspepsia* when there are no manifestations of severe inflammatory trouble. In these cases the chief symptoms are a coated, flabby tongue, anorexia, fulness and distress for some time after eating, deficient secretion of hydrochloric acid, flatulence, and constipation. In such types of indigestion the tincture of nux vomica (4 or 5 drops, before meals), with or without hydrochloric acid, often acts most happily. It is also of service in *dilatation of the stomach* due to atony. It is a trustworthy remedy in those cases of *marked flatulence*, common in elderly people, dependent upon a lack of tone in the walls of the stomach and bowel.

Strychnin is contraindicated in gastric affections when there is hypersecretion of hydrochloric acid, and when there are evidences of well-marked inflammation.

Other Uses.—In *hyperemesis* not dependent upon organic disease of the stomach, the tincture of nux vomica (1–2 drops) alone, or with wine of ipecac ($\frac{1}{2}$–1 drop), every hour or two, will often afford relief. Good results are sometimes obtained with strychnin as a prophylactic remedy in *uterine inertia*. It has been recommended by Murrell and Brunton to check the *night-sweats of phthisis.* It is useful in *sexual impotence* due to lowered nervous tone. Dana and others speak very favorably of strychnin in heroic doses in *tic douloureux* occurring in anemic and exhausted patients, and when the duration of the disease has not been more than one or two years. The drug is given hypodermically once a day, and the dose is gradually increased from $\frac{1}{30}$ to $\frac{1}{5}$ of a grain (0.002 to 0.013 gm.), ten to twenty days being required to reach this maximum.

Administration.—Strychnin sulphate and extract of nux vomica are suitable preparations for use in pills. When a liquid preparation is desired, the tincture will be found reliable. In paralytic affections it is best to inject a solution of strychnin sulphate directly into the affected muscles. In these cases the drug should be given once a day, at first in small doses ($\frac{1}{50}$ gr.—0.0013 gm.), the amount being rapidly increased to the limit of tolerance. Nervous excitement, restlessness, and twitching of the muscles are indications that the limit of physiologic action has been reached. The nitrate of strychnin is sometimes preferred to the sulphate for hypodermic use, but it has no special advantage over the latter.

Incompatibles.—Strychnin is incompatible with tannic acid, alkalies, chlorids, iodids, and bromids.

SPINAL CORD DEPRESSANTS.

Spinal cord depressants are drugs which lessen the functional activity of the spinal cord. They may diminish reflex activity directly by depressing the lower motor or sensory neurons, or indirectly by stimulating Setschenow's inhibitory centers in the medulla oblongata. In very large doses many of them not only abolish reflex action, but they also cause more or less complete loss of motor power. Some spinal depressants have a special affinity for the sensory neurons, and this is manifested in the poisoned animal by the persistence of voluntary movement after the abolition of reflex action.

The following drugs in full therapeutic doses are spinal cord depressants:

Physostigma.	Chloral.
Nitrites.	Ether.
Bromids.	Chloroform.
Gelsemium.	Hyoscin.

A number of other drugs in *toxic doses* have a similar action on the spinal cord:

Cimicifuga.	Tartar emetic.
Hydrastinin.	Potassium salts.
Alcohol.	Cocain.
Opium.	Antipyrin.
Hydrocyanic acid.	Quinin.
Veratrum viride.	Digitalis.

Physostigma, nitrites, gelsemium, chloral, hyoscin, hydrastinin, and *veratrum viride* depress especially the motor centers of the spinal cord.

Bromids, ether, chloroform, cocain, and *tartar emetic* depress both sensory and motor neurons, but the former more powerfully than the latter.

Digitalis and *quinin* diminish reflex activity first by stimulating Setschenow's inhibitory centers in the medulla oblongata, and later by depressing the spinal cord.

Spinal cord depressants are useful in combating symptoms

resulting from irritation of the neurons of the lower motor segment. Thus, they are of value in controlling the paroxysms of laryngismus stridulus, the spasms of tetany, and the tremors of paralysis agitans. They are also of service in allaying the hyperexcitability of the lower motor neurons caused by certain poisons, as strychnin and that of tetanus. Finally, spinal cord depressants are not without value in the treatment of convulsive seizures which have a cerebral origin; for by depressing the lower motor neurons they impair their power of transmitting excessive discharges from the brain to the muscles, and by depressing the lower sensory neurons they afford the brain more or less protection from the effects of external stimuli.

PHYSOSTIGMA, U. S. P.

(Calabar Bean.)

Physostigma is the seed of *Physostigma venenosum*, a perennial climber, growing in West Africa. It contains two alkaloids, *physostigmin* or *eserin*, and *calabarin;* the latter is less important than the former, and resembles strychnin in its action.

PREPARATIONS.	DOSE.
Tinctura Physostigmatis, U. S. P. . . .	5–30 min. (0.3–1.8 c.c.).
Extractum Physostigmatis, U. S. P . . .	$\frac{1}{8}$–$\frac{1}{4}$ gr. (0.008–0.02 gm.).

Physostigmin, or eserin, fairly represents the active properties of the bean. It occurs in yellowish-white crystals, deliquescent, odorless, bitter, sparingly soluble in water, but readily so in alcohol and dilute acids, and turning pink on exposure. The dose is from $\frac{1}{120}$–$\frac{1}{30}$ gr. (0.0005–0.002 gm.). Two salts are official:

> Physostigminæ sulphas.
> Physostigminæ salicylas.

The sulphate is very soluble in water, and the salicylate is soluble in 150 parts of water. The dose of either salt is from $\frac{1}{100}$–$\frac{1}{30}$ gr. (0.0006–0.002 gm.).

Physiologic Action. — Circulatory System. — Small doses do not affect the circulation; large doses slow the pulse, probably by stimulating the peripheral vagi, and raise the arterial pressure by stimulating the heart or its contained ganglia, and probably also by stimulating the muscular fibers of the arterioles. Lethal doses paralyze the heart.

Nervous System.—The dominant action of the drug is upon the motor neurons of the spinal cord, which are paralyzed by large doses. Upon the sensory neurons and the peripheral nerves it has but a feebly depressant influence. It does not seem to affect the brain in any dose. The convulsions occasionally seen after the ingestion of poisonous doses are probably due to calabarin.

Respiratory System.—Therapeutic doses do not affect respiration, but toxic doses kill by paralyzing the respiratory center.

Muscles.—In toxic doses it acts as an irritant to voluntary muscles, causing fibrillary tremors and tetanoid contractions.

Alimentary Canal.—It increases peristalsis, probably by directly stimulating the unstriped muscle-fibers of the intestines.

Secretions.—It increases to some extent the secretions of the salivary, sweat-, intestinal, and mammary glands.

Eye.—Physostigmin powerfully contracts the pupil, the myosis being due to stimulation of the peripheral fibers of the oculomotor nerve and to paralysis of the peripheral filaments of the sympathetic nerve. It also causes accommodative spasm and diminishes intra-ocular tension.

Toxicology.—Physostigmin-poisoning is characterized by vertigo, great muscular relaxation, tremors and twitchings of the muscles, abolition of reflex action, contraction of the pupils, slowing of the pulse, salivation, and sometimes colicky pains with nausea and vomiting.

Treatment of Poisoning.—This consists in evacuating the stomach by an emetic, in administering atropin as a physiologic, and tannic acid as a chemical, antidote, and in combating respiratory and cardiac failure with such stimulants as ammonia, strychnin, and alcohol.

Therapeutics.—Physostigma has a very limited range of usefulness ; it is chiefly employed to depress the spinal cord, to contract the pupils, to lessen intra-ocular tension, and to stimulate involuntary muscles, especially those of the stomach and intestine.

It has been used to depress the spinal cord in *tetanus*, *strychnin-poisoning*, and *chorea*, but in these affections we have far more reliable remedies.

The extract sometimes makes a useful addition to cathartic pills intended to relieve *constipation* dependent upon atony of the intestinal walls. Ringer has recommended it in *chronic bronchitis* with dyspnea, due to weakness of the bronchial muscles. It has found favor with some practitioners in the treatment of *atonic dilatation of the stomach.*

Eye-affections.—Physostigmin sulphate is used to counteract *mydriasis* produced by atropin, to lessen intra-ocular tension in the early stages of *glaucoma*, and sometimes alternately with atropin to break up adhesions in *iritis*. In *peripheral ulceration of the cornea*, in the absence of iritis, it often acts better than atropin.

Untoward Effects.—Strong solutions instilled into the eye sometimes cause pain in the eye and head, spasm of the ocular muscles, vertigo, and faintness.

Administration.—Physostigma may be given internally in the form of the extract, the tincture, or the alkaloid. For hypodermic use physostigmin sulphate is preferable on account of its ready solubility in water. For the same reason the sulphate should be employed in affections of the eye. For continued use it is generally prescribed in solutions containing from $\frac{1}{8}$–1 gr. (0.008–0.065 gm.) to the ounce (30.0 c.c.), but when a powerful and prompt effect upon the eye is desired much stronger solutions (2–3 gr.—0.13–0.2 gm.) to the ounce (30.0 c.c.) may be employed. On standing solutions of eserin turn red and lose some of their effectiveness.

AMYL NITRIS.

(Amyl Nitrite, $C_5H_{11}NO_2$.)

Amyl nitrite is prepared by distilling equal volumes of pure amylic alcohol and nitric acid.

It appears as a yellowish, highly volatile liquid having a strong ethereal odor. The dose is from 1–3 min. (0.06–0.2 gm.), although it is rarely employed internally.

Physiologic Action.—The inhalation of nitrite of amyl is speedily followed by flushing of the face, a sense of fulness in the head, quickening of the pulse, and a pronounced fall of blood-pressure. The flushing of the face and the fall of blood-pressure are chiefly due to dilatation of the arterioles, which is brought about by the direct paralyzant effect of the drug on the muscular coats of the blood-vessel walls. The quickening of the pulse results from a depression of the cardiac inhibitory centers in the medulla, from widening of the blood-paths throughout the body, and probably also from stimulation of the heart itself. While the drug primarily acts as a cardiac stimulant, this action is transient, and soon gives way to depression, especially when large doses are employed.

Nervous System.—The ringing in the ears, throbbing, and headache which follow the inhalation of large amounts of the drug are due to cerebral congestion. Its dominant action on the nervous system is lessening of reflex activity, which is due to depression of the motor neurons of the spinal cord. Upon the brain, sensory neurons of the spinal cord, and peripheral nerves amyl nitrite has but little if any effect.

Respiratory System.—Small doses quicken the respiration; toxic doses kill by paralyzing the respiratory center.

Blood.—Inhalations of nitrite of amyl impart a chocolate hue to the blood. This change of color is due to the conversion of oxyhemoglobin into methemoglobin, a compound having less oxidizing power.

Uterus.—The uterine muscle, like unstriated muscle generally, is sensitive to the relaxing influence of the drug.

Temperature.—Large doses of amyl nitrite produce a decided fall of the body-temperature, which is probably chiefly due to lessened oxidation.

Local Action.—In concentrated form it is a paralyzant to all highly organized tissues.

Elimination.—In part it is eliminated with great rapidity through the kidneys, and in part it is oxidized in the body. Its escape through the kidneys is sometimes attended with glycosuria and polyuria.

Therapeutics.—Nitrite of amyl is employed chiefly for two purposes: To relax spasm (by depressing the lower motor neurons) and to lower arterial tension (by dilating the arterioles). Owing to the rapidity with which it is absorbed and eliminated, it is of service only when a very prompt but transitory effect is required.

As an Antispasmodic.—It is used extensively to relax spasms, both general and local. In the convulsions of epilepsy, tetanus, uremia, and strychnin-poisoning, inhalations of the drug often prove very efficient. When the epileptic seizures are preceded by an aura the patient may carry the remedy in a small bottle containing the exact dose, or in *perles* made especially for the purpose. While it has been recommended in puerperal eclampsia, its use in this affection is not unattended with danger, since serious flooding may result from its relaxing influence on the uterine muscle.

As a Vasodilator.—On account of the fugaciousness of its action amyl nitrite is not so useful as nitroglycerin or erythrol tetranitrate in circulatory diseases associated with persistent high arterial tension. In attacks of *angina pectoris*, however, it often acts most happily. Curiously enough, the drug is just as likely to do good in this disease when the pulse is soft as when it is hard. Brunton explains this fact in the following way: The excessive work to be done, which is the cause of the pain, may be due either to the resistance in the vessels being increased above the normal, or to the power of the heart to overcome the resistance being lessened below the normal. In either case the resistance becomes too great for the heart.

The vasodilator action of the drug has led to its use in *malaria*, to cut short the first stage of the paroxysms. It is said to be of service in *migraine*, when the pain is associated with constriction of the peripheral vessels.

Administration.—Amyl nitrite is almost always given by inhalation: 2–5 min. (0.1–0.3 c.c.) may be dropped upon a handkerchief, or one of the small glass capsules containing the requisite dose may be crushed in a handkerchief, and the vapor inhaled. According to Osler it sometimes acts better in angina pectoris when given by the mouth, combined with the tincture of capsicum and peppermint-water. As a rule, however, nitroglycerin is better suited for internal administration.

NITROGLYCERINUM.

(Nitroglycerin, Glonoin, Trinitrin, $C_3H_5(NO_3)_3$.)

Nitroglycerin is prepared by gradually adding dehydrated glycerin to nitric acid, or a mixture of nitric and sulphuric acids. It is official in the form of a 1 per cent. (by weight) alcoholic solution (*Spiritus Glonoini*, U. S. P.), the dose of which is from 1–3 min. (0.06–0.2 c.c.), cautiously increased. The spirit should be kept in well-stoppered tin-cans, and should be stored in a cool place, remote from lights or fire. Care must be exercised in handling it, since a dangerous explosion may occur if any considerable quantity be spilled.

Physiologic Action.—The action of nitroglycerin is similar to that of amyl nitrite, except that it is not quite so prompt, and is more persistent. Its effects are recognizable within a few minutes, and last, according to the dose, from one to three hours.

Therapeutics.—In *fatty and fibroid degeneration of the myocardium* nitroglycerin is often a useful remedy, especially when there is precordial pain or oppression. In *angina pectoris* it is often of value when given between the paroxysms, alone or in conjunction with potassium iodid. In *chronic valvular disease* it is sometimes advantageous to give nitroglycerin along with digitalis in order to counteract the tendency of the latter to constrict the peripheral vessels. In such cases, however, the two drugs should not be combined, since the action of the one is very prompt, but fugacious, while that of the other is slow, but persistent. In the heart-failure of *croupous pneumonia*, while nitroglycerin occasionally does good, it often does harm. The indications for its use cannot be definitely stated. Each case must be considered on its own merits, and the remedy used tentatively.

No remedy equals nitroglycerin in relieving the high arterial tension, dyspnea, palpitation, headache, vertigo, and insomnia of *chronic nephritis.*

In small, frequently repeated doses it sometimes proves effective in controlling persistent *hiccough.* Wade recommends it in the *abdominal palpitations* frequently seen in nervous women. He believes that it acts by lowering the blood-pressure in the splanchnic area. Good results occasionally follow its use in *migraine* and *sciatica.* Excellent results are said to have been obtained from its use in the asphyxial stage of *Raynaud's disease.*

Untoward Effects.—In certain individuals the smallest dose of nitroglycerin induces flushing of the face, headache, tinnitus aurium, and vertigo.

Administration.—It may be given by the mouth or subcutaneously. For internal use, the available preparations are the official spirit and the tablets, which are commonly found in the market, and which contain from $\frac{1}{500}-\frac{1}{50}$ gr. (0.0001–0.0012 gm.). For subcutaneous injection either the spirit or soluble hypodermic tablets may be employed. It is very important that the preparation should be fresh, as both solutions and tablets deteriorate with age. On account of the extreme variation in individual susceptibility to the drug, very small doses should be administered at first, and these gradually increased until the therapeutic or physiologic effect (flushing, etc.) is secured. According to Wood, a single drop has caused insensibility; on the other hand, Whittaker has given as much as $8\frac{1}{3}$ gr. (0.56 gm.) in twenty-four hours without untoward effects. Toleration of the drug is rapidly acquired, and this is another reason for making the initial doses the smallest that will prove effective.

ERYTHROL TETRANITRATE.
(Tetranitrol, $(CH_2.ONO_2)_2(CH.ONO_2)_2.$)

Erythrol tetranitrate occurs in the form of large scales, which are soluble in alcohol, insoluble in water, and which readily explode on percussion. In the form of chocolate or cocoa-butter tablets, however, it may be handled with perfect safety. The dose is from $\frac{1}{2}$–2 gr. (0.03–0.13 gm.).

Physiologic Action and Therapeutics.—The action of erythrol tetranitrate closely resembles that of nitroglycerin, but it is milder and decidedly more prolonged. As in the case of the nitrites, patients vary considerably in their susceptibility to the drug: $\frac{1}{2}$ gr. (0.03 gm.) has caused intense headache and great malaise lasting thirty-six hours; on the other hand, so much as $\frac{1}{2}$ dr. (2.0 gm.) in twenty-four hours has been given with good results.

Erythrol tetranitrate is indicated in the same class of cases as nitroglycerin. On account of its prolonged action it is especially useful in *arteriosclerosis, angina pectoris, chronic interstitial nephritis,* and *Raynaud's disease.*

SODII NITRIS, U. S. P.

(Sodium Nitrite, NaNO$_2$.)

Sodium nitrite occurs in the form of white, opaque, fused masses, or colorless, transparent crystals, odorless, and of a mild saline taste. It is soluble in 1.5 parts of water, and sparingly soluble in alcohol. The dose is from 2–3 gr. (0.13–0.2 gm.).

Therapeutics.—The action of sodium nitrite is identical with that of amyl nitrite, except that it is less rapid and more lasting. As a vasodilator it is not so reliable as erythrol tetranitrate, although it resembles that drug in the permanency of its effects. It is indicated in the same class of cases as erythrol tetranitrate and nitroglycerin.

POTASSII BROMIDUM, U. S. P.

(Potassium Bromid, KBr.)

Potassium bromid occurs in colorless or white cubical crystals, odorless, and having a pungent saline taste. It is soluble in 1.6 parts of water and in 200 parts of alcohol. The dose is from 10–60 gr. (0.65–4 gm.).

Physiologic Action.— Circulatory System.—Small doses have no influence, but large doses depress the heart and vasomotor center.

Nervous System.—The dominant action of potassium bromid is on the nervous system; it depresses first the cerebral cortex, next the sensory neurons of the spinal cord and sensory nerves, then the motor neurons of the spinal cord, and, finally, the motor nerves. It has a special affinity for the sexual centers in the spinal cord, and, even in moderate doses, it distinctly lessens the sexual appetite.

Wright and Crisafulli have found after poisonous doses marked degenerative changes (vacuolation and atrophy) of the cortical nerve-cells.

Respiratory System.—The respiratory centers share with

other parts of the nervous system in the depression caused by large doses of bromids, and in fatal poisoning respiration may fail before the heart ceases to beat.

Alimentary Canal.—In concentrated form potassium bromid acts as an irritant to the stomach, and may occasion nausea and vomiting.

Temperature.—Toxic doses cause a marked fall of the body-temperature, probably from vasomotor paresis and the consequent increased dissipation of heat.

Absorption and Elimination.—Potassium bromid is rapidly absorbed, and may appear unchanged in the urine within fifteen minutes after its ingestion. As elimination is not effected quite so rapidly as absorption, there is a tendency for the drug to accumulate in the body.

Bromism.—The continuous use of bromids is usually followed by a group of symptoms to which the term *bromism* has been applied. This condition is characterized by anemia, a general eruption of acne, fetor of the breath, gastric disturbance, mental depression, somnolence, failure of memory, abolition of the sexual appetite, diminution of the reflexes, impairment of tactile sensibility, unsteady gait, and muscular weakness.

Exceptional Effects of Bromids.—The continued exhibition of bromids in large doses is occasionally followed by one of the following symptoms: Paresis of the limbs, ptosis, depression of the heart, intense irritability of temper, depression of spirits, and suicidal or homicidal impulses.

Acute Poisoning.—The chief symptoms resulting from the ingestion of a single toxic dose of potassium bromid are severe headache, inco-ordination, somnolence, dilatation of the pupils, burning in the stomach, and collapse. No fatal case of bromid-poisoning has been reported.

Action of Potassium Bromid Compared with Other Bromids.—On account of the relatively large amount of potassium (44 per cent.), this salt is the most powerful depressant of all the bromids.

Therapeutics.—Potassium bromid is used to control convulsions, to allay nervous excitement, to promote sleep, and to relieve certain forms of headache and neuralgia.

To Control Convulsions.—The bromids are by far the most useful drugs we possess in the treatment of *epilepsy.* They not only exert a direct sedative influence on the cerebral cortex, but through their action on the sensory neurons of the spinal cord they protect the brain from the disturbing influence of external stimuli. While they rarely effect a cure, they very often lessen materially the frequency of the paroxysms.

The amount required varies with the severity of the case and the susceptibility of the individual, and must be determined experimentally in each case. Saturation is indicated by a loss of the palatal reflex, mental hebetude, and gastric disturbance. The daily dose usually ranges between 1 and 2 dr. (4.0–8 0 gm.). When larger doses than these are required, it is better to try a combination of other remedies with the bromid than to depend upon the latter alone. Sometimes small doses of potassium iodid, or moderate doses of antipyrin or acetanilid, will serve to lessen the amount of bromid required. A combination of several bromids often acts more satisfactorily than a single bromid. In *nocturnal epilepsy* the occasional exhibition of chloretone or sulphonal with the bromid often proves very efficacious. When the circulation is weak a combination of a bromid with digitalis or of adonis venalis, as recommended by Bechterew, will be found useful. It has been shown by Richet and Toulouse that the deprivation of salt from food very decidedly increases the efficacy of the bromid treatment. The addition of a drop or two of Fowler's solution to each dose of the bromid often prevents the outbreak of acne.

In some cases of epilepsy it is necessary to withhold the bromids entirely, since the general disturbance which they occasion more than counterbalances the good they accomplish in diminishing the number of seizures.

The bromids are of considerable value also in other *convulsive disorders* besides epilepsy, such as puerperal eclampsia, uremia, infantile convulsions, strychnin-poisoning, and tetanus. In severe cases of *chorea* with incessant movements the bromids may be employed, but they are less efficient than chloral. In local spasmodic affections—*whooping-cough, laryngismus stridulus*, and *essential asthma*—bromids are sometimes of service.

To Allay Nervous Excitement.—In nervous irritability of all kinds potassium bromid is useful. It is an excellent sedative in the excitement of *overwork, worry*, and *anxiety*, and also in that which attends hysteria. In the varied nervous disturbances which attend the *menopause* it is our most reliable remedy. In fully developed *delirium tremens* it is of little value, but in the stage of excitement which precedes the latter it is very efficient.

Potassium bromid is a powerful anaphrodisiac; by diminishing the irritability of the sexual centers in the spinal cord it renders excellent service in *priapism, spermatorrhea*, and *nymphomania*.

Given for several days before the expected voyage, in doses

of from 10–20 gr. (0.65–1.3 gm.), three times a day, it often prevents *sea-sickness.* It may also be tried in the *vomiting of pregnancy.* Da Costa found that from 30–60 gr. (2.0–4.0 gm.) of potassium bromid, administered an hour before a dose of opium, would prevent the disagreeable after-effects of the latter.

To Promote Sleep.—As pure hypnotics the bromids are feeble compared with opium and chloral, but in insomnia due to nervous excitement they frequently produce refreshing sleep.

To Relieve Headache and Neuralgia.—Potassium bromid is often useful in attacks of *migraine, trifacial neuralgia,* and the *headache of cerebral congestion.* In these affections it may be given alone or in combination with antipyrin, phenacetin, caffein, or a salicylic compound.

Administration.—Potassium bromid should be given in solution, well diluted, after meals. In epilepsy it is best to divide the daily dose according to the time at which the seizures are likely to occur. Thus in nocturnal epilepsy comparatively small doses may be given early in the day, and a very large dose ($\frac{1}{2}$–1 dr.—2.0–4.0 gm.) at supper-time. The bromid may be combined with antipyrin and Fowler's solution, as in the following formula suggested by Wood:

```
R  Potassii bromidi,
   Ammonii bromidi,               aa   ℥iij (12.0 gm.);
   Liquoris potassii arsenitis,        f℥j (4.0 c.c.);
   Antipyrini,                         ℥j (4.0 gm.);
   Aquæ menthæ piperitæ,     q. s. ad f℥ vj (180.0 c.c.).—M.
Sig.  Tablespoonful in water night and morning.
```

In insomnia from nervous excitement potassium bromid may be combined with chloral, as in the following formula:

```
R  Chloralis,                         gr. xl (2.6 gm.);
   Potassii bromidi,                  ℥iss (6.0 gm.);
   Aquæ et syrupi aurantii,   aa q. s. ad f℥ij (60.0 c.c.).—M.
Sig.  Tablespoonful in water at bedtime.
```

Incompatibles.—Potassium bromid is incompatible with acids and acid salts. It also precipitates certain alkaloids—morphin, quinin, and strychnin—from neutral solutions.

AMMONII BROMIDUM, U. S. P.

(Ammonium Bromid, NH_4Br.)

Bromid of ammonium occurs in the form of white crystals or as a yellowish-white powder having a disagreeable salty taste. It dissolves in 1½ parts of water and in 30 of alcohol. Its dose is from 20–60 gr. (1.3–4.0 gm.), well diluted.

Physiologic Action.—The dominant action of ammonium bromid, like the corresponding potassium salt, is on the nervous system. It depresses the cortical brain-cells, and both the sensory and motor neurons of the spinal cord, especially the former. It differs in its action from potassium bromid in being less depressing to the heart, muscles, and peripheral nerves.

Therapeutics.—It may be substituted with advantage for the potassium salt in all the conditions in which a bromid is indicated.

Incompatibles.—The same as in the case of potassium bromid, and, in addition, spirit of nitrous ether.

SODII BROMIDUM, U. S. P.

(Sodium Bromid, NaBr.)

Sodium bromid occurs in colorless or white cubical crystals, or as a white granular powder, odorless, and of a saline, bitter taste. It is soluble in 1.2 parts of water and in 13 of alcohol. It contains 78 per cent. of bromin. The dose is from 20–60 gr. (1.3–4.0 gm.), well diluted.

Physiologic Action and Therapeutics.—Its action closely resembles that of potassium bromid, but it is less irritating to the stomach and somewhat less depressing. It may be used as a substitute for the potassium salt in the various conditions in which a bromid is indicated.

STRONTII BROMIDUM, U. S. P.

(Strontium Bromid, SrBr$_2$.6H$_2$O.)

Strontium bromid occurs in colorless, very deliquescent, hexagonal crystals, odorless, and of a bitter, saline taste. It is readily soluble in water and in alcohol. The dose is from 30–60 gr. (2.0–4.0 gm.).

Therapeutics.—Strontium bromid has an action on the nervous system similar to that of the other bromids, but it is less powerful. At the same time it is less depressing, less prone to cause acne, and, instead of disturbing the stomach, it tends rather to allay gastric irritation. It may be used as a succedaneum for other bromids when only a mild effect is required or when there is irritability of the stomach. In other cases it may be combined with potassium or ammonium bromid to mitigate their unpleasant by-effects. Germain Sée and others have spoken favorably of its action in *gastralgia* and *acid, dyspepsia.*

LITHII BROMIDUM, U. S. P.

(Lithium Bromid, LiBr.)

Lithium bromid is a white, granular, deliquescent salt, odorless, of a sharp, bitter taste. It is freely soluble in water and in alcohol. The dose is from 20–60 gr. (1.3–4.0 gm.).

Therapeutics.—Lithium bromid has a therapeutic value about equal to that of sodium bromid.

ACIDUM HYDROBROMICUM DILUTUM, U. S. P.

(Diluted Hydrobromic Acid, HBr.)

Diluted hydrobromic acid is an odorless, colorless liquid, of acid properties, composed of 10 per cent. by weight of absolute hydrobromic acid and 90 per cent. of water. The dose is from 1–4 fl. dr. (4.0–15.0 c.c.), well diluted.

Therapeutics.—Hydrobromic acid has very much the same action as the bromids. It differs from potassium bromid in being less powerful, less likely to cause acne and general depression, and in being more irritating to the stomach.

It may be given alone or in combination with bromids in *epilepsy* and other convulsive disorders. It is especially valuable in *tinnitus aurium* resulting from aural disease or following the administration of quinin or salicylic compounds. It has been employed as a sedative in *febrile excitement, whooping-cough*, and *nervous headache.* The following combination of dilute hydrobromic acid and horse-nettle is sometimes efficacious in the milder forms of epilepsy:

```
R  Acidi hydrobromici diluti,
     Extracti solani carolinensis fluidi,    aa f℥iss (45.0 c.c.);
     Syrupi aurantii,                        f℥j (30.0 c.c.);
     Aquæ,                       q. s. ad f℥vj (180.0 c.c.).—M.
   Sig.  A tablespoonful in water after meals.
```

GELSEMIUM, U. S. P.

(Yellow Jasmine.)

Gelsemium is the rhizome and root of *Gelsemium semper-virens*, a climber indigenous to the southern United States. It contains gelsemin, gelseminin, and gelseminic acid. Gelsemin, which represents to a considerable extent the active properties of the crude drug, occurs in the form of a white amorphous powder, of a bitter taste, insoluble in water, but slightly soluble in alcohol. The dose is from $\frac{1}{60}$–$\frac{1}{30}$ gr. (0.001–0.002 gm.).

PREPARATIONS.	DOSE.
Tinctura Gelsemii, U. S. P.	10–20 min. (0.6–1.2 c.c.).
Extractum Gelsemii Fluidum, U. S. P.	5–10 min. (0.3–0.6 c.c.).

Physiologic Action. — **Circulatory System.** — Small doses do not affect the circulation, but large doses lessen both the force and the frequency of the pulse, probably by directly depressing the heart itself.

Nervous System.—In large doses gelsemium depresses the spinal cord, especially its motor neurons. It also paralyzes the peripheral filaments of the cranial nerves, the third or oculomotor nerve being the most sensitive to its influence. Upon other motor nerves, the sensory nerves, and the brain the drug has but little influence.

Respiratory System.—Toxic doses kill by paralyzing the respiratory centers.

Eye.—Gelsemium is a powerful but slowly acting mydriatic. This action is probably due to paresis of the oculomotor nerve. If a solution of sufficient strength be employed (8 gr. of gelsemin to the ounce—0.5 gm.-30.0 c.c.), accommodation also is ultimately paralyzed.

Toxicology.—Gelsemium-poisoning is characterized by vertigo, frontal headache, disordered vision, dilatation of the pupils, ptosis, falling of the jaw, failure of articulation, extreme muscular weakness, slow respiration, a weak, thready pulse, and, finally, collapse. Consciousness is usually retained until near the close of life. Convulsions have been observed in animals, but not in man.

Treatment.—Restorative measures consist in evacuating the stomach, administering tannic acid, maintaining the body-temperature, and combating collapse with such stimulants as strychnin, whisky, and ammonia.

Therapeutics.—Gelsemium has been employed as an antispasmodic in *laryngismus stridulus, whooping-cough, asthma,* and *chorea,* but in these affections there are far more efficient remedies. At one time gelsemin was recommended as a mydriatic, but homatropin and atropin are superior, being more prompt and less dangerous. The drug is sometimes distinctly useful in *spasmodic affections of the muscles,* like torticollis, and in *obstinate neuralgia,* especially of the trifacial nerve. In the latter, a combination of butyl-chloral hydrate (5 gr.—0.3 gm.) and gelsemin ($\frac{1}{100}$ gr.—0.0006 gm.), as recommended by Murrell, is particularly efficacious. Ringer has found it useful in some cases of *Ménière's disease.* On account of the variability of the preparations, they should be given in small doses, gradually increased until such symptoms as vertigo or dimness of vision appear. In persistent torticollis gelsemin may be injected directly into the muscle.

OTHER SPINAL CORD DEPRESSANTS.

Chloral.—The depressant action of chloral on the motor tracts of the spinal cord is secondary in importance only to its action on the cerebrum. In *tetanus* and *strychnin-poisoning* it renders valuable service in allaying the irritability of the spinal motor neurons.

Chloroform and Ether.—These drugs depress both the sensory and motor neurons of the spinal cord, but the former more powerfully than the latter. They are useful in checking convulsions which endanger life, whether the motor discharges emanate from the brain or from the spinal cord. As spinal cord depressants they are of value in cerebral convulsions, by lessening the power of the sensory columns to transmit to the brain stimuli from without, and also by lessening the power of the motor columns to convey discharges from the brain to the muscles.

Hyoscin.—The action of hyoscin on the motor centers of the spinal cord is subordinate to its action on the brain. Its influence on the sexual centers, however, is very pronounced, in consequence of which it is a valuable remedy in all forms of *sexual excitement.*

MOTOR NERVE DEPRESSANTS.

The most important depressants of the peripheral motor nerves are:

Conium.	Curare.
Lobelia.	Pelletierin.
Belladonna.	Bromids.

Of the above group, however, only conium, belladonna, and perhaps lobelia are employed for their effect on the motor nerves. *Curare* is a paralyzant of all highly organized tissue, the motor nerves being particularly sensitive to its influence. It has been used in chorea, epilepsy, and tetanus, but without much success. The action of the *bromids* on the motor nerves is entirely subordinate to their action on the brain, spinal cord, and sensory nerves. *Pelletierin,* the active principle of pomegranate, affects the motor nerves only in toxic doses. The action and uses of *belladonna* have been fully considered on page 70.

Motor nerve depressants are useful in local spasmodic affections like whooping-cough, torticollis, and asthma.

CONIUM, U. S. P.

(Hemlock.)

Conium is the fruit of the *Conium maculatum*, gathered while yet green. It contains a volatile oil and a yellowish, oily, liquid alkaloid, *coniin*, which represents the active properties of the drug. Coniin forms with acids crystallizable salts which are soluble in water. The dose of the alkaloid or its salts is from $\frac{1}{20}$–$\frac{1}{6}$ gr. (0.003–0.01 gm.).

PREPARATIONS.	DOSE.
Extractum Conii Fluidum, U. S. P.	2–5 min. (0.1–0.3 c.c.).
Extractum Conii, U. S. P.	$\frac{1}{2}$–2 gr. (0.03–0.13 gm.).
Tinctura Conii	$\frac{1}{2}$–1 fl. dr. (2.0–4.0 c.c.).

Physiologic Action.—Toxic doses of conium produce a staggering gait, muscular relaxation, paralysis, tremors, vertigo, ptosis and dilatation of the pupils (from paresis of the oculo-motor nerve), and failure of circulation. Consciousness and sensibility are preserved until near the end.

Nervous System.—The dominant action of conium is on the peripheral motor nerves, which it paralyzes. On the spinal cord and sensory nerves it acts as a feeble depressant. On the brain it exerts no special influence.

Circulatory and Respiratory Systems.—Respiration and circulation are not affected except by very large doses, which exert a depressing influence on both.

Therapeutics.—Conium has been used as an antispasmodic in *chorea, paralysis agitans, asthma,* and *whooping-cough,* but in these affections it is of less value than other remedies. In the form of vapor it is sometimes useful in allaying the cough of *bronchitis.* Spitzka and others strongly recommend it in the mental and physical excitement of *acute mania.* On account of the uncertain strength of the preparations, it is necessary to give the drug in small doses rapidly increased until some effect is produced.

Added to poultices, it is a useful sedative for *painful ulcers, cancer,* etc.

LOBELIA, U. S. P.

Lobelia is the leaves and tops of *Lobelia inflata,* a weed growing in the wild places of Canada and the United States. It contains *lobelic acid* and a yellow liquid alkaloid, *lobelin.* The latter forms with acids non-crystalline salts, which are soluble in water. The dose of the sulphate of lobelin is from $\frac{1}{6}$–1 gr. (0.01–0.06 gm.).

10

<table>
<tr><td align="center">PREPARATIONS.</td><td align="center">DOSE.</td></tr>
<tr><td>Tinctura Lobeliæ, U. S. P. . . . .</td><td>As an expectorant, 10–30 min. (0.6–2.0 c.c.).
As an emetic, 1–2 fl. dr. (2.0–4.0 c.c.).</td></tr>
<tr><td>Extractum Lobeliæ Fluidum, U. S. P. . .</td><td>As an expectorant, 2–5 min. (0.1–0.3 c.c.).
As an emetic, 20–30 min. (1.2–2.0 c.c.).</td></tr>
</table>

Physiologic Action.—The ingestion of a large dose of lobelia (a teaspoonful of the tincture) is generally followed by nausea, or even vomiting and faintness. Toxic doses cause severe burning pain in the esophagus and stomach, persistent vomiting, cold sweats, weakness of the pulse, extreme prostration, and, finally, death from paralysis of the respiratory center. In frogs toxic doses of lobelia are speedily followed by a loss of voluntary power, which is the result of paralysis of the peripheral motor nerves. In mammals, however, this action of lobelia on the motor nerves is probably subordinate to its depressant influence on the respiratory center and the fibers of the pneumogastric, which are distributed to the bronchial muscles.

The treatment of *lobelia-poisoning* consists in the administration of tannic acid and of diffusible stimulants, and in the use of artificial heat to maintain the bodily temperature.

Therapeutics.—Lobelia was formerly used as an *emetic*, but as such it has been almost entirely displaced by less depressing drugs. It is, however, an exceedingly useful antispasmodic in *asthma* and in *bronchitis* associated with spasmodic dyspnea. To secure the best results, it is necessary to give the drug in increasing doses until slight nausea is produced.

SENSORY NERVE DEPRESSANTS.

A drug when taken internally, if the dose has been sufficiently large, may depress the peripheral sensory nerves, or it may do so only when applied directly to them. The following drugs when taken in large doses exert a sedative influence on the sensory nerves:

Bromids.	Antipyrin.
Belladonna.	Phenacetin.
Aconite.	Acetanilid.

When applied locally a drug may lessen or destroy sensibility through a direct action on the peripheral nerves, or through an indirect action, as is the case with ice, ether, and ethyl chlorid, by rendering the part intensely cold.

Sensory nerve depressants are employed locally to relieve pain and to allay itching.

The following agents act as local anodynes or as local anesthetics:

Cocain.	Carbolic acid.
Eucain.	Iodoform.
Holocain.	Camphorated chloral.
Tropacocain.	Chloroform.
Orthoform.	Aconite.
Chloretone.	Ether.
Menthol.	Ice.
Ethyl chlorid.	

Many of the *aromatic oils*, like oil of cloves and oil of peppermint, also possess anodyne properties.

Cocain (see p. 78).—The application of a cocain solution of moderate strength (6 per cent.) to a mucous membrane is followed in a few moments by pallor and anesthesia. The pallor is due to constriction of the blood-vessels, and is subsequently replaced by congestion; the anesthesia is due to the direct action of the drug upon the sensory nerves. Applied to cutaneous surfaces it is without effect, since absorption through the skin is very slight. In the eye it causes, in addition to anesthesia, mydriasis and a slight impairment of accommodation.

The uses of cocain as a local anesthetic are various. *In operations on the eye, nose, throat, urethra, and rectum*, its solutions are indispensable in preventing or in lessening pain.

When injected under the skin it is also of established value in many *minor operations involving the cutaneous structures*, such as circumcision, opening small abscesses, amputating fingers, and excising small growths.

In *inflammations* and *ulcerations of the nose, pharynx*, and *larynx* it may be employed alone or in combination with antiseptic sprays and powders. In *acute coryza* and *hay-fever* it gives much relief by lessening the sensibility and the turgescence of the tissues. In *tuberculous laryngitis* it serves a useful purpose in lessening the intense pain and dysphagia. It is also efficacious in diminishing the pain caused by the application of caustics.

For anesthetizing mucous membranes solutions are employed varying in strength from 2–20 per cent. On the nose and throat from 4–6 per cent. solutions are usually sufficient.

In the larynx somewhat stronger solutions are required. Solutions stronger than 4 per cent. should not be used in the eye on account of the danger of inducing degenerative changes in the corneal epithelium. When the drug is employed subcutaneously after the method of Schleich, it is not safe to allow more than 1 grain to remain in the tissues.

Eucain.—Eucain is a synthetic compound allied chemically to cocain. Two forms are upon the market, which are known respectively as Alpha-eucain or eucain hydrochlorate "*α*" and Beta-eucain or eucain hydrochlorate "*β*." The former, on account of its irritant properties, has been almost completely supplanted by the latter. Beta-eucain occurs as a white, neutral, crystalline powder, soluble in about 30 parts of cold water, making a 3 per cent. solution. It is somewhat less toxic than cocain, and, moreover, its solutions are more stable, and can be sterilized by boiling without losing in efficacy. A disadvantage of eucain is that it has no constricting influence on the blood-vessels, in consequence of which it causes hyperemia rather than ischemia. When injected subcutaneously it is somewhat more liable than cocain to induce sloughing.

For general surgical purposes a 2 per cent. solution (9 gr.–1 oz.—0.6 gm.–30.0 c.c.) is generally employed.

Holocain.—Holocain is a synthetic compound related chemically to phenacetin. The hydrochlorate, in which form it is always employed, occurs in colorless crystals, having a bitter taste and slightly soluble in cold water.

It is very sensitive to alkalies—even the small amount of alkali dissolved out of the glass on boiling a solution of the drug in a test-tube is sufficient to decompose it.

Experimentally holocain has been found to be more actively poisonous than cocain, and therefore it is not suitable for subcutaneous use. It has qualities, however, which make it a very valuable anesthetic in ophthalmic surgery. The instillation into the eye of a few drops of a 1 per cent. solution of holocain causes slight burning, and is followed in from twenty to thirty seconds by complete anesthesia, lasting for from ten to fifteen minutes. It is without effect upon the pupil, the ciliary muscle, the intra-ocular tension, or the corneal epithelium. It is more rapid in its action than cocain, and it has, moreover, some bactericidal power. As it has no constricting effect upon the blood-vessels, it lacks the power of cocain to control hemorrhage.

Since it acts so quickly and does not cause mydriasis, it is especially valuable in *minor operations on the eye*, such as the removal of foreign bodies. Norris has found it very efficacious in *sluggish corneal ulcers*, and de Schweinitz and others have spoken favorably of it in *keratitis*.

A 1 per cent. solution is usually employed. When the eye is inflamed, the effectiveness of both holocain and cocain is distinctly enhanced by the previous application of a solution (1 : 10,000) of adrenalin.

Tropacocain Hydrochlorate.—Tropacocain is an alkaloid obtained from the small Java coca leaves. It forms white needles which are readily soluble in water. It resembles cocain in its action, but it is much less toxic, and when applied to mucous membranes it does not cause ischemia.

It anesthetizes the eye without affecting the pupil or accommodation, but it has no advantages over holocain. On account of its lesser toxicity it may replace cocain in solutions intended for infiltration anesthesia, especially when it is necessary to employ large doses of the anesthetic. Block recommends the following solution :

Tropacocain hydrochlorate,	3 gr. (0.2 gm.) ;
Sodium chlorid,	3 gr. (0.2 gm.) ;
Sterilized distilled water,	3 fl. oz. (100.0 c.c.).

It has also been used successfully in the production of medullary anesthesia, from $\frac{2}{5}$–$\frac{4}{5}$ gr. (0.26–0.05 gm.) being injected for this purpose into the subarachnoid space.

Orthoform.—This synthetic compound is the methyl-ester of meta-amido-para-oxybenzoic acid. It occurs as a white, voluminous powder, without odor or taste, and sparingly soluble in water. Orthoform hydrochlorid is freely soluble in water, but it is too irritating for general use. Orthoform has pronounced analgesic, antiseptic, and desiccant properties, and for this reason it has been recommended as a dressing for painful wounds and ulcers.

When applied pure or in the form of an ointment (10–20 per cent.) to exposed sensory nerve-endings it causes a slight burning sensation, which is soon followed by complete analgesia lasting from ten to twenty hours. The prolonged action of the drug is due in large part to the slowness with which it is absorbed. On account of its acid reaction it is too irritating to be applied to sensitive mucous membranes like the eye or to be injected hypodermically. Unfortunately it is not altogether free from toxic properties, and its effects should be carefully watched. Brocq, Asam, Yonge, and others have seen extensive erythema, urticaria, eczema, and even gangrene follow its use. A study of the published records indicates that these accidents are more likely to occur from the use of the drug in an ointment than when it is applied directly in the form of a powder.

Orthoform has been used successfully as a local remedy in *painful wounds*, in *burns*, especially of the third degree, in *cancerous* and *tuberculous ulcerations*, and in *fissures* and *excoriations* of the mucocutaneous junctions. It will be found very efficacious in relieving the pains of *tuberculous laryngitis* and of *fissured nipples*. In the former the powdered drug may be used; in the latter, a saturated alcoholic solution, or an ethereal solution like the following, recommended by Bardet:

℞	Orthoform,	gr. lxxx (5.0 gm.);
	Ætheris,	q. s. ad solv.;
	Olei amygdalæ expressi,	f ℥j (30.0 c c.).

Added to arsenical pastes (orthoform, 1 part; acacia, 1 part; arsenous acid, 2 parts) it materially lessens the painfulness of their caustic action. Its internal use in doses of from 5–10 gr. (0.3–0.65 gm.) has been recommended in *gastric ulcer* and *cancer*, but Epstein has observed vomiting and collapse after its administration by the mouth.

Orthoform is incompatible with silver nitrate, antipyrin, and bismuth subnitrate.

Chloretone (see p. 101).—This is a compound formed by the addition of potassium hydrate to equal weights of chloroform and acetone. It appears as a white, crystalline powder, with a camphoraceous odor and taste. In addition to its usefulness as a hypnotic it has a definite value as an antiseptic and a local anesthethic. It has been used with considerable success as a dusting-powder in *painful wounds*. It has also been employed in the form of a saturated solution for producing infiltration anesthesia, but the results have not been uniformly favorable.

Menthol (peppermint camphor).—Menthol is a stearopten obtained from the essential oil of peppermint. It occurs as colorless prismatic or acicular crystals, having the odor of mint and a camphoraceous taste. It is sparingly soluble in water, but freely so in alcohol, ether, and chloroform. The dose is from 1–3 gr. (0.06–0.2 gm.), in pills, capsules, or in alcoholic solution.

Menthol is employed as a local anesthetic and an antiseptic. It makes a useful application in *frontal headache* and in *neuralgia of the superficial nerves*. For use in these affections it is best dissolved in chloroform or ether. When equal parts of chloral or menthol are heated together in a water-bath, an oily liquid is formed (chloral menthol), which is efficacious in *toothache*. One part of menthol to 10 of olive oil makes a soothing application for *burns*. Dissolved in collodion in the proportion of 1 : 4 it forms a useful dressing for *small contusions*. The inhalation of menthol in the form of a vapor or spray affords considerable relief from the disagreeable symptoms of acute *coryza*. The following solution will be found useful in both coryza and acute laryngitis :

    ℞ Menthol,   gr. vj (0.4 gm.) ;
      Eucalyptol,   ℥v (0.3 c.c.) ;
      Petrolati liquidi,   f℥j (30.0 c.c.).—M.
    Sig. To be used as a spray several times a day.

Liquid petrolatum containing from 5–10 per cent. of menthol makes an excellent vehicle for various remedies employed in the treatment of *chronic rhinitis* and *chronic laryngitis*.

The following mixture in the form of a spray has been recommended as a local anesthetic in minor surgical operations :

    ℞ Menthol,   gr. x (0.65 gm.) ;
      Chloroformi,   f℥v (18.5 c.c.) ;
      Ætheris,   f℥j (30.0 c.c.).—M.

Internally, menthol is sometimes used in persistent *vomiting* and *gastralgia*, but it is rarely successful.

Ethyl Chlorid.—This compound is produced by the action of dry hydrochloric acid gas on absolute alcohol. It is a colorless, highly volatile, inflammable liquid, having a strong ethereal odor and boiling at 54° F. On account of the intense cold caused by its rapid evaporation it is employed as a local anesthetic. It is usually sold in sealed glass tubes, the ends of which are drawn to a point, so that they can be readily broken and the liquid expelled as a spray by the heat of the hand.

Ethyl chlorid is a convenient local anesthetic for use in *minor operations* requiring but a single incision, such as opening boils and aspirating pleural or abdominal effusions. It is seldom efficient in operations upon the deeper structures requiring dissection or curetting.

Dethlefsen has reported good results in *lupus vulgaris* from repeated freezings with ethyl chlorid. The drug has also been used to some extent as a *general anesthetic*, but according to the statistics of Lotheisseni is distinctly more dangerous than chloroform.

Carbolic Acid (see p. 383).—As a local anesthetic carbolic acid is chiefly employed to allay itching in *jaundice* and *pru-*

ritic skin diseases like eczema and urticaria. It may be used in the strength of 1½–2 dr. (5.8–7.8 gm.) to the pint (0.5 L) of water.

Iodoform (see p. 312).—On account of its analgesic properties iodoform is a valuable topical remedy in various *ulcerations of a painful character.* It is particularly useful in relieving the pain and dysphagia of tuberculous laryngitis, in which affection it is employed by insufflation, either in a pure form (2–3 gr.—0.13–0.2 gm.) or mixed with morphin ($\frac{1}{16}$ – $\frac{1}{12}$ gr.—0.004–0.005 gm.). Suppositories of iodoform are often of service in painful hemorrhoids and in fissure of the anus.

Camphorated Chloral.—This is a syrupy liquid made by rubbing together equal parts of camphor and chloral. It is soluble in alcohol, ether, glycerin, and oils, but it is decomposed by water. It is used as a local anesthetic in *neuralgia, toothache*, and *pruritus.* In local pruritus the following ointment is sometimes efficacious:

> ℞ Pulveris camphoræ,
>   Pulveris chloralis,   āā ʒj (4.0 gm.);
>   Trit. et add.
>   Unguenti aquæ rosæ,   ʒj (30.0 gm.).—M.

Chloroform (see p. 109).—Chloroform is used locally for its anesthetic effect in *neuralgic affections of the superficial nerves* and in *skin diseases* attended with *itching.* In neuralgia it is usually employed in the form of a liniment, like the following:

> ℞ Tincturæ aconiti,
>   Chloroformi,   āā f℥ss (15.0 c.c.);
>   Linimenti saponis,   f℥j (30.0 c.c.).—M.

Bartholow has spoken enthusiastically of deep injections of chloroform (5–10 min.—0.3–0.6 c.c.) in the neighborhood of the painful nerve, but serious sloughing may follow this method of treatment. In urticaria chloroform may be used as a lotion with alcohol in the proportion of 1 : 8.

Aconite (see p. 59).—The tincture of aconite, rubbed into the affected part, is sometimes useful in relieving the pain of *neuralgia* and in allaying the itching of *chilblains.* Ointments containing aconitin (5–10 gr. to the ounce—0.3–0.6 gm. to 30.0 gm.) are also prescribed in neuralgia, but on account of the extreme virulence of the alkaloid they must be used with great caution. Liniments containing aconite and chloroform are employed with benefit in *muscular rheumatism.*

Ether (see p. 105).—The application of ether in the form of a spray is one of the oldest methods of producing local anesthesia. The sensory nerves are benumbed by the intense cold which is occasioned by the rapid evaporation of the drug. This method may be employed preliminary *to opening small abscesses,* or *performing paracentesis.*

MYDRIATICS.

The size of the pupil is regulated by the unstriped muscle-fibers of the iris, of which there are probably two sets—concentrically arranged constrictor fibers and radiating dilator fibers—although the existence of the latter is denied by some observers. Two antagonistic nerves control these muscles: the oculomotor, which when stimulated contracts the pupil through the agency of the constrictor fibers, and the sympathetic, which when stimulated dilates the pupil probably through the agency of the dilator fibers.

Mydriasis or dilatation of the pupil may be brought about in a variety of ways: By paralyzing the oculomotor center in the corpora quadrigemina; by paralyzing the peripheral filaments of the oculomotor nerve; by paralyzing the constrictor muscle of the iris; or by stimulating the sympathetic center in the medulla; by stimulating the peripheral fibers of the sympathetic nerve, or possibly by stimulating the radiating muscular fibers of the iris itself, if such fibers exist. The mode of action of the various mydriatic drugs has not been definitely determined. *Atropin* appears to dilate the pupil by paralyzing the peripheral filaments of the oculomotor nerves and by stimulating the peripheral filaments of the sympathetic nerves. *Cocain* causes mydriasis by stimulating the peripheral ends of the sympathetic nerves.

All mydriatics impair more or less the *accommodative power* of the eye, that is, its power of adjusting itself to vision at different distances. The agency through which the adjustment is effected is the ciliary muscle. In accommodating for near objects this muscle contracts, the suspensory ligament relaxes, the lens, owing to its inherent elasticity, becomes more convex. Of the mydriatics, *atropin* and *duboisin* are the most powerful *cycloplegics* or paralyzants of the ciliary muscle, and *cocain* is the least powerful.

Another property of mydriatics is their power of *increasing the intra-ocular tension,* and so favoring the development of glaucoma when a tendency to that affection already exists. Dilatation of the pupil mechanically increases the intra-ocular pressure by narrowing the angle between the iris and the cornea, thus impeding the escape of humor from the eye through the small openings communicating with the canal of Schlemm. Of all mydriatics, *cocain* has the least effect on the intra-ocular tension.

Mydriatics are employed to facilitate ophthalmoscopic examination, to paralyze accommodation when estimating refractive errors, to rest the iris and to prevent or break loose adhesions in iritis, to enlarge the field of vision in nuclear cataract, when the periphery of the lens is still clear, and to allay irritation in inflammation of the cornea.

The most important local mydriatics are:

Atropin.	Scopolamin.
Homatropin.	Hyoscyamin.
Cocain.	Hyoscin.
Duboisin.	Daturin.
Euphthalmin.	

Atropin Sulphate (see p. 70).—Atropin dilates the pupil by paralyzing the peripheral ends of the oculomotor nerves and by stimulating the peripheral ends of the sympathetic nerves. It also destroys the power of accommodation and increases the intra-ocular tension. After the instillation into the eye of a drop or two of a solution of atropin (4 gr.–1 fl. oz.—0.26 gm.–30.0 c.c.) mydriasis begins in fifteen minutes, and attains its maximum in about half an hour. Accommodation is not affected so quickly, paralysis not being complete within an hour and a half. On the other hand, mydriasis persists somewhat longer than the suspension of accommodation. The effect of an atropin solution of the strength indicated does not usually disappear completely for about ten days.

Atropin may be used as a simple mydriatic to *facilitate ophthalmoscopic examination*, but euphthalamin, cocain, and homatropin are as efficient for this purpose and cause less inconvenience to the patient. As it paralyzes accommodation it is a reliable mydriatic for general use in *refraction work*, but many ophthalmologists prefer cycloplegics which have a less persistent action. In *iritis* it is indispensable in preventing and in breaking up adhesions between the iris and the capsule of the lens. In *acute keratitis* it is also very useful in allaying ciliary irritation.

As a simple mydriatic ¼ gr. (0.016 gm.) to the ounce (30.0 c.c.) is sufficient, and the mydriasis from this solution does not last longer than four or five days. As a cycloplegic 4 gr. (0.26 gm.) to the ounce (30.0 c.c.) should be employed. A solution of the same strength is generally used in iritis. In keratitis a solution containing 1–2 gr. (0.065–0.13 gm.) to the ounce (30.0 c.c.) will be effective.

Homatropin Hydrobromate.—Homatropin is an artificial alkaloid obtained by heating tropin (a component of atropin) with either oxytolinic acid or mandelic acid, in the presence of dilute hydrochloric acid. It is employed in the form of the hydrobromate, which is freely soluble in water.

Like atropin, it dilates the pupil and paralyzes accommodation. Its action is complete within an hour, and lasts from two to four days. It has an advantage over atropin in the shorter duration of its effects. It is, however, several times more costly than the natural alkaloid. To paralyze the accommodation a solution of 4–6 gr. (0.26–0.4 gm.) to the ounce (30.0 c.c.) must be instilled in the conjunctival sac five or six times at intervals of five or ten minutes. In facilitating ophthalmoscopic examination a solution of 1–2 gr. (0.05–0.13 gm.) to the ounce (30.0 c.c.) is sufficient.

Cocain Hydrochlorate (see p. 78).—The instillation into the eye of a few drops of a 4 per cent. solution of cocain causes, in from ten to fifteen minutes, in addition to local anesthesia, marked dilatation of the pupil. The mydriasis attains its maximum in about an hour, and persists for from twelve to twenty-four hours. It is of peripheral origin, and is probably caused by stimulation of the sympathetic nerves. Cocain only slightly impairs accommodation, but in strong solutions it has an injurious effect upon the corneal epithelium.

As the effects of cocain on the pupil are of short duration, and as it does not seriously disturb accommodation, it is a convenient mydriatic for *retinal examinations.* In iritis atropin is distinctly preferable on account of its forcible action and lasting effect. The mydriasis induced by cocain is readily overcome by the instillation into the eye of a few drops of a $\frac{1}{2}$ per cent. solution of eserin. Stronger solutions of cocain than 4 per cent. should not be used in the eye on account of the danger of causing degenerative changes in the corneal epithelium.

Duboisin Sulphate (see p. 103).—The effects of duboisin sulphate on the eye resemble those of atropin; like the latter it dilates the pupil, paralyzes accommodation, and increases intra-ocular tension. Its action, however, is more intense, more prompt, and of shorter duration than that of atropin. Its chief drawback is its greater tendency to excite constitutional disturbance. Since it is nearly twice as powerful as atropin, a solution of the sulphate containing 2 gr. to the ounce (0.13 gm.–30.0 c.c.) is sufficiently active. In this strength it is perhaps less irritating to the eye than a 4-grain solution of atropin.

Euphthalmin.—This synthetic alkaloid is the hydrochlorid of the mandelic acid derivative of eucain B. It occurs in the form of a white crystalline powder, soluble in cold water. Its solution can be sterilized by boiling, and may thus be kept for a long time. If a few drops of a 4 per cent. solution be placed in the conjunctival sac, mydriasis begins in a few minutes, attains its maximum in about half an hour, and passes away within six or seven hours. While its mode of action has not been fully determined, it is probable that the drug dilates the pupil by paralyzing the peripheral ends of the oculomotor nerve. Euphthalmin does not injure the corneal epithelium; it does not irritate the conjunctiva; it gives rise to no constitutional disturbance; and it only slightly impairs accommodation. Contrary to what has been stated, however, it does increase to some extent the intra-ocular tension, and Knapp has observed glaucoma follow the use of a 7.5 per cent. solution. On account of the brief duration of its action on the pupil, and its slight cycloplegic influence, it is perhaps the best agent we possess for *simple ophthalmoscopic purposes.* Solutions varying from 4 to 6 per cent. are usually employed. According to Jackson the following solution is more active in dilating the pupil and at the same time less persistent in its effect than a stronger solution of euphthalmin alone:

Euphthalmin,	1 part;
Cocain hydrochlorate,	1 "
Distilled water,	100 parts.

Scopolamin Hydrobromate.—This alkaloid is obtained from the roots of *Scopolia carniolica,* growing in Southern Europe. It occurs in the form of colorless, hygroscopic crystals, which are soluble in water. By some authorities it is regarded as being identical with hyoscin, but the physiologic action of the two alkaloids is somewhat different. H. C. Wood, Jr., finds that while the action of the scopolia alkaloids resembles that of the alkaloids of belladonna, it is distinctly more powerful.

Interest in scopolamin centers around its action as a mydriatic. Two instillations, at intervals of an hour, of a drop or two of a $\frac{1}{5}$ per cent. solution will dilate the pupil and paralyze accommodation, the effect completely disappearing in from five to eight days. Like atropin it increases intra-ocular tension. It has an advantage over atropin in the rapidity with which the power of accommodation is restored. Its effect is more prolonged, however, than that of homatropin, and it has a greater tendency than the latter to excite constitutional disturbance.

It is commonly used in solutions of from $\frac{1}{10}-\frac{1}{5}$ per cent. ($\frac{9}{20}-\frac{9}{10}$ gr. to the ounce—0.03–0.06 gm. to 30.0 c.c.).

Hyoscyamin Hydrobromate.—The action of this drug on the eye is very similar to that of duboisin.

Hyoscin Hydrobromate.—This drug is a powerful mydriatic and cycloplegic, resembling atropin in its action, but on account of its liability to cause intoxicating effects it is not often used by ophthalmologists. It is effective in solutions of from $\frac{1}{2}-1$ gr. to the ounce (0.03–0.065 gm.—30.0 c.c.).

Daturin Hydrochlorate.—According to Ladenburg this preparation, which is obtained from stramonium, is a mixture of atropin and hyoscyamin. On account of the uncertainty of its composition it is rarely used as a mydriatic.

MYOTICS.

Myosis or contraction of the pupil may result from causes the reverse of those mentioned as being capable of bringing about mydriasis. *Morphin,* when taken internally, contracts the pupil by stimulating the oculomotor center. *Physostigmin* acts locally probably by paralyzing the peripheral fibers of the sympathetic nerve, and by stimulating the peripheral ends of the oculomotor nerve, and perhaps the constrictor muscle of the iris itself. *Pilocarpin* acts locally by stimulating the peripheral ends of the oculomotor nerve. Physostigmin also stimulates the ciliary muscle, and in strong solutions causes spasm of accommodation. Pilocarpin has a similar but a less powerful action on the ciliary muscle.

Myotics lessen intra-ocular tension by widening the angle between the iris and the cornea, and thus favoring the escape of humor from the eye.

Myotics are employed to overcome the effects of mydriatics; to lessen intra-ocular tension in beginning glaucoma and staphyloma; and alternately with mydriatics in marginal ulcers of the cornea.

The most important local myotics are:

Physostigmin.   Pilocarpin.   Arecolin.

Physostigmin or Eserin Sulphate (see p. 131).—This is the most commonly used myotic. When a gentle, continuous action is desired a solution containing from ⅛–1 gr. (0.008–0.065 gm.) to the ounce (30.0 c.c.) is sufficient, but when a prompt and powerful action is necessary, a solution containing from 2–3 gr. (0.13–0.2 gm.) to the ounce (30.0 c.c.) may be employed. The maximum effect of these solutions is secured in from one-half to three-quarters of an hour, and persists from a few hours to several days. On standing, solutions of physostigmin turn red and lose some of their effectiveness.

Pilocarpin Hydrochlorate (see p. 252).—The action of pilocarpin on the eye is very much like that of physostigmin, but decidedly less powerful. It is employed in solution of from 1–4 gr. (0.065–0.26 gm.) to the ounce (30.0 c.c.).

Arecolin Hydrobromate.—This is the hydrobromate of a liquid alkaloid obtained from *areca catechu*, or betel-nut. It is a white crystalline salt, soluble in water. In the form of ½ or 1 per cent. solution (2¼–4½ gr. to the ounce—0.14–0.3 gm. to 30.0 c.c.) it is a powerful myotic. Contraction of the pupil begins in from three to five minutes, reaches its maximum in from ten to fifteen minutes, and is accompanied by spasm of the ciliary muscle. The pupil returns to its normal condition within an hour or two. Its action is more rapid, but at the same time more transient, than that of physostigmin. Its power of lowering tension in *glaucoma* is apparently equal to that of eserin, and in Bietti's hands it proved successful in a case of glaucoma in which physostigmin had failed.

Arecolin has been used with asserted success as an anthelmintic, and, according to Frohner and Clemesma, it is, when injected hypodermically, a more powerful sialagogue than pilocarpin.

EMETICS.

Vomiting is a reflex act consisting of a forcible spasmodic contraction of the abdominal muscles and diaphragm. While the contraction of the muscular coat of the stomach may assist in the act, it is not an essential factor, for Magendie found that when the stomach was removed and was replaced by a bladder filled with water, emesis could still occur. The afferent impulses which excite vomiting travel from the stomach through the sensory fibers of the vagus to a center in the medulla closely connected with the respiratory center, and thence the efferent impulses are conducted through the phrenics, the spinal nerves, and the vagus to the muscles concerned in the act.

Theoretically, a drug may cause vomiting either by directly affecting the center in the medulla or by indirectly affecting it through irritation of the sensory nerve-endings in the stomach. It is not always an easy matter to determine the exact mode of action of an emetic. If the drug be injected into a vein, it may act directly on the vomiting center, or, being eliminated through the stomach, it may act reflexly by irritating the sensory nerves of that organ; or again, if the drug be given by the mouth, it may irritate the stomach and provoke emesis through a reflex action, or, being absorbed, it may excite the vomiting center directly. If, however, a drug acts more promptly when injected subcutaneously than when it is given by the mouth, and if vomiting follows its injection after the stomach has been replaced by a bladder filled with water, or after all the arteries leading to the stomach have been ligated, it may be assumed that its action is chiefly a direct one on the center in the medulla. The action of *apomorphin* appears to be in the main a direct one; *ipecac* and *antimony* undoubtedly act both directly and indirectly; and the action of *copper sulphate, zinc sulphate, yellow sulphate of mercury, alum,* and *mustard* is chiefly an indirect or reflex one.

Emetics may be employed for one of three purposes: To expel irritating food or poisons from the stomach; to expel foreign bodies, false membrane, or excessive secretion from the respiratory tract; and to expel mucus and bile from the gall-ducts in catarrhal jaundice.

Emetics should be avoided or used with extreme caution in advanced pregnancy and in persons suffering from aneurysm, atheroma, or hernia.

Apomorphin Hydrochlorate (see p. 265).—This drug is the most prompt and reliable emetic that we possess; moreover, it causes little nausea, and it is effective when administered hypodermically. It is very useful in *poisoning*, especially when swallowing is impossible, or when the state of the stomach is such as to prohibit the use of a mechanical or irritant emetic. In *acute alcoholism* it is exceedingly effectual. In some cases of narcotic poisoning, on account of the decreased sensitiveness of the medullary center, apormorphin, like other emetics, may prove inactive. As an emetic the drug should be administered hypodermically, although it will provoke vomiting when given by the mouth if the dose be sufficiently large. It must be borne in mind that in the very young and infirm considerable depression may attend its action. The dose is from $\frac{1}{12}$–$\frac{1}{6}$ gr. (0.005–0.01 gm.). Solutions are most conveniently prepared from tablets containing the requisite dose. Both tablets and solution are liable to deteriorate with age and on exposure to light and air.

Ipecacuanha (see p. 263).—This drug is a safe and fairly prompt emetic. Its action is not so vigorous or so depressing as that of the mineral emetics. It is especially adapted for children, in whom it may be employed to unload the stomach of irritant food or to expel tenacious mucus from the air-passages. The syrup and the wine are eligible preparations, and either may be given to children in doses of from ½–2 fl. dr. (2.0–8.0 c.c.).

Tartar Emetic (see p. 63).—The use of antimonial preparations as emetics is fortunately obsolescent. As their action is slow and is attended with prolonged nausea and depression, apomorphin, ipecacuanha, zinc sulphate, or mustard should always be selected in their stead.

Zinc Sulphate (see p. 363).—As an emetic zinc sulphate is employed chiefly in *narcotic poisoning*. The dose is from 10–30 gr. (0.65–2.0 gm.), repeated in fifteen or twenty minutes, if necessary.

Copper Sulphate (see p. 362).—This drug is a more prompt and powerful emetic than zinc sulphate. In *phosphorus poisoning* it serves a double purpose: it not only unloads the stomach, but it also acts as a partial antidote by coating the phosphorus with an insoluble phosphid of copper. The dose of copper sulphate is from 5–10 gr. (0.3–0.65 gm.), and when this amount has been administered without effect, it is best not to repeat it, but to resort to some other emetic.

Mercury Subsulphate (Turpeth Mineral).—This drug was formerly used in doses of 2–3 gr. (0.13–0.2 gm.) as an emetic in croup. Although certainty, quickness of action, tastelessness, and small bulk are in its favor, it cannot be recommended on account of its exceedingly irritant properties.

Alum.—Powdered alum is a safe but somewhat uncertain emetic. It may be given to children in doses of a teaspoonful, in syrup, and repeated once or twice if vomiting does not ensue.

Mustard.—Mustard flour is a prompt and energetic emetic. It may replace zinc sulphate or copper sulphate in *narcotic poisoning*. It is contraindicated when there is acute gastritis. The dose is a tablespoonful stirred up in a glass of water, and repeated in ten or fifteen minutes if necessary.

ANTI-EMETICS.

The treatment of vomiting must be modified according to the cause. In the vomiting of intestinal obstruction it is only in rare instances that anything short of operative interference brings relief. Severe vomiting is sometimes an early manifestation of chronic Bright's disease, and when such is the case, gastric sedatives are of little avail; our main reliance must be on eliminant measures. Reflex vomiting is not uncommon, and may be associated with lesions in various parts of the body, as the bowel, gall-bladder, kidney, lung, internal ear, or the uterus or its appendages. In this type of vomiting treatment, to be really curative, must be directed to the original cause.

The exact etiology of the vomiting of pregnancy is still undetermined, and hence its treatment must be in a measure empiric. There is a growing conviction, however, that hyperemesis gravidarum is dependent upon several co-operative factors, the most important of which are the presence in the blood of irritant toxins; reflex irritation from uterine displacement or from inflammatory lesions of the uterus or its appendages; circulatory disturbances in the medulla; and, finally, in certain cases, a neurotic or hysteric element. In this form of vomiting remedies directed solely to the stomach generally prove ineffective. To rid the body of noxious materials warm baths, mild laxatives, and saline injections into the bowel or under the skin are useful remedies. Saline transfusions are particularly efficacious in that they dilute the poisons circulating in the blood, increase elimination, serve to maintain the arterial pressure, and help to nourish the patient. Pelvic irritation should receive careful attention. In some cases drugs that raise the arterial pressure on the medulla, like strychnin, act beneficially; in others, vasodilators, like nitroglycerin, give better results. Drugs that directly depress the vomiting center are sometimes serviceable; chloral and bromids are the best of this class; morphin, as a rule, is badly borne. In neurotic subjects firmness tempered with kindliness and encouragement is essential to success. When the hysterical element is very pronounced, the Weir-Mitchell treatment can be recommended with considerable confidence.

When the measures which have been indicated, after a thorough trial prove futile, and the life of the patient is jeopardized by the incessant vomiting, there ought to be no question in the mind of the practitioner as to the advisability of terminating the pregnancy.

11

A curious form of vomiting is that which is known, for want of a better name, as nervous vomiting. It is not associated with any anatomic lesion of the stomach or with changes in the quantity or quality of the food, nor is it apparently of reflex character. In many instances it is evidently a symptom of hysteria. In this condition anti-emetics—cerium oxalate, menthol, hydrocyanic acid, and bismuth—are sometimes useful, but they more often fail. Suppositories of valerian, asafetida, and belladonna may prove efficacious. Lavage followed by the administration of silver nitrate (gr. $\frac{1}{4}$—0.016 gm.) often serves to allay the irritability of the gastric nerves. Hot compresses, sinapisms, or blisters over the epigastrium are indicated. A spray of ethyl chlorid along the spine and over the stomach has given good results. Galvanism with a current of low density, one pole (negative) over the back of the neck and the other (positive) within the stomach, may afford prompt relief In many cases it is necessary to keep the patient perfectly quiet in the recumbent position, and to feed exclusively by the rectum. In intractable cases the Weir-Mitchell treatment should be given a thorough trial.

Vomiting dependent upon cancer of the stomach is sometimes controlled temporarily or lessened by anti-emetics. Cerium oxalate, ice-cold carbonated water, chloroform, hydrocyanic acid, and bismuth may be tried. Morphin hypodermically or by the rectum is also useful. When there is obstruction of the pylorus, lavage is of great service.

When the vomiting of ulcer is not relieved promptly by gastric sedatives, the patient should be kept at complete rest for a few days and nourished by rectal injections. Counter-irritation by means of small blisters over the epigastrium is useful. Inhalations of vinegar will often allay the nausea and vomiting resulting from etherization.

From the foregoing it is evident that while anti-emetics are often exceedingly useful, they should be regarded as only palliative remedies or as adjuvants to more important curative measures. Anti-emetics may accomplish their purpose directly by acting on the center in the medulla, or indirectly by allaying irritability of the sensory nerve-endings in the stomach. To the first class belong:

Opium.　　　　　Bromids.　　　　　Chloral.

To the second class belong:

Carbonated water.	Chloroform.
Ice.	Silver nitrate.
Bismuth subnitrate.	Menthol.
Hydrocyanic acid.	Lime-water.
Carbolic acid.	Ipecac (minute doses).
Cocain.	Cerium oxalate.

All these agents, with the exception of cerium oxalate, are considered under other headings.

CERII OXALAS, U. S. P.

(Cerium Oxalate; $Ce_2(C_2O_4)_3 + 9H_2O$.)

Cerium oxalate is a white, odorless, and tasteless powder, insoluble in water or alcohol, but soluble in diluted hydrochloric acid or sulphuric acid. Its dose is from 3–5 gr. (0.2–0.3 gm.).

It has been employed chiefly as an anti-emetic in the *vomiting of pregnancy*, but it is also of value in vomiting dependent upon *organic disease of the stomach*.

Bechterew has strongly recommended it in the *gastric crises* of locomotor ataxia. It has also been used, but without much success, as a sedative in *chorea* and *epilepsy*. The manner of its action has not yet been determined. It may be administered in pill, or, better, in the form of a dry powder.

GASTRIC ANTACIDS.

Gastric antacids are drugs employed to neutralize acidity of the stomach-contents. Some drugs, like the carbonate and bicarbonate of sodium, have the power of lessening not only the acidity of the stomach-contents, but also that of the urine. Others, like ammonia and ammonium carbonate, serve as antacids in the stomach, but are eliminated in such a form that they tend to increase rather than to diminish the acidity of the urine. Again, there are other drugs, like the vegetable salts of potassium and lithium, which have very little effect upon the acidity of the stomach, but, being eliminated as alkaline carbonates, render the urine less acid.

Gastric antacids are employed to neutralize the organic acids (lactic, butyric, and acetic) resulting from fermentation of the stomach-contents, and which excite eructations, heartburn, and gastralgia; to lessen the superacidity in cases of hyperchlorhydria; and to antagonize poisons of an acid character.

The question whether the administration of alkalies before meals does or does not stimulate gastric secretion has not yet been satisfactorily settled.   Claude Bernard taught, and Ringer and Lemoine agree with him, that alkalies increase acid secretion; on the other hand, Reichmann, as the result of many experiments, doubts whether they have any such power. Those who deny that alkalies excite secretion attribute their good effects in diseases of the stomach to their power of neutralizing excessive acidity (hydrochloric acid or organic acids), removing excess of mucus, and of imparting tone to the gastric mucous membrane (Reichmann).

The most important antacids are:

Sodium bicarbonate.	Bismuth subnitrate.
Lime-water.	Calcium carbonate.
Magnesia.	Ammonia.
Magnesium carbonate.	

SODII BICARBONAS, U. S. P.

(Sodium bicarbonate; $NaHCO_3$.)

Sodium bicarbonate is a white, opaque, odorless powder, having a cooling, faintly alkaline taste.   It is soluble in 11.3 parts of water and insoluble in alcohol.   The dose is from 5–30 gr. (0.3–2.0 gm.).

PREPARATIONS.	DOSE.
Mistura Rhei et Sodæ, U. S. P.	1–2 fl. dr. (2.0–8.0 c.c.).
Pulvis Effervescens Compositus (Seidlitz Powder), U. S. P.	
Trochisci Sodii Bicarbonatis, U. S. P., contain 3 gr. (0.2 gm.) each.	

Therapeutics.—Sodium bicarbonate is extensively used in diseases of the stomach to neutralize abnormal acids or hydrochloric acid when in excess.   Given before meals in small doses (3–5 gr.—0.2–0.3 gm.), with some bitter stomachic, it often affords prompt relief in *mild forms of indigestion.*   Given an hour or two after meals it allays the burning pain, eructations, and palpitation due to the *acids of fermentation.*   Given at the height of digestion in large doses it relieves the painful crises of *hyperchlorhydria.*   Large doses of sodium bicarbonate are also useful in neutralizing the hydrochloric superacidity encountered in *gastric ulcer.*   In the *chronic gastro-intestinal catarrh of childhood*, characterized by capricious appetite, tympanites, eructations, troubled sleep, and hard, lumpy, mucous stools, sodium bicarbonate with a bitter and a mild laxative often gives excellent results.

In daily doses of from 5–10 dr. (20.0–40.0 gm.) it has been employed successfully by Huchard and others in averting *diabetic coma*. Intravenous injections of alkalies have been tried in *fully developed acetonemia*, but without encouraging results.

Sodium bicarbonate is sometimes used externally as a sedative dressing in *superficial burns*. A 5 to 10 per cent. solution is an excellent remedy for *thrush*. It is employed as a detergent in many washes designed for *chronic catarrhal affections of the nasopharynx*, and it is an ingredient of the well-known Dobell's solution, which may be prescribed as follows:

   R Sodii bicarbonatis,
     Sodii boratis,   āā ʒj (4.0 gm.);
     Acidi carbolici,   gr. xxx (2.0 gm.);
     Glycerini,   fʒj (30.0 c.c.);
     Aquæ,   Oij (1.0 L.).—M.

Sodium bicarbonate is incompatible with acids, metallic salts, and alkaloids.

LIQUOR CALCIS, U. S. P.

(Solution of Calcium Hydrate; Lime-water.)

Lime-water is a saturated aqueous solution of calcium hydrate made by slaking lime with water. It is a clear, colorless liquid, odorless, and having a saline, slightly caustic taste. It contains about 0.17 per cent. of calcium hydrate. The dose is from $\frac{1}{2}$–2 fl. oz. (15.0–60.0 c.c.).

PREPARATIONS.

Linimentum Calcis, U. S. P. (lime liniment, Carron oil), contains equal volumes of lime-water and linseed oil.

Lotio Hydrargyri Flava (yellow-wash), $\frac{1}{2}$ dr. (2.0 gm.) of corrosive sublimate to 1 pint (0.5 L.) of lime-water.

Lotio Hydrargyri Nigra (black-wash), 1 dr. (4.0 gm.) of calomel to 1 pint (0.5 L.) of lime-water.

Therapeutics.—In all conditions in which cows' milk is the chief article of diet, lime-water may be added to the milk, to prevent the formation in the stomach of hard curds.

In the *diarrhea of children* and in *typhoid fever* it is especially useful when employed in this way. It is sometimes a useful remedy in *obstinate vomiting*, particularly when the latter is due to a high degree of acidity. In *chronic gastric catarrh* with excessive secretion of mucus, washing out the stomach with a weak solution of lime-water (1 : 10) before breakfast may be practised with great benefit.

Inhalations of atomized lime-water have been recommended for their solvent effect in *diphtheria* and *fibrinous bronchitis.* As an alkaline astringent lime-water is sometimes employed as an injection in *vaginitis, leucorrhea,* and *urethritis.* It may be used also as an injection in *seat-worms,* although it is inferior to an infusion of quassia. Carron oil is used as a soothing application for *burns* and *scalds.* It has received its name from being so extensively used by the workmen in the foundries at Carron, Scotland.

OTHER ANTACIDS.

Magnesia (see p. 193).—Magnesia or magnesium oxid is official in two forms : light or calcined magnesia (*Magnesia,* U. S. P.) and heavy magnesia (*Magnesia Ponderosa,* U. S. P.). Both forms occur as white, insoluble powder, odorless, and having a slightly earthy taste. Heavy differs from light magnesia not only in its weight, but also in its not readily uniting with water to form a gelatinous hydrate. The dose of either oxid is from 10–60 gr. (0.65–4.0 gm.).

Magnesia combines the usefulness of an antacid with that of a mild laxative. When there is constipations, it may be given instead of sodium bicarbonate or combined with that drug in all conditions in which an antacid is indicated. In the *diarrhea* of childhood, when there is unnatural acidity of the *primæ viæ,* it is an excellent laxative for removing sour, indigestible food. In *ulcer of the stomach* it may be advantageously combined with bismuth to control constipation.

Magnesium Carbonate (see p. 193).—This drug is used in the same doses and for the same purposes as magnesia.

Bismuth Subnitrate (see p. 370).—Bismuth subnitrate is but a feeble antacid, but owing to its sedative, astringent, and antiseptic properties it is an invaluable remedy in many diseases of the stomach. Combined with sodium bicarbonate it affords prompt relief in *heart-burn* and *pyrosis* due to unnatural acidity of the stomach-contents. In *hyperchlorhydria* and *peptic ulcer* it should be given in large doses (20–30 gr.—1.3–2.0 gm.), combined with sodium bicarbonate, or, in case there is constipation, with magnesia.

Calcium Carbonate (see p. 373).—Calcium carbonate or chalk is an unirritating but feeble antacid and astringent. It is very useful in *diarrhea* with acidity of the primæ viæ, and it may be combined with bismuth subnitrate in the treatment of *gastric ulcer* when there is a marked tendency to looseness of the bowels. It is a chemical antidote to all the *poisonous acids*.

Ammonia (see p. 40).—Aromatic spirit of ammonia, in doses of from 20–30 min. (1.2–2.0 c.c.), may be used instead of sodium bicarbonate to relieve the headache, heart-burn, and pyrosis arising from acid fermentation in the stomach.

STOMACHICS.

Stomachics are drugs which sharpen the appetite and promote the functional activity of the stomach. The most important members of this class of remedies are:

Gentian.	Condurango.
Quassia.	Taraxacum.
Calumba.	Eupatorium.
Nux vomica.	Serpentaria.
Cinchona.	Cascarilla.
Chirata.	Wild cherry.
Chamomile.	Alcohol.
Hydrastis.	Orexin.

All stomachics, with the exception of the last two in the foregoing list, contain a bitter principle. It is generally admitted that bitters strengthen the muscular movements of the stomach, but whether or not they increase the glandular secretion is a moot point. Tschelzoff and Jaworski conclude from their experiments on dogs and human beings that they actually lessen the amount of the secretion. On the other hand, Marcone and Wirszubski believe that bitters increase secretion, and that they bring about this result chiefly through stimulation of the endings of the vagi in the stomach, thus producing a reflex effect upon the secreting glands. Riegel reports that in a series of experiments on animals he was unable to demonstrate an invariable result with any one drug upon gastric secretions. With the exception of cinchona and nux vomica the direct influence of the bitter stomachics probably does not extend beyond the digestive organs. As a class, stomachics are useful in atony of the stomach, in chronic gastritis when the mucous membrane is not very irritable, and in the early stages of cancer. They are contraindicated in all conditions involving the stomach in which there is marked irritability or hypersecretion.

GENTIANA, U. S. P.

(Gentian.)

Gentian is the root of *Gentiana lutea*, a perennial herb growing in the mountainous districts of Central and Southern Europe. It contains an extremely bitter glucosid, *gentio-picrin*, and *gentisic* or *gentianic acid*, which is inert.

PREPARATIONS.	DOSE.
Tinctura Gentianæ Composita, U. S. P. (contains also bitter orange-peel and cardamom)	1–2 fl. dr. (4.0–8.0 c.c.).
Extractum Gentianæ Fluidum, U. S. P.	½–1 fl. dr. (2.0–4.0 c.c.).
Infusum Gentianæ Compositum, U. S. P.	½–2 fl. oz. (15.0–60.0 c.c.).
Extractum Gentianæ, U. S. P.	2–5 gr. (0.13–0.3 gm.).

Therapeutics.—Gentian is one of the most reliable of the bitter tonics. It is often of great value in the milder forms of *chronic indigestion*, in the *dyspepsia of phthisis*, and in the early stages of *cancer*. It is an excellent restorative in the *convalescence of acute fevers*. In indigestion it may be advantageously combined with an alkali, as in the following formula:

℞ Sodii bicarbonatis,   ʒij (8.0 gm.);
    Infusi gentianæ compositi,   fʒvj (180.0 c.c.).—M.
Sig. A tablespoonful before meals.

As a restorative, it may be combined with hydrochloric acid, as in the following formula:

℞ Acidi hydrochlorici diluti,   fʒij (8.0 c.c.)
    Tincturæ gentianæ compositæ, q. s. ad fʒiij (90.0 c.c.).—M.
Sig. A teaspoonful in water after meals.

QUASSIA, U. S. P.

Quassia is the wood of *Picræna excelsa*, a tree resembling the common ash, growing in the West Indies. It contains a bitter crystalline principle, *quassin*, but no tannic acid. The dose of quassin is from $\frac{1}{30}-\frac{1}{3}$ gr. (0.002–0.02 gm.).

PREPARATIONS.	DOSE.
Tinctura Quassiæ, U. S. P.	½–1 fl. dr. (2.0–4.0 c.c.).
Infusum Quassiæ, U. S. P.	½–2 fl. oz. (15.0–60.0 c.c.).
Extractum Quassiæ Fluidum, U. S. P.	10–20 min. (0.6–1.0 c.c.).
Extractum Quassiæ, U. S. P.	1–3 gr. (0.06–0.2 gm.).

Therapeutics.—As a stomachic, quassia may be used in the same class of cases as gentian.

The rectal injection of an infusion made by adding an ounce (31.0 gm.) of quassia chips to a pint (0.5 L.) of water is an efficient anthelmintic against *seat-worms* (*Oxyuris vermicularis*).

To secure the best results the bowel should first be thoroughly cleansed by means of a simple soap-and-water enema.

As quassia does not contain tannic acid, its preparations are not incompatible with the salts of iron.

CALUMBA, U. S. P.

(Columbo.)

Calumba is the root of *Jateorhiza palmata*, growing in Eastern Africa. It contains an alkaloid, *berberin*, a neutral principle, *columbin*, and *columbic acid*, all of which are bitter.

PREPARATIONS.	DOSE.
Tinctura Calumbæ, U. S. P.	½–2 fl. dr. (2.0–8.0 c.c.).
Extractum Calumbæ Fluidum, U. S. P.	10–30 min. (0.6–2.0 c.c.).
Extractum Calumbæ	2–5 gr. (0.13–0.3 gm.).
Infusum Calumbæ	½–2 fl. oz. (15.0–60.0 c.c.).

Therapeutics.—Calumba is a pure, unirritating bitter, and may be prescribed with confidence in the various conditions in which stomachics are indicated. It may be combined with either acids or alkalies, and since it contains no tannin, it does not form a black, unsightly mixture with the salts of iron.

CHIRATA, U. S. P.

(Chiretta.)

Chirata is the entire plant of *Swertia Chirata*, growing in the mountains of Northern India. It contains a bitter crystalline glucosid, *chiratin*, and *ophelic acid*, which is also bitter.

PREPARATIONS.	DOSE.
Tinctura Chiratæ, U. S. P.	½–1 fl. dr. (2.0–4.0 c.c.).
Extractum Chiratæ Fluidum, U. S. P.	10–30 min. (0.6–2.0 c.c.).
Infusum Chiratæ, U. S. P.	½–2 fl. oz. (15.0–60.0 c.c.).

Therapeutics.—In addition to its virtues as a bitter stomachic, chirata is supposed to possess decided cholagogue properties. It is useful in those forms of indigestion which are associated with so-called bilious attacks. Like quassia and calumba, it is free from tannic acid; it is therefore not incompatible with iron salts.

ANTHEMIS, U. S. P.

(Chamomile.)

Chamomile is the flower-heads of *Anthemis nobilis*, a perennial plant cultivated in Western Europe, and to some extent in the United States. It contains a volatile oil, a bitter principle, anthemin, and a small amount of tannin. Chamomile is a mild stomachic and carminative, and is usually employed in the form of an infusion (unofficial), of which 1–2 fl. oz. (30.0–60.0 c.c.) may be given.

CONDURANGO.

Condurango is the bark of *Gonolobus Condurango*, a climbing vine indigenous to Ecuador. It contains tannin and one or two bitter glucosids. It is usually employed in the form of the fluid extract in doses of from 20–60 min. (1.0–4.0 c.c.).

Therapeutics.—Condurango was first recommended by Friedreich in 1874 as a curative remedy in cancer of the stomach. Later, Immermann and Riess reported cases in support of this view; but the testimony of numerous impartial observers has proved conclusively that the drug is without specific action. Notwithstanding the adverse opinion as to its curative power, nearly all who have used the drug agree that it often acts most beneficially on the concomitant catarrh, sharpening the appetite, lessening the pain, and promoting digestion. In some cases of *simple gastric catarrh* it is also efficacious. It may be prescribed with hydrochloric acid, as in the following formula:

> R Acidi hydrochlorici diluti,   f℥iss (6.0 c.c.);
> Extracti condurango fluidi,
> Glycerini,   aa f℥vj (22.5 c.c.);
> Vini xerici,   q. s. ad f℥iv (120.0 c.c.).—M.
> Sig. A dessertspoonful in water after meals.

TARAXACUM, U. S. P.

(Dandelion.)

Taraxacum is the root of *Taraxacum officinale*, growing in Europe and North America. It contains a crystalline bitter principle, *taraxacin*.

PREPARATIONS.	DOSE.
Extractum Taraxaci Fluidum, U. S. P. . . ½–2 fl. dr. (2.0–8.0 c.c.).
Extractum Taraxaci, U. S. P. . . . . . . 5–30 gr. (0.3–2.0 gm.).

Therapeutics.—Taraxacum is a mild stomachic, possessing feeble laxative properties.

EUPATORIUM, U. S. P.

(Boneset; Thoroughwort.)

Eupatorium is the leaves and flowering tops of *Eupatorium perfoliatum*, a perennial plant growing in North America. It contains a bitter crystalline glucosid, *eupatorin*, tannin, and a volatile oil.

PREPARATIONS.	DOSE.
Extractum Eupatorii Fluidum, U. S. P.	½–1 fl. dr. (2.0–4.0 c.c.).
Infusum Eupatorii, U. S. P.	1–4 fl. oz. (30.0–120.0 c.c.).

Therapeutics.—Eupatorium is employed as a stomachic and diaphoretic. Large doses excite vomiting and purging. The infusion, taken while hot, has long been a popular household remedy in *colds* and *muscular rheumatism.*

SERPENTARIA, U. S. P.

(Virginia Snakeroot.)

Serpentaria is the rhizome and roots of *Aristolochia Serpentaria*, a perennial herb growing in the woods of the Eastern and Southern United States. It contains a bitter principle, *aristolochin*, a resin, tannin, and a volatile oil.

PREPARATIONS.	DOSE.
Tinctura Serpentariæ, U. S. P.	½–2 fl. dr. (2.0–8.0 c.c.).
Extractum Serpentariæ Fluidum, U. S. P.	10–30 min. (0.6–2.0 c.c.).
Tinctura Cinchonæ Composita, U. S. P., contains 2 per cent. serpentaria	1–4 fl. dr. (4.0–15.0 c.c.).
Infusum Serpentariæ	½–2 fl. oz. (15.0–60.0 c.c.).

Therapeutics.—Serpentaria is believed to possess, in addition to its properties as a bitter tonic, the power of increasing the secretions, especially the sweat and the urine. Large doses are capable of causing nausea and vomiting. Serpentaria has been extensively employed as a stomachic in *atonic dyspepsia*, and, combined with more potent remedies, like nux vomica and cinchona, as a tonic in the convalescent stage of *acute fevers.* Garrod believes it to be a remedy of some power in *chronic rheumatism.*

CASCARILLA, U. S. P.

Cascarilla is the bark of *Croton Eleuteria*, a shrub growing in the Bahama Islands. It contains a bitter principle, *cascarillin*, a volatile oil, a resin, and tannin.

PREPARATIONS.	DOSE.
Tinctura Cascarillæ	½–2 fl. dr. (2.0–8.0 c.c.).
Extractum Cascarillæ Fluidum	10–30 min. (0.6–2.0 c.c.).
Infusum Cascarillæ	½–2 fl. oz. (15.0–60.0 c.c.).
Extractum Cascarillæ	5–10 gr. (0.3–0.65 gm.).

Therapeutics.—The action of cascarilla resembles somewhat that of serpentaria. For two centuries it was a favorite remedy in Germany and France in low fevers, diarrhea, and dysentery. At present it is rarely employed except as a bitter tonic. Mineral acids should not be given with the tincture or the fluid extract, since they precipitate the resin.

PRUNUS VIRGINIANA, U. S. P.

(Wild Cherry.)

Prunus virginiana is the bark of *Prunus serotina*, a large tree growing in North America. It contains tannic acid, a bitter principle, a non-crystalline glucoside—*amygdalin*, and a ferment—*emulsin.* Hydrocyanic acid is formed from the union of the last two constituents in the presence of water.

PREPARATIONS.	DOSE.
Extractum Pruni Virginianæ Fluidum, U. S. P.	½–1 fl. dr. (2.0–4.0 c.c.).
Syrupus Pruni Virginianæ, U. S. P.	1–4 fl. dr. (4.0–15.0 c.c.).
Infusum Pruni Virginianæ, U. S. P.	½–2 fl. oz. (15.0–60.0 c.c.).

Therapeutics.—Wild cherry is a bitter tonic and a feeble nerve-sedative. It has been extensively employed in *phthisis* to allay irritable cough, its efficacy being attributed to the hydrocyanic acid. The amount of acid present, however, is so small that it can scarcely be of therapeutic value. It has also been used on the recommendation of Allbutt to combat the dyspnea and dyspepsia of *heart-disease,* but Allbutt himself admits that it is only in the very early stages of the disease that it exerts any decided influence. On account of its agreeable taste it is an excellent *vehicle* for unpalatable drugs.

OREXIN.

(Phenyl-dihydro-chin-azolin.)

Orexin is a complex synthetic compound derived from chinolin. The tannate of orexin being freest from irritating properties is the best preparation for medicinal use. This salt appears as a yellowish-white, odorless powder, having a chalky taste. It is insoluble in water, but readily soluble in dilute hydrochloric acid. The dose is from 5–8 gr. (0.3–0.5 gm.), in capsules or tablets.

Therapeutics.—Orexin possesses the power to a certain extent of sharpening the appetite, stimulating the secretion of hydrochloric acid, and increasing the motor activity of the stomach. It is sometimes useful in restoring the appetite in the *early stages of phthisis,* in *chlorosis,* in *neurasthenia,* and in *convalescence* from acute diseases. Penzoldt and Hermanni have spoken very favorably of it in the *vomiting of pregnancy.* It is contraindicated in organic diseases of the stomach—cancer, ulcer, and gastritis—and in cases of hyperacidity.

It should be given two or three times a day, two hours before meals, and followed by a draft of half a glass of water.

OTHER STOMACHICS.

Alcohol (see p. 43).—In small quantities dilute alcohol whets the appetite and favors digestion. Its presence in the stomach tends to retard digestion, but this effect is more than counterbalanced by the increased flow of saliva and of gastric juice, and the improvement in the motor power of the stomach which the drug calls forth. A little dry sherry, light claret, or pure whisky, well diluted, is often of service in *dyspepsia associated with atony of the stomach,* particularly in old people; but the greatest caution is necessary in ordering it on account of the danger of inducing the alcohol habit. All alcoholic preparations are contraindicated when there is hyperesthesia of the stomach or supersecretion of acid.

Nux Vomica (see p. 126).—On account of its general tonic properties nux vomica has a decided advantage over many other bitters. It not only sharpens the appetite and increases secretion, but it probably has more effect than any other stomachic on the muscular movement of the stomach and intestines. In *chronic indigestion,* when there are no evidences of decided irritation, it is often very efficient. In *dilatation of the stomach* the result of atony it is our most valuable drug, and should be given in full doses. It is also useful in those cases of *marked flatulence,* common in elderly people, and dependent upon a lack of tone in the walls of the stomach and bowel.

Cinchona (see p. 409).—Cinchona to a certain extent shares with nux vomica the advantage of being a general tonic as well as a stomachic. The compound tincture of cinchona (Huxham's tincture) is specially serviceable as a bitter tonic in the *convalescence of acute febrile diseases.* The indigestion resulting from overwork is often promptly relieved by a combination of quinin in small doses with nux vomica, as in the following formula:

℞. Quininæ sulphatis, gr. x (0.65 gm.);
 Ferri lactatis, gr. xx (1.3 gm.);
 Extracti nucis vomicæ, gr. v (0.3 gm.);
 Pulveris ipecacuanhæ, gr. iij (0.2 gm.);
 Oleoresinæ capsici, gr. ij (0.13 gm.).—M.
Fiant pilulæ No. xx. Sig. One after meals.

Hydrastis (see p. 375).—Hydrastis is a useful bitter tonic in *asthenic gastritis* with motor insufficiency. In addition to its virtues as a stomachic, it is an active stimulant to intestinal peristalsis. According to the researches of Rutherford and of Cerna, it increases the flow of bile. The drug may be employed in the form of the fluid extract ($\frac{1}{2}$–1 fl. dr.—2.0–4.0 c.c.) or of its chief alkaloid, hydrastin ($\frac{1}{6}$–$\frac{1}{2}$ gr.—0.01–0.03 gm.).

DIGESTANTS.

Of the agents which take part in the process of natural digestion the most important from the standpoint of the therapeutist are hydrochloric acid, pepsin, and the enzymes contained in the pancreatic secretion—trypsin, amylopsin, and steapsin. *Hydrochloric acid*, the natural acid of the stomach, is the most valuable of all digestants. It aids digestion not only directly, but also indirectly by stimulating the activities of the stomach, by preventing fermentation, and by assisting in the conversion of pepsinogen into the active pepsin and of labzymogen into the active labferment. As a remedy, hydrochloric acid is indicated in all conditions associated with inacidity or marked hypoacidity.

Pepsin, being absent from the gastric juice far less frequently than hydrochloric acid, is distinctly less useful as a remedy than the latter. It is of service, however, in conditions associated with apepsia or hypopepsia, such as the advanced stages of chronic gastritis, atrophy of the gastric glands, and carcinoma.

Of the *pancreatic enzymes*, trypsin digests proteids, amylopsin acts upon starches, and steapsin decomposes fats. Since these enzymes act best in alkaline or neutral solutions, and the contents of the stomach are naturally acid, it is argued by some authorities that artificial pancreatic extracts can possess no medicinal value; but it should be remembered that the contents of the stomach do not normally become acid until from ten to fifteen minutes after the ingestion of the food, that in indigestion the secretion of acid is often delayed, and that at least one of the constituents of the pancreatic secretion—trypsin, a more powerful proteolytic enzyme than pepsin—retains its effectiveness even in solutions that are somewhat acid.

The digestants in common use are:

Hydrochloric acid.	Pancreatin.	Pepsin.
Diastase.	Papayotin.	

ACIDUM HYDROCHLORICUM, U. S. P.
(Hydrochloric or Muriatic Acid; HCl.)

The official acid is a colorless, fuming liquid, of a pungent odor and an intensely acid taste. It contains 31.9 per cent. by weight of absolute hydrochloric acid. The dose is from 3–6 min. (0.2–0.4 c.c.) well diluted.

PREPARATION.	DOSE.
Acidum Hydrochloricum Dilutum, U. S. P., contains 10 per cent. of absolute hydrochloric acid	5–20 min. (0.3–1.2 c.c.).

Toxicology.—In overdoses, hydrochloric acid, like other mineral acids, acts as an irritant poison, causing intense burning pain, vomiting and purging of mucous and bloody material, and collapse. The treatment consists in antagonizing the poison by soda, lime, chalk, soap, or other alkalies; in protecting the stomach by demulcents; in combating collapse with diffusible stimulants, administered hypodermically; and in relieving pain with morphin.

Therapeutics.—Hydrochloric acid is useful in certain conditions of the stomach associated with inacidity or hypo-acidity, such as *chronic asthenic gastritis, atrophy of the gastric glands,* and *cancer of the stomach.* It is contraindicated, even though the normal acid is wanting, when there is any evidence of active inflammation. In the *continued fevers* it sometimes acts beneficially as a refrigerant and digestant, the gastric juice in these cases being generally deficient. .

The strong acid is sometimes employed externally as an *escharotic.*

Administration.—In indigestion the acid should be given after meals, well diluted, care being taken to rinse the mouth thoroughly after its exhibition. It may be prescribed with a bitter tonic, as in the following formula :

℞. Strychninæ sulphatis,	gr. j (0.065 gm.) ;
Acidi hydrochlorici diluti,	f℥ij (8.0 c.c.) ;
Pepsini,	ℨiss (6.0 gm.) ;
Tincturæ cardamomi compositæ,	f℥ss (15.0 c.c.) ;
Aquæ,	q. s. ad f℥iv (120.0 c.c.).—M.

Sig. A teaspoonful in water after meals.

Incompatibles.—Hydrochloric acid is incompatible with oxids, alkalies, carbonates, and hydrates, with many metallic salts, with albumin, and with some glucosids (glycyrrhizin). In the above formula the hydrochloric acid sets free insoluble carminic acid, but its formation is of no moment.

PEPSINUM, U. S. P.
(Pepsin.)

Pepsin is a proteolytic ferment obtained from the glandular layer of fresh stomachs from healthy pigs. To be up to the official standard, it should be capable of digesting, under favorable conditions, not less than 3000 times its own weight of freshly coagulated and disintegrated egg-albumen. It occurs as a yellowish-white, amorphous powder, or as yellow translucent scales, having a faint odor and slightly acidulous or saline taste. It is soluble in about 100 parts of water, more so in water containing hydrochloric acid, and insoluble in alcohol. The dose is from 5–20 gr. (0.3–1.3 gm.).

PREPARATION.	DOSE.
Pepsinum Saccharatum, U. S. P. (10 per cent. pepsin and 90 per cent. sugar of milk)	20–60 gr. (1.3–4.0 gm.).

Therapeutics.—Pepsin is sometimes of service in *chronic gastritis, atrophy of the gastric glands,* and *cancer,* but, as a rule, it is less efficacious than pancreatin, and both are inferior to dilute hydrochloric acid and bitters. Care should be taken to secure a reliable preparation. Saccharated pepsin has very feeble digestive power. Pepsin should be given in full doses, with hydrochloric acid, after meals.

Most substances destroy or impair the proteolytic action of the drug, especially alcohol and alkalies.

PANCREATINUM, U. S. P.
(Pancreatin.)

Pancreatin is a mixture of enzymes naturally existing in the pancreas of warm-blooded animals, and usually obtained from the fresh pancreas of the hog. It appears as a yellowish-white, amorphous powder, having a faint odor and a meat-like taste. The dose is from 5–20 gr. (0.3–1.3 gm.).

In alkaline solution pancreatin has the power of emulsifying fats, and of converting proteids into diffusible peptones, and starches into sugars. In the presence of an acid it soon becomes inert.

Pancreatin is extensively used for peptonizing milk and other foods, the process being as follows: Mix 5 gr. (0.3 gm.) of pancreatin and 20 gr. (1.3 gm.) of sodium bicarbonate in a small teacupful of cool water, and pour into a bottle containing a pint (0.5 L.) of fresh milk. Place the bottle in water so hot that the hand can be held in it without discomfort for a minute. As thoroughly digested milk has an unpleasant bitter taste, it is well to arrest digestion at the end of fifteen or twenty minutes by raising the milk for a few seconds to the boiling-point or by placing the bottle on ice.

Pancreatin is a useful aid to digestion in conditions associated with hypoacidity. It should be combined with sodium bicarbonate and given immediately before or during meals, so that its digestive powers can be exercised in the stomach before the contents become distinctly acid.

. DIASTASE.

Diastase is an amylolytic ferment obtained from malted grain—wheat, barley, oats, and rice. It is a yellowish-white or brownish-yellow amorphous powder, tasteless, soluble in water, but insoluble in alcohol; 1 part, under favorable conditions, should convert 2000 parts of starch into dextrin and maltose. The dose is from 3–5 gr. (0.2–0.3 gm.).

Extracts of malt containing a variable quantity of diastase are extensively employed. They are marketed in 3 forms: A thin liquid closely resembling beer; a thick brown, syrupy liquid, containing much saccharine material; and a dry powder, containing, in addition to diastase, dextrin, dextrose, and the salts of barley. These extracts are used as digestants and as general tonics, but their value has been much overrated.

Chittenden finds that although diastase will act in neutral or alkaline solutions, it is most effective in solutions that are slightly acid. It may be given before meals when there is difficulty in digesting starchy foods.

PAPAYOTIN.

(Papain ; Papoid ; Caroid.)

Papayotin is an albuminous ferment obtained from *Carica Papaya*, true papaw, or melon tree, growing in the tropics. It is a grayish-white amorphous powder, tasteless, and without odor. It is quite soluble in water and glycerin, but insoluble in alcohol and ether. The dose is from 5–10 gr. (0.3–0.6 gm.).

It is claimed for papayotin that it will convert proteids into peptone, starch into maltose, and emulsify fats; and that, although it will act in neutral or acid solutions, it is most active in solutions of an alkaline reaction. Clinical experience with the preparations on the market indicates that the drug is of doubtful value.

12

CARMINATIVES.

Carminatives are drugs that aid in the expulsion of gas from the stomach or intestines. The origin of the word is somewhat obscure; it is probably derived from *carmen*, a card for wool—that is, a cleanser, but according to some authorities it is derived from *carmen*, a charm. The gases found in the alimentary canal are swallowed with the food, are formed by the action of acid upon the carbonates contained in the saliva and food, or they are generated through fermentation or putrefaction of food. The most common cause, however, of abnormal accumulation of gas is fermentation. Excessive accumulations of flatus distend the viscus, excite spasm of its muscular fibers, and thus occasion distress or actual pain. Carminatives prove effective probably by relaxing the spasm, especially at the orifices, thus allowing the gas to escape. A form of distention occurs that is dependent upon atony of the walls of the stomach rather than upon an excessive accumulation of gas. In this condition carminatives are of little use. Although carminatives are chiefly valuable in expelling gas already formed, they also possess the power, to a very limited extent, of preventing the formation of flatus, for by quickening the gastric circulation and probably by increasing glandular activity, they play the part of stomachics, thus aiding digestion and lessening fermentation.

As a class, carminatives are of service also in allaying spasmodic contraction of the intestine caused by irritants other than flatus; thus they are effective in relieving colicky pains caused by irritant food or drugs, provided the irritation has not been sufficient to set up an inflammatory condition, in which case drugs of this class should be avoided. Carminatives are frequently combined with purgatives to prevent the griping pains that the latter are likely to induce when administered alone.

With the exception of alcohol, ether, and chloroform, carminatives are aromatics containing volatile oils as their chief active ingredients. The most important drugs of this class are:

Capsicum.	Ginger.	Fennel.
Pepper.	Cinnamon.	Coriander.
Peppermint.	Cajuput.	Asafetida.
Spearmint.	Anise.	Ether.
Cardamom.	Pimenta.	C o m p o u n d
Cloves.	Sassafras.	spirit of ether
Nutmeg.	Caraway.	Chloroform.
	Turpentine.	

Carminatives are usually administered by the mouth, but asafetida and turpentine are often exceedingly efficacious in removing intestinal flatus when they are administered in the form of enemata.

CAPSICUM, U. S. P.
(Cayenne Pepper; African Pepper.)

Capsicum is the fruit of *Capsicum fastigiatum*, a small shrub growing in tropical America, in Asia, and in Africa. Its active principle is probably *capsaicin*, which appears in the form of colorless crystals of an exceedingly pungent character. The dose of the powdered drug is from 1–3 gr. (0.06–0.2 gm.).

Preparations.	Dose.
Tinctura Capsici, U. S. P.	10–20 min. (0.6–1.2 c.c.)
Extractum Capsici Fluidum, U. S. P.	1–2 min. (0.06–0.12 c.c.).
Oleoresina Capsici, U. S. P.	½–1 min. (0.016–0.06 c.c.).
Emplastrum Capsici, U. S. P.	

Physiologic Action and Therapeutics.—Externally applied, capsicum excites burning and redness, and in concentrated form even vesication. When taken internally in small doses, it produces a sense of warmth in the stomach, stimulates peristalsis, and aids in the expulsion of flatus. Large doses produce severe irritation of the gastro-intestinal and the genito-urinary tract, characterized by pain, vomiting, purging, and dysuria, with scanty, dark-colored urine. It is employed chiefly as a rubefacient, stomachic, and carminative. In the form of a liniment it is sometimes an excellent remedy in *wry-neck, muscular rheumatism*, and *sprains*. Capsicum plaster is an efficient counterirritant in *pleurodynia* and *bronchitis*. Combined with cantharides (tincture of cantharides, tincture of capsicum, and alcohol, of each, ½ fl. oz.—15 c.c.) it is frequently employed as a stimulating lotion in *alopecia areata*. A gargle consisting of half an ounce (15 c.c.) of the tincture in half a pint (235 c.c.) of water has been much used in domestic medicine in *sore throat* with relaxation of the uvula, but the remedy is inferior to many others, and in severe inflammations it is calculated to do harm.

In the *gastric catarrh* following an alcoholic debauch, and characterized by fetid breath, a heavily furred tongue, anorexia, nausea, and a sinking sensation in the epigastrium, the tincture of capsicum in ten-drop doses is an invaluable stomachic. In the *flatulent dyspepsia* of the aged, especially of old gourmands, it is a useful adjuvant to other stomachics. Cayenne pepper in small doses may be employed to promote the absorption of certain drugs, notably of quinin. In *obstinate constipation* due to deficient peristalsis a small amount of the oleoresin may be added advantageously to a cathartic pill.

PIPER, U. S. P.
(Black Pepper.)

Black pepper is the unripe fruit of *Piper nigrum,* a climbing vine cultivated in the East Indies. It contains a crystalline neutral principle, *piperin,* an aromatic volatile oil, and a pungent resin.

PREPARATIONS.	DOSE.
Piperinum, U. S. P.	1–5 gr. (0.06–0.3 gm.).
Oleoresina Piperis, U. S. P.	½–2 min. (0.03–0.1 c.c.).

Therapeutics.—Pepper is used chiefly as a condiment. Owing to its carminative properties, the oleoresin is sometimes added to cathartic pills. Piperin was at one time considered to be of value in malarial fever, but for lack of satisfactory evidence of its efficacy its use in this affection has been entirely abandoned.

MENTHA PIPERITA, U. S. P.
(Peppermint.)

Peppermint is the leaves and tops of *Mentha piperita,* a perennial herb growing in wet places throughout the Temperate Zone. Its active principle is a greenish-yellow volatile oil, having a strong, characteristic odor and a pungent taste, followed by a sense of coolness when air is drawn into the mouth. When subjected to refrigeration, it yields the stearopten, *menthol* (p. 150).

PREPARATIONS.	DOSE.
Oleum Menthæ Piperitæ, U. S. P.	1–5 min. (0.06–0.3 c.c.).
Spiritus Menthæ Piperitæ, U. S. P. (10 per cent.)	10–30 min. (0.6–2.0 c.c.).
Aqua Menthæ Piperitæ, U. S. P.	1–4 fl. dr. (4.0–15.0 c.c.).
Trochisci Menthæ Piperitæ, U. S. P.	

. The oil also enters into the official compound pills of rhubarb.

Therapeutics.—Peppermint has long been employed as a carminative for relieving *flatulence* and *colic,* especially in young children. It is particularly useful in covering the taste of unpalatable medicines. The oil, or better still menthol, is sometimes effective, when applied locally, in relieving the milder forms of *neuralgia*.

MENTHA VIRIDIS, U. S. P.
(Spearmint.)

Spearmint is the leaves and tops of *Mentha viridis*, a perennial herb growing wild in Europe and North America. Its active principle is a volatile oil.

PREPARATIONS.	DOSE.
Oleum Menthæ Viridis, U. S. P.	1–5 min. (0.06–0.3 c.c.).
Spiritus Menthæ Viridis, U. S. P. (10 per cent.)	10–30 min. (0.6–2.0 c.c.).
Aqua Menthæ Viridis, U. S. P.	1–4 fl. dr. (4.0–15.0 c.c.).

Spearmint is the therapeutic equivalent of peppermint.

CARDAMOMUM, U. S. P.
(Cardamom.)

Cardamom is the fruit of *Elettaria repens*, a perennial herb cultivated in the mountainous regions of India. Its active principle is a volatile oil, of which it contains about 5 per cent.

PREPARATIONS.	DOSE.
Tinctura Cardamomi, U. S. P.	1–2 fl. dr. (4.0–8.0 c.c.).
Tinctura Cardamomi Composita, U. S. P. (contains also cinnamon, caraway, cochineal, and glycerin)	1–2 fl. dr. (4.0–8.0 c.c.).
Pulvis Aromaticus, U. S. P. (contains also ginger, cinnamon, and nutmeg)	10–30 gr. (0.6–2.0 gm.).

Cardamom also enters into the compound extract of colocynth, compound tincture of gentian, tincture of rhubarb, and sweet tincture of rhubarb.

Therapeutics.—It is used as an agreeable aromatic for disguising the taste of other drugs, and as an adjuvant to simple bitters in *flatulent dyspepsia*. When free acids are added to the compound tincture of cardamom, they separate from the cochineal insoluble carminic acid, but this incompatibility is of no particular importance.

CARYOPHYLLUS, U. S. P.
(Cloves.)

Cloves are the unexpanded flowers of *Eugenia aromatica,* an evergreen indigenous to the East Indian Islands, and cultivated to some extent in South America and Africa. Their active constituent is a yellow volatile oil of a characteristic odor and a pungent, spicy taste.

PREPARATION.   DOSE.
Oleum Caryophylli, U. S. P. . . . . 1–5 min. (0.06–0.3 c.c.).

They also enter into compound tincture of lavender, aromatic tincture of rhubarb, and wine of opium.

Therapeutics.—Cloves are used as a counterirritant, local anesthetic, and carminative. They are one of the active ingredients of the *spice-poultice*, which consists of powdered cloves, ginger, and cinnamon, of each, one or two teaspoonfuls; flour, a tablespoonful; whisky, sufficient to make a mass moist enough to spread on soft flannel. In this form they are a useful rubefacient for applying to the abdomen of children suffering from diarrhea. As a carminative, the oil is a useful addition to laxative pills, aiding materially in preventing *griping*. In *intestinal colic* brought on by exposure a drop or two of the oil in a teaspoonful of paregoric, repeated at short intervals, often gives speedy relief. The oil is also a popular remedy for *toothache*, being applied on a pledget of cotton to the cavity of the tooth.

MYRISTICA, U. S. P.
(Nutmeg.)

Nutmeg is the seed of *Myristica fragrans*, an evergreen tree growing in the Molucca Islands and neighboring East India Islands. It contains a volatile oil, upon which its aromatic properties depend, and a fixed oil.

PREPARATIONS.   DOSE.
Oleum Myristicæ, U. S. P. . . . . 1–5 min. (0.06–0.3 c.c.).
Spiritus Myristicæ, U. S. P. . . . . ½–1 fl. dr. (2 0–4.0 c.c.).

Nutmeg also enters into aromatic powder, compound tincture of lavender, aromatic tincture of rhubarb, and troches of sodium bicarbonate.

In large doses nutmeg acts as a narcotic poison, producing headache, vertigo, delirium, stupor, coma, and, finally, death from respiratory paralysis. Like cloves, it is employed as a carminative and a local anodyne.

ZINGIBER, U. S. P.
(Ginger.)

The rhizome of *Zingiber officinale*, a perennial herb growing in tropical countries. It contains a volatile oil having the odor of ginger, and a viscid resinous principle having a hot, pungent taste.

PREPARATIONS.	DOSE.
Tinctura Zingiberis, U. S. P.	20–60 min. (1.2–4.0 c.c.).
Extractum Zingiberis Fluidum, U. S. P.	10–30 min. (0.6–2.0 c.c.).
Syrupus Zingiberis, U. S. P.	½–2 fl. dr. (2.0–8.0 c.c.).
Oleoresina Zingiberis, U. S. P.	½–2 min. (0.03–0.1 c.c.).
Trochisci Zingiberis, U. S. P.	

Ginger also enters into compound powder of rhubarb and aromatic powder.

Therapeutics.—It is employed as a carminative and as a flavoring agent.

CINNAMOMUM.

(Cinnamon.)

The United States Pharmacopœia recognizes 3 varieties of cinnamon: *Cinnamomum Cassia* (Cassia cinnamon, or Chinese cinnamon); *Cinnamomum Saigonicum* (Saigon cinnamon), a species cultivated in the neighborhood of Saigon, the capital of French Cochin-China, and *Cinnamomum Zeylanicum* (Ceylon cinnamon). Cinnamon contains a volatile oil and a small amount of tannin. The oil consists chiefly of cinnamic aldehyd, which on exposure is oxidized into resin and *cinnamic acid*.

PREPARATIONS.	DOSE.
Oleum Cinnamomi, U. S. P.	1–5 min. (0.06–0.3 c.c.).
Spiritus Cinnamomi, U. S. P. (10 per cent. of oil)	5–30 min. (0.3–2.0 c.c.).
Tinctura Cinnamomi, U. S. P. (10 per cent. of cinnamon)	½–2 fl. dr. (2.0–8.0 c.c.).
Aqua Cinnamomi, U. S. P.	½–1 fl. oz. (15.0–30.0 c.c.).

Cinnamon also enters into aromatic fluid extract, infusion of digitalis, compound tincture of cardamom, compound tincture of lavender, compound tincture of catechu, wine of opium, and aromatic powder.

Therapeutics.—Cinnamon possesses carminative and feebly astringent properties. The oil is an active antiseptic. Cinnamon-water is an agreeable vehicle for *diarrhea* mixtures. The oil has been used to some extent as an antiseptic in surgical dressings. Very favorable results have been reported from the use of cinnamon in *influenza*.

Cinnamic acid and its sodium salt have been highly extolled by Landerer, Heusser, and others in the treatment of *tuberculosis*. It is claimed that intravenous injections of sodium cinnamate (⅛ gr.—0.008 gm.—every forty-eight hours, gradually increased to ⅓ gr.—0.02 gm.) induce a marked hyperleukocytosis and favor the formation of cicatricial tissue around the tuberculous area. A careful review of the reported cases, however, does not warrant the belief that this method of treatment is any more effective than other methods that are far less painful.

OLEUM CAJUPUTI, U. S. P.
(Oil of Cajuput.)

Oil of cajuput is a thin, bluish-green, volatile oil distilled from the leaves of *Melaleuca Leucadendron,* a small tree growing in the East India Islands. It has a camphoraceous odor and an aromatic, bitter taste. The dose is from 2–10 min. (0.12–0.6 c.c.) in emulsion, in capsules, or on sugar.

Therapeutics.—Cajuput oil is one of the most efficient of the carminatives. It is an excellent adjunct to other remedies in the *flatulent dyspepsia* of the aged. Combined with opium it is undoubtedly useful in *diarrhea* with choleraic symptoms. It has been used also as a rubefacient in muscular rheumatism and as a parasiticide in ringworm.

ANISUM, U. S. P.
(Anise.)

Anise is the fruit of *Pimpinella Anisum,* a small plant cultivated in Southern Europe and North America. It contains a volatile oil having the characteristic odor of anise and a sweetish, aromatic taste.

PREPARATIONS.	DOSE.
Oleum Anisi, U. S. P.	2–5 min. (0.12–0.3 c.c.).
Aqua Anisi, U. S. P.	2–8 fl. dr. (8.0–30.0 c.c.).
Spiritus Anisi, U. S. P. (10 per cent. of oil)	1–2 fl. dr. (4.0–8.0 c.c.).

Anise also enters into sweet tincture of rhubarb, camphorated tincture of opium, compound spirit of orange, elixir of phosphorus, compound syrup of sarsaparilla, and troches of opium and licorice.

Therapeutics.—Although anise is an effective carminative, it is used chiefly as a flavoring agent.

PIMENTA, U. S. P.
(Allspice.)

Allspice is the nearly ripe fruit of *Pimenta officinalis,* an evergreen tree growing in the West Indies and South America. It contains a volatile oil.

PREPARATION.   DOSE.

Oleum Pimentæ, U. S. P. . . . . 1–5 min. (0.06–0.3 c.c.).

Oil of pimenta enters into spirit of myrcia, or bay-rum.

SASSAFRAS.

Sassafras is official as the bark (*sassafras*) and as the pith (*sassafras medulla*) of *Sassafras variifolium*, a shrub or tree growing in Eastern North America. It contains a fragrant, aromatic, volatile oil and a fair amount of tannin (6 per cent.).

PREPARATIONS.   DOSE.

Oleum Sassafras, U. S. P. . . . . 1–5 min. (0.06–0.3 c.c.).
Mucilago Sassafras Medullæ, U. S. P. (2 per
   cent.) . . . . 1–8 fl. dr. (4.0–30.0 c.c.).

Sassafras also enters into compound decoction of sarsaparilla, compound fluid extract of sarsaparilla, and compound syrup of sarsaparilla.

CARUM, U. S. P.

(Caraway.)

Caraway is the fruit of *Carum Carvi*, an herb indigenous to Asia, and cultivated in Europe and North America. Its active ingredient is a volatile oil.

PREPARATION.   DOSE.

Oleum Cari, U. S. P. . . . . 1–5 min. (0.06–0.3 c.c.).

Caraway also enters into compound tincture of cardamom and compound spirit of juniper.

FŒNICULUM, U. S. P.

(Fennel.)

Fennel is the fruit of *Fœniculum capillaceum*, grown chiefly in Southern Europe. It contains an aromatic volatile oil.

PREPARATIONS.   DOSE.

Oleum Fœniculi, U. S. P. . . . . 1–5 min. (0.06–0.3 c.c.).
Aqua Fœniculi, U. S. P. . . . . 1–8 fl. dr. (4.0–30.0 c.c.).

Fennel also enters into compound licorice powder, compound spirit of juniper, and compound infusion of senna.

CORIANDRUM, U. S. P.
(Coriander.)

Coriander is the fruit of *Coriandrum sativum*, an herb grown in all parts of Europe and the United States. It contains a volatile oil.

PREPARATION.	DOSE.
Oleum Coriandri, U. S. P.	1–5 min. (0.06–0.3 c.c.).

Coriander also enters into confection of senna, syrup of senna, and compound spirit of orange.

The carminative properties of *asafetida*, *ether*, *compound spirit of ether*, *chloroform*, and *turpentine* are discussed elsewhere in this volume.

CATHARTICS.

Cathartics are drugs that are employed in medicine to cause evacuation of the bowels. It is conceivable that they may act either by increasing the velocity of the peristaltic movements or by increasing the natural secretions. That many of them increase the rate of the intestinal contractions seems to be well established both by clinical observation and by experimental research. Whether some of them act also by increasing secretion is still an open question. It is true that active purgation is always accompanied by an increased amount of fluid in the fecal discharges; but by some authors this is attributed to the rapid evacuation of the intestinal contents, whereby reabsorption of the natural secretions is materially lessened, or, in cases in which the action of the drug has been more or less violent, to the development of an actual inflammatory process, with the formation of a serous exudation. Those who believe that the natural secretions are not increased base their belief chiefly upon the investigations of Thiry and Radjiejewski, who observed no increase in the secretions when various purgatives were applied to a small section of the intestine that had been isolated without injury to its supplying blood-vessels or nerves. Other pharmacologists, however, who have reinvestigated the subject, notably Moreau, Vulpian, and Brunton, have observed a very decided increase in the intestinal fluids from the application of certain drugs. Although conclusions based on experiments of this character, which modify so seriously natural conditions, must be accepted with considerable reservation, a careful study of all the evidence strongly favors the view that cathartics increase not only the rate of the peristaltic movements, but also the quantity of intestinal secretion.

Although many studies have been undertaken to determine the effect of cathartic drugs upon the biliary secretion, there is very little concerning the matter that can be stated with absolute certainty. It is generally accepted as a fact that certain drugs, like mercury and podophyllin, have the power of stimulating the cells of the liver to secrete a greater amount of bile, and, in consequence, these drugs are known as *cholagogues* or as *cholagogue purgatives.* Not much evidence, however, can be adduced in support of this assumption. It is true that the group of symptoms to which the term "biliousness" is applied often disappears as by magic after the administration of a full dose of calomel, but there is no proof that these symptoms are due to inactivity of the liver; on the contrary, there is much to support the view that they are due solely to some derangement of the stomach and intestines. Nor can the dark-green color of the stools that is observed after the administration of certain cathartics be claimed as an evidence of increased secretion, for it is doubtless due in large part to the rapid flow of bile into the lower bowel and its escape from decomposition and reabsorption. Rutherford ligated the common bile-duct, inserted a cannula into it, and then noted the effect that the injection of certain drugs into the duodenum had upon the flow of bile. From his experiments he concluded—and his conclusions have been generally accepted as trustworthy—that the following drugs cause a decided increase in the secretion of bile:

Podophyllin.	Jalap.
Aloes.	Sodium sulphate.
Rhubarb.	Sodium phosphate.
Colchicum.	Rochelle salt.
Iridin.	Euonymin.
Colocynth.	Ipecac.

Corrosive sublimate.

Contrary to what was expected, calomel, unless it was combined with a minute quantity of corrosive sublimate, had no effect whatever. Unfortunately, Rutherford's experiments were too few in number and his results too often contradictory to justify his conclusions, and, moreover, recent investigations, especially those of Stadelmann and Pfaff, fail completely to confirm the facts embodied in Rutherford's report. It is evident, therefore, that at the present time no positive statement can be made concerning the action of cathartics upon the hepatic secretion.

The colicky pains induced by cathartics are probably the result of scybala interrupting the regular peristaltic movements, and throwing the muscular fibers into violent spasm. This view is confirmed by the experiments of Cash, who found that the slightest weight sufficed to check the onward movement of the substance in the intestine and to set up contractions of a painful character.

Classification.—Cathartics have been variously classified according to the intensity of their action and the character of the stools they produce. Since the effect of any cathartic is dependent, to a great extent, upon the dose, the time of administration, the susceptibility of the patient, and the state of the bowel, it is evident that a classification based upon physiologic action must necessarily be an imperfect one; nevertheless, when its limitations are fully recognized, such a classification is the most convenient one that can be adopted. Cathartics, therefore, may be divided into laxatives, purgatives, drastics, and hydragogues.

Laxatives are the least irritating of all the cathartics, and ordinarily produce stools that are nearly normal in appearance and consistence. Many articles of food, such as molasses, bran bread, figs, prunes, apples, etc., serve as laxatives. The most important drugs belonging to this class are:

Manna.	Cassia fistula.
Tamarind.	Euonymus.
Cascara sagrada.	Butternut.
Frangula.	Iris.
Magnesia.	Leptandra.
Sulphur.	Ox-gall.

Purgatives are more powerful than laxatives, and usually produce one or more copious stools of a semiliquid consistence. The difference between the action of a laxative and that of a purgative is mainly one of degree: laxatives in large doses act as purgatives, and the latter in small doses act as laxatives. The most important purgatives are:

Aloes.	Castor oil.
Rhubarb.	Calomel.
Senna.	Blue mass.

Drastics have a violent action, and in overdoses produce the symptoms of acute enteritis. The most important are:

Croton oil.	Podophyllum.
Colocynth.	Jalap.
Gamboge.	Elaterium.
Scammony.	Bryonia.

Hydragogues produce large watery stools and give rise to but slight irritation. The principal ones are:

Magnesium sulphate.	Potassium and sodium tartrate.
Sodium sulphate.	Sodium phosphate.
Magnesium citrate.	

Some of the drastics, especially jalap, elaterium, and bryonia, in appropriate doses, are efficient hydragogues.

Indications.—*To Relieve Constipation.*—In *simple acute constipation* the prompt administration of a cathartic or the employment of an enema is nearly always advisable. In *chronic constipation the result of atony of the bowel* a resort should first be had to dietetic and hygienic measures, but, these failing, a laxative should be prescribed. As a rule, the use of laxatives in these cases, even when habitual, is less baneful than the persistent constipation. Indeed, mild vegetable cathartics are often taken for indefinite periods without producing harmful effects. Brodie speaks of a man, eighty-six years of age, who for threescore years took an aloetic pill every night.

To Remove Irritants from the Bowel.—In the beginning of *acute diarrhea* it is generally well to administer a dose of castor oil, calomel, or Epsom salts, to rid the bowel of undigested material, ptomains, and micro-organisms. In *poisoning*, if the irritant has escaped from the stomach into the bowel, a purgative should take the place of an emetic.

To Promote Absorption.—Hydragogue cathartics, especially the salines, are often useful in *cardiac* and *renal dropsy*. They not only remove directly from the body a certain amount of fluid, but, by depleting the blood, promote the reabsorption of the lymph that has accumulated in the tissues.

In *serous effusions* of an inflammatory character, such as are met with in *pleurisy* and *pericarditis*, cathartics are rarely serviceable.

Certain drugs, such as digitalis and quinin, may prove ineffective until the bowel has been unloaded and the portal system depleted by the administration of a brisk cathartic.

To Remove Excrementitious Substances from the Blood.—In *uremia* and *puerperal eclampsia* the administration of cathartics is a measure of the greatest value in securing the elimination of noxious material.

To Relieve Cerebral Congestion.—In *acute cerebral congestion* and *mania* cathartics serve a useful purpose, since, by inviting blood to the bowel, they tend to deplete the brain. In *cerebral hemorrhage*, also, they may do good by preventing further extravasation.

Administration.—Cathartics are most commonly administered by the mouth. The salines are given in solution; the vegetable preparations, most conveniently in the form of pills. In chronic constipation a combination of several drugs usually gives rise to less pain and is more effective than a single drug. A little of the extract of belladonna or a drop of one of the aromatic oils is frequently employed as an adjuvant to prevent griping. When atony of the bowel is pronounced, extract of nux vomica or physostigma may be added to the pill to enhance its stimulating effect. The time of administration is a factor to be considered in prescribing any cathartic. Pills of the vegetable cathartics are likely to cause less inconvenience if administered after the evening meal or on going to bed. Salines or saline mineral waters act more promptly and powerfully when given before breakfast.

Very often an enema is the most satisfactory means of unloading the bowel. A metal syringe with a piston or one made entirely of soft rubber may be employed for the purpose. The syringe should be filled with the fluid; all the air should be expelled from it before the nozzle is inserted. The patient should lie near the edge of the bed, on the left side, with the knees drawn up. A towel should be held against the anus on withdrawing the nozzle, so that the fluid may be retained in the bowel for several minutes. An enema may act by distending the bowel (mechanically), by softening the intestinal contents, or by directly irritating the intestinal walls.

A simple enema is one composed of cool or tepid water or of soap and water. For an infant, 1 or 2 ounces (30.0–60.0 c.c.) may be employed; for a child, from 4 to 8 ounces (120.0–240.0 c.c.); and for an adult, from 1 to 2 pints (0.5–1.0 L.). The injection may be made more powerful by adding, in the case of an adult, 1 ounce (30.0 c.c.) of castor oil, $\frac{1}{2}$ ounce (15.0 c.c.) of molasses, or from 1 to 2 drams (4.0–8.0 c.c.) of turpentine.

The following enema, official in the British Pharmacopeia, is very useful :

Magnesium sulphate . . . . 1 ounce (30.0 c.c.).
Olive oil . . . . 1 ounce (30.0 c.c.).
Mucilage of starch . . . . 15 ounces (445.0 c.c.).

A small enema of glycerin (1–2 drams—4.0–8.0 c.c.) is generally very efficacious. In cases of fecal impaction an enema of warm salad or linseed oil (6–8 ounces—180.0–240.0 c.c.) or an infusion of ox-gall often proves serviceable. In some cases of chronic indigestion with constipation flushing out the lower bowel with from 6 to 8 pints (3.0–4.0 L.) of tepid water, two or three times a week, by means of a soft-rubber tube passed well up into the sigmoid flexure, is followed by excellent results.

It is well to remember that bulky enemata containing hard soap are occasionally followed, after the lapse of a few hours, by a scarlatiniform, measly, or urticarial rash.

The introduction into the rectum of suppositories made of yellow soap, glycerin, or gluten is another means frequently employed to unload the bowel.

It has been shown that certain drugs—aloin, colocynthin, and cathartic acid—will produce catharsis when given hypodermically. The irritation caused by the injections of these substances, however, is so severe as to forbid the employment of this method of administration except under unusual circumstances.

MANNA, U. S. P.

Manna is a concrete saccharine exudation from *Fraxinus Ornus*, a small tree a native of Sicily and other Mediterranean islands. It occurs in the form of yellowish-white or brownish-white three-edged pieces or fragments, having a crystalline structure, a honey-like odor, and a sweetish, faintly acrid taste. Its chief constituent is *mannite* (50–80 per cent.), a sweet crystalline principle soluble in water. In doses of from 1 to 2 ounces (30.0–60.0 gm.) manna acts as a mild laxative. It is usually given in combination with other cathartics.

PREPARATION.   DOSE.

Infusum Sennæ Compositum, U. S. P. (Black
    Draft) . . . . 1–3 fl. oz. (30.0–90.0 c.c.).

TAMARINDUS, U. S. P.
(Tamarind.)

Tamarind is the preserved pulp of the fruit of *Tamarindus indica*, a large tree indigenous to Africa, and cultivated in

the West Indies.   It is a gentle laxative, about equal in power
to the fig and prune.   Its aperient properties are due chiefly
to the potassium salts of tartaric, citric, malic, and acetic acids,
of which it contains from 8 to 12 per cent.   The dose is from
1 to 8 drams (4.0–30.0 gm.).

PREPARATION.	DOSE.
Confectio Senna, U. S. P.	1–2 dr. (4.0–8.0 gm.).

RHAMNUS PURSHIANA, U. S. P.
(Cascara Sagrada; Chittem Bark.)

Cascara sagrada is the bark of *Rhamnus Purshiana*, a small
tree growing in Northern California, Oregon, and Washington.
It is allied to *Rhamnus Frangula* (buckthorn) and to *Rhamnus cathartica* (buckthorn).   It contains *purshianin*, a crystalline glucosid that is four or five times as active as the crude
drug.

PREPARATIONS.	DOSE.
Extractum Rhamni Purshianæ Fluidum, U. S. P.	10–30 min. (0.6–2.0 c.c.).
Extractum Rhamni Purshianæ	2–8 gr. (1.3–0.5 gm.).

Therapeutics.—Cascara sagrada is used exclusively as a
tonic laxative, and in *habitual constipation* due to torpor of the
bowel it is a most reliable remedy.   It possesses a great
advantage over many cathartics in not readily losing its effect
when frequently taken; indeed, in many cases the dose can
be diminished gradually and still give satisfactory results.

In most cases a single dose at bedtime will afford relief, but
sometimes small doses (10 drops) after each meal are more
effective.   On account of its unpleasant bitter taste the fluid
extract is best given in some aromatic syrup or cordial, as in
the following formula :

℞   Extracti rhamni purshianæ fluidi,
   Extracti sarsaparillæ fluidi compositi,
   Glycerini,   aa f℥j (30.0 c.c.).—M.
Dose : A teaspoonful or more.

FRANGULA, U. S. P.
(Buckthorn.)

Frangula is the bark of *Rhamnus Frangula*, a shrub growing in Europe and Northern Asia.   When fresh, it is a powerful irritant, but its action is modified by age, and therefore the
Pharmacopeia specifies that it should be at least one year old

before being used. It contains a crystalline glucosid, *frangulin.*

PREPARATION.   DOSE.
Extractum Frangulæ Fluidum, U. S. P. . . . ½–1 fl. dr. (2.0–4.0 c.c.).

Frangula has an action not unlike that of senna. In this country it has been almost entirely displaced by *Rhamnus Purshiana.*

MAGNESIA, U. S. P.

(Light Magnesia; Calcined Magnesia; Magnesium Oxid; MgO.)

Light magnesia is a. white, very light, odorless powder, having an earthy taste. It is almost insoluble in water, insoluble in alcohol, but soluble in dilute acids. The dose as a laxative is from 30–60 gr. (2.0–4.0 gm.).

PREPARATIONS.   DOSE.
Pulvis Rhei Compositus, U. S. P. (65 per cent.) . . . . 30–60 gr. (2.0–4.0 gm.).
Ferri Oxidum Hydratum cum Magnesia, U. S. P. (1.25 per cent.) . . . . 2–4 fl. oz. (60.0–120.0 c.c.).
Massa Copaibæ, U. S. P. (6 per cent.) . . . 10–60 gr. (0.65–4.0 gm.).

Therapeutics.—Magnesia is partly converted by the acids of the stomach into soluble salts that have a laxative action; it therefore combines the properties of an antacid with those of a mild aperient. It is not a suitable drug for habitual use in simple chronic constipation, since, when not thoroughly acidified, there is a possibility of its forming intestinal concretions. It may replace sodium bicarbonate, however, when *excessive acidity exists with constipation.* In *ulcer of the stomach* it may be combined with bismuth subnitrate to control constipation. It may be given with syrup of rhubarb in the early stages of *acute diarrhea* to rid the bowel of any sour, acrid material that may give rise to irritation. It is also an antidote in *arsenic poisoning*, and is doubly efficient in the form of ferric hydrate with magnesia.

MAGNESIA PONDEROSA, U. S. P.

(Heavy Magnesia; Heavy Calcined Magnesia; Magnesium Oxid; MgO.)

Heavy magnesia is three and a half times heavier than light magnesia, and, unlike the latter, does not readily unite with water to form a gelatinous hydrate. It has the same properties and uses as light magnesia.

13

MAGNESII CARBONAS, U. S. P.

(Magnesium Carbonate; $(MgCO_3)_4.Mg(OH)_2+5H_2O$.)

Magnesium carbonate occurs in light, white, friable masses or as a light white powder, free from odor, and having a slightly earthy taste. It is almost insoluble in water, insoluble in alcohol, but soluble in dilute acids with active effervescence. The dose as a laxative is from 20–60 gr. (1.3–4.0 gm.).

Magnesium carbonate has the same therapeutic value as magnesia, but when there is much acid in the stomach, it is likely to cause unpleasant eructations of gas.

SULPHUR.

(S.)

The United States Pharmacopeia recognizes 3 forms of sulphur:

> Sulphur Sublimatum.
> Sulphur Lotum.
> Sulphur Præcipitatum.

Sublimed sulphur (flowers of sulphur) is a fine yellow powder having a slightly characteristic odor and a faintly acid taste. It is insoluble in water, but partially soluble in absolute alcohol, ether, chloroform, boiling solutions of alkaline hydrates, oil of turpentine, and many other oils. It usually contains small quantities of sulphuric acid, arsenic sulphid, and other impurities. The dose is from 10 to 120 gr. (0.65–8.0 gm.).

Washed sulphur is prepared by digesting for three days sublimed sulphur in weak ammonia water.

There are 2 official preparations of washed sulphur:

> Unguentum Sulphuris (30 per cent.).
> Pulvis Glycyrrhizæ Compositus (8 per cent.).

Precipitated sulphur (milk of sulphur) is a fine, almost white powder, without odor or taste. It is prepared by precipitating calcium sulphid with diluted hydrochloric acid. The dose is from 10 to 120 gr. (0.65–8.0 gm.).

Physiologic Action and Therapeutics.—When sulphur is taken internally, a considerable portion passes out unchanged; some, however, is converted by the alkaline secretions of the bowel into hydrogen sulphid and other sulphids. It is to the latter compounds that sulphur owes its active properties. They are the cause of the mild catharsis and the offensive flatus that follow the ingestion of sulphur. Being partially absorbed, they escape from the lungs and skin as sulphids, and from the kidneys as sulphates.

Sulphur may be employed as a laxative when a very mild action is desirable, as in *pregnancy, hemorrhoids, fissure of the anus*, and *prolapse of the bowel*. The dose as a laxative is from 1 to 2 dr. (4.0–8.0 gm.) in syrup or molasses. Sulphur has a long-standing reputation, both as an internal remedy and as a local application, in *chronic articular rheumatism*. Externally, it is much used as a stimulant and parasiticide in certain chronic diseases of the skin. It is often useful in *eczema, comedo, acne*, and *psoriasis*, and it may be prescribed in an ointment having the strength of from ½ to 2 dr. (2.0–8.0 gm.) to the ounce (30.0 gm.). It is absolutely contraindicated in the acute stages of any skin disease. In parasitic diseases, like *scabies* and *ringworm*, it is a reliable remedy.

CASSIA FISTULA, U. S. P.
(Purging Cassia.)

Purging cassia is the fruit of *Cassia Fistula*, a tree extensively cultivated in all tropical countries. It contains about 60 per cent. of sugar, with albuminoids and organic salts. The dose is from 1–4 dr. (4.0–15.0 gm.). It has a mild laxative action, but on account of its tendency to cause griping, it is prescribed only in combination with other remedies.

PREPARATION.	DOSE.
Confectio Sennæ, U. S. P. (16 per cent.) . .	1–2 dr. (4.0–8.0 gm.).

EUONYMUS, U. S. P.
(Wahoo.)

Euonymus is the bark of the root of *Euonymus atropurpurens*, growing in Eastern North America. It contains a soluble, amorphous glucosid, *euonymin*. The dose of the latter is from ½–3 gr. (0.03–0.2 gm.).

PREPARATION.	DOSE.
Extractum Euonymi, U. S. P.	1–5 gr. (0.065–0.3 gm.).

The action of euonymus resembles that of podophyllum, although it is less powerful. It has been credited with cholagogue as well as laxative properties, but although it undoubtedly hastens the elimination of bile and prevents the reabsorption of that secretion, there is not sufficient proof to warrant the belief that the drug actually increases the functional activity of the liver.

Combined with cathartics, euonymus is a useful remedy in the condition popularly known as "biliousness."

JUGLANS, U. S. P.
(Butternut.)

Butternut is the bark of the root of *Juglans cinerea*, a large tree belonging to the walnut family, and growing in North America.

PREPARATION.	DOSE.
Extractum Juglandis, U. S. P.	5–10 gr. (0.3–0.65 gm.).

Its action somewhat resembles that of rhubarb, for which it has been used as a substitute. It is seldom employed at the present day.

IRIS, U. S. P.
(Blue Flag.)

Blue flag is the rhizome and roots of *Iris versicolor*, a perennial herb growing in the swampy places of North America. It contains an acrid resin and possibly an amorphous alkaloid.

PREPARATIONS.	DOSE.
Extractum Iridis Fluidum, U. S. P.	10–20 min. (0.6–1.3 c.c.).
Extractum Iridis, U. S. P.	1–4 gr. (0.06–0.25 gm.).

Iris in sufficient dose produces active catharsis with bilious stools. Its action resembles that of podophyllum, though milder. At one time held in high esteem, it has at present fallen into disuse.

LEPTANDRA, U. S. P.
(Culver's Root.)

The rhizome and roots of *Veronica virginica*, a perennial herb growing in the eastern portion of the United States. It contains a crystalline glucosid, *leptandrin*, which is to be distinguished from an impure resin also known as leptandrin.

PREPARATIONS.	DOSE.
Extractum Leptandræ Fluidum, U. S. P.	10–60 min. (0.6–4.0 c.c.).
Extractum Leptandræ, U. S. P.	1–5 gr. (0.06–0.3 gm.).
Pilulæ Catharticæ Vegetabiles, U. S. P.	1–3 pills.

In the fresh state leptandra is a violent gastro-intestinal irritant. Preparations made of the dried root have a laxative or purgative effect, according to the dose, but they are somewhat uncertain in their action. Leptandra has been largely superseded by more reliable remedies.

FEL BOVIS, U. S. P.
(Oxgall.)

Purified oxgall (*Fel Bovis Purificatum,* U. S. P.) only is used in medicine. It is prepared by evaporating to a pilular consistence fresh oxgall that has been subjected for several days to the action of alcohol. It is a yellowish-green, soft solid, having a peculiar odor and a partly sweet and partly bitter taste. The dose is from 5–15 gr. (0.3–1.0 gm.).

Physiologic Action and Therapeutics.—Bile is an active cholagogue and an uncertain laxative. That it is a true hepatic stimulant, increasing both the liquid and the solids secreted by the liver, has been amply confirmed by Rosenberg, Stadelmann, Joslin, and other investigators who have experimented on both animals and human beings with permanent biliary fistulæ. When bile is not present in the intestines, the digestion of fat is materially aided by the administration of oxgall by the mouth. In the test-tube bile suspends peptic digestion by precipitating the soluble proteids (proteoses and peptones) and probably also pepsin; but in the living subject, when the fat is excessive in the stools, it has been shown that it promotes the digestion of proteid food.

Careful bacteriologic studies do not indicate that bile possesses the pronounced antiseptic properties that are generally attributed to it. It is probable that the fetid character acquired by the stools in the absence of bile is due to impaired intestinal digestion, and, therefore, to better opportunities for the action of bacteria, rather than to the withdrawal of any antiseptic influence exerted directly by the bile itself.

Bile is too uncertain in its action to be employed as an ordinary laxative. In *fecal impaction,* however, a solution of oxgall, containing an ounce to a pint of water (30.0 gm.–0.5 L.), often makes a very serviceable enema. When, for any reason, the biliary secretion is lacking in the intestine, oxgall may be used as an *adjuvant to certain cathartics*—podophyllum, jalap, rhubarb, and scammony—that are ordinarily inactive when bile is not present.

In cases of *biliary fistula* or of *obstruction of the common duct* it may be used to promote the digestion of fats and proteids, and to prevent, indirectly, putrefactive changes in the intestinal contents. It is best administered in capsules, about two hours after meals.

ALOE.

(Aloes.)

Aloes is the inspissated juice prepared from the leaves of several species of the genus *Aloe,* a familiar example of which is the American century plant. It thrives in nearly all hot, dry countries. The United States Pharmacopeia recognizes 2 varieties: Barbadoes aloes (*Aloe barbadensis*), obtained from *Aloe vera,* and Socotrine aloes (*Aloe Socotrina*), obtained from *Aloe Perryi.* The latter is generally mixed with a large amount of Curaçoa aloes, a cheaper but equally active variety. Aloes contains a neutral crystalline principle, *aloin,* a resin, and a trace of volatile oil. The purgative principle is probably *emodin,* which is obtained from aloin, and which is found also in senna and rhubarb. The dose of aloes is from 1–10 gr. (0.065–0.65 gm.).

PREPARATIONS.	DOSE.
Aloinum, U. S. P.	$\frac{1}{4}$–2 gr. (0.016–0.13 gm.).
Aloe Purificatum, U. S. P.	1–10 gr. (0.065–0.65 gm.).
Tinctura Aloes, U. S. P.	$\frac{1}{2}$–2 fl. dr. (2.0–8.0 c.c.).
Tinctura Aloes et Myrrhæ, U. S. P. (10 per cent. of each, with glycyrrhizæ)	$\frac{1}{2}$–2 fl. dr. (2.0–8.0 c.c.).
Pilulæ Aloes, U. S. P. (about 2 gr.—0.13 gm.)	1–5 pills.
Pilulæ Aloes et Asafœtidæ, U. S. P. (about $1\frac{1}{3}$ gr.—0.08 gm.—of each)	1–5 pills.
Pilulæ Aloes et Ferri, U. S. P. (about 1 gr.—0.065 gm.—each of aloes and ferrous sulphate)	1–3 pills.
Pilulæ Aloes et Mastiches, U. S. P. (Lady Webster's pills; about 2 gr.—0.13 gm.—of aloes, with mastich and red rose)	1–5 pills.
Pilulæ Aloes et Myrrhæ (2 gr.—0.13 gm.)	1–4 pills.
Pilulæ Rhei Compositæ ($1\frac{1}{2}$ gr.—0.1 gm.—each)	1–3 pills.

Aloes also enters into compound extract of colocynth (50 per cent.), compound tincture of benzoin (2 per cent.), compound cathartic pills, and vegetable cathartic pills.

Physiologic Action and Therapeutics.—Aloes is a slowly acting but brisk purgative. A dose of 2 or 3 grains ordinarily produces one or two copious, dark-brown stools in the course of ten or twelve hours. The evacuation is generally attended with more or less griping pain. Large doses congest the pelvic viscera and irritate the rectum. It has been shown that a part of that which is ingested escapes from the body in the urine, and, in the case of nursing women, in the milk. Aloes also causes catharsis when introduced into the system through channels other than the mouth, such as the skin and the rectum. When injected into the bowel, however, it is active only when combined with a solvent, like oxgall or glycerin. Aloin is slower in its action and somewhat less certain in its effects than aloes. Aloes is rarely used singly as a cathartic, but in *simple, persistent constipation* it is very efficacious in combination with other remedies, particularly with nux vomica, belladonna, ipecac, rhubarb, or podophyllum. *Chlorosis* with constipation often yields to pills of aloes and iron. Contrary to what was formerly believed, the use of aloes is not contraindicated by the existence of hemorrhoids unless inflammation has developed; indeed, the drug often benefits *indolent piles* by overcoming the sluggishness of the bowel that led to their development. Since it congests the pelvic organs, aloes is sometimes serviceable as an emmenagogue in *amenorrhea* associated with troublesome constipation, and that is not dependent upon inflammation of the uterus or adnexa.

When dysentery, cystitis, or inflammation of any pelvic organ is present, aloes should be avoided. On account of the possibility of its causing abortion it is best not to use it during pregnancy.

Aloes is usually administered in the form of pills. The liquid preparations, on account of their disagreeable taste, are rarely employed. Meyer has shown that the action of the drug is increased by the addition of potassium carbonate or ferrous sulphate.

RHEUM, U. S. P.
(Rhubarb.)

Rhubarb is the root of *Rheum officinale*, a perennial herb resembling garden rhubarb, but of larger growth, and a native of China, Thibet, and other Asiatic countries. It contains *chrysophanic acid, emodin, tannic acid*, and several principles of a somewhat resinous character. Chrysophanic acid is not a purgative, but emodin is; it is doubtful, however, whether the latter represents all the active properties of the plant. The dose of rhubarb is from 5–30 gr. (0.3–2.0 gm.).

PREPARATIONS.	DOSE.
Tinctura Rhei, U. S. P. (10 per cent., with cardamom)	1–2 fl. dr. (4.0–8.0 c.c.).
Tinctura Rhei Aromatica, U. S. P. (20 per cent., with aromatics)	$\frac{1}{2}$–1 fl. dr. (2.0–4.0 c.c.).
Tinctura Rhei Dulcis, U. S. P. (10 per cent., with glycyrrhizæ, anise, and cardamom)	1–2 fl. dr. (4.0–8.0 c.c.).
Extractum Rhei Fluidum, U. S. P.	10–30 min. (0.6–2.0 c.c.).
Extractum Rhei, U. S. P.	5–10 gr. (0.3–0.65 gm.).
Mistura Rhei et Sodæ, U. S. P. (15 per cent., with sodium bicarbonate, fluid extract of ipecac, and spirits of peppermint)	$\frac{1}{2}$–2 fl. dr. (2.0–4.0 c.c.).
Syrupus Rhei, U. S. P. (10 per cent. of fluid extract)	1–6 fl. dr. (4.0–22.5 c.c.).

<table>
<tr><td>PREPARATIONS.</td><td>DOSE.</td></tr>
</table>

Syrupus Rhei Aromaticus, U. S. P. (spiced
syrup of rhubarb: 15 per cent. of aromatic
tincture) . . . . 1–6 fl. dr. (4.0–22.5 c.c.).
Pulvis Rhei Compositus, U. S. P. (Gregory's
powder: rhubarb, 25; magnesia, 65; gin-
ger, 10) . . . . 20–60 gr. (1.3–4.0 gm.).
Pilulæ Rhei, U. S. P. (3 gr.—0.2 gm.) . . . 1–5 pills.
Pilulæ Rhei Compositæ, U. S. P. (rhubarb, 2
gr.—0.13 gm.; aloes, 1½ gr.—0.1 gm.;
myrrh; and oil of peppermint) . . . . . 1–5 pills.

Physiologic Action and Therapeutics.

In appropriate doses rhubarb acts as a purgative and stomachic. It affects the bowel more promptly than aloes, a full dose of from 20–30 gr. (1.3–2.0 gm.) usually producing evacuation in from four to eight hours. On account of the tannic acid that it contains it frequently causes constipation as a secondary effect. Doses of from 1–3 gr. (0.065–0.2 gm.) often exert no cathartic influence, but act upon the stomach as a tonic and mild astringent. It is eliminated in the various secretions—urine, milk, and sweat—and imparts to them a yellowish color.

Rhubarb is an excellent remedy for removing irritant material from the bowel in the beginning of *acute diarrhea*. In the *dyspeptic diarrhea of childhood* the withdrawal of all food for several hours and the administration of a few doses of the aromatic syrup with magnesia generally affords prompt relief:

R   Magnesiæ,   gr. xl (2.6 gm.);
    Syrupi rhei aromatici,   f℥vj (22.5 c.c.);
    Aquæ menthæ piperitæ,   q. s. ad f℥j (30.0 c.c.).—M.
Sig. A teaspoonful repeated once or twice for a child two years old.

Rhubarb alone is not a suitable remedy for *chronic constipation*, but benefit is often derived from a combination of small doses with other cathartics. *Mild attacks of indigestion* the result of overindulgence, frequently yield to the union of rhubarb with an antacid:

R   Pulveris nucis vomicæ,   gr. xii (o 8 gm.);
    Pulveris thei,   gr. xxiv (1.5 gm.)
    Sodii bicarbonatis,   ℨj (4.0 gm.).—M.
    Fiant chartulæ No. xii.
Sig. One before meals.

SENNA, U. S. P.

Senna is the leaflets of *Cassia acutifolia* and of *Cassia angustifolia*, small shrubs growing respectively in Africa and India. Its active principle is an acid glucosid known as *cathartinic acid* or *cathartin*. It also contains chrysophanic acid and one or two bitter principles. The dose is from 1–3 dr. (4.0–12.0 gm.).

PREPARATIONS.	DOSE.
Extractum Sennæ Fluidum, U. S. P.	1–2 fl. dr. (4.0–8.0 c.c.).
Syrupus Sennæ, U. S. P. (25 per cent.)	1–4 fl. dr. (4.0–16.0 c.c).
Infusum Sennæ Compositum, U. S. P. (Black Draft: senna, 6; fennel, 2; manna, 12; Epsom salts, 12)	1–4 fl. oz. (30.0–120.0 c.c.).
Confectio Sennæ, U. S. P. (10 per cent., with cassia fistula, tamarind, prune, fig, sugar, and coriander oil)	1–2 dr. (4.0-8.0 gm.).
Pulvis Glycyrrhizæ Compositus, U. S. P. (18 per cent. of senna, with licorice, sulphur, sugar, and fennel oil)	½–2 dr. (2.0–8.0 gm.).

Senna (1.5 per cent.) also enters into compound syrup of sarsaparilla.

Physiologic Action and Therapeutics.—Senna is an energetic purgative. A full dose produces in from four to six hours one or two yellowish, loose, or even watery, stools, the evacuation being attended with considerable griping and flatulence. Its action is more irritating than that of rhubarb, and more prompt and powerful than that of aloes. Very large doses cause nausea, vomiting, violent purging, and depression. Cathartinic acid also acts when injected subcutaneously. Like rhubarb, senna imparts to the urine a deep yellow or red color, the pigmentation being due to the presence of chrysophanic acid. Senna is a safe and reliable purgative for unloading the bowel in *simple, acute constipation*. Its tendency to cause griping is in a measure corrected by giving it with an aromatic or a saline. Compound licorice powder, in doses of a teaspoonful, more or less, at bedtime, is a popular household remedy in *habitual costiveness*.

OLEUM RICINI, U. S. P.
(Castor Oil.)

Castor oil is a fixed oil expressed from the seed of *Ricinus communis*, a plant indigenous to India, but extensively cultivated in other countries having a warm or temperate climate. It is a pale-yellowish, viscid oil, having a faint odor and a slightly acrid, offensive taste. It is freely soluble in alcohol. It is composed chiefly of *ricinolein*, the glycerid of ricinoleic acid, which is the purgative principle. The seeds themselves are never used for medicinal purposes. They contain *ricin*, an intensely poisonous proteid body, which, when injected into the blood of an animal, causes, after the lapse of several days, anorexia, vomiting, diarrhea, and profound prostration. In fatal poisoning the chief postmortem lesions are extensive ecchymoses of the mucous and serous membranes, tumefaction of the abdominal lymph-glands, and numerous areas of necrosis in the various organs. The dose of castor oil for an infant is from 1–2 dr. (4.0–8.0 c.c.); for an adult, $\frac{1}{2}$–1 ounce (15.0–30.0 c.c.).

Physiologic Action and Therapeutics.—Castor oil is a mild purgative, unloading the bowels thoroughly in from four to six hours, without causing much colic or flatulence. While its fate in the body has not been definitely determined, it is probable that it escapes from the stomach unchanged, and that in the presence of the intestinal juices saponification occurs with the liberation of ricinoleic acid, which is subsequently converted into ricinoleates. The latter induce catharsis by stimulating the muscular coat of the bowel, and are probably absorbed, since the oil is known to impart its purgative properties to the milk when given to nursing women.

Castor oil is not a suitable remedy for habitual constipation, but on account of its mild and speedy action it is perhaps the best remedy we possess to remove irritant material from the bowel in the beginning of *acute inflammatory diarrhea.* It is also an excellent laxative for use during *pregnancy* or *labor.*

When it is desirable to add an oil to a lotion intended for the scalp, castor oil is generally selected on account of its solubility in alcohol.

The leaves of the castor oil plant are said to promote the flow of milk when applied to the breasts.

Many substances have been suggested to disguise the disagreeable taste of castor oil, which is the main objection to its employment; those in common use are the oils of peppermint, gaultheria, and cinnamon. Wood speaks favorably of a combination of equal parts of glycerin and castor oil flavored with a drop or two of one of the above-named oils. It is often given in the froth of porter, beer, or soda-water. It may be prescribed in emulsion, but in this form it is bulky and not so effective. The best method of administering it is in flexible capsules containing a dram (30.0 c.c.) or more.

HYDRARGYRI CHLORIDUM MITE, U. S. P.

(Mild Mercurous Chlorid; Calomel; Hg_2Cl_2.)

Calomel is a white, odorless, tasteless powder, insoluble in all ordinary menstrua. The dose is from $\frac{1}{10}$–10 gr. (0.0065–0.65 gm.).

PREPARATIONS.   DOSE.

Pilulæ Antimonii Compositæ, U. S. P. (Plummer's Pills: ⅓ gr.—0.04 gm.—of calomel and sulphurated antimony with guaiac and castor oil) . . . . 1–2 pills.

Pilulæ Catharticæ Compositæ, U. S. P. (1 gr.—0.06 gm.— of calomel, with gamboge, compound extract of colocynth, and extract of jalap) . . . . 1–3 pills.

Calomel acts as a laxative or as a purgative according to the method of its administration and the susceptibility of the patient. Unlike other cathartics, its effect does not increase in direct ratio with the dose. Fractional doses repeated every half hour until 1 or 2 gr. (0.065–0.13 gm.) have been taken generally operate more freely than a dose of 10 gr. (0.65 gm.) taken at once. Calomel produces a thorough evacuation, the stools being large and loose, and commonly charged with undecomposed bile. Its action is not usually accompanied by much griping, but in some persons the drug excites more or less nausea. The change that calomel undergoes in the body is still a matter of dispute. It was formerly supposed that at least a portion was converted into mercuric chlorid by the hydrochloric acid of the stomach, but the experiments of Bucheim, Winkler, and Jeannel prove that such a conversion is impossible at the temperature of the body. It is more probable, as Jeannel suggests, that in the presence of the alkaline juices of the intestines the drug is changed into a gray oxid, a compound that is freely soluble in oily alkaline mixtures, such as are naturally present in the duodenum. Whether or not calomel directly increases the secretion of bile is another question that has evoked considerable discussion, and to which no definite answer can be given at the present time. There can be no doubt that the amount of bile discharged in the stools is actually increased under its influence, but the evidence, experimental and clinical, does not favor the view that calomel increases the quantity of bile formed by the liver; on the contrary, it suggests that the drug, through its stimulant action on the duodenum, merely speeds the flow of bile through the intestines and prevents its reabsorption.

No remedy is so useful as calomel in the condition known as "*biliousness*," which is characterized by a thickly coated tongue, fetid breath, heavy urine, headache, and depression of spirits. A sixth of a grain (0.01 gm.) may be given every fifteen or twenty minutes until 1 gr. (0.065 gm.) has been taken. If the bowels do not move freely, the mercurial may be followed by Epsom salts or a Seidlitz powder. Attacks of *indigestion associated with clay-colored stools*, in either children or adults, are often relieved by a few small doses

of calomel.   In *dyspeptic diarrhea* it is not quite so effica-
cions as castor oil in removing undigested food from the
bowel, but it has advantages in its tastelessness and small
bulk.   In the beginning of *acute disease*, especially the specific
infections, it is an excellent cathartic for unloading the bowels
without exciting irritation.   In *chronic heart disease* with
venous engorgement of the digestive organs calomel is of the
utmost value in relieving the tension in the portal circulation,
and without its aid digitalis often proves ineffectual.

Massa Hydrargyri, U. S. P. (Blue Mass) is made by
triturating mercury, 33; glycyrrhize, 5; althæa, 25; glycerin,
3; and honey of rose, 34.   It is employed for the same purposes as calomel, although it is less active.   The dose is from
$\frac{1}{4}$–15 gr. (0.016–1.0 gm.).

Hydrargyrum cum Creta, U. S. P. (mercury with chalk;
gray powder).—This is an intimate mixture of mercury, honey,
and chalk, containing 38 per cent. of mercury.   It is weaker
than blue mass.   The dose is from 1–10 gr. (0.065–0.65 gm.).

OLEUM TIGLII, U. S. P.
(Croton Oil.)

Croton oil is a fixed oil expressed from the seeds of *Croton
Tiglium*, a small tree, indigenous to China, but extensively cultivated in India and the Philippine Islands.   It is a yellowish,
viscid oil, having a faint odor and an acrid, burning taste.   In
addition to several inactive fatty acids it contains *crotonoleic
acid*, which is present both as a free acid and as a glycerid.
The dose of the oil is from $\frac{1}{2}$–2 min. (0.03–0.12 c.c.).

Physiologic Action and Therapeutics.—When applied
to the skin, croton oil causes redness and burning, followed by
a copious eruption of pustules.   When taken internally, it acts
as a violent drastic cathartic, causing in from one to two hours
several copious movements, which are partly formed and
partly watery.   The evacuations are usually attended with
considerable pain, and not infrequently with nausea.   Large
doses produce all the symptoms of a severe gastro-enteritis.
As croton oil contains a certain amount of free crotonoleic
acid, it causes, when swallowed without a demulcent, a burning sensation in the fauces and stomach, but its maximum irritant effect is not manifested until it reaches the intestine, where
the bulk of the acid is liberated from the oil by saponification.

On account of the ease with which it can be administered
croton oil is a well-adapted cathartic for cases in which deglutition is seriously affected.   Its prompt irritant action on the

bowel makes it also a useful revulsant in cases of cerebral congestion. Thus, it may be given with advantage in *uremia*, *apoplexy*, and *acute mania.*

It is also useful in very *obstinate constipation*, such as occurs in chronic lead-poisoning. As it acts promptly when simply placed on the back of the tongue, it is often selected as a cathartic for the insane. Its use is contraindicated whenever there is inflammation of the gastro-intestinal tract.

The drug is best administered in a little olive oil or in a pill with bread-crumb as an excipient.

Croton oil is sometimes used externally as a counterirritant. It is rarely applied to the skin in the pure state, a mixture with from 2 to 4 parts of some indifferent oil generally being preferred. A liniment is official in the British Pharmacopœia that is made of 2 parts of the oil with 7 parts each of oil of cajuput and alcohol. This liniment may be applied to the chest in *chronic bronchitis, phthisis,* or *fibrinous pleurisy.*

A mixture of croton oil (1 part) and tincture of iodin (2 parts) employed as a pigment often acts very well in *neuritis.*

COLOCYNTHIS, U. S. P.

(Colocynth.)

Colocynth is the fruit of *Citrullus Colocynthis* deprived of its rind. The plant grows in arid places in Asia, Africa, and Southern Europe. The chief cathartic principle is *colocynthin,* a bitter glucosid.

PREPARATIONS.	DOSE.
Extractum Colocynthidis, U. S. P.	2–5 gr. (0.13–0.3 gm.).
Extractum Colocynthidis Compositum, U. S. P. (extract of colocynth, 16; aloes, 50; resin of scammony, 14; cardamom, 6; soap, 14) . .	5–20 gr. (0.3–1.3 gm.).
Pilulæ Catharticæ Compositæ, U. S. P. (each pill contains about 1¼ gr.—0.08 gm.—of compound extract of colocynth; 1 gr.—0 06 gm.—of calomel; ½ gr.—0.03 gm.—of extract of jalap; ¼ gr.—0.015 gm.—of gamboge)	1–3 pills.
Pilulæ Catharticæ Vegetabiles, U. S. P. (each pill contains about 1 gr.—0.06 gm.—of compound extract of colocynth; ½ gr.—0.03 gm.—of extract of jalap; ½ gr.—0.03 gm.—of extract of hyoscyamus; ¼ gr.—0.015 gm.—of extract of leptandra; ¼ gr.—0.015 gm.—of extract of podophyllum; oil of peppermint)	1–3 pills.

Physiologic Action and Therapeutics.—Colocynth is a powerful drastic cathartic, producing in full doses copious watery stools, which are often accompanied by griping. Very large doses excite intense inflammation of the whole alimentary tract, which may prove fatal. The drug is too irritant to be used alone. At the present time it is prescribed in combination with other remedies only in *obstinate chronic constipation.* It is not a suitable cathartic for habitual use, but it may be employed now and then to secure a thorough evacuation of the bowels.

CAMBOGIA, U. S. P.
(Gamboge.)

Gamboge is a gum-resin from *Garcinia Hanburii,* a laurel-like tree growing in the East Indies. Its active principle is *cambogic acid.* The dose is from ½–5 gr. (0.03–0.3 gm.).

<table>
<tr><td>PREPARATION.</td><td>DOSE.</td></tr>
<tr><td>Pilulæ Catharticæ Compositæ, U. S. P. (about ¼ gr.—0.015 gm.—in each pill) . . . .</td><td>1–3 pills.</td></tr>
</table>

Physiologic Action and Therapeutics.—Gamboge is an extremely irritating drastic cathartic, fully capable in large doses of causing fatal gastro-enteritis. It is never used alone, but in combination with less powerful cathartics it is sometimes employed in *obstinate chronic constipation.*

SCAMMONIUM, U. S. P.
(Scammony.)

Scammony is a resinous exudation from the living root of *Convolvulus Scammonia,* a perennial herb growing in Western Asia. Its active principle is a resin, *scammonin,* which appears to be identical with jalapin.

<table>
<tr><td>PREPARATIONS.</td><td>DOSE.</td></tr>
<tr><td>Resina Scammonii, U. S. P. . . . .</td><td>2–8 gr. (0.13–0.5 gm.).</td></tr>
<tr><td>Extractum Colocynthidis Compositum, U. S. P. (14 per cent.) . . . .</td><td>5–20 gr. (0.3–1.3 gm.).</td></tr>
</table>

Physiologic Action and Therapeutics.—Scammony has been used as a drastic cathartic since the time of Hippocrates. It is more powerful than jalap, but less irritating than gamboge. It is rarely employed except in the form of the " compound cathartic pill."

PODOPHYLLUM, U. S. P.
(May Apple; Mandrake.)

Podophyllum is the rhizome and roots of *Podophyllum peltatum,* a perennial herb growing in moist shady places in Canada and Northern United States. It contains two iso-

meric glucosids—*podophyllotoxin* and *picropodophyllin*—and
podophylloresin. Picropodophyllin appears to be inert; the
other two principles are active cathartics.

PREPARATIONS.	DOSE.
Resina Podophylli, U. S. P. (Podophyllin)	$\frac{1}{8}$–$\frac{1}{2}$ gr. (0.008–0.03 gm.).
Extractum Podophylli, U. S. P.	1–3 gr. (0.065–0.2 gm.).
Extractum Podophylli Fluidum, U. S. P.	5–20 min. (0.3–1.2 c.c.).

The resin also enters into compound cathartic pills.

Physiologic Action and Therapeutics.—In full doses
podophyllum is an active, irritant cathartic. It produces its
effects slowly, however—rarely earlier than ten or twelve hours
after its administration—the evacuations being copious and
liquid, and generally attended with much griping pain. It is
said to act as a cathartic when applied to an open wound or
administered subcutaneously. It is believed by many practitioners to possess the power of directly increasing the
secretion of bile, but very little evidence can be adduced to
support this claim.

Podophyllum is especially prized as a cathartic in *habitual
constipation*, associated with so-called "bilious attacks." To
many persons the action of the drug is very agreeable, but in
others it causes severe pain and depression. When combined
with other cathartics only those should be selected which are
also slow in action, such as aloes and calomel. Extract of
belladonna or hyoscyamus may be added to prevent griping.
The resin is the most reliable preparation. A combination
like the following often makes a useful dinner pill:

R Resinæ podophylli,	gr. iv (0.26 gm.);
Aloes purificatæ,	gr. xx–xl (1.3–2.6 gm.) ;
Extracti nucis vomicæ,	
Extracti belladonnæ,	aa gr. iv (0.26 gm.).—M.
Fiant pilulæ No. xx.	
Sig. One at bedtime.	

JALAPA, U. S. P.
(Jalap.)

Jalap is the tuberous root of *Ipomœa Jalapa*, a perennial
herb growing in Mexico. It contains two active glucosids,
convolvulin and *jalapin.* The dose is from 5–30 gr. (0.3–
2.0 gm.).

PREPARATIONS.	DOSE.
Extractum Jalapæ, U. S. P.	3–10 gr. (0.2–0.65 gm.).
Resina Jalapæ, U. S. P.	1–5 gr. (0.065–0.3 gm.).
Pulvis Jalapæ Compositus, U. S. P. (35 per cent. of jalap and 65 per cent. of potassium bitartrate)	15–60 gr. (1.0–4.0 gm.).

PREPARATIONS.   DOSE.

Pilulæ Catharticæ Compositæ, U. S. P. ($\frac{1}{2}$ gr.—0.03 gm.
  —of extract) . . . . 1–3 pills.
Pilulæ Catharticæ Vegetabiles, U. S. P. ($\frac{1}{2}$ gr.—0.03 gm.
  —of the extract) . . . . 1–3 pills.

Physiologic Action and Therapeutics.—Jalap is a powerful hydragogue cathartic. It acts generally within three or four hours, and produces copious watery stools. It not infrequently excites nausea and colic, but its action is less harsh than that of either gamboge or scammony. According to Stadelmann, it is inactive in the absence of bile, the latter probably serving as a solvent. Except as an ingredient of "compound cathartic pills" or "vegetable cathartic pills" jalap is not employed in simple constipation. It is used especially for the removal of *dropsical effusions*, and for this purpose it is frequently combined with a saline, as in the official compound jalap powder. Associated with calomel jalap is sometimes a valuable depletive in *chronic heart disease* with congestion of the portal circulation.

ELATERIUM.

Elaterium is a substance deposited by the juice of *Ecballium Elaterium*, or squirting cucumber, a vine growing on the shores of the Mediterranean Sea. It is of uncertain strength and is not official, but its active properties are fully represented by a neutral principle, *elaterin* (Elaterinum, U. S. P.), which is official. The latter occurs in minute white scales or crystals, without odor, and having an acrid, bitter taste. It is almost insoluble in water, but readily soluble in chloroform and hot alcohol. The dose of elaterium is from $\frac{1}{8}$–$\frac{1}{4}$ of a gr. (0.008–0.016 gm.).

PREPARATIONS.   · DOSE.

Elaterinum, U. S. P. . . . . $\frac{1}{30}$–$\frac{1}{10}$ gr. (0.002–0.0065 gm.).
Trituratio Elaterini, U. S. P. (10 per cent.,
  with sugar of milk) . . . . $\frac{1}{4}$–1 gr. (0.016–0.065 gm.).

Physiologic Action and Therapeutics.—Elaterium is a decided irritant to all tissues. Given internally it is, perhaps, the most powerful of the hydragogue cathartics, producing large watery stools, with some griping, and, occasionally, nausea. Its action is more vigorous than that of jalap and less harsh than that of gamboge or colocynth. In overdoses it is a violent, acrid poison.

In appropriate doses elaterium is often very useful in reducing *local serous effusions* (hydrothorax, hydropericardium, ascites) and the *general dropsy* of cardiac or renal disease. On account of its prompt action it may be employed also as a derivative in *cerebral congestion* and *uremia*. Elaterium itself, probably on account of its adulteration, is unreliable, and, therefore, only official preparations should be prescribed.

BRYONIA, U. S. P.
(Bryony.)

Bryonia is the root of *Bryonia alba* and of *Bryonia dioica*, a perennial climbing plant growing in Central and Southern Europe. Its active principle is probably *bryonin*, an intensely bitter, soluble glucosid.

PREPARATION.   DOSE.
Tinctura Bryoniæ, U. S. P. . . . . 1–2 fl. dr. (4.0–8.0 c.c.).

Bryonia is a drastic, hydragogue cathartic. It was formerly a much esteemed remedy in general dropsy and local effusions, but at present it is very rarely used.

MAGNESII SULPHAS, U. S. P.
(Magnesium Sulphate; Epsom Salt; $MgSO_4 + 7H_2O$.)

Magnesium sulphate occurs in small, colorless, rhombic prisms or acicular crystals, odorless, and having an unpleasant, bitter, saline taste. It is soluble in 1.5 parts of water, and insoluble in alcohol. The dose is from 1–8 dr. (4.0–30.0 gm.).

Physiologic Action.—The saline cathartics, of which magnesium sulphate may be taken as a type, produce evacuation of the bowels by increasing the amount of fluid in the intestine, not like the vegetable cathartics, by directly stimulating peristalsis. They hinder the absorption of fluid taken with food, and they also abstract fluid directly from the tissues and blood. The increased bulk and weight of the intestinal contents give an impetus to peristalsis and purgation follows, but there is no direct irritation of the muscular coat of the bowel, as occurs with the vegetable cathartics. Under ordinary conditions saline cathartics are absorbed from the intestine very slowly and in but small quantities, the greater part escaping from the body in the stools. Their action, therefore, appears to be a purely local one, and this view is supported by the fact that they do not produce catharsis when injected intravenously. It is a noteworthy fact that when a saline cathartic is administered to an anhydremic subject, instead of the blood yielding its water to the intestine, the process is reversed, and the blood attracts fluid from the bowel, the salt is in large part absorbed, and purgation does not ensue.

Saline cathartics, by concentrating the blood, primarily lessen the secretion of urine; subsequently, however, they may be absorbed from the intestine in sufficient quantity to exert a diuretic effect.

In full doses and in concentrated solution magnesium sulphate is an active hydragogue cathartic, producing in the course of a few hours copious watery stools without much pain or systemic disturbance. Its chief drawback is its tendency to excite nausea. Dilute solutions are much less active as purgatives, but, being more readily absorbed, they increase the flow of urine. According to Hay, Epsom salt, when administered by intravenous injection, is intensely toxic, paralyzing both the heart and the respiration. When taken by the mouth, the drug does not ordinarily produce untoward symptoms because it is absorbed slowly and only in small quantities. That large doses, however, in the absence of purgation, are capable of producing systemic poisoning is evidenced by the case reported by Christison, in which two ounces proved fatal to a lad ten years of age.

Therapeutics.—Since Epsom salt is free from irritant properties, it is an excellent cathartic for removing undigested material from the bowel in *acute enteritis* and *colitis*. On account of its unpleasant taste and large bulk, however, it is not so convenient for children as calomel, or even castor oil. As it increases the fluidity of the intestinal contents it is well suited for use in cases of *fecal accumulation*. In *chronic constipation* a saline taken in small dose before breakfast is sometimes more efficacious than a vegetable cathartic. It is an excellent remedy for reducing *dropsical effusions*, both local and general. Recently several competent observers have testified to the value of Epsom salt in the treatment of *tropical dysentery*. Doses of from 1–2 dr. (4.0–8.0 gm.) are recommended to be given in some aromatic water, and continued until the stools lose their mucous character.

Magnesium sulphate is also used as an antidote to *acute lead-poisoning* and to *carbolic acid poisoning*. With sugar of lead it forms an insoluble sulphate, and with carbolic acid an innocuous sulphocarbolate.

Administration.—When a full hydragogue effect is desired, the drug should be given before breakfast, in one large dose, and with the smallest quantity of water that will dissolve it. In cases of general dropsy the restriction of liquids after the administration of the salt makes its action still more effective. Solutions of Epsom salt are rendered more acceptable to the stomach by the addition of magnesium carbonate (20–30 gr.—1.3–2.0 gm.). Cathartic enemata are rendered more active by the addition of Epsom salt. The following enema, recommended by Noble, will be found very efficacious in cases of marked intestinal torpor:

℞	Magnesii sulphatis,	℥ij (60.0 gm.);
	Olei terebinthinæ,	f℥ss (15.0 c.c.);
	Glycerini,	f℥j (30.0 c.c.);
	Aquæ,	q. s. ad f℥iv (120.0 c.c.).—M.

Incompatibles.—Magnesium sulphate is incompatible with lead acetate, silver nitrate, alkaline carbonates, and lime-water. From solutions of sodium phosphate it precipitates the insoluble magnesium phosphate, and from solutions of Rochelle salt, after a time, the insoluble magnesium tartrate.

SODII SULPHAS, U. S. P.

(Sodium Sulphate; Glauber's Salt; $Na_2SO_4 + 10H_2O$.)

Sodium sulphate occurs in large, colorless, transparent prisms, or granular crystals, odorless, and of a bitter, saline taste. It is soluble in 2.8 parts of water. The dose is from 2–8 dr. (8.0–30.0 gm.).

Therapeutics.—Glauber's salt is a powerful hydragogue cathartic, producing large watery stools, accompanied by griping and borborygmi. Epsom salt, being far less irritating, has largely superseded it. It enters into the mixture known as artificial Carlsbad salt, which makes an efficient mild saline aperient:

℞	Sodii sulphatis,	℥v (150.0 gm.);
	Sodii bicarbonatis,	℥ij (60.0 gm.);
	Sodii chloridi,	℥j (30.0 gm.).—M.
	Sig. A teaspoonful in a tumblerful of hot water half an hour before breakfast.	

The sulphates of sodium and magnesium are the active ingredients in certain natural mineral waters, such as Hunyadi Jànos, Friedrichshall, Carlsbad, Rubinat, and Pullna.

POTASSII ET SODII TARTRAS, U. S. P.

(Potassium and Sodium Tartrate; Rochelle Salt; $KNaC_4H_4O_6 + 4H_2O$.)

Rochelle salt occurs in colorless, transparent, prismatic crystals, or as a white powder, odorless, and having a cooling, saline taste. It is soluble in 1.4 parts of cool water and in less than 1 part of boiling water. The dose is from 2–4 dr. (8.0–16.0 gm.).

PREPARATION.   DOSE.

Pulvis Effervescens Compositus, U. S. P. (Seidlitz powder) . 1 powder.
(The *blue paper* contains sodium bicarbonate, 40 gr.—2.6 gm.—and potassium and sodium tartrate, 120 gr.—8.0 gm. The *white paper* contains tartaric acid, 35 gr.—2.3 gm. The contents of each paper should be dissolved separately, the two solutions mixed, and the whole taken while effervescing.)

Therapeutics.—Rochelle salt may be used as a hydragogue cathartic in the same class of cases in which magnesium sulphate is indicated. It is less active than the latter salt, but more agreeable to take. Seidlitz powder is a very mild saline cathartic. The carbonic acid evolved during effervescence not only makes the solution palatable, but exerts a sedative influence on the stomach.

SODII PHOSPHAS, U. S. P.

(Sodium Phosphate; $Na_2HPO_4 + 12H_2O$.)

Sodium phosphate occurs in large, colorless, prismatic crystals, odorless, and having a cooling, saline taste. It is soluble in 5.8 parts of water, and insoluble in alcohol. The dose for a young child is from 1–10 gr. (0.065–0.65 gm.); for an adult, 1–4 dr. (4.0–16.0 gm.).

Therapeutics.—Sodium phosphate in small doses acts as a laxative; in large doses, as a purgative. Its mild action and agreeable taste commend it especially for children, to whom it may be given in milk or other food. It is useful also in the less severe forms of constipation occurring in adults. In *chronic gastric catarrh with constipation* small doses in hot water before meals often have a very happy effect not only on the bowel, but also on the stomach. Taken in the same way, sodium phosphate is a valuable depletive in *simple catarrhal jaundice* secondary to duodenitis.

MAGNESII CITRAS.

(Magnesium Citrate; $Mg_3 2C_6H_5O_7 + 14H_2O$.)

Magnesium citrate is official as the *solution of magnesium citrate* (liquor magnesii citratis) and as *effervescent magnesium citrate* (magnesii citras effervescens). The first preparation is an effervescing solution of magnesium citrate containing a small quantity of sugar and free citric acid. In doses of from 6–12 ounces (180.0–360.0 c.c.) it is a very pleasant, but somewhat irritating, cathartic. It is suitable for occasional use, when it is desired simply to evacuate the alimentary canal. It should not be employed when there is any inflammation of the gastro-intestinal tract.

Effervescent magnesium citrate is a dry mixture of magnesium carbonate, sodium bicarbonate, citric acid, and sugar. It appears as a white, granular, deliquescent salt, which effervesces on the addition of water. The dose is from 1–3 dr. (4.0–12.0 gm.) in solution while effervescing. It is neither so pleasant nor so efficacious as the official solution.

DIURETICS.

The secretory activity of the kidneys varies with the quantity of blood flowing through them, and also with the composition of the blood.

Quantity of Blood.—A moderate rise in the general arterial pressure, other conditions remaining the same, augments the flow of urine; but Heidenhain, Munk, and Senator have shown that the increased secretion depends less on the greater pressure within the glomeruli than on the more rapid flow of blood through them. The flow of urine is not augmented when the rise in arterial pressure is accompanied by a corresponding constriction of the renal arteries; indeed, when the latter condition is very pronounced, it more than counteracts any favorable effect on the secretion exerted by the rise in blood pressure. The kidneys receive a larger quantity of blood, and in consequence their secretory activity is increased, when, (1) there is unusual vascular fulness; (2) when the action of the heart is stimulated; (3) when the blood-vessels in other vascular areas are constricted; (4) when the arteries of the kidneys are dilated by direct or reflex depression of their vasoconstrictor nerve-fibers or stimulation of their vasodilator nerve-fibers.

Composition of the Blood.—The inorganic salts, urea, and other soluble proteids normally present in the blood may be regarded as the physiologic stimuli of the secretory cells of the kidneys. Many other substances, not natural components of the blood, increase the secretion of urine by exerting a direct and specific influence on the cells of the glomeruli and convoluted tubes.

Diuretics are agents that increase the flow of urine. They may act indirectly by driving a greater volume of blood through the kidneys, or directly by stimulating the renal epithelium. A number of them act in both ways. The following circulatory stimulants act also as diuretics:

Digitalis.	Convallaria.
Strophanthus.	Squill.
Caffein.	Scoparius.
Theobromin.	Apocynum.

Of these drugs, *digitalis* seems to have no special influence on the secreting cells of the kidneys, its action as a diuretic depending almost entirely upon its influence on the cardiovascular system. The others appear to have a direct action on the renal epithelium, but undoubtedly their influence on the circulation contributes very largely to their efficiency as diuretics, especially in conditions associated with a low arterial tension.

Copious draughts of *water* excite diuresis by swelling the volume of blood. In acute Bright's disease, in the absence of dropsy, water may be drunk freely, since it not only increases the quantity of urine, but serves to flush the kidneys and to lessen irritation. The injection of warm *normal salt solution* under the skin, into a vein, or into the bowel, also increases the renal secretion by distending the arterial system, and also by exerting, through the salt, a specific effect on the secreting cells of the kidneys.

While vasodilators, like *nitroglycerin* and *nitrite of sodium,* are not in themselves especially useful as diuretics, they sometimes increase the diuretic action of digitalis and other cardiovascular stimulants by counteracting the constricting effect of these drugs on the renal arteries.

The following salts increase the flow of urine probably by exerting a specific influence on the cells of the kidneys:

Potassium bicarbonate.	Potassium nitrate.
Potassium carbonate.	Potassium chlorate.
Potassium bitartrate.	Lithium carbonate.
Potassium acetate.	Lithium citrate.
Potassium citrate.	Lithium benzoate.

The vegetable salts of potassium not only increase the quantity of urine, but, being eliminated in large part as carbonates, they tend to make it alkaline in reaction.

Finally, there are several drugs that, under certain conditions, produce a copious flow of urine, but the mode of action of which has not been clearly determined. It is supposed, however, that they act directly on the renal epithelium. To this group belong:

Pilocarpin.   Calomel.   Lactose.

Concerning the influence of drugs upon the solids eliminated by the kidneys, we have at present very little definite knowledge. Large quantities of *water* frequently increase the solids of the urine, but it has been shown that this is due to the flushing out of waste-products and not to any increase of proteid metabolism. It is commonly believed that the *alkaline diuretics* hasten oxidation in the tissues, and in consequence increase nitrogenous elimination, but it has not been satisfactorily demonstrated that these drugs have such an action. Many of the investigations undertaken to decide the question have been attended by numerous opportunities for error, and those that have not been have given such conflicting results that no conclusions can be drawn from them.

Digitalis, strophanthus, and *caffein* have no pronounced effect on the excretion of the solids normally found in the urine, but the last sometimes induces a slight glycosuria. Pilocarpin, probably in health, and certainly in disease, increases the elimination of urea.

Certain drugs which have no special activity as diuretics notably increase the solid constituents of the urine. Thus, *salicylic acid* increases the elimination of urea, uric acid, and sulphur compounds; *colchicum* probably increases the excretion of uric acid, and *thyroid extract,* by increasing the destruction of the proteids of the tissues, very decidedly augments the output of nitrogen and phosphates.

Diuretics are employed for the following purposes:

1. *To Remove Excrementitious Matters From the Blood.*—They are useful for this purpose when the secretory function of the kidneys is impaired or suspended, as it often is in acute febrile diseases, in passive congestion of the kidneys, and in some forms of Bright's disease.

2. *To Promote the Absorption and Excretion of Dropsical Effusion.*—For obvious reasons diuretics are less efficacious in dropsy resulting from organic disease of the kidneys than in that resulting from lowered arterial tension.

3. *To Lessen Irritation of the Genito-urinary Tract.*—For this purpose alkaline diuretics are useful when the urine is too concentrated or is excessively acid. They also relieve the distressing symptoms occasioned by uric acid gravel and uric

acid stones, not by exercising a solvent action on the concretions, but by producing a more copious secretion of urine.

DIGITALIS, U. S. P.

Digitalis has very little, if any, influence on the renal epithelium; it produces diuresis mainly through its stimulant action
on the heart and blood-vessels.

In the *dropsy* of *heart disease* it is preëminent as a diuretic
when used in doses just sufficient to maintain the proper
degree of tension in the renal blood-vessels. We must remember, however, that by giving it too freely we may cause excessive constriction of the arterioles, and thereby defeat the object
for which we have employed it.

Various opinions are held as to the advisability of using
digitalis to excite the secretory function of the kidneys in *acute
Bright's disease.* In deciding this question a very important
factor to be considered is the state of the circulation. We
believe that when the pulse is strong, when the second sound
of the heart at the aortic cartilage is accentuated, and when
dropsy is absent or is slight, the drug should not be employed.
On the other hand, when there are evidences of lowered
arterial tension and when dropsy is marked, it is often of great
service in reëstablishing elimination through the kidneys. In
chronic parenchymatous nephritis it may be used to increase the
flow of urine and to reduce the dropsy, but, like all other
diuretics, it is a very uncertain remedy. In *chronic interstitial
nephritis* so long as the arterial tension is high and the urine
is abundant, there is, of course, no indication for diuretics;
but late in the disease, when dilatation of the heart supervenes
and occasions oliguria, dropsy, dyspnea, and a weak, irregular
pulse, digitalis is our most reliable remedy. In *dropsy of hepatic origin* the drug may be combined with mercury, as in the
well-known Niemeyer's pill (p. 51), but its action is often disappointing. In *serous effusions of an inflammatory character*,
such as are encountered in pleurisy and pericarditis, it is of
doubtful utility.

STROPHANTHUS, U., S. P.

Strophanthus excites diuresis by increasing the quantity of
blood flowing through the kidneys, and also by stimulating
directly the epithelial cells of the kidneys. Notwithstanding
this double action, it is a less reliable diuretic than digitalis.
It may be employed, however, in the same class of cases in
which digitalis has been recommended.

CAFFEINA, U. S. P.
(Caffein.)

Caffein is a very powerful diuretic, acting on the renal epithelium and, in a less degree, on the circulation. Its effect on the elimination of nitrogenous matters has been studied repeatedly, but with most discordant results. After the ingestion of much sugar-forming food it sometimes gives rise to a slight glycosuria. It may be employed in any form of *general dropsy* except that of acute nephritis. In *cardiac dropsy* it not infrequently surpasses digitalis in its power of producing diuresis, and a combination of digitalis and caffein may prove more effective than either drug singly. In *chronic parenchymatous nephritis* also caffein may succeed when digitalis has failed. Its tendency to cause insomnia must always be borne in mind.

THEOBROMINA.
(Theobromin.)

Theobromin is an alkaloid closely related to caffein, and obtained from the seeds of *Theobroma Cacao*, or chocolate tree, extensively grown in South America and the West Indies. It occurs in minute whitish crystals, of a bitter taste, and sparingly soluble in water and alcohol. The dose is from 5–10 gr. (0.3–0.65 gm.). On account of its greater solubility, *Sodiotheobromin salicylate*, a combination containing about 50 per cent. of theobromin, is preferred to the pure alkaloid. *Diuretin* is the trade name of a proprietary remedy, composed of sodio-theobromin salicylate. The double salts of theobromin are readily decomposed on exposure. The dose of sodiotheobromin salicylate or diuretin is from 10–20 gr. (0.65–1.3 gm.), in capsules or in solution.

The action of theobromin resembles that of caffein, but it has less influence than the latter on the brain and circulatory system. It produces diuresis mainly through its influence on the renal epithelium. It is eliminated in the urine partly unchanged and partly in the form of xanthin bodies, to which, like caffein, it is closely allied.

Theobromin may be used as a diuretic in the same class of cases as caffein. On account of its inferior rank as a circulatory stimulant it is not so generally useful as digitalis in the dropsy of cardiac origin, but it will be found an excellent adjuvant to that drug in some cases. In *chronic nephritis*, when degeneration of the kidneys is far advanced, like other diuretics, it often proves ineffectual in removing the dropsy.

CONVALLARIA, U. S. P.

The action of convallaria resembles that of digitalis. While it has advantages in not having any cumulative effect and in not disturbing the stomach, it is distinctly inferior to digitalis both as a circulatory stimulant and as a diuretic. It occasionally succeeds, however, in cases of *cardiac dropsy* in which digitalis has been tried without avail.

SCILLA, U. S. P.

(Squill.)

Squill is the bulb of the *Urginea maritima*, a perennial herb growing on the shores of the Mediterranean Sea. According to Merck, it contains three active constituents, *scillitin, scillitoxin (scillain)*, and *scillin*. No principle, however, that has yet been isolated represents fully the properties of the crude drug. The dose of powdered squill is from 1–3 gr. (0.065–0.2 gm.).

Preparations.	Dose.
Acetum Scillæ, U. S. P.	10–30 min.
Extractum Scillæ Fluidum, U. S. P.	1–3 min. (0.06–0.2 c.c.).
Tinctura Scillæ, U. S. P.	5–20 min. (0.3–1.2 c.c.).
Syrupus Scillæ, U. S. P.	½–1 fl. dr. (2.0–4.0 c.c.).
Syrupus Scillæ Compositus, U. S. P. (fl. ext. of squill, 8 per cent. ; fl. ext. of senega, 8 per cent. ; tartar emetic, 2 per cent.)	10–60 min. (0.6–4.0 c.c.).

Physiologic Action and Therapeutics.—Squill has an influence on the circulation somewhat similar to digitalis, but less powerful. It also produces diuresis, partly by increasing the quantity of blood flowing through the kidneys, and partly, there is reason to believe, by directly stimulating the renal epithelium. Large doses cause vomiting and purging, and, in addition, if much of the drug be absorbed, albuminuria and hematuria.

Squill is employed chiefly as a stimulating expectorant (see p. 281) and as a diuretic. For the latter purpose it is rarely given by itself, but in combination with digitalis. Reference has already been made to the value of a combination of digitalis, squill, and blue-mass in some cases of *cardiac* and *hepatic dropsy*. It is contraindicated in *acute nephritis*. As a diuretic it is usually prescribed in substance and in pill form.

SCOPARIUS, U. S. P.

(Broom.)

Scoparius is the tops of *Cytisus Scoparius*, a shrub indigenous to Western Asia and Southern Europe, and cultivated in the

United States. It contains a liquid alkaloid, *spartein*, and a neutral principle, *scoparin*. The salts of spartein, of which the sulphate is official, are crystalline solids. The dose of spartein sulphate is from ¼–1 gr. (0.016–0.065 gm.). Scoparin is rarely used; its dose by the mouth is from 5–15 gr. (0.3–1.0 gm.).

Preparations.	Dose.
Extractum Scoparii Fluidium, U. S. P.	½–1 fl. dr. (2.0–4.0 c.c.).
Decoctum Scoparii (1 oz.–1 pint—30.0 gm.– 0.5 L.)	1–3 fl. oz. (30.0–90.0 c.c.).

Physiologic Action and Therapeutics.—The chief effect of moderate doses of scoparius is an increase in the flow of urine. The diuresis is probably due in large part to the action of scoparin on the renal epithelium, although it is possible that spartein may contribute, through its influence on the circulation (see p. 54). Large doses cause vomiting and purging. Scoparius, alone or with digitalis, is sometimes efficacious in *cardiac dropsy*. The decoction is generally employed.

APOCYNUM, U. S. P.

(Canadian Hemp.)

Apocynum is the root of *Apocynum cannabinum*, a perennial plant growing in North America, from Canada to Florida. In addition to tannin and a bitter extractive, it contains an indifferent resinous principle, *apocynin*, and an active glucosid, *apocynein*.

Preparation.	Dose.
Extractum Apocyni Fluidum, U. S. P.	5–10 min. (0.3–0.6 c.c.).

Physiologic Action and Therapeutics.—Apocynum, through its glucosid, apocynein, has an action on the circulation resembling that of digitalis. Like the latter, also, it increases the flow of urine. In large doses it is an active irritant to the gastro-intestinal tract, causing severe vomiting and purging. Apocynum has been used as a diuretic in *cardiac* and *renal dropsy*, but it is unreliable and prone to disturb the stomach.

VEGETABLE SALTS OF POTASSIUM.

Physiologic Action.—In very large doses potassium is a depressant to both the nervous system and the circulatory system, especially to the latter. In the case of the vegetable salts of potassium, however, the action of the base is entirely subordinate to that of the acid radical, and a depressant effect is never observed except after an enormous dose of the salt or after its prolonged administration. The chief effect of a moderate dose of any one of the vegetable salts of potassium is an increased secretion of urine. The mineral salts of the urine, both of potassium and sodium, are increased, but the evidence is not convincing that the salts of potassium hasten oxidation in the tissues, and in consequence augment nitrogenous elimination. In large doses they impart an alkaline reaction to the urine. In concentrated solution they act upon the bowel and produce free watery discharges. The claim made by some authorities that they increase the secretion of bile has not been substantiated by recent investigations.

Therapeutics.—It is an established fact that *acute rheumatism* is favorably influenced by the administration of the vegetable salts of potassium, but until we have a clearer understanding of the nature of this disease it would be useless to speculate upon the *modus operandi* of the remedies employed in its treatment. Thirty grains (2.0 gm.) of the alkali should be given every three or four hours until the urine becomes neutral or slightly alkaline. Excellent results are often obtained by combining an alkaline salt with a salicylic compound.

Alkalis in copious drafts of water are efficient adjuvants to colchicum in the treatment of *gout;* but it is not known whether they act by hindering the formation of uric acid, by promoting the solution of uric acid compounds, or simply by benefiting gastric catarrh, relieving constipation, and stimulating the activity of the kidneys.

Alkalis or mineral waters containing them are apparently of service in *acute catarrh of the bile-ducts* and in *cholelithiasis*, probably by relieving coexisting duodenal catarrh, and not by exerting any direct influence on the bile itself or its concretions. They are also useful in relieving *dysuria* that is due to *excessive acidity of the urine*, and through their diluent action they tend to mitigate the suffering occasioned by *uric acid gravel* and *uric acid calculi.* In *acute Bright's disease* they serve to maintain the excretion through the kidneys and to render the urine less irritating.

Finally, the vegetable salts of potassium are efficient sedative expectorants in the beginning of *acute bronchitis*, especially when the secretion is viscid and scanty.

Incompatibles.—The vegetable salts of potassium are incompatible with acids, mineral salts, and alkaloidal salts.

POTASSII BICARBONAS, U. S. P.

(Potassium Bicarbonate; $KHCO_3$.)

Potassium bicarbonate occurs in colorless prismatic crystals, odorless, and of a saline, alkaline taste. It is soluble in 3.2 parts of water, and almost insoluble in alcohol. The dose is from 10–30 gr. (0.65–2.0 gm.) in solution well diluted. On account of its unpleasant taste it is less frequently prescribed than the citrate and acetate of potassium. Potassium bicarbonate is used in making the following preparations:

Liquor potassii arsenitis (Fowler's solution), U. S. P.
Liquor potassii citratis, U. S. P.
Liquor magnesii citratis, U. S. P.

POTASSII CARBONAS, U. S. P.

(Potassium Carbonate; K_2CO_3.)

Potassium carbonate is a white, granular, deliquescent powder, alkaline in reaction, and caustic to the taste. It is freely soluble in water, but insoluble in alcohol. The dose is from 10–30 gr. (0.65–2.0 gm.), but on account of its irritant properties it is rarely used internally. From potassium carbonate the following preparations are made:

Potassa sulphurata, U. S. P.
Mistura ferri composita, U. S. P.
Pilulæ ferri carbonatis (Blaud's Pills), U. S. P.

POTASSII BITARTRAS, U. S. P.

(Potassium Bitartrate; Cream of Tartar; $KHC_4H_4O_6$.)

Potassium bitartrate occurs in colorless or slightly opaque, rhombic crystals, or as a white powder, odorless, and of an acidulous taste. It is soluble in about 201 parts of water, and very sparingly soluble in alcohol. The dose is from 15–60 gr. (1.0–4.0 gm.). Unlike the other vegetable salts of potassium, the bitartrate resists oxidation in the body, and therefore much of it is eliminated unchanged. It is also a more active cathartic than the other salts of potassium. The double tartrate of potassium and sodium (Rochelle salt) is formed by adding potassium bitartrate to a hot solution of sodium carbonate.

PREPARATION.	DOSE.
Pulvis Jalapæ Compositus, U. S. P. (jalap, 35; potassium bitartrate, 65)	15–60 gr. (1.0–4.0 gm.).

POTASSII ACETAS, U. S. P.

(Potassium Acetate; $KC_2H_3O_2$.)

Potassium acetate occurs as a white powder or as crystalline masses, very deliquescent, odorless, and of a saline taste. It is soluble in 0.36 part of water and in 1.9 parts of alcohol. The dose is from 15–60 gr. (1.0–4.0 gm.). On account of its ready solubility, pleasant taste, and freedom from irritant properties, potassium acetate is a favorite alkaline diuretic.

POTASSII CITRAS, U. S. P.

(Potassium Citrate; $K_3C_6H_5O_7 + H_2O$.)

Potassium citrate occurs in transparent, prismatic crystals or white, granular powder, deliquescent on exposure to air, odorless, and having a pleasant saline taste. It is soluble in 0.6 part of water, but sparingly soluble in alcohol. The dose is from 15–60 gr. (1.0–4.0 gm.).

PREPARATIONS.	DOSE.
Liquor Potassii Citratis, U. S. P. (neutral mixture)	½–1 fl. oz. (15.0–30.0 c.c.).
Potassii Citras Effervescens, U. S. P.	30–90 gr. (2.0–6.0 gm.).

The citrate of potassium has properties closely resembling those of acetate of potassium, and the two salts may be used interchangeably. The solution of potassium citrate is used in *mild febrile conditions* to promote the secretion of the skin and kidneys.

POTASSII NITRAS, U. S. P.

(Potassium Nitrate; Saltpeter; KNO_3.)

Potassium nitrate occurs as colorless, transparent, rhombic prisms, or as a crystalline powder having a cooling, saline taste. It is permanent in the air. It is soluble in 3.8 parts of water, and sparingly soluble in alcohol. The dose is from 10–30 gr. (0.65–2.0 gm.), in solution, well diluted.

PREPARATION.

Charta Potassii Nitratis, U. S. P.

Physiologic Action.—Beyond being more irritating to the stomach, the effect of potassium nitrate in moderate doses is not materially different from that of the vegetable salts of potassium. Like the latter, it increases the flow of urine through its direct action on the secreting cells of the kidneys. In large doses potassium nitrate is not only more irritating than the vegetable salts, but, owing to the relatively greater influence of the metallic base,—potassium,—it is distinctly more depressing to the heart and nervous system. The drug is partly eliminated in the urine and saliva unchanged ; a greater part undergoes reduction in the tissues, but its educts are at present unknown.

Toxicology.—Toxic doses produce the symptoms of acute gastro-enteritis, profound muscular weakness, and collapse. In some instances diuresis is replaced by oliguria or anuria.

Therapeutics.—Potassium nitrate may be employed as a diuretic in *general dropsy*, but its utility is no greater than that of the acetate or bitartrate of potassium, which are less irritating to the gastro-intestinal tract. Inhalation of the smoke of burning niter paper is sometimes effective in *asthma*.

POTASSII CHLORAS, U. S. P.

(Potassium Chlorate ; $KClO_3$.)

Potassium chlorate occurs in the form of colorless, crystalline plates, odorless, and of a cooling saline taste. It is soluble in 16.7 parts of water, and sparingly soluble in dilute alcohol. The dose is from 5–20 gr. (0.3–1.3 gm.), in solution, well diluted.

PREPARATION. DOSE.

Trochisci Potassii Chloratis, U. S. P. (each contains about
5 gr.—0.3 gm.) . . . . 1–4 troches.

Physiologic Action.—When applied in dilute form to mucous membranes, potassium chlorate produces a stimulating and alterative effect; in concentrated form it is a decided irritant. When taken internally it does not, as was formerly believed, yield its oxygen to the blood, but it passes out of the body unchanged. Most of it—more than 90 per cent.— escapes in the urine, but, according to Isambert, small quantities can be recovered from the saliva, the tears, and the milk of nursing women. The only appreciable effect of moderate doses, well diluted, is an increase in the flow of urine. Concentrated solutions, even of therapeutic doses, give rise to a burning sensation in the stomach, followed by nausea and vomiting. Jacobi[1] was the first to point out that the practice of giving the drug internally in large doses was a dangerous one, and to prove that many deaths that had been attributed to other causes were in reality due to potassium chlorate poisoning.

Toxicology.—The symptoms of potassium chlorate poisoning are the resultant of the drug's action on the heart,

[1] *Medical Times*, April, 1861.

digestive tract, kidneys, and blood. Through its basic radical it depresses the circulation, and through its acid radical it irritates the stomach, intestines, and kidneys, and also effects a peculiar change in the blood—namely, the conversion of hemoglobin into methemoglobin, a compound very tenacious of its oxygen. The first symptoms of poisoning are thirst and abdominal pain, followed by vomiting and purging. Cyanosis is almost constantly present, and is accompanied by dyspnea and the usual phenomena of cardiac failure. The urine is scanty, dark colored, and albuminous, and on microscopic examination reveals numerous pigmented tube-casts and granular detritus derived from the red blood-cells. Jaundice is often present, and in many cases death is preceded by coma and convulsions. The immediate cause of death may be heart failure, asphyxia, or uremia, according as the action of the poison has been most pronounced on the circulation, blood, or kidneys. Apart from gastro-enteritis and nephritis, the most striking postmortem change is that which involves the blood. The color of the blood is changed to a chocolate hue, oxyhemoglobin is replaced by methemoglobin, and many of the red cells are decolorized and more or less disintegrated. The organs generally may be chocolate colored from the accumulation in the tissues of liberated blood-pigment. In the recorded cases the minimum fatal dose for an adult has been 4 drams; for a child of four years, 3 drams; and for an infant, 1 dram.

The *treatment of poisoning* consists in evacuating the stomach and in administering demulcents and diluents. Subcutaneous injections of normal salt solution may be tried.

Therapeutics.—Potassium chlorate makes an excellent local application in inflammatory affections of the mouth and throat. Thus, it is valuable in the various forms of *stomatitis* and in *acute pharyngitis.* In these affections a solution of from 10–20 gr. (0.65–1.3 gm.) to the ounce (30.0 c.c.) may be used as a wash or gargle. In acute pharyngitis tannic acid or some preparation containing it is frequently added to the solution.

In *scarlatina* and *diphtheria* it is generally advisable to avoid solutions of potassium chlorate, and to select an agent that, if swallowed, will not contribute to the existent renal irritation. In *ulcerous stomatitis* this salt is almost a specific, and may be used internally as well as locally. The dose for a child of three years is from 2–3 gr. (0.13–0.2 gm.), well diluted, every three hours. Benefit from its internal administration is to be attributed to its continuous elimination in the saliva. Tincture of ferric chlorid may be used as an adjuvant, as in the following formula :

> ℞. Potassii chloratis,   gr. xl (2.6 gm.);
> Tincturæ ferri chloridi,   ♏ xxxij (2.0 c.c.);
> Syrupi zingiberis,   f℥vj (22.5 c.c.);
> Aquæ,   q. s. ad f℥ij (60.0 c.c.).—M.
> Sig. A teaspoonful in water every three hours.

It is generally believed that *salivation* is less likely to result from the continuous use of mercury when a solution of potassium chlorate is used at the same time as a mouth-wash. In some cases of *syphilis* troches of the salt are useful as preventives of both mercurial stomatitis and of mucous patches.

Incompatibles.—Owing to rapid oxidation or reduction violent explosion may occur when potassium chlorate is triturated with such substances as tannic acid, phosphorus, pulverized charcoal, sulphur, sulphids, sugar, hyposulphites, hypophosphites, and ammonium chlorid. Mixtures of potassium chlorate, tincture of ferric chlorid, and glycerin, when warm, are liable to explode. The salt reacts with strong hydrochloric acid, setting free chlorin gas; with potassium iodid it may form the exceedingly irritant potassium iodate, and from syrup of ferrous iodid it may precipitate free iodin.

LITHII CARBONAS, U. S. P.

(Lithium Carbonate; Li_2CO_3.)

Lithium carbonate occurs as a white powder, odorless, and of an alkaline taste. It is soluble in 80 parts of water and insoluble in alcohol. Its dose is from 5–20 gr. (0.3–1.3 gm.).

Physiologic Action and Therapeutics.—The effect of the vegetable salts of lithium is very similar to that of the corresponding salts of potassium; like the latter, they increase the quantity of urine and lessen its acidity. Lithium salts were originally recommended in the treatment of *gout* and the *uric acid diathesis* because it was found that outside of the body lithium united with uric acid to form a more soluble salt than did either sodium or potassium; but as it has been clearly demonstrated that alkaline medication is without influence on the urates present in the blood or tissues, it is very doubtful whether the salts of lithium are any more potent in these diseases than the vegetable salts of potassium. The natural mineral waters are extensively employed in the treatment of gout and nephrolithiasis, but any efficacy that they may possess is probably dependent upon the depurant action of the water itself and not upon the small amount of salt they contain.

LITHII CITRAS, U. S. P.

(Lithium Citrate; $Li_3C_6H_5O_7$.)

Lithium citrate is a white, deliquescent powder, odorless, and of a cooling, alkaline taste. It is soluble in two parts of water and insoluble in alcohol. Its dose is from 5–30 gr. (0.3–2.0 gm.).

PREPARATION.	DOSE.
Lithii Citras Effervescens, U. S. P.	1–2 dr. (4.0–8.0 gm.).

Lithium citrate is the therapeutic equivalent of lithium carbonate, and may be substituted for the latter when a solution is preferred to a powder.

LITHII BENZOAS, U. S. P.

(Lithium Benzoate; $LiC_7H_5O_2$.)

Lithium benzoate occurs in the form of a white powder or small shining scales, permanent in the air, odorless, and of cooling, sweetish taste. The dose is from 5–30 gr. (0.3–2.0 gm.).

This drug was recommended in the treatment of *gout* on the ground that it would prove especially active as a uric acid solvent, the acid radical uniting with proteid derivatives to form the soluble hippuric acid, and the basic radical uniting with free uric acid to form a readily soluble urate. This reasoning, however, has been shown to be fallacious by recent studies on uric acid formation and elimination, and, moreover, in practice, lithium benzoate has been found to be less efficacious in gouty conditions than the other salts of lithium.

OTHER DIURETICS.

Pilocarpin (see p. 252).—Small doses of pilocarpin ($\frac{1}{16}$–$\frac{1}{12}$ gr.—0.004–0.005 gm.), especially if the patient be lightly wrapped, cause well-marked diuresis. The manner of the drug's action is not definitely known, but as albuminuria and even strangury have followed its use in a few instances, it is surmised that it acts directly on the secreting structure of the kidneys. The excretion of urea is also increased by pilocarpin. The drug is sometimes useful as a diuretic in the *dropsy of subacute and chronic nephritis.*

Calomel (see p. 323).—Several theories have been advanced to explain the diuretic action of this drug; the most tenable one is that it acts directly on the renal epithelium. It is especially useful in the *dropsy of cardiac disease*, but it is not without value in serous effusions of hepatic or renal origin. A few large doses (2–3 gr.—0.13–0.2 gm.) may be given, opium being added, when necessary, to prevent purging; or small doses (¼ gr.—0.016 gm.) may be employed as adjuvants to digitalis and squill.

Lactose (see p. 465).—This drug has been highly recommended by See and others as a powerful diuretic in *cardiac dropsy.* The manner of its action has not been determined. From 2–4 ounces (30.0–125.0 gm.) may be given daily, dissolved in milk.

Spirit of Nitrous Ether (see p. 256).—Sweet spirit of niter is an agreeable remedy of feeble diuretic power. Each of its components—ethyl nitrite, ether, and alcohol—probably contributes to its activity. It is useless in general dropsy, but it is very efficacious in relieving *oliguria* that is due to *febrile disease* or *active congestion of the kidneys.* The usual dose as a diuretic is from 1–2 fl. dr. (4.0–8.0 c.c.).

Incompatibles.—Spirit of nitrous ether is incompatible with antipyrin, iodids, ferric sulphate, tincture of guaiacum, mucilage of acacia, and tannic acid.

STIMULANTS TO THE GENITO-URINARY TRACT.

Under this heading will be considered those drugs that stimulate the mucous membrane of the whole genito-urinary tract. All these act also upon the secretory cells of the kidneys, and increase more or less the quantity of urine, but this influence is subordinate to that exerted upon the mucous membranes over which the urine flows. Many of them contain antiseptic principles that are eliminated by the kidneys, and that are powerful enough to inhibit the growth of bacteria and to prevent decomposition of the urine. In large doses these drugs are capable of producing acute inflammation of the bladder and urethra, and even of the kidneys themselves.

The most important members of this group are:

Copaiba.	Oil of erigeron.
Cubeb.	Buchu.
Matico.	Uva ursi.
Oil of sandalwood.	Chimaphila.
Turpentine.	Pareira.
Juniper.	Cantharides.

The combined stimulant and antiseptic properties of these drugs make them useful remedies in subacute and chronic pyelitis, cystitis, and urethritis. They are contraindicated when the inflammation is of an acute character.

COPAIBA, U. S. P.

(Balsam of Copaiba; Copaiva.)

Copaiba is the oleoresin of *Copaiba Langsdorffii* and other species of *Copaifera*, a small tree growing in the northern states of South America. It is a yellowish or brownish, viscid liquid, having a peculiar, aromatic odor and a bitter, acrid taste. It contains a volatile oil (40.9 per cent.), resins, a bitter principle, and copaivic acid. The dose of copaiba is from 10–30 min. (0.6–2.0 c.c.) in capsule or emulsion.

PREPARATIONS.	DOSE.
Massa Copaibæ, U. S. P. (copaiba, 94 per cent.; magnesia, 6 per cent.)	15–30 gr. (1.0–2.0 gm.).
Oleum Copaibæ, U. S. P.	5–15 min. (0.3–1.0 c.c.).
Resina Copaibæ, U. S. P.	5–15 gr. (0.3–1.0 gm.).

Physiologic Action and Therapeutics.—In therapeutic doses copaiba acts as a stimulant to mucous membranes. As its active principles are eliminated chiefly by the kidneys and lungs, it particularly influences the mucous membrane of the genito-urinary tract and bronchi. The drug also excites diuresis, and imparts to the urine decided antiseptic properties. The resins and most of the oil are excreted by the kidneys, but a part of the oil escapes through the lungs and skin. The oil appears in the urine in part unchanged, and in part in combination with glycuronic acid. Urine containing the resins of copaiba responds to the nitric acid test for albumin, but the resinous precipitate can be distinguished by its rapid disappearance on the addition of alcohol. Fehling's solution may also be reduced by urine containing the educts of copaiba. In large doses copaiba acts as an irritant, causing burning in the stomach, vomiting and purging, lumbar pains, frequent micturition, and even strangury. In some persons the administration of the drug is followed by an eruption on the skin, urticarial, scarlatinoid, or morbilliform in type.

In inflammatory diseases of the genito-urinary tract, such as *pyelitis, cystitis*, and *urethritis*, copaiba is a useful remedy; it should not be used, however, until the most acute symptoms have subsided. In *gonorrhea*, after the discharge has become well established, the drug affords considerable relief and tends to prevent complications. It may be combined advantageously with other drugs, like cubeb and salol, which tend to render the urine sterile. As a simple diuretic it has been well recommended in *dropsy* due to *cirrhosis of the liver.* In *subacute and chronic bronchitis* with profuse purulent expectoration it is sometimes beneficial, but generally it is less useful than oil of eucalyptus or the preparations of guaiacol. Copaiba has a long-standing reputation as a stimulant application for *indolent ulcers.*

The oleoresin and oil are the most eligible preparations. On account of their unpleasant taste they should always be prescribed in capsules or as an emulsion.

CUBEBA, U. S. P.

(Cubeb.)

Cubeb is the unripe fruit of *Piper cubeba*, a perennial climber indigenous to Borneo and the neighboring islands. It contains a volatile oil (5–15 per cent.), a resin, cubebic acid, and cubebin, which is inert. The dose of powdered cubeb is from $\frac{1}{2}$–2 dr. (2.0–8.0 gm.).

PREPARATIONS.	DOSE.
Extractum Cubebæ Fluidum, U. S. P.	10–30 min. (0.6–2.0 c.c.).
Oleoresina Cubebæ, U. S. P.	5–15 min. (0.3–1.0 c.c.).
Oleum Cubebæ, U. S. P.	5–15 min. (0.3–1.0 c.c.).
Tinctura Cubebæ, U. S. P.	$\frac{1}{2}$–2 fl. dr. (2.0–8.0 c.c.).
Trochisci Cubebæ, U. S. P. (each contains about $\frac{2}{3}$ gr. (0.04 gm.) of oleoresin)	1–5 troches.

Physiologic Action and Therapeutics.—Cubeb possesses properties closely resembling those of copaiba; hence it may be employed to meet the same indications as the latter. A combination of the two is often more effective than either alone. Cubeb is somewhat less liable than copaiba to disturb digestion or to cause cutaneous rashes. Troches of cubeb are useful in relieving hoarseness and fatigue of the larynx resulting from prolonged use of the voice. The oleoresin is the most active preparation; it should be given in capsules or as an emulsion.

MATICO, U. S. P.

Matico is the leaves of *Piper angustifolium*, a shrub growing in Mexico and South America. It contains a volatile oil, a resin, tannin, and artanthic acid.

PREPARATIONS.	DOSE.
Tinctura Matico, U. S. P.	$\frac{1}{2}$–2 fl. dr. (2.0–8.0 c.c.).
Extractum Matico Fluidum, U. S. P.	$\frac{1}{2}$–2 fl. dr. (2.0–8.0 c.c.).

Physiologic Action and Therapeutics.—Matico somewhat resembles cubeb in its action, and has been given in the same class of cases. The leaves, on account of their hairy surfaces, favor coagulation and arrest bleeding when applied to small wounds. The drug is rarely employed at the present time.

OLEUM SANTALI, U. S. P.

(Oil of Santal; Oil of Sandalwood.)

Oil of santal is a volatile oil distilled from the wood of *Santalum album*, a small tree growing in India and the East Indian Islands. It is a thick, pale-yellow liquid, having a strong, aromatic odor and a pungent, spicy taste. It is soluble in alcohol. The dose is from 5–20 min. (0.3–1.2 c.c.), in capsules or as an emulsion.

Oil of santal resembles copaiba in its action, but is less irritant.

TEREBINTHINA, U. S. P.

(Turpentine.)

Turpentine is a concrete oleoresin obtained from *Pinus palustris* and other species of *Pinus* growing in the Southern United States, especially in North Carolina. When subjected to distillation it yields a volatile oil, *oleum terebinthinæ*, and a solid residue, *resin*.

OLEUM TEREBINTHINÆ, U. S. P.

(Oil of Turpentine; Spirit of Turpentine.)

Oil of turpentine is a limpid, colorless, highly inflammable liquid, having a characteristic odor and taste. It contains several terpene hydrocarbons. The rectified oil is made by distilling the ordinary oil with lime-water, and is the preparation always selected for internal use.

PREPARATIONS.   DOSE.

Oleum Terebinthinæ Rectificatum, U. S. P. . . 5–20 min. (0.3–1.2 c.c.).
Linimentum Terebinthinæ, U. S. P. (oil of turpentine, 35; resin cerate, 65).
Terpin hydrate (see p. 281) is a crystalline compound obtained by the interaction of oil of turpentine, alcohol, and nitric acid.
Terebene (see p. 281) is a liquid hydrocarbon made by oxidizing oil of turpentine with strong sulphuric acid.

Physiologic Action.—When applied to the skin, oil of turpentine acts as an irritant, producing redness and burning, and, if the contact be prolonged, vesication. The drug has decided antiseptic properties, and is capable, even in dilute form, of preventing fermentation and putrefaction. Internally, in full doses, it causes a sense of warmth in the stomach, quickened respiration, and an increase in the rate and tension of the pulse. It escapes from the body through the respiratory tract and the kidneys, and imparts to the urine an agreeable odor resembling that of violets. In overdoses it produces abdominal pain, nausea, vomiting, embarrassed respiration, a rapid, feeble pulse, great muscular relaxation, lumbar pains, dysuria, scanty bloody urine, delirium, and finally coma.

In therapeutic doses oil of turpentine acts as a circulatory stimulant and diuretic. In the alimentary canal it plays the part of a carminative and an antiseptic. On leaving the body it imparts its peculiar stimulant and antiseptic properties to the urine and the secretions of the respiratory tract. In susceptible persons an erythematous or papular eruption may result from either its internal or its external use.

Therapeutics.—Externally oil of turpentine is used as a rubefacient in various inflammatory affections, such as *bronchitis, pleurisy, pneumonia, gastritis,* and *enteritis.* It is best applied in these cases in the form of a stupe made by sprinkling freely with the oil a piece of flannel which has first been soaked in boiling water and then wrung dry. It may be allowed to remain on the affected part from ten to twenty minutes, according to the sensitiveness of the skin. Turpentine liniment makes an excellent application in *muscular rheumatism* and *chilblains.*

Oil of turpentine is a reliable carminative, and may be administered either by the mouth or by the rectum. Troublesome *tympanites* will often yield to the application of a stupe to the abdomen and the rectal injection of a pint of soapy water containing a tablespoonful of the oil.

The oil is a valuable adjunct to alcohol in certain cases of *typhoid fever.* It is useful when the tongue is dry, brown, and fissured; when there is a tendency to muttering delirium; and when there is marked abdominal distention. It sometimes acts favorably as a stimulant and antiseptic in subacute and chronic inflammations of the genito-urinary tract, such as *cystitis* and *urethritis.* It may be employed as an expectorant in *subacute and chronic bronchitis,* when the expectoration is copious and purulent, but generally terebene will be found more agreeable. With some practitioners it has acquired a reputation as a hemostatic in *passive hemorrhage* from the stomach, intestines, lungs, or kidneys. Some cases of *purpura hæmorrhagica* are favorably influenced by the oil.

Oil of turpentine, ½–1 fl. oz. (15.0–30.0 c.c.), with castor oil, has been used as an anthelmintic to destroy *tape-worm*, but there are other remedies equally effective and decidedly less dangerous.

A mixture of 3 parts of oil of turpentine and 2 of ether, on the recommendation of Durande, of Dijon, was at one time largely employed as a solvent of gall-stones, but it has been abandoned from the lack of evidence as to its efficacy.

Oil of turpentine is a powerful deodorant. In the form of vapor it may be employed to overcome the fetid odor of the breath in *bronchiectasis* and *gangrene of the lung*. It is serviceable for removing the odor from the hands after postmortem examinations. Madden has found it efficacious in destroying the offensive character of the discharges in *cancer of the uterus*. He recommends a douche composed of a tablespoonful each of the oil and magnesia to a quart of boiling water; the mixture is to be cooled and thoroughly stirred before being used. The penetrating odor of iodoform is lost to a great extent in the presence of turpentine.

Whatever may have been the merits of old preparations of the oil in *phosphorus-poisoning*, it is certain that the present article of commerce is entirely without efficacy.

Contraindications.—It should not be prescribed when there is acute inflammation of the stomach, intestines, or genito-urinary tract.

Administration.—Oil of turpentine may be administered on lumps of sugar, in capsules, or in emulsion. The following formula illustrates its use in emulsion:

℞ Olei terebinthinæ,		f℥ij (8.0 c.c.);
Pulveris acaciæ,	q. s.	
Olei cinnamomi,		♏ xij (0.75 c.c.);
Aquæ et syrupi,	q. s. ad	f℥iij (90.0 c.c.).—M.
Misce et fiat emulsum.		
Sig. A teaspoonful every three hours.		

Incompatibles.—Bromin, iodin, nitric acid, and strong sulphuric acid act with violence on oil of turpentine.

RESINA, U. S. P.

(Resin; Colophony.)

Resin is the hard, transparent, amber-colored residue left after distilling off the volatile oil from turpentine.

PREPARATIONS.

Ceratum Resinæ, U. S. P. (resin, 35 parts; yellow wax, 15 parts; lard, 50 parts).

Emplastrum Resinæ, U. S. P. (resin, 14 parts; lead-plaster, 80 parts; yellow wax, 6 parts).

Therapeutics.—Resin cerate is chiefly employed as a stimulating application for *indolent ulcers*. Resin plaster or lead adhesive plaster is commonly used for the fixation of surgical dressings. As it is less irritating to the skin than the rubber adhesive plaster, it is to be preferred to the latter for strapping the chest in fracture of the ribs and in pleurisy, and for making compression over indolent leg-ulcers.

JUNIPERUS.

(Juniper.)

Juniper is the fruit of *Juniperus communis*, an evergreen shrub growing in Northern Europe, Asia, and America. It contains a volatile oil, resin, and a non-crystalline principle, juniperin.

PREPARATIONS.	DOSE.
Oleum Juniperi, U. S. P.	3–5 min. (0.2–0.3 c.c.).
Spiritus Juniperi, U. S. P. (5 per cent. of oil of juniper)	½–1 fl. dr. (2.0–4.0 c.c.).
Spiritus Juniperi Compositus, U. S. P. (equivalent to gin)	1–4 fl. dr. (4.0–15.0 c.c.).

Physiologic Action and Therapeutics.—Juniper is a more active diuretic than most of the stimulants to the genito-urinary tract. The compound spirit is useful in increasing the flow of urine in *passive congestion of the kidneys* resulting from chronic heart disease. Its combination with a vegetable salt of potassium enhances its effect. In subacute and chronic inflammatory diseases of the genito-urinary tract it is not generally so useful as copaiba, oil of santal, or buchu.

OLEUM ERIGERONTIS, U. S. P.

(Oil of Erigeron; Oil of Fleabane.)

Oil of erigeron is a volatile oil distilled from the fresh, flowering herb of *Erigeron canadense*, an annual shrub growing in waste places in North America. It is a pale-yellow liquid, having a persistent aromatic odor and taste. The dose is from 5–15 min. (0.3–1.0 c.c.), on sugar, in capsules, or as an emulsion.

Oil of erigeron resembles oil of turpentine in its action, but it is less powerful. It has been used to some extent as a substitute for copaiba and oil of santal in the treatment of *cystitis* and *gonorrhea*. It has an established reputation as an *internal hemostatic* in hemorrhage from the uterus, lungs, intestines, and kidneys, especially when the bleeding is slight but persistent.

BUCHU, U. S. P.

Buchu is the leaves of *Barosma betulina* and *Barosma crenulata*, growing in Southern Africa. Its active principle is a volatile oil, from which is derived *barosma camphor* or *diosphenol*.

PREPARATION.   DOSE.

Extractum Buchu Fluidum, U. S. P. . . . . . ½–1 fl. dr. (2.0–4.0 c.c.).

Physiologic Action and Therapeutics.—In therapeutic doses buchu acts as a stimulant to the mucous membrane of the genito-urinary tract. It is a feeble diuretic. The volatile oil is eliminated by the kidneys, and imparts to the urine an aromatic odor. Its presence also makes the urine more or less antiseptic. Overdoses may produce vomiting, purging, and strangury.

Buchu is useful in the less severe forms of *subacute and chronic cystitis*. It is especially efficacious in *chronic irritability of the bladder*, manifested by frequent desire to urinate. In *subacute cystitis* it may be advantageously combined with an alkali, as in the following formula :

R   Extracti buchu fluidi,   f℥vj (22 5 c.c.) ;
    Potassii citratis,   ℥ss (15.0 c.c.) ;
    Spiritus ætheris nitrosi,   f℥j (30.0 c.c.) ;
    Syrupi limonis,   f℥ij (60 0 c.c.) ;
    Aquæ,   q. s. ad f℥vj (180.0 c.c.).—M.
Sig. A tablespoonful in water every three hours.

UVA URSI, U. S. P.

(Bearberry.)

Uva ursi is the leaves of *Arctostaphylos Uva-ursi*, an evergreen shrub growing in the northern parts of Europe, Asia, and North America. It contains two glucosids,—*arbutin* and *ericolin*,—tannic acid, and an inert, resinous principle, urson. Its virtues depend very largely on arbutin, which appears as white, crystalline needles, of a bitter taste, and freely soluble in alcohol and hot water, but sparingly soluble in cold water. The dose of arbutin is from 2–5 gr. (0.13–0.3 gm.).

PREPARATIONS.   DOSE.

Extractum Uvæ Ursi Fluidum, U. S. P. . . . 1–4 fl. dr. (4.0–15.0 c.c.).
Extractum Uvæ Ursi, U. S. P. . . . . . . 5–30 gr. (0.3–2.0 gm.).

Physiologic Action and Therapeutics.—Uva ursi in therapeutic doses stimulates the mucous membrane of the genito-urinary tract, increases the activity of the kidneys, and renders the urine slightly antiseptic. It is excreted by the kidneys partly unchanged and partly as hydroquinon. Both uva ursi and arbutin may impart to the urine a dark-green color that is probably due to the presence of pigments derived from the oxidation of hydroquinon. As uva ursi contains a considerable quantity (6–8 per cent.) of tannic acid, it has well-marked astringent properties. For the same reason its preparations are incompatible with spirit of nitrous ether. Uva ursi may be employed to meet the same indications as buchu, but it is generally less efficient.

CHIMAPHILA, U. S. P.

(Pipsissewa.)

Chimaphila is the leaves of *Chimaphila umbellata*, a perennial indigenous to North America, Northern Europe, and Northern Asia. It contains tannic acid, arbutin, ericolin, urson, and a neutral, crystalline principle, *chimaphilin.*

PREPARATION. DOSE.
Extractum Chimaphilæ Fluidum, U. S. P. . . . ½–1 fl. dr. (2.0–4.0 c.c.).

Chimaphila has the same action and uses as uva ursi.

PAREIRA, U. S. P.

(Pareira Brava.)

Pareira is the root of *Chondrodendron tomentosum*, a climbing tree growing in Brazil and Peru. It contains tannic acid and an alkaloid, *buxin.*

PREPARATION. DOSE.
Extractum Pareiræ Fluidum, U. S. P. . . . . . ½–2 fl. dr. (2.0–8.0 c.c.).

Pareira has been used in the same class of cases as buchu.

CANTHARIS, U. S. P.

(Cantharides; Spanish Flies.)

Cantharides (see p. 432) in moderate doses is an active stimulant to the mucous membrane of the genito-urinary tract. In large doses it is a powerful irritant, causing intense abdominal pain, vomiting and purging, frequent micturition, dysuria, priapism, and scanty, albuminous, and at times bloody urine. The tincture, in doses of from 1–5 min. (0.06–0.3 c.c.), has been used with advantage in certain cases of *chronic parenchymatous nephritis.* The greatest caution, however, must be exercised in in its administration, since if there be any remnant of an active inflammatory process, or if the drug be used too freely, serious harm may ensue.   Henderson claims to have used cantharides successfully in a case of persistent *hematuria*, in which other remedies had been tried without avail.   The drug has also been recommended in *chronic pyelitis, cystitis*, and *urethritis*.

SEDATIVES TO THE GENITO-URINARY TRACT.

Certain drugs play indirectly the part of sedatives to the genito-urinary tract by removing the causes of irritation. Thus, the alkaline diuretics afford relief from dysuria and frequent micturition, when these symptoms are due to excessive acidity of the urine.   Mineral waters, through their diluent action, soothe the mucous membrane when uric-acid calculi are present.   Again, boric acid, salol, and urotropin, through their antiseptic and acidifying properties, subdue the distressing symptoms resulting from ammoniacal fermentation of the urine.   But apart from these drugs, which allay irritation by changing the character of the urine, there are others which exert a direct sedative influence on the urinary passages themselves.   In this class are belladonna (see p. 70), hyoscyamus (see p. 75), and perhaps zea.   These remedies may be employed for their sedative effect when there is acute inflammation or excessive irritability of the bladder or urethra.

ZEA, U. S. P.

(Corn-silk.)

Zea is the styles and stigmas of *Zea Mays*, or Indian corn. Its active constituents are a resin and *maizenic acid*.

PREPARATIONS.	DOSE.
Extractum Zeæ Fluidum, U. S. P.	½–2 fl. dr. (2.0–8.0 c.c.).
Decoctum Zeæ (5 per cent.)	1–2 fl. oz. (30.0–60.0 c.c.).

Physiologic Action and Therapeutics.—Corn-silk acts as a feeble diuretic and as a sedative to the mucous membrane of the genito-urinary tract.   It is very useful in *acute cystitis* and in the various forms of *vesical irritability*.

URIC-ACID SOLVENTS.

Under this heading will be considered certain drugs that are employed to prevent the precipitation of insoluble urates in the body, or to effect the solution of uratic concretions that have already been formed. The *alkalis* and the *alkaline mineral waters* were among the first agents to be used for these purposes. That these remedies are often beneficial in gout cannot be denied, but that they owe their efficacy to a solvent action is exceedingly doubtful. Uric acid does not, as was formerly supposed, exist as such in the blood or tissues, but is present in the blood chiefly as a soluble quadriurate, and in the tissues as a solid and stable biurate. The only way in which alkalis could affect the quadriurate in the blood would be by hindering its conversion into the insoluble biurate, but there is sufficient experimental evidence to show that these remedies have no such retarding effect. Nor can it be assumed that their usefulness depends upon their influence on the alkalinity of the blood, for it has been shown by Klemperer, Magnus-Levy, and others that the alkalinity of the blood in gout is very little, if at all, diminished, and that, moreover, considerable variations in alkalinity are met with in health.

We are able, by the administration of alkalies, to keep the uric acid in the urine from precipitating, and possibly to redissolve uric-acid gravel. This is accomplished, according to von Noorden, even with a urine that is still weakly acid, by the removal of the monophosphate of sodium, which, in particular, renders the urine insolvent, the biphosphate of sodium remaining as an active solvent. There is, however, very little evidence to support the belief that uratic deposits in the tissues or well-formed calculi in the pelvis of the kidney or bladder can be redissolved by the administration of alkalis. Finally, it has not been proved that the constitutional symptoms of gout, which are so often ameliorated by alkaline medication, are due to the formation of uric acid; on the contrary, there is much testimony to substantiate the view that uric acid itself is not peculiarly toxic. Until we possess a more complete knowledge of the pathogenesis of gout, no adequate explanation of the favorable action of the alkalis can be given. Doubtless much of the good that accrues to gouty patients from sojourns at watering-places is to be attributed to strict regimen, temperate living, agreeable exercise in the open air, and the liberal use of pure water, rather than to the specific action of any salts that may be contained in the waters.

Knowing that *benzoic acid* was eliminated in the urine chiefly as the soluble hippuric acid, Ure assumed that the drug would be efficacious as a uric-acid solvent in gout and nephrolithiasis. In consequence, the acid itself and its salts have been repeatedly recommended in these conditions. But the assumption is opposed to the results of more recent investigations, which have shown that the hippuric acid is formed in the kidney, and not in the blood, and, moreover, that its formation is not attended by a corresponding decrease in the excretion of uric acid.

Piperazin, or **diethylendiamin,** has been highly recommended for its solvent effects in gout and nephrolithiasis, but in the experience of many observers it has proved to be entirely inert. It is prepared by the action of ammonia on ethylene bromid or chlorid, and appears as colorless, acicular crystals, very hygroscopic, and freely soluble in water. Its dose is from 5–10 gr. (0.3–0.65 gm.), well diluted. The only appreciable effect of such doses is a moderate increase in the quantity of urine. The excretion of uric acid, even in the presence of gouty concretions, is not increased. Large doses may cause nausea, malaise, stupor, muscular weakness, tremors, and incoordination. Solutions of 1 or 2 per cent. in normal urine will readily dissolve fragments of uric acid calculi in the test-tube, but the amount of unoxidized piperazin appearing in the urine after ordinary doses have been administered by the mouth has been shown to be without solvent action. Gouty tophi have been removed by the local injections of the drug, but the treatment is said to be very painful. *Lycetol,* or dimethylpiperazin tartrate, and *lysidin,* made by the interaction of ethylendiamin hydrochlorid and sodium acetate, are claimed to be more powerful solvents of uric acid than piperazin.

Tunnicliffe and Rosenheim believe that the *tartrate of piperidin,* a derivative of piperin, in doses of 15 grains (1.0 gm.), is an excellent solvent of gouty concretions. Weiss, Fränkel, Ewald, and others have recently advocated the use of quinic acid, in combination with lithium (urosin) or with piperazin (sidonal). According to Weiss, quinic acid decreases the amount of uric acid in the urine, hippuric acid appearing in its stead.

APHRODISIACS AND ANAPHRODISIACS.

Aphrodisiacs are agents which stimulate the sexual appetite and increase the virile power. Sexual impotence results from a variety of causes; in consequence its treatment, to be successful, must vary accordingly. In one instance the symptom is due to diabetes, in another to locomotor ataxia, and in still another to neurasthenia. The last is by far the most common cause of acquired impotence, and its diverse etiology must be carefully considered in formulating any special line of treatment. General hygienic measures should never be neglected. Judicious physical exercise, fresh air, and careful feeding are necessary to improve the nutritive functions. Abstinence from all sexual excitement is absolutely essential. Iron and arsenic are indicated when there is anemia. Hydrotherapy, massage, and electricity are valuble adjuvants in the treatment. Electricity, locally applied, often proves especially effective. A large electrode—the positive—should be applied over the lumbar spine, while the negative electrode is applied over the spermatic cord or perineum. A current of from 5 to 10 milliampères is sufficient. Hirt has repeatedly seen good effects follow the introduction of a metal-tipped sound into the urethra as far as the fossa navicularis, while the positive pole (anode) is placed over the lumbar cord, the current at the negative pole being made and broken several times. In many instances the psychic element is the all-important factor, and when such is the case, treatment must be directed to the mind rather than to the body. Advice calculated to promote confidence, to give encouragement, and to distract the patient's thoughts from his infirmity will be the most effectual.

The chief drugs for which special aphrodisiac action has been claimed are:

Nux vomica.	Cantharides.
Phosphorus.	Yohimbin.

In certain cases of impotence *nux vomica* proves very useful, not only through its general tonic influence, but also through its stimulant action on the sexual centers of the spinal cord.

Phosphorus is generally believed to possess the power of improving the nutrition of the central nervous system, and, in consequence, to be of value in the various manifestations of nervous exhaustion. Whether this be true or not, small doses of phosphorus are often beneficial in sexual neurasthenia. The drug may be prescribed as phosphorus in doses of from $\frac{1}{100}$–$\frac{1}{50}$ gr. (0.00064–0.0013 gm.), or as phosphid of zinc, in doses of from $\frac{1}{20}$–$\frac{1}{10}$ gr. (0.003–0.006 gm.).

Cantharides in large doses, by irritating the bladder and urethra, sometimes induces sexual emotions and erections, but in doses that can be regarded as safe the drug is entirely without aphrodisiac influence.

Yohimbin is an alkaloid or a mixture of alkaloids obtained

from the bark of the cameroon tree—johimbehe.   According
to Löwey, it stimulates the sexual appetite and produces a
marked congestion of the testes.   The dose is from $\frac{1}{12}-\frac{1}{6}$ gr.
(0.005–0.01 gm.).   Berger and Eulenberg have employed it
with asserted excellent results in sexual neurasthenia.   Krav-
koff, however, who has very recently made a careful study of
yohimbin, finds that the congestion of the sexual organs is
not due to any specific action of the drug, but to its general
vasodilating effect, and that while it possesses no aphrodisiac
influence, it is, in the doses recommended, distinctly toxic.

Orchitic and ovarian extracts, when first introduced, enjoyed
for a time the reputation of being aphrodisiacs.

Anaphrodisiacs are drugs which allay sexual desires. The
most important members of this group are:

Hyoscin.	Camphor.
Bromids.	Lupulin.

Anaphrodisiacs are useful in satyriasis, nymphomania, spermatorrhea, priapism, and chordee.   Hyoscin and the bromids
are the most reliable of all drugs for subduing morbid excitability of the genital organs.   Camphor in itself is of little value,
but the monobromated camphor is often efficacious.

Spermatorrhea and impotence are not infrequently associated
symptoms in neurasthenia, and when such is the case, an anaphrodisiac, such as hyoscin, may be given at bedtime to overcome the irritable weakness of the sexual apparatus, while
general roborant measures are employed to combat the nervous
exhaustion.   In some instances seminal incontinence is the
result of a local cause, such as persistent constipation, hemorrhoids, prostatic hypertrophy, phimosis, extreme sensibility of
the prostatic urethra, or loading of the urine with crystals
of uric acid or calcium oxalate.   It is needless to say that
under these circumstances the removal of the local irritation
is of paramount importance.   The local application of cold
water, or even of ice, avoidance of stimulating food, sleeping
on a hard mattress without too much covering, the adjustment
of some mechanical device to prevent the patient from sleeping
on his back, are measures calculated to lessen the sensibility
of the genital organs.   When the emissions occur toward
morning and are excited by distention of the bladder, it is a
good plan to have the patient form the habit of voiding his
urine at 1 or 2 o'clock in the morning.

EMMENAGOGUES.

Emmenagogues are remedies that promote the menstrual flow. They may act indirectly by removing the remote pathologic cause of amenorrhea, or they may act directly by stimulating the uterus itself. There are very few indications for the employment of direct emmenagogues. In a large class of cases the suppression of the menstrual flow is dependent upon some general cause which has resulted in constitutional enfeeblement; thus it may be due to anemia, persistent constipation, tuberculosis, Bright's disease, extreme obesity, or myxedema. In such cases treatment should be directed to the primary disease, there being rarely any indication to address remedies to the amenorrhea as such. In conjunction with measures calculated to improve the general nutrition *iron* will be found efficacious in chlorosis, purgatives, especially *aloes*, when there is obstinate constipation, and *thyroid extract* when myxedema is the primary factor. In some cases absence of menstruation is dependent upon a pronounced defect in the development of the genital organs or upon atresia of the genital passages. Under such circumstances, emmenagogues would, of course, be of no avail. Again, amenorrhea may be a symptom of uterine or ovarian disease, and when such is the case, it is manifestly improper to goad the affected organs with stimulating drugs. Occasionally suppression is associated with some psychic disturbance, such as grief, anxiety, or fear. In these cases restoration of the catamenia usually occurs after a time without recourse to special medication. Acute suppression from exposure to cold should be treated by rest in bed, hot sitz-baths, hot drinks, and hot applications to the abdomen and genitals. Constipation should be relieved by a laxative dose of aloes, and febrile reaction controlled with such remedies as aconite or spirit of nitrous ether.

The only indications for treating amenorrhea as such by the use of direct uterine stimulants are : (1) When the suppression itself is the cause of painful symptoms—that is, when there is a menstrual molimen without bloody discharge; (2) when the menses do not return upon the restoration of the general health, as is sometimes the case in girls during the developmental period.

The following drugs are the most important direct emmenagogues :

16

Salts of manganese.	Myrrh.
Apiol.	Pennyroyal.
Senecio.	Tansy.
Cantharides.	Savine.
Oxalic acid.	Rue.
Pulsatilla.	

In addition to these drugs, *electricity*, particularly faradism, locally applied, has been highly extolled as a direct uterine stimulant in amenorrhea dependent upon imperfect development of the internal genitals or upon atrophy caused by superinvolution. One pole should be applied over the sacrum and the other above the pubes, or, better, within the uterine cavity.

MANGANUM.

(Manganese; Mn.)

The finding of traces of manganese in the red-blood cells and other tissues led to the recommendation of its salts in the treatment of chlorosis and other forms of anemia, but in the judgment of competent observers the drug is valueless as a hematinic. Ringer was the first to recommend potassium permanganate as an emmenagogue, and there is sufficient evidence at present fully to establish its worth in this respect. It probably acts directly on the uterus, and not indirectly through any influence exerted on the blood. It is most efficacious when given a few days before the expected menstrual period. It should be prescribed in doses of from 1–3 gr. (0.065–0.2 gm.), in pills made with kaolin or soft paraffin, and these should be taken after meals in large drafts of water. As potassium permanganate is a highly irritating salt, and, moreover, as it is at once decomposed in the stomach into the black oxid of manganese, the latter preparation (mangani dioxidum, U. S. P), which is free from irritant properties, may be substituted with advantage. Its dose is from 2–5 gr. (0.1–0.3 gm.).

APIOL.

Apiol is an oleoresin obtained from *Apium Petroselinum*, or garden parsley. It is a greenish oily liquid from which a crystalline stearoptene is obtained, known as *white apiol*. The dose of apiol is from 5–10 min. (0.3–0.6 c.c.), in capsules. When given for a few days before the expected period it is a fairly reliable emmenagogue. It was formerly recommended as a substitute for quinin in *malaria*, but it proved to be useless.

SENECIO.

(Ragwort; Life Root.)

Senecio 'is the entire plant of *Senecio aureus*, a perennial herb growing in the northern and western parts of the United States. In the form of a fluid extract or tincture it has been highly recommended as a corrective of *functional amenorrhea and dysmenorrhea.* Murrell compares its action to that of potassium permanganate. The dose of the fluid extract is from ½–1 fl. dr. (2.0–4.0 c.c.); of the tincture, 1–2 fl. dr. (4.0–8.0 c.c.).

CANTHARIS, U. S. P.

(Cantharides.)

The stimulant influence of cantharides on the uterus is probably the indirect effect of its irritant action on the urinary passages. Small doses of the tincture—2–3 min. (0.12–0.2 c.c.)—are generally prescribed in combination with other members of the same class, as in the well-known emmenagogue mixture of Dewees, which Wood regards as the most effective combination ever made in *atonic amenorrhea.* The formula of Dewees' mixture is as follows:

 ℞ Tincturæ ferri chloridi,   f℥iij (11.0 c.c.);
   Tincturæ cantharidis,   f℥j (4.0 c.c.);
   Tincturæ aloes,   f℥ss (15.0 c.c.);
   Tincturæ guaiaci ammoniatæ, f℥iss (45.0 c.c.);
   Syrupi,   q. s. ad f℥vj (180.0 c.c.).—M.
  Sig. A tablespoonful three times a day.

ACIDUM OXALICUM.

(Oxalic Acid; $H_2C_2O_4 + H_2O$.)

Oxalic acid occurs as transparent, prismatic crystals of a very acid taste. It is readily soluble in water and in alcohol. **Physiologic Action and Therapeutics.**—Oxalic acid is a powerful uterine stimulant, in full doses quite capable of exciting labor pains and of expelling the product of conception. As Poulet first demonstrated, it is an efficient emmenagogue if given in doses of ¼–⅓ gr. (0.016–0.02 gm.), thrice daily. Larger doses should not be given. Talley has reported a case of acute poisoning in an anemic girl from the administration of three doses of ½ gr. (0.03 gm.) each, at intervals of four hours. On account of its irritant properties it should be taken well diluted, after meals.

Oxalic acid is an active germicide, and its saturated solution has been highly recommended by Kelly and others as a *disinfectant for the hands.*

Toxicology.—The poisonous properties of oxalic acid do not depend solely upon its corrosive action, for the neutral oxalates may also cause death by paralyzing the central nervous system. The resemblance of the drug to Epsom salts has led to many fatal mistakes. When swallowed in a concentrated form it causes the usual symptoms of an irritant poison; when taken in dilute form it is absorbed and induces muscular weakness, cyanosis, coma, and collapse. The chemical antidote is chalk or lime, which acts by forming an insoluble and inert calcium oxalate. The salts of potassium and sodium cannot be used as antidotes since they form soluble and poisonous oxalates.

MYRRHA, U. S. P.

(Myrrh.)

Myrrh is a gum-resin obtained from *Commiphora Myrrha*, a small tree growing in eastern Africa and Arabia. It appears in the form of brownish-red irregular-shaped tears, having an agreeable aromatic odor and a bitter acrid taste.

PREPARATIONS.	DOSE.
Tinctura Myrrhæ, U. S. P.	10–30 min. (0.6–2.0 c.c.).
Tinctura Aloes et Myrrhæ, U. S. P. (10 per cent. of each)	½–2 fl. dr. (2.0–8.0 c.c.).
Mistura Ferri Composita, U. S. P. (Griffith's mixture; ferrous sulphate, 6; myrrh, 18; sugar, 18; potassium carbonate, 8; spirit of lavender, 60; rose-water, to make 1000)	4–8 fl. dr. (15.0–30.0 c.c.).
Pilulæ Aloes et Myrrhæ, U. S. P. (aloes, 2 gr.—0.13 gm.; myrrh, 1 gr.—0 06 gm.)	1–4 pills.
Pilulæ Rhei Compositæ, U. S. P. (aloes, 1½ gr.—0.1 gm.; rhubarb, 2 gr.—1.3 gm.; myrrh, 1 gr.—0.06 gm.).	1–5 pills.

Therapeutics.—Myrrh is used as an emmenagogue and as a stimulant to mucous membranes. When employed as an emmenagogue, it is usually associated with iron or aloes. The tincture, diluted with water, or with a weak solution of borax or potassium chlorate has been extensively employed as a local application in *ptyalism, ulcerative stomatitis*, and *spongy gums*. Myrrh was also used at one time as a stimulant expectorant in *chronic bronchitis*, but it has been replaced by more efficient remedies.

HEDEOMA, U. S. P.

(Pennyroyal.)

Pennyroyal is the leaves and tops of *Hedeoma pulegioides*, an annual herb indigenous to North America. It contains a volatile oil, a bitter principle, and tannin.

<pre>
PREPARATIONS.   DOSE.
Oleum Hedeomæ, U. S. P. . . . . 3–10 min. (0.2–0.6 c.c.).
Infusum Hedeomæ . . . . 1–2 fl. oz. (30.0–60.0 c.c.).
</pre>

Therapeutics.—Pennyroyal, in the form of a hot infusion, has a popular reputation as an emmenagogue which is not altogether unmerited. It is especially employed in acute suppression brought on by exposure.

TANACETUM, U. S. P.

(Tansy.)

Tansy is the leaves and tops of *Tanacetum vulgare*, a perennial herb growing wild in Europe and Asia, and naturalized in North America. Its active principle is a volatile oil.

<pre>
PREPARATION.   DOSE.
Oleum Tanaceti . . . . 1–3 min. (0.06–0.2 c.c.).
</pre>

Physiologic Action and Therapeutics.—Tansy has long been used as a domestic remedy in *amenorrhea*, the herb itself being generally employed in doses of from ½–1 dr. (2.0–4.0 gm.), in decoction. The oil has also been taken frequently in order to produce abortion, many times with ill success, but with grave or even fatal results to the mother. Large doses of the oil cause severe abdominal pain, vomiting, epileptiform convulsions, coma, and collapse. Tansy was formerly employed as an anthelmintic, but this use of the drug is now obsolete.

SABINA, U. S. P.

(Savine.)

Savine is the tops of *Juniperus Sabina,* an evergreen shrub growing in northern Europe, Asia, and America. Its active constituent is a volatile oil.

<pre>
PREPARATIONS.   DOSE.
Oleum Sabinæ, U. S. P. . . . . 3–6 min. (0.2–0.4 c.c.).
Extractum Sabinæ Fluidum, U. S. P. . . . . . 5–15 min. (0.3–1.0 c.c.).
</pre>

Physiologic Action and Therapeutics.—The action of oil of savine resembles that of oil of juniper, but it is more irritant. Large doses cause abdominal pain, vomiting, purging, anuria, unconsciousness, and collapse. In pregnant women the intoxication generally culminates in abortion. Owing to the fact that expulsion of the fetus does not occur until the irritation of the intestinal canal and kidneys is sufficiently violent to endanger life, many cases of fatal poisoning have resulted from the use of the drug as an abortifacient. Oil of savine has long borne a reputation in *amenorrhea* and in certain forms of *menorrhagia*, but at the present time it is rarely prescribed.

RUTA.

(Rue.)

Rue is the leaves of *Ruta graveolens*, a perennial plant indigenous to Southern Europe, and cultivated in this country. It contains a volatile oil. Its action is similar to that of savine. The dose of the oil is from 3–6 min. (0.2–0.4 c.c.).

PULSATILLA, U. S. P.

Pulsatilla (see p. 66) has been recommended in sudden suppression of menstruation induced by chill, but its use as an emmenagogue is superfluous.

OXYTOCICS, OR ECBOLICS.

Oxytocics are drugs that tend to accelerate parturition by increasing the expulsive force of the uterus. It is conceivable that they may act directly by stimulating the unstriped muscle of the uterus, or indirectly by influencing a center for uterine contraction supposed to exist in the lumbar spinal cord. The chief oxytocics are:

Ergot.	Hydrastis.
Quinin.	Cotton-root bark.
Corn smut.	

Indications.—Oxytocics may be employed during labor to hasten the expulsion of the fetus; at the close of labor, to prevent or to check postpartum hemorrhage; and in the puerperium to overcome certain forms of subinvolution.

ERGOTA, U. S. P.

(Ergot; Ergot of Rye; Secale Cornutum.)

Ergot is the sclerotium of a parasitic fungus, *Claviceps purpurea*, which replaces the grain of common rye, *Secale cereale*. The sclerotium represents a stage in the life-history of the fungus which is intermediate between that of the mycelium or spawn, and that of the spore-bearing thallus. Ergot is obtained principally from Russia, Germany, and Spain. The official preparation is horn-shaped, about an inch long (2 or 3 cm.) and ⅛ inch (3 mm.) thick; externally it is purplish-black, and internally, whitish with pinkish striæ. It has a peculiar heavy odor and an oily, disagreeable taste. The chemistry of ergot has not been definitely determined. Kobert has separated in an impure state three bodies—*ergotinic acid, cornutin,* and *sphacelinic acid.* None of these constituents represents completely the activities of the drug. Ergotinic acid is a protoplasmic poison, and when injected intravenously, produces inflammation of serous and mucous membranes, disintegration of the red blood-cells, and wide-spread ecchymoses; cornutin excites the central nervous system and causes general convulsions; and sphacelinic acid induces gangrene. Jacobi has recently isolated a pure body, *sphacelotoxin,* which he believes represents the active principle of sphacelinic acid.

Preparations.	Dose.
Vinum Ergotæ, U. S. P. (15 per cent.)	1–8 fl. dr. (4.0–30.0 c.c.).
Extractum Ergotæ Fluidum, U. S. P.	½–1 fl. dr (2.0–4.0 c.c.).
Extractum Ergotæ, U. S. P.	5–15 gr. (0.3–1.0 gm.).
Ergotin (Bonjean's Ergotin: a purified hydro-alcoholic extract of ergot)	3–10 gr. (0.2–0.65 gm.).

Physiologic Action.—Considerable difference of opinion exists concerning the physiologic action of ergot. This is in part due, no doubt, to the fact that the activities of the drug are materially affected by age and by the season of the year in which it is gathered. We know with certainty, however, that it increases the parturitive action of the uterus; that in sufficient doses it constricts the arterioles in certain parts of the body, in consequence of which the arterial pressure is somewhat raised; and that its prolonged administration is often followed by gangrene.

Circulatory Action.—The subcutaneous injection of ergot in moderate doses is followed by constriction of the arterioles, with an increase in the arterial pressure, which is due to the stimulant effect of the drug (sphacelotoxin) on the vasomotor center, and probably, also, to its direct action on the muscular coats of the vessels. When the injection is made intravenously, the blood-pressure is first lowered and then raised, the primary fall, according to Wood, being due to temporary depression of the heart from direct contact with the drug. After toxic doses there may be a marked and continuous fall in the blood-pressure from depression of both the vasomotor center and the heart. The rise of blood pressure is often accompanied by slowing of the heart, which seems to be due to stimulation of the inhibitory center in the medulla. The gangrene observed in chronic poisoning with ergot is generally attributed to prolonged constriction of the arterioles, but it is probably due more directly to degenerative changes in the vessel-walls and the consequent formation of hyaline thrombi (Von Recklinghausen, Grigorjeff, Winogradow, and Grün-feld).

Nervous System.—Ordinary doses of ergot, if not long con-tinued, have no appreciable influence on the nervous system. In one form of chronic poisoning, however, nervous phenomena are pronounced. These include paresthesia, anesthesia, con-vulsions, tonic contractions of the limbs, ataxia, mental dulness, and dementia. Grigorjeff has found postmortem in animals slowly poisoned with ergot degenerative changes in the central nervous system, especially in the posterior columns of the spinal cord.

Alimentary Canal.—Ergot increases the peristaltic move-ments of both the stomach and intestine, probably through its direct action on their muscular coats. Even in therapeutic doses it may excite nausea and vomiting.

Uterus.—In the parturient woman small doses of ergot induce uterine contractions which often closely resemble the rhythmic contractions observed in normal labor. After full doses, however, the contractions usually become very power-ful, and uninterrupted or tetanic. The preparations of the drug on the market vary so much in activity that it is quite difficult to predict the effect of an average dose in a given case. The influence of ergot on the gravid but non-parturient uterus is less marked, but there is ample evidence to show that in animals it is capable of originating expulsive movements that culminate in abortion. It is not definitely known whether the ecbolic action of ergot is due to direct stimulation of the unstriped muscle of the uterus itself, or to an influence exerted on the uterine center in the spinal cord. It is probably in part, at least, of centric origin.

Muscles.—On the voluntary muscle ergot has no effect, but it seems to have a stimulant influence on the involuntary muscle throughout the body.

Toxicology.—In *acute ergot poisoning* the symptoms most commonly observed are nausea, vomiting, intense thirst, cold-ness of the surface, headache, stupor, and generalized purpura, The pulse may be unduly slow, or, after very large doses. rapid and feeble.

Chronic ergot poisoning has not been uncommon. In nearly every instance it has resulted from eating bread made of ergotized grain. It assumes two forms: Spasmodic ergotism and gangrenous ergotism. The first is characterized by malaise, inordinate craving for food, vomiting, diarrhea, headache, disordered vision, tinnitus aurium, formication, numbness of the extremities, ataxia, convulsive seizures, tonic contractions of the muscles, and mental hebetude, followed by delirium or acute dementia.

Gangrenous ergotism is characterized by exhaustion, coldness of the surface, cyanosis, aching of the limbs, and the development of gangrenous areas in various parts of the body, especially in the fingers and toes. Cataract has also been frequently observed.

Therapeutics.—Ergot may be used during parturition, cautiously and in small doses, to hasten the expulsion of the fetus, when the cause of the delay is simply *uterine inertia*, and not mechanical obstruction, but it is very rarely necessary. Its use in full doses, especially when dystocia is the result of contracted pelvis, rigid os uteri, displacement of the uterus, abnormal presentation, or other physical abnormity, may prove disastrous by exciting tetanic contraction and consequent asphyxia of the child or even rupture of the uterus. At the close of labor, after the head is born, ergot is very useful in *preventing postpartum hemorrhage*. It may be used also after the removal from the uterus of any abnormal contents—retained placenta, clots, etc.—to control hemorrhage that has already occurred. It is a useful adjuvant in the treatment of certain cases of *subinvolution of the uterus*, but the possibility of its affecting the child through the mother's milk must be borne in mind. The treatment of *uterine fibroids* with repeated injections of ergot, as recommended by Hildebrandt in 1872, has met with a very limited success. By contracting the blood-vessels of the uterus, however, the drug may temporarily control hemorrhage, and in rare instances check the growth of the tumor, or even bring about its involution. The treatment is adapted only to small intramural fibroids that are soft and vascular.

While the efficacy of ergot in certain forms of metrorrhagia has been conclusively proved, there is no satisfactory evidence that the drug is of value in other forms of hemorrhage, such as hemoptysis, hematuria, or hematemesis. On the contrary, our knowledge of its physiologic action indicates that it may be productive of harm. Granting that ergot can lessen the caliber of eroded, degenerated, or seriously wounded vessels, there is no reason for supposing that its influence is restricted to the vessels concerned in the bleeding, and if such is not the case, the universal constriction of the arterioles induced by the drug, and the consequent rise in the general blood pressure must favor the escape of blood from the injured vessels and militate against the formation of an occluding thrombus. Ergot has been used with some success in *diabetes insipidus* (Da Costa, Ringer), in *colliquative sweats* (Wood), and in *headaches of a migraine type* (Thomson). It has also been recommended in a variety of other affections, such as diabetes mellitus, acute meningitis, myelitis, splenic enlargements, acute dysentery, and hyperemic skin diseases, but it has not borne the test of experience.

Administration.—The preparations of ergot on the market are exceedingly variable in strength; some are absolutely inert. The best preparation for internal use is a good fluid extract. It is less liable to excite nausea when taken after meals. In urgent cases the drug should be given hypodermically, notwithstanding its tendency to produce local inflammation. The injections should not be made directly beneath the skin, but deeply into a muscle. An aqueous solution of the extract, freshly prepared and filtered, may be employed. Bonjean's ergotin is unreliable.

OTHER OXYTOCICS.

Quinin (see p. 410).—This drug has some power of increasing the rapidity and energy of the uterine contractions when they have once begun; its action, however, is inconstant, and is comparable to that of strychnin. Unlike ergot, quinin causes intermittent and not tetanic contractions, and probably effects its object through its general stimulant properties rather than through any special influence exerted upon the uterus itself. As it has no power of originating labor pains, it is in no sense an abortifacient, and the cases of abortion that have been reported as following its administration have no doubt been due either to an idiosyncrasy or to the original disease for which the quinin was prescribed. According to some obstetricians, its use during labor distinctly increases the tendency to postpartum hemorrhage.

Hydrastis (see p. 375).—Both the natural alkaloid of this drug, *hydrastin*, and its artificial derivative, *hydrastinin*, are of considerable value in controlling hemorrhage from the uterus. It is definitely known that both drugs induce constriction of the peripheral vessels, but whether their efficacy in hemorrhage is due, in part, to their power of exciting contraction of the uterine muscle is a moot question. Many writers, however, maintain that these alkaloids have pronounced ecbolic properties, and are capable, in sufficient dose, of producing abortion.

As a uterine hemostatic hydrastinin is preferable to hydrastin. It is particularly efficacious in *simple menorrhagia* of young girls and in *metrorrhagia* following abortion or resulting from disease of the adnexa. Hemorrhages resulting from the presence of cancer or fibroids are not much affected by the drug. The dose of the official hydrastinin hydrochlorate is from $\frac{1}{2}$–1 gr. (0.03–0.065 gm.). When a prompt effect is required, it should be given hypodermically in the form of a 10 per cent. aqueous solution.

Cotton-root Bark.—This drug (*Gossypii Radicis Cortex*, U. S. P.) is said to have been much valued among the slaves of the South as an abortifacient. Some obstetricians have claimed for it an action resembling that of ergot. The dose of the official fluid extract as an ecbolic is from $\frac{1}{2}$–2 fl. dr. (2.0–8.0 c.c.).

Corn Smut.—This drug (*Ustilago Maydis*) is a fungus growing on the stems and tassels of *Zea. Mays*, or Indian corn. According to some obstetricians, the fluid extract, in doses of from $\frac{1}{2}$–1 fl. dr. (2.0–4.0 c.c.), induces in the parturient uterus vigorous but intermittent contractions, and is, therefore, a useful stimulant when the labor pains are feeble from exhaustion.

DIAPHORETICS, OR SUDORIFICS.

Diaphoretics are agents which promote the secretion of sweat. They may act directly by stimulating the special nerve-fibers supplying the sweat-glands, indirectly by stimulating the cells in the central nervous system (sweat-centers) from which those fibers originate, or reflexly by irritating peripheral sensory nerves. The activity of the sweat-glands is increased by dilatation of the peripheral vessels, external heat, muscular exercise, dyspnea, strong emotion, nausea, and various drugs.

Although dilatation of the cutaneous vessels promotes perspiration when other conditions are not unfavorable, the mechanism of sweat secretion is in large measure independent of the blood-flow. Thus, in fever the flushed skin may be unnaturally dry, while in collapse, with pallid skin, there may be free perspiration. External heat may act partly by inducing hyperemia of the skin, but that it acts chiefly by stimulating the sweat-fibers reflexly through the cutaneous sensory nerves appears evident from the fact that exposure of a part to a high temperature does not cause sweating if all the nerves leading to the part be severed. Muscular exercise probably acts by increasing the flow of blood to the glands and by stimulating the peripheral nerves and nerve-centers. Dyspnea acts by increasing the proportion of venous blood, the latter being a powerful stimulant of the sweat-centers. Strong emotions affect secretion probably also by action on the nerve-centers. The sweating that attends nausea is, no doubt, induced reflexly by stimulation of the sweat-centers through the sensory nerves of the stomach. Among drugs, alcohol probably acts mainly by increasing the quantity of blood flowing through the cutaneous vessels; opium, by stimulating the sweat-centers, and pilocarpin, by stimulating directly the endings of the nerve-fibers in the glands. The drugs most commonly employed as diaphoretics are:

Pilocarpus (pilocarpin).   Spirit of nitrous ether.
Opium.   Ammonium acetate.
                  Warburg's tincture.

Indications.—Diaphoretics are given: (1) To promote the absorption of dropsical fluid; (2) to rid the system of peccant matters; and (3) to arrest the development of local congestions or inflammations that have resulted from exposure to cold.

PILOCARPUS, U. S. P.

(Jaborandi.)

According to the Pharmacopeia pilocarpus is the leaflets of *Pilocarpus Selloanus* and of *Pilocarpus Jaborandi*, shrubs growing in South America. The pilocarpus of the market, however, consists very largely of the leaflets of an allied plant —*Pilocarpus microphyllus*. It contains three alkaloids—*pilocarpin, pilocarpidin*, and *isopilocarpin*—which produce similar effects, but pilocarpin is the most active, and is present in larger quantity ($\frac{1}{4}$–$\frac{1}{2}$ per cent.) than the others. Jowett, in a recent investigation, was unable to find in the leaves at present obtainable any trace of an alkaloid answering to the description of *jaborin*, discovered by Harnack and Meyer, and which is said to resemble atropin in its action.

Pilocarpin yields soluble crystalline salts, of which the nitrate and the hydrochlorate (Pilocarpinæ Hydrochloras, U. S. P.) are most commonly employed. The dose of either of these salts is from $\frac{1}{12}$–$\frac{1}{2}$ gr. (0.005–0.03 gm.).

PREPARATION.   DOSE.

Extractum Pilocarpi Fluidum, U. S. P. . . 20–60 min. (1.0–4 0 c.c.).

Physiologic Action.—Secretion.—The dominant action of pilocarpin is on the secretory glands of the body. Under its influence the sweat and saliva are enormously increased. It increases also, but to a less extent, the lacrimal, nasal, bronchial, gastric, intestinal, and renal secretions. Milk and bile are probably not increased. As its effect on secretion is completely suspended by atropin, it would seem to be due to stimulation of the peripheral fibers of the nerves supplying the glands.

Circulatory System.—The action of pilocarpin on the circulation has not been definitely determined. In certain animals, like the frog, cat, and rabbit, it appears first to stimulate and then to depress the peripheral fibers of the vagus, in consequence of which the pulse is primarily slowed and then quickened. After large doses the pulse again becomes slow, and the blood pressure falls considerably below normal, owing to weakness of the heart and vasomotor center. In man the circulation appears to be affected in a somewhat different manner; the pulse-rate is usually increased, the peripheral vessels are dilated, and the blood pressure is lowered. Not infrequently, however, there is a distinct rise in the arterial pressure, especially at first. Although the evidence, derived from experimental sources, is in some respects contradictory, clinical experience is in accord with the view that pilocarpin, at least in full doses, is a circulatory depressant.

Nervous System.—In therapeutic doses pilocarpin has no action on the central nervous system. In certain animals toxic doses induce convulsions which are probably of spinal origin.

Respiration.—Small doses are without effect; toxic doses depress the respiratory center. In the presence of a weakened circulation the drug tends to induce pulmonary edema.

Muscles.—Pilocarpin excites contractions in the unstriped muscle of the iris, stomach, intestines, bladder, uterus, and spleen, probably by stimulating the terminations of the nerve-fibers supplying the muscle.

Alimentary Canal.—Nausea and vomiting are sometimes induced by jaborandi or pilocarpin, especially when the drug is given by the mouth. This effect is probably caused by its irritant action on the peripheral nerves of the stomach. It also increases both gastric and intestinal peristalsis.

Eye.—Pilocarpin, especially when applied locally, contracts the pupil by stimulating the terminal filaments of the oculomotor nerve. It also decreases the intra-ocular tension.

Kidneys.—In small doses it increases the quantity of urine.

The urea, certainly in disease, and probably in health, is also increased.

Blood.—According to Horbaczewski, it increases the number of leukocytes in the blood, and in consequence the quantity of uric acid in the urine.

Temperature.—Coincident with the profuse diaphoresis which the drug excites, and probably dependent upon it, there is a well-marked fall in the body-temperature.

Toxicology.—The chief phenomena of pilocarpin poisoning are salivation, profuse sweating, nausea and vomiting, serous purging, contraction of the pupils, muscular tremors, and collapse. Atropin is an effective physiologic antidote.

Idiosyncrasy and Untoward Effects.—The action of pilocarpin in man is by no means constant. Occasionally even small doses are followed by severe cardiac depression or pulmonary edema. Swelling of the salivary glands, profuse salivation, and strangury have also been observed after its administration. On the other hand, there may be a peculiar insusceptibility to its action, as in a case reported by Hare, in which a woman of thirty years received $\frac{3}{4}$ gr. (0.05 gm.) of pilocarpin hydrochlorate hypodermically in half an hour without any effect. According to Ringer, children are far less susceptible than adults.

Therapeutics.—Pilocarpin in doses of from $\frac{1}{16}$–$\frac{1}{12}$ gr. (0.004–0.005 gm.) is very useful in promoting moderate diaphoresis in *acute and chronic parenchymatous nephritis*. It tends to avert uremia by relieving the congestion of the kidneys and by ridding the blood of poisonous matters. Its action should be supplemented by external heat, and any depression or faintness should be combated with strychnin or ammonia. The remedy may be repeated every other day or even every day. Samuel West believes that there is no more useful drug in *chronic interstitial nephritis* than pilocarpin in small doses by the mouth. He finds that headache, irritability, digestive disturbances, and dry skin often yield rapidly to a dose or two. In *acute uremia* larger doses, $\frac{1}{6}$–$\frac{1}{3}$ gr. (0.01–0.02 gm.) may be employed in conjunction with hot-air or vapor baths. In *puerperal eclampsia* the drug is very often badly borne. In *general dropsy*, especially that due to Bright's disease, pilocarpin is a valuable aid to absorption. In local effusions, however, such as occur in the pleura and pericardium, it is rarely effective. It is sometimes employed for its diaphoretic effect in *muscular rheumatism, acute coryza*, and *acute sthenic infections*.

Daily injections of pilocarpin ($\frac{1}{6}$ gr.—0.01 gm.), as first employed by Politzer in 1880, have given gratifying results in some forms of *acute and subacute labyrinthine disease*, characterized by deafness, tinnitus aurium, and vertigo.

The action of pilocarpin on the eye is much like that of eserin, but it is decidedly less powerful. It has been used to lessen intra-ocular tension in *glaucoma*, but eserin is more reliable. When, however, a gentle but persistent stimulant effect on the ciliary muscle is desirable, as in cases of *accommodative asthenopia*, pilocarpin will be found very serviceable. As an adjuvant to mixed treatment daily injections of pilocarpin have sometimes proved efficacious in intractable cases of *syphilitic keratitis*. There is sufficient evidence to prove that the internal administration of pilocarpin promotes the growth of hair, and at the same time renders it darker and coarser. Favorable results from the hypodermic use of the drug in *extensive alopecia* and *alopecia areata* have been reported by Simon, Pick, Morris, Pringle, and others, but the practice of applying the remedy in the form of lotions to the scalp in premature baldness is of doubtful utility.

Contraindications.—The existence of grave cardiac or pulmonary lesions prohibits its use. Caution should be exercised in giving pilocarpin to women advanced in pregnancy, as the drug possesses some power as an abortifacient.

Administration.—On account of the uncertain alkaloidal strength of the fluid extract and its greater tendency to cause vomiting, one of the salts of pilocarpin is usually preferable. The latter may be given by the mouth, but when a prompt effect is desired, it should be given hypodermically. Its diaphoretic action is greatly facilitated by the application of heat to the patient. Circulatory failure should be combated with ammonia, strychnin, or alcohol, and pulmonary edema with hypodermic injections of atropin. According to Tyson, a freshly prepared infusion may be injected into the rectum with prompt results. Four ounces (120.0 c.c.) of hot water should be poured on a dram (4.0 gm.) of jaborandi leaves, and, when sufficiently cool, strained and injected. To secure a diuretic effect small doses of pilocarpin ($\frac{1}{20}$ gr.—0.003 gm.) should be given by the mouth every three or four hours, and the patient kept lightly covered.

OTHER DIAPHORETICS.

Opium (see p. 85).—As a diaphoretic, opium is usually employed in the form of Dover's powder, or powder of ipecac and opium. This preparation, in doses of from 5–10 gr. (0.3–0.6 gm.), produces a mild sudorific effect, and is useful in breaking up *acute catarrhal conditions of the respiratory organs* and in relieving so-called *muscular rheumatism.* It should be borne in mind that large doses not infrequently cause nausea or even vomiting when the stomach is sensitive.

Spirit of Nitrous Ether (see p. 227).—Sweet spirit of niter (Spiritus Ætheris Nitrosi, U. S. P.) is an alcoholic solution of ethyl nitrite, yielding, when freshly prepared, not less than eleven times its own volume of nitrogen dioxid, the equivalent of about 4 per cent. of pure ethyl nitrite. It is a clear, volatile, inflammable liquid, having a pale yellowish or greenish-yellow tint, a pleasant ethereal odor, and a sharp, burning taste. On exposure to light and air it rapidly deteriorates. The dose for an adult is from ½–2 fl. dr. (2.0–8.0 c.c.); for a child, 10–30 min. (0.6–2.0 c.c.).

In medicinal doses nitrous ether acts as a mild diaphoretic, diuretic, and antispasmodic. When the patient is kept well covered after its administration, its diaphoretic effect is more pronounced than its diuretic effect, and the reverse is true when the patient is lightly covered. Toxic doses produce symptoms resembling poisoning by amyl nitrite and other nitrites. Sweet spirit of niter is a useful remedy in the *mild febrile affections of children;* it is best given in small doses, well diluted with hot water, at frequent intervals. As a diuretic it is efficacious in relieving *oliguria* due to *fever* or *acute congestion of the kidneys.*

Incompatibles.—Iodids, ferric sulphate, mucilage of acacia, tannic acid, tincture of guaiacum, and antipyrin, acetanilid, etc.

Ammonium Acetate.—This compound is employed in the form of its official solution, the spirit of mindererus (Liquor Ammonii Acetatis, U. S. P.). According as the patient is kept warm or cool it acts as a feeble diaphoretic or diuretic. It is often used in *febrile diseases* as a vehicle for spirit of nitrous ether or aconite. Its dose is from 1–4 fl. dr. (4.0–15.0 c.c.).

Warburg's Tincture (see p. 416).—This preparation in doses of ½ fl. oz. (15.0 c.c.) is capable of inducing copious sweating. To be effective, it should be given undiluted, and drinks should be withheld. It has been highly commended in the *malignant types of malarial fever.*

ANTIHYDROTICS.

Antihydrotics are agents that check the secretion of sweat. Theoretically, they may act by lessening the irritability of the sweat-centers; by depressing the terminal fibers of the secreting nerves, or, possibly, the secretory gland-cells themselves; or by stimulating the respiratory center. Brunton has called attention to the close relation existing between respiratory depression and the occurrence of colliquative sweats. He states that when the respiratory center is exhausted from any cause it responds less readily than the sweat-centers to the stimulating influence of venous blood, in consequence of which profuse sweating results. This teaching accords with the well-known fact that respiratory stimulants are frequently efficacious in controlling the night-sweats of phthisis. The most important antihydrotics are:

Atropin.	Agaricin.
Sulphuric acid.	Picrotoxin.
Camphoric acid.	Tellurium compounds.
Thallium acetate.	

Gallic acid, ergot, and *zinc oxid* have also been employed, but they are of little value. In addition to the internal remedies, certain external applications are useful in controlling excessive perspiration; thus, sponging the surface with *vinegar* and water or a hydro-alcoholic solution of *tannic acid* or *alum* sometimes affords relief. A weak solution of *formalin,* as recommended by Hirschfeld, is a powerful antihydrotic, but its irritant effects on the eyes and respiratory tract forbid its use when the sweating is general. Recently, Strasburger has substituted for the formalin solution a dusting-powder of *tanno-form,* a condensation product of tannin and formaldehyd. The application, at bedtime, of a powder composed of 1 part of tannoform to 2 parts of talcum, is said to render excellent service in preventing the night-sweats of phthisis.

In local hyperidrosis internal remedies are rarely of benefit; the best results are obtained from external applications, the most effective being dusting-powders containing tannoform, salicylic acid, or boric acid, or a pigment of formalin and alcohol (equal parts).

Atropin (see p. 70).—This is the most powerful of the antihydrotics. It checks the secretion of the sweat-glands by paralyzing the terminations of the secretory nerves. In doses of from $\frac{1}{200}-\frac{1}{100}$ gr. (0.00032–0.00064 gm.) at bedtime it often effectually controls the exhausting *night-sweats of phthisis;* unfortunately, however, its repeated use is frequently attended by untoward symptoms, such as dryness of the throat, thirst, and dimness of vision. The tincture of belladonna is sometimes employed as a lotion in *local hyperidrosis* occurring about the hands and feet.

17

Sulphuric Acid (see p. 441).—Aromatic sulphuric acid has been used longer and perhaps more extensively than any of the other antihydrotics. Its power, however, compared with that of drugs more recently introduced, is very feeble. It should be given in doses of from 5–10 min. (0.3–0.6 c.c.) largely diluted with cool water.

ACIDUM CAMPHORICUM.

(Camphoric Acid; $C_8H_{14}(COOH)_2$.)

Camphoric acid is a crystalline powder made by oxidizing camphor with nitric acid. It is odorless, of a slightly acid taste, soluble in 160 parts of cold water, and readily soluble in alcohol and ether. The dose is from 20–30 gr. (1.3–2.0 gm.). On account of its sparing solubility in water it is best prescribed in powders, in capsules, or in an elixir.

Therapeutics.—Camphoric acid is an antiseptic and antihydrotic. As an antiseptic it has been used both internally and externally, but in this respect it is inferior to many other drugs in common use. It is of much greater service in checking *excessive perspiration*, but the manner of its action is unknown. As it is rapidly eliminated in the urine it is best given about two hours before the time at which the sweating is likely to occur. Its drawback is its tendency to cause nausea and vomiting.

AGARICIN.

Agaricin is a crystalline body obtained from *Polyporus officinalis*, a fungus growing on the larch tree. The fungus itself (white agaric or purging agaric), although an active irritant, was recommended for excessive sweating by de Haen as early as 1768. Agaricin is an impure form of *agaric* or *agaricic acid*, which appears as a white, lustrous powder, odorless, almost tasteless, and slightly soluble in water and alcohol. The dose of agaricin is from $\frac{1}{4}$–1 gr. (0.016–0.065 gm.), and of agaric acid, $\frac{1}{6}$–$\frac{1}{2}$ gr. (0.01–0.03 gm.). The latter is the less irritating and more reliable preparation.

Physiologic Action and Therapeutics.—When administered by the mouth in large doses agaricin acts as an irritant and induces severe vomiting and purging. According to Hofmeister, when it is administered to animals subcutaneously or intravenously in sufficient dose it first stimulates and then depresses the centers of the medulla oblongata, death ultimately resulting through cardiac or respiratory paralysis. The drug is chiefly interesting to the therapeutist on account of its effect on the perspiration. Like atropin, it arrests the secretion of sweat probably by paralyzing the nerve-endings in the sudoriferous glands. Unlike atropin, however, it has no influence on the salivary or lacrimal secretions.

Agaric acid is preferable to agaricin, being less liable to cause nausea. As it is absorbed slowly it should be given several hours before the time of the expected sweating. It may be prescribed in the form of a pill or powder.

PICROTOXINUM, U. S. P.

(Picrotoxin.)

Picrotoxin is a neutral principle obtained from fish berries— *Anamirta cocculus* and *Anamirta paniculata*, growing in India and the Malayan Islands. It is a colorless, shiny, crystalline powder, odorless, and of an intensely bitter taste. It is soluble in alcohol and in acidulous or alkaline solutions, but only sparingly so in water. The dose is from $\frac{1}{80}-\frac{1}{30}$ gr. (0.0008–0.002 gm.).

Physiologic Action and Therapeutics.—The dominant action of picrotoxin appears to be upon the centers in the medulla oblongata. In animals toxic doses produce violent clonic convulsions, acceleration of the respirations, slowing of the pulse, a marked rise in the arterial pressure, and vomiting. Death often results from asphyxia induced by the convulsive seizures.

Picrotoxin has been recommended in a variety of conditions, but it is mainly useful in controlling the *night-sweats of phthisis*. The manner of its action is unknown; but it may act indirectly, as Cushny suggests, by increasing the respirations and thus preventing the stimulant effect of partial asphyxia on the sweat-centers. As an antihydrotic it often succeeds when atropin and camphoric acid have failed. It is best given by the mouth, at bedtime, in the form of a pill or tablet.

The tincture of Cocculus indicus (*Anamirta paniculata*) is a reliable remedy for *head lice*.

TELLURIUM.

The most important tellurium compounds from a therapeutic standpoint are the tellurates of sodium and potassium. These appear as white crystalline powders, soluble in water. The dose of either salt is from $\frac{1}{3}-\frac{3}{4}$ gr. (0.02–0.05 gm.).

Physiologic Action and Therapeutics.—In warm-blooded animals large doses of any one of the tellurium compounds cause vomiting, diarrhea, somnolence, weakening of the reflexes, paralysis, unconsciousness, respiratory failure, convulsions, and death. A characteristic feature of the poisoning is the development of a strong garlicky odor in the breath a few minutes after the administration of the drug. This odor has been determined by Hofmeister to be due to the formation of methyl tellurid. In man full doses of sodium or potassium tellurate arrest perspiration, impart an alliaceous odor to the breath, and impair slightly the digestion. Tellurium compounds undergo partial reduction in the body, and, according to Gies, are eliminated in the metallic form in the feces and as methyl tellurid in the breath, urine, feces, and epidermal secretions. As the antihydrotic action of tellurium is offset by pilocarpin (Czapek and Weil), it is probably due to depression of the peripheral secretory nerves.

The tellurates have been recommended by Neusser, Combemale, and Dubiquet, and others for their antihydrotic effects in advanced phthisis, but the garlicky odor of the breath, often persisting several weeks after treatment has been suspended, is a strong objection to their use.

THALLII ACETAS.

(Thallium Acetate ; $TlC_2H_3O_2$.)

Thallium acetate occurs in the form of white deliquescent crystals which are readily soluble in water and alcohol. The dose is from 1–3 gr. (0.065–0.2 gm.) in pills.

Physiologic Action and Therapeutics.—Although thallium possesses decided toxic properties, these are not manifest when its salts are administered in therapeutic doses and for short periods. Combemale, Huchard, and others have found the acetate very effective in controlling the *night-sweats of phthisis*. Unfortunately, however, its continued use is not infrequently followed by alopecia. This untoward effect was observed in 8 of 34 cases in which Vossaux employed it as an antihydrotic. Huchard also has reported a case of total baldness following the administration of a dozen pills. The drug should be given about an hour before the expected sweating, and not repeated for more than four successive nights.

EXPECTORANTS.

Expectorants are drugs that modify the secretion of the air-passages, affect its quantity, and facilitate its expulsion. The exact manner of their action is obscure. Those that promote secretion and render it less viscid in character, and, therefore, more easy of removal, are generally termed *seda-tive expectorants*. Most of the latter are also centric emetics and, no doubt, owe their power of increasing secretion very largely to their nauseating properties. As a rule, they are employed as expectorants in much smaller doses than required to produce vomiting, but occasionally vomiting itself renders valuable service in expelling mechanically excessive accumulations of tenacious mucus from the bronchi. The increased secretion which these drugs induce may be of service in depleting the congested bronchi, in protecting their surfaces from the air, in diluting and washing out the irritant, and, possibly, in exercising a bactericidal influence. The sedative expectorants in common use are:

Ipecac.	Antimony (tartar emetic).
Apomorphin.	Lobelia.
Quebracho.	Potassium citrate.

Sedative expectorants are indicated in *acute bronchitis*, particularly in the early stage, when there is hard cough with little or no sputum.

Drugs that stimulate the mucous membrane of the respiratory tract and lessen the quantity of sputum are known as *stimulant expectorants*. Some of them appear to act by imparting tone to the relaxed membrane, and others also by exercising an antiseptic influence on the secretions. The most important stimulant expectorants are:

Ammonium chlorid.	Sanguinaria.
Ammonium carbonate.	Garlic.
Benzoin (benzoic acid).	Grindelia.
Balsam of Peru.	Creasote (guaiacol).
Balsam of tolu.	Tar.
Ammoniac.	Squill.
Eucalyptus.	Oil of sandalwood.
Oil of myrtle (myrtol).	Copaiba.
Senega.	Cubeb.

Oil of turpentine (terebene; terpin hydrate).

Stimulant expectorants are indicated in subacute and chronic bronchitis, especially when there is free expectoration.

Adjuvants to Expectorants.—When cough is disproportionate to expectoration and is in itself a source of distress, drugs may be combined with expectorants which lessen the excitability of the respiratory center or of the afferent and efferent nerves of lung.   The most reliable drugs for this purpose are *opium (morphin, codein, heroin), dilute hydrocyanic acid, chloroform, bromids, hyoscyamus,* and *cannabis indica.*   Sometimes cough is dependent to a great extent upon excessive irritability of the fauces; if this be the case, demulcents—*licorice, glycerin, acacia*—will be found useful adjuvants.   When the respiratory center has become weakened through persistent cough, *strychnin* will prove beneficial.   The same remedy also renders valuable aid when the pulmonary circulation is sluggish, owing to insufficient cardiac power.   When bronchial catarrh excites asthma, antispasmodics—*belladonna, Hoffmann's anodyne, chloroform, bromids*—should be associated with expectorants.   *Lobelia* is particularly useful in these cases, because it is both an expectorant and a depressomotor.

Administration.—Expectorants, as a rule, are most effective when administered by the mouth.   Sometimes, however, inhalations act beneficially in allaying cough, facilitating expectoration, and lessening fetor of the breath.   Except in the case of very volatile substances, it is doubtful whether vapors or sprays penetrate beyond the trachea and the main bronchial divisions.   There are several methods of inhalation intended to influence the mucous membrane of the respiratory tract.   The simplest plan consists in breathing deeply the warm vapor arising from the surface of boiling water. For this purpose a volatile expectorant—compound tincture of benzoin, terebene, creasote—may be dropped into a wide-mouthed jug nearly filled with hot water, and the vapor conducted to the nose and mouth through a cone made of stiff paper or a folded napkin.   Inhalations of this class are rarely useful, except in acute bronchitis.   More complicated, but somewhat more serviceable, are the various hand-ball and steam atomizers designed to reduce medicated fluids to a nebular form.   In chronic bronchitis the cold spray has an advantage over the steam spray in being less relaxing in its effect on the throat and trachea.

Another method of administering medicated vapors is by the so-called respirator.   One of the most useful of the latter contrivances is the oronasal respirator devised by Dr. Burney Yeo, which consists of a perforated zinc mask, shaped to fit over the nose and mouth, and held in position by tapes passed around the ears.   This mask incloses a sponge which is charged from time to time with the volatile inhalant.   Creasote (with an equal amount of spirit of chloroform), turpentine, terebene, eucalyptol, and menthol are the drugs most frequently used in the respirator.

The intralaryngeal injection of drugs.into the trachea some-times affords great relief in chronic bronchitis, in bronchi-ectasis, and even in phthisis. If they are skilfully made, they rarely excite cough or other discomfort. Some bland oil, like olive oil, is usually made the vehicle for the active drug—guaiacol (1 : 100 or 1 : 50) or menthol (1 : 50).

SEDATIVE EXPECTORANTS.

IPECACUANHA, U. S. P.

(Ipecac.)

Ipecac is the root of *Cephaëlis Ipecacuanha*, a perennial shrub growing in Brazil and other South American States. According to Paul and Cownley, it contains three alkaloids—*emetin*, *cephaëlin*, and *psychotrin*. The dose of the powdered drug as an expectorant is from $\frac{1}{2}$–2 gr. (0.03–0.13 gm.); as an emetic, 15–30 gr. (1.0–2.0 gm.).

PREPARATIONS.	DOSE.
Extractum Ipecacuanhæ Fluidum, U. S. P.	As an expectorant, 2–5 min. (0.1–0.3 c.c.); as an emetic, 15–30 min. (1.0–2.0 c.c.).
Syrupus Ipecacuanhæ, U. S. P.	As an expectorant, 10–60 min. (0.6–4.0 c.c.); as an emetic, 2–4 fl. dr. (8.0–15.0 c.c.).
Vinum Ipecacuanhæ, U. S. P.	As an expectorant, 10–60 min. (0.6–4.0 c.c.); as an emetic, 2–4 fl. dr. (8.0–15.0 c.c.).
Tinctura Ipecacuanhæ et Opii, U. S. P. (1 gr.—0.065 gm.—each of ipecac and powdered opium in 10 min.—0.6 c.c.)	5–15 min. (0.3–1.0 c.c.).
Pulvis Ipecacuanhæ et Opii, U. S. P. (Dover's powder : ipecac, 1 part; powdered opium, 1 part; sugar of milk, 8 parts)	5–10 gr. (0.3–0.6 gm.).
Trochisci Ipecacuanhæ, U. S. P. ($\frac{1}{3}$ gr.—0.02 gm.—in each).	
Trochisci Morphinæ et Ipecacuanhæ, U. S. P. (morphin, $\frac{1}{40}$ gr.—0.0016 gm.; and ipecac, $\frac{1}{15}$ gr.—0.005 gm.).	

Physiologic Action.—Ipecac is a powerful local irritant. When rubbed into the skin it causes erythema and vesication; when applied to mucous membranes it produces redness, swelling, and an increase of secretions. In some persons the mucous membrane of the respiratory tract is so sensitive to the drug that the inhalation of an exceedingly small quantity of the powder is sufficient to excite lacrimation, sneezing, running from the nose, and asthmatic breathing. The ingestion of a full dose of ipecac is followed in a short time by nausea, vomiting, and a marked increase of the secretions of the nose, mouth, and bronchi. These effects are due to cephaëlin and emetin. Both alkaloids are active emetics, but the former is decidedly more powerful in this respect than the latter. The emesis is partly of peripheral origin, and partly, also, of central origin, since it follows the hypodermic administration of the drug and its local application to the medulla.

The subcutaneous injection of a toxic dose of emetin is followed by nausea, vomiting, purging, and collapse. It is well-established clinically that small doses of ipecac promote both the cutaneous and bronchial secretions.

Therapeutics.—Ipecac is chiefly employed as a sedative expectorant, an emetic, an anti-emetic, an antidysenteric, and a diaphoretic.

Expectorant.—Ipecac is an excellent expectorant in *acute bronchitis* before secretion has been established. In *chronic bronchitis*, when the sputum is scanty and viscid, it is also serviceable. In the form of a spray the wine, diluted with twice its bulk of water, as recommended by Murrell, is often very efficacious in " winter cough " and bronchial asthma.

Emetic.—Ipecac, though not very prompt in its action, is a certain and safe emetic. It is especially useful in children, in whom it is frequently employed to unload the stomach of irritant material. In *capillary bronchitis*, when the bronchi become filled with mucus and asphyxia is threatened, ipecac may be given in doses sufficient to excite emesis, with the hope of forcibly expelling the secretion. In poisoning more prompt emetics, like apomorphin or zinc sulphate, are preferable.

Anti-emetic.—Ipecac in minute doses sometimes succeeds in controlling *persistent vomiting* that is not dependent upon any organic disease of the stomach. The wine is the best preparation; it should be given in doses of from $\frac{1}{2}$–1 minim (0.03–0.06 c.c.) every half-hour or hour. In some instances its efficacy is enhanced by the addition of an equal quantity of the tincture of nux vomica.

Antidysenteric.—Ipecac has a long-standing reputation in *acute tropical dysentery;* a large dose of from 30–40 gr. (2.0–2.6 gm.) should first be given to induce vomiting, and after this has occurred smaller doses of from 3–5 gr. (0.2–0.3 gm.) should be given at intervals of two hours. To insure the retention of the latter doses, 5 min. (0.3 c.c.) of laudanum may be administered fifteen minutes before each dose of ipecac. A successful result is usually indicated by the appearance within twelve hours of copious tarry stools. In the sporadic dysentery of temperate zones this treatment is of little value.

Diaphoretic.—The mild and agreeable diaphoretic effect of ipecac in the form of Dover's powder has made the latter a favorite remedy in the initial stage of *acute coryza*, in so-called *muscular rheumatism*, in *influenza*, and in *acute suppression of the menses.*

Other Uses.—Ipecac has been recommended as a hemostatic since Trousseau's time, but its use in hemorrhage is at present obsolescent. Rutherford and Vignal concluded, from their experiments on dogs, that ipecac is a direct hepatic stimulant, but the work of other investigators has served to throw considerable doubt on this conclusion. In minute doses ($\frac{1}{8}$–$\frac{1}{4}$ gr.—0.008–0.016 gm.) it is an excellent adjuvant to strychnin in *indigestion dependent upon motor insufficiency.* In *atony of the bowel,* also, it often makes a useful addition to cathartic pills.

Administration.—For producing emesis either the wine or the syrup may be selected; in children the syrup, $\frac{1}{2}$–2 fl. dr. (2.0–8.0 c.c.), is preferable. Paul and Cownley recommend cephaëlin as a powerful and certain emetic, free from depressing effects when given in doses of from $\frac{1}{10}$–$\frac{1}{5}$ gr. (0.006–0.01 gm.), but it is too costly for ordinary use. The preparations and alkaloids of ipecac are all too irritating to be administered hypodermically. As an expectorant the wine, syrup, or fluid extract may be employed. A combination of ipecac with an alkaline expectorant, as in the following formula, is often effective in the first stage of bronchitis:

℞ Vini ipecacuanhæ,		f℥iss (10.0 c.c.);
Potassii citratis,		℥iij (12.0 gm.);
Tincturæ opii camphoratæ,		
Syrupi acaciæ,	āā	f℥j (30.0 c.c.);
Aquæ,	q. s. ad	f℥vj (180.0 c.c.).—M.

Sig. A tablespoonful every four hours.

APOMORPHINÆ HYDROCHLORAS, U. S. P.

(Apomorphin Hydrochlorate.)

Apomorphin is an artificial alkaloid obtained by abstracting from morphin a molecule of water It is prepared by maintaining at a high temperature for several hours, in a sealed retort, a mixture (1 : 20) of morphin and strong hydrochloric acid. The hydrochlorate of this alkaloid appears as minute, grayish-white, acicular crystals, having a faintly bitter taste, and acquiring a greenish tint on exposure to light and air. It is soluble in about 50 parts of water or alcohol. The dose as an expectorant is from $\frac{1}{30}-\frac{1}{20}$ gr. (0.002–0.003 gm.); as an emetic, from $\frac{1}{12}-\frac{1}{6}$ gr. (0.005–0.01 gm.).

Physiologic Action.—In man the subcutaneous injection of $\frac{1}{10}$ gr. (0.006 gm.) of apomorphin is followed within fifteen minutes by nausea and vomiting. The emesis is accompanied by muscular relaxation, quickening of the pulse, free perspiration, and a marked increase in the nasal, faucial, and bronchial secretions. The vomiting is of centric origin.

On the central nervous system apomorphin appears to have first a stimulant, and then a paralyzant, effect. The heart is not influenced by therapeutic doses, but toxic doses induce paralysis. The rapid pulse observed after emetic doses is due, according to Reichert, to stimulation of the accelerators.

Therapeutics.—Apomorphin is employed as an emetic and as a sedative expectorant.

Emetics (see p. 160).—The certainty and promptness of its action and its freedom from irritant properties when administered hypodermically make it almost an ideal emetic in *poisoning*, especially when swallowing is impossible or the state of the stomach is such as to forbid the use of emetics having a peripheral action. In young children and infirm subjects considerable caution must be exercised in its employment, since its action may be attended with profound exhaustion or even collapse.

Expectorant.—Apomorphin is a useful expectorant in both *acute* and *chronic bronchitis*, when the sputum is scanty and viscid. As it is eliminated rather rapidly, it should be given at comparatively short intervals—that is, every two or three hours. Its continuous use is sometimes followed by a soporific effect, which is no doubt due to the contamination of the preparation with traces of unconverted morphin.

Administration.—As an emetic apomorphin should be administered hypodermically. Solutions can be conveniently prepared from tablets containing the requisite dose. Both tablets and solution are liable to deteriorate with age and on exposure to light and air. As an expectorant it should be given by the mouth in the form of a solution, pill, or capsule.

Incompatibles.—Alkalis, potassium iodid and ferric chlorid. It is said to be incompatible also with heroin.

Apocodein Hydrochlorid.—This compound is prepared from codein after the manner of preparing apomorphin from morphin. It is an amorphous, yellow powder, soluble in water and alcohol. It has been recommended chiefly as an expectorant, in doses of from $\frac{3}{4}-1$ gr. (0.05–0.065 gm.).

ASPIDOSPERMA, U. S. P.

(Quebracho.)

Quebracho is the bark of *Aspidosperma Quebracho-blanco*, an evergreen tree growing in South America. It contains several alkaloids, the most important of which is *aspidospermin*. The aspidospermin of commerce, however, is really a mixture of the various alkaloids, and may be given in doses of from $\frac{1}{2}$–1 gr. (0.03–0.065 gm.).

<table>
<tr><td>PREPARATION.</td><td>DOSE.</td></tr>
<tr><td>Extractum Aspidospermatis Fluidum, U. S. P. . .</td><td>20–60 min. (1.2–4.0 c.c.).</td></tr>
</table>

Physiologic Action and Therapeutics.—Several of the alkaloids of quebracho resemble apomorphin in their action in that they first stimulate and then depress the central nervous system. Like apomorphin also they induce nausea and increase the secretions of the throat and bronchi. Only one alkaloid, however, is an active emetic, and that is aspidosamin. Aspidospermin in small dose is apparently an active respiratory stimulant.

Because of its influence on respiration and bronchial secretion quebracho has been found useful in *bronchitis associated with asthma or emphysema*.

OTHER SEDATIVE EXPECTORANTS.

Tartar Emetic (see p. 63).—Antimonial preparations, although not used so much as formerly, often afford prompt relief in the early stages of *acute bronchial catarrh*, when given in small but frequently repeated doses. The dose of tartar emetic as an expectorant is from $\frac{1}{16}$–$\frac{1}{12}$ gr. (0.004–0.005 gm.). The wine of antimony, which contains 2 gr. (0.13 gm.) of tartar emetic to the ounce (30.0 c.c.), is the preparation most commonly employed. It may be prescribed with a mild febrifuge, as in the following formula:

R Vini antimonii, f℥j (4 0 c.c.);
Tincturæ aconiti, m vj (0.4 c.c.);
Syrupi tolutani, f℥ij (60.0 c.c.);
Liquoris ammonii acetatis, q. s. ad f℥iv (120.0 c.c.).—M.
Sig. A teaspoonful every two hours for a child of two years.

Tartar emetic also enters into compound syrup of squill (0.2 per cent.) and compound licorice mixture or brown mixture.

Lobelia (see p. 145).—Lobelia, like other nauseants, increases the bronchial secretions. In large dose it also depresses the fibers of the vagus distributed to the bronchial muscles. Owing to this combined action, the drug is especially valuable in *bronchitis with asthma.* The dose of the tincture of lobelia as an expectorant is from 10–30 min. (0.6–2.0 c.c.); of the fluid extract, from 2–5 min. (0.1–0.3 c.c.).

Potassium Citrate (see p. 222).—This and other vegetable salts of potassium are useful expectorants in the early stages of *acute bronchitis.* The citrate is generally preferred to the other salts, owing to its less unpleasant taste. It may be prescribed in doses of from 20–30 gr. (1.3–2.0 gm.), in combination with a nauseant like ipecac or antimony.

STIMULANT EXPECTORANTS.
AMMONII CHLORIDUM, U. S. P.
(Ammonium Chlorid; Sal Ammoniac; NH_4Cl.)

Ammonium chlorid is a white crystalline powder, odorless, of a cooling, saline taste, and permanent in the air. It is soluble in 3 parts of cold water, but almost insoluble in alcohol. The dose is from 5–20 gr. (0.3–1.3 gm.).

PREPARATION.

Trochisci Ammonii Chloridi, U. S. P. (each contains ammonium chlorid, 1½ gr.—0.1 gm.; licorice extract, 4 gr.—0.25 gm.; with tragacanth, sugar, and syrup of tolu).

Physiologic Action.—The chief effect of ammonium chlorid when administered by the mouth in moderate doses is to increase and render less viscid the mucous secretion of the bronchi, and perhaps, also, of the stomach and bowel. While it lacks the stimulant influence of ammonium carbonate on the circulation and respiration, its action on the mucous membranes is much more prolonged than that of the latter salt. Very large doses have an irritant action and excite the symptoms of gastro-enteritis. Blood dyscrasia with great muscular weakness has been observed after the prolonged use of ammonium salts in large doses. Ammonium chlorid is eliminated in the urine for the most part unchanged.

Therapeutics.—Ammonium chlorid is particularly useful as an expectorant in *acute bronchial catarrh* after the symptoms of the initial stage have passed. It is sometimes successfully employed also in *chronic bronchitis* when the sputum is chiefly thick, tenacious mucus. In *subacute and chronic pharyngitis* or *laryngitis* lozenges of ammonium chlorid or a warm spray of the salt (5 gr.–1 oz.—0.3 gm.–30.0 c.c.) may prove beneficial. The drug has found favor with some practitioners in *subacute gastro-intestinal catarrh* and in *catarrhal jaundice.*

A solution of ammonium chlorid—5 gr. (0.3 gm.) to ½ ounce (15.0 c.c.) each of alcohol and water—makes an excellent lotion in *contusions with much ecchymosis*.

Administration.—The drug is generally given in aqueous solution to which licorice is added to disguise its unpleasant salty taste. It may also be prescribed in the form of compressed tablets to be taken after meals in several ounces of water:

R Ammonii chloridi, ℈iiss (10.0 gm.);
Syrupi Scillæ, f℥v (18.5 c.c.);
Tincturæ opii deodorati, ♏ xl (2 5 c.c.);
Extracti glycyrrhizæ, ℈j (4.0 gm.);
Glycerini, f℥ss (15.0 c.c.);
Aquæ, q. s. ad f℥iv (120.0 c.c.).—M.
Sig. A dessertspoonful in water four times a day.

In chronic bronchitis with inspissated secretions an iodid may be combined with ammonium chlorid, as in the following formula:

R Ammonii iodidi, ℈ss (2.0 gm.);
Ammonii chloridi, ℈ij (8.0 gm.);
Syrupi pruni Virginianæ, f℥ij (60.0 c.c.);
Aquæ, q. s. ad f℥iv (120.0 c.c.).—M.
Sig. A dessertspoonful in water every four hours.

Incompatibles.—Alkalis, mineral acids, tartaric acid, and soluble salts of silver and lead.

AMMONII CARBONAS, U. S. P.

(Ammonium Carbonate; $NH_4HCO_3 . NH_4NH_2CO_2$.)

The ammonium carbonate of commerce is a mixture of ammonium bicarbonate (acid carbonate) and ammonium carbamate. It occurs in white, translucent masses having a strong ammoniacal odor and a sharp saline taste. It is soluble in about 5 parts of cold water, and is decomposed by boiling water, with the elimination of carbon dioxid and ammonia. On exposure to air it is converted into an opaque white powder of ammonium bicarbonate. The dose is from 5–10 gr. (0.3–0.8 gm.).

PREPARATION.	DOSE.
Spiritus Ammoniæ Aromaticus, U. S. P. (ammonium carbonate, 3.4; ammonia water, 9; aromatic oils; water to make 100 parts) . .	20–60 min. (1.0–4.0 c.c.).

Physiologic Action.—In therapeutic doses ammonium carbonate acts as a heart stimulant, a respiratory stimulant, and an expectorant. Its action is prompt but evanescent. In large doses it is an irritant and capable of provoking emesis. It is eliminated in the urine probably partly as nitric acid (Bence Jones) and partly as urea.

Therapeutics.—Ammonium carbonate may be substituted for the water or one of the spirits of ammonia in combatting *syncope, collapse*, and *other forms of sudden heart failure.* As a stimulant expectorant it is especially useful in *bronchitis* associated with cardiac or respiratory weakness. In *pneumonia*, both croupous and catarrhal, it often acts beneficially in facilitating expectoration and in stimulating the circulation. In *pulmonary emphysema*, when the dyspnea and cyanosis are greatly exaggerated by attacks of acute bronchial catarrh, it acts admirably in conjunction with strychnin and digitalis.

Mixed with half its bulk of stronger water of ammonia and scented, it constitutes the " smelling-salts," a popular remedy for syncope.

Administration.—On account of the rapidity with which ammonium carbonate is decomposed after absorption it is necessary, to secure the desired effect, to administer it at intervals of not more than two hours. It is always given in solution, the favorite vehicle being syrup of acacia.

Incompatibles.—It is incompatible with acids, acid salts, alkaloidal salts, and ferric salts.

BENZOINUM, U. S. P.

(Benzoin; Gum Benzoin.)

Benzoin is a balsamic resin obtained from *Styrax Benzoin*, a large tree growing in Sumatra, Java, Borneo, and Siam. It contains several resins, benzoic acid (15–20 per cent.), a volatile oil, cinnamic acid, and vanillin. It is soluble in about 5 parts of warm alcohol and in solutions of the fixed alkalis. Benzoic acid is soluble in 2 parts of alcohol and very sparingly soluble in water. The salts of benzoic acid—ammonium, lithium, and sodium—are readily soluble in water.

PREPARATIONS.	DOSE.
Tinctura Benzoini, U. S. P.	20–60 min. (1.2–4.0 c.c.).
Tinctura Benzoini Composita, U. S. P.	
(Friar's or Turlington's Balsam : 12 per cent., with aloes, tolu, and storax)	20–60 min. (1.2–4.0 c.c.).
Adeps Benzoinatus, U. S. P. (2 per cent.).	
Acidum Benzoicum, U. S. P.	5–20 gr. (0.3–1.3 gm.).
Ammonii Benzoas, U. S. P.	5–30 gr. (0.3–2.0 gm.).
Lithii Benzoas, U. S. P.	5–30 gr. (0.3–2.0 gm.).
Sodii Benzoas, U. S. P.	5–30 gr. (0.3–2.0 gm.).

Benzoic acid (0.4 per cent.) also enters into camphorated tincture of opium, or paregoric.

Physiologic Action.—Benzoic acid is an antiseptic of considerable power and a stimulant to the mucous membranes. In health moderate doses do not affect the functions of the body beyond increasing, to a variable degree, the nitrogenous output and lessening the quantity of ethereal sulphates and of indican in the urine. It escapes from the body rapidly, chiefly through the kidneys, and for the most part in the form of hippuric acid—a compound of benzoic acid with the proteid derivative, glycocoll. It is generally believed that benzoic acid increases the acidity of the urine, but according to W. W. Ashhurst, who has investigated the subject very carefully, sodium benzoate, at least, has little or no influence on the acidity of urine that is already acid, but that it does render alkaline urine acid; not so much by its presence, however, as by its power of arresting ammoniacal fermentation. In large doses benzoic acid and benzoates have an irritant action on the stomach, and in consequence excite nausea and vomiting.

Therapeutics.—The compound tincture of benzoin painted over the part and allowed to dry is a useful *protectant for small wounds.* A combination of the compound tincture with 4 parts of glycerin makes an efficient application in *chapped nipples, hands,* or *lips.* A dram (4.0 c.c.) to a pint (0.5 L.) of boiling water may be used as an inhalant in *acute laryngitis* and *bronchitis.* The tincture may be prescribed internally also in *subacute* and *chronic bronchitis* when expectoration is very viscid. Benzoic acid itself often affords prompt relief in *phosphaturia,* and in *cystitis with ammoniacal urine.* Sodium benzoate, in doses of from 5–10 gr. (0.3–0.6 gm.), is a very useful remedy in *acute pharyngitis* and *tonsillitis.*

Owing to the fact that benzoic acid is eliminated as the soluble hippuric acid, it was hoped that the benzoates would prove of value in gout, but experience has shown that they are of no avail. Senator and others have recommended the benzoates in rheumatism, but they have been found to be far less efficacious than the salicylates.

The treatment of *phthisis* by intravenous injections of cinnamic acid (see p. 183), so highly recommended by Landerer, is very painful and does not seem to have given any better results than can be obtained by other means.

Administration.—Benzoic acid is best administered in capsules or pills. The benzoates may be administered in solution or in powders. The tinctures are incompatible with aqueous preparations.

BALSAMUM PERUVIANUM, U. S. P.

(Balsam of Peru.)

Balsam of Peru is a brownish-black, syrupy liquid, obtained by bruising the bark of *Toluifera Pereira,* a large tree growing in Central America. It has a smoky, vanilla-like odor and a bitter, persistent taste. It is soluble in alcohol, chloroform, and glacial acetic acid. It contains resins, cinnamic and benzoic acids, and traces of vanillin. The dose is from 5–30 min. (0.3–2.0 c.c.).

Therapeutics.—Balsam of Peru is chiefly employed as a protectant and parasiticide. It was at one time used as a stimulant expectorant in *chronic bronchitis,* but it has been entirely replaced by more elegant remedies. It is still occasionally employed as a stimulant protectant dressing in *bedsores* and *other ulcers.* Alone, or in combination with sulphur, it is a reliable remedy in *scabies,* in which disease it may be prescribed as follows:

<pre>
R   Sulphuris præcipitati,
    Balsami Peruviani,   aa ʒij (8.0 gm.);
    Olei olivæ,   fℨss (15.0 c.c.);
    Adipis,   q. s. ad ℨij (60.0 gm.).—M.
</pre>

BALSAMUM TOLUTANUM, U. S. P.

(Balsam of Tolu.)

Balsam of tolu is a yellowish-brown, semiliquid mass, obtained by incising the bark of *Toluifera Balsamum,* an evergreen tree growing in South America. It has a vanilla-like odor and a mild, aromatic taste, and is soluble in alcohol, ether, and chloroform, but is insoluble in water. It contains resin, a volatile oil, vanillin, and cinnamic and benzoic acids. The dose is from 5–30 gr. (0.3–2.0 gm.).

PREPARATIONS.	DOSE.
Syrupus Tolutanus, U. S. P.	1–4 fl. dr. (4.0–15.0 c.c.).
Tinctura Tolutana, U. S. P.	½–1 fl. dr. (2.0–4.0 c.c.).
Tinctura Benzoini Composita, U. S. P. (4 per cent.)	20–60 min. (1.2–4.0 c.c.).

Therapeutics.—Balsam of tolu has but feeble medicinal properties. It is chiefly employed on account of its agreeable flavor as a vehicle in cough-mixtures.

AMMONIACUM, U. S. P.

(Ammoniac.)

Ammoniac is a gum-resin obtained by puncturing the stems of *Dorema Ammoniacum*, a plant growing in Persia and Turkestan. The dose is from 10–30 gr. (0.65–2.0 gm.).

PREPARATIONS.	DOSE.
Emulsum Ammoniaci, U. S. P. (4 per cent.)	1–6 fl. dr. (4.0–22.0 c.c.).
Emplastrum Ammoniaci cum Hydrargyro, U. S. P.	

Therapeutics.—Ammoniac was once much used as a stimulant expectorant and a counterirritant, but it has been largely superseded by more efficient remedies.

EUCALYPTUS, U. S. P.

Eucalyptus is the leaves of *Eucalyptus globulus*, or gum-tree, a native of Australia, and cultivated in Europe and Southern United States. It contains a volatile oil (6 per cent.), from which is obtained a neutral body, *eucalyptol.* The latter is a colorless liquid, having a camphoraceous odor and a pungent, spicy taste, and on exposure to a temperature below 30° F. solidifying into a mass of needle-shaped crystals. It is insoluble in water, but freely so in alcohol.

PREPARATIONS.	DOSE.
Extractum Eucalypti Fluidum, U. S. P.	. 20–60 min. (1.2–4.0 c.c.).
Oleum Eucalypti, U. S. P.	. 3–10 (0.2–0.6 c.c.).
Eucalyptol, U. S. P.	. 3–10 (0.2–0.6 c.c.).

Physiologic Action.—Oil of eucalyptus is a local irritant. When applied to the skin, evaporation being prevented, it causes redness and even vesication. When taken internally in large doses (20 min.—1.2 c.c.), it causes a sense of burning in the throat and stomach, nausea, looseness of the bowels, and some mental exhilaration, which in turn is followed by a feeling of calm. Toxic doses depress the spinal cord, medulla, and heart, and in consequence induce extreme muscular weakness, a fall of temperature, a feeble pulse, shallow respiration, and finally death from asphyxia. The drug is eliminated by all the emunctories, and, like turpentine, it imparts the odor of violets to the urine.

Both eucalyptol and the oil have pronounced antiseptic properties.

Therapeutics.—Eucalyptus was at one time used as a sub-

18

stitute for quinin in malaria, but its value in this disease was never established, and to-day it is rarely, if ever, employed. In *subacute* and *chronic bronchitis* attended with copious expectoration the oil is an exceedingly useful expectorant. As an inhalant, also, oil of eucalyptus is sometimes efficacious in diminishing profuse expectoration in *bronchiectasis* and *phthisis*. A dram (4.0 c.c.) may be added to an ounce (30.0 c.c.) of liquid petrolatum and used in an atomizer; or eucalyptol alone or in conjunction with terebene may be used in an inhalation respirator.

Eucalyptol in the form of an oily spray acts very favorably in *subacute* and *chronic rhinitis*, if used after the nares have been thoroughly cleansed with an alkaline wash. The following formula, suggested by Douglass, is an excellent one:

℞	Thymol,	gr. x (0.65 gm.);
	Menthol,	gr. xx (1.3 gm.);
	Eucalyptol,	gtt. xx (0.6 c.c.);
	Olei cubebæ,	gtt. xl (1.2 c.c.);
	Benzoinol,	fℨvj (175.0 c.c.).—M.

Oil of eucalyptus is sometimes used as a urinary antiseptic in *cystitis* and *urethritis*.

Administration.—The oil and eucalyptol are the best preparations for internal use. They may be given on sugar, in capsules, or in an emulsion.

OLEUM MYRTI.

(Oil of Myrtle; Myrtol.)

Oil of myrtle is a greenish-yellow, volatile oil, distilled from the leaves and flowers of *Myrtus communis*, the common European myrtle. It has properties almost identical with those of eucalyptol. The dose is from 3–10 min. (0.2–0.6 c.c.). It is an excellent stimulant expectorant in *bronchorrhea, bronchiectasis*, and *fibroid phthisis*. Eichhorst has used it with satisfaction in *pulmonary gangrene*.

SENEGA, U. S. P.

Senega is the root of *Polygala Senega*, a perennial herb growing in woody places in North America. It contains a white, amorphous body—senegin—belonging to the saponin group of glucosids.

PREPARATIONS.	DOSE.
Extractum Senegæ Fluidum, U. S. P.	5–20 min. (0.3–1.2 c.c.).
Syrupus Senegæ, U. S. P.	½–2 fl. dr. (2.0–8.0 c.c.).
Syrupus Scillæ Compositus, U. S. P. (fl. ext. of squill, 8 per cent.; fl. ext. of senega, 8 per cent.; tartar emetic, 2 per cent.)	10–60 min. (0.6–4.0 c.c.).

Physiologic Action and Therapeutics.—Senega is a local irritant and emetic. It is chiefly used, however, as a stimulant expectorant in *subacute* and *chronic bronchitis* with copious expectoration. It is often prescribed in combination with squill.

SANGUINARIA, U. S. P.

(Bloodroot.)

Sanguinaria is the rhizome of *Sanguinaria canadensis*, a perennial herb growing in the woods of North America. It contains an alkaloid, *sanguinarin*.

PREPARATIONS.	DOSE.
Tinctura Sanguinariæ, U. S. P.	10–30 min. (0.6–2.0 c.c.).
Extractum Sanguinariæ Fluidum, U. S. P.	1–5 min. (0.06–0.3 c.c.).

Physiologic Action and Therapeutics.—Sanguinaria is a powerful irritant. In large doses it excites nausea and vomiting. Toxic doses of sanguinarin cause salivation, violent intestinal peristalsis, spinal convulsions, and, finally, cardiac and respiratory failure. Sanguinaria has been employed as an emetic and a stimulant expectorant, but it is now mostly replaced by more efficient and less dangerous drugs.

ALLIUM, U. S. P.

(Garlic.)

Garlic is the fresh bulb of *Allium sativum*, a plant indigenous to Southern Europe and cultivated in North America. Its active principle is a volatile oil.

PREPARATION.	DOSE.
Syrupus Allii, U. S. P.	1–4 fl. dr. (4.0–15.0 c.c.).

Therapeutics.—Garlic was at one time a favorite stimulant expectorant, especially for children; it has been superseded, however, by more agreeable remedies.

GRINDELIA, U. S. P.

Grindelia is the leaves and flowering tops of *Grindelia robusta* and *Grindelia squarrosa*, perennial herbs growing in North America, west of the Rocky Mountains. It contains resin, a bitter principle, and a volatile oil of a terebinthinate odor.

PREPARATION.   DOSE.

Extractum Grindeliæ Fluidum, U. S. P.   ₰15–60 min. (1.0–4.0 c.c.).

Therapeutics.—Grindelia is a very useful remedy in *bronchitis complicated with asthma.* It may be prescribed alone or in combination with other antiasthmatics, as in the following formula:

R   Sodii iodidi,   ʒj (4.0 gm.);
     Extracti grindeliæ fluidi,   fʒss (15.0 c.c.);
     Tincturæ lobeliæ,
     Tincturæ belladonnæ,   aa fʒij (8.0 c.c.);
     Extracti glycyrrhizæ fluidi,   fʒiij (11.0 c.c.);
     Syrupi tolutani,   q. s. ad fʒiij (90.0 c.c.).—M.
  Sig.   A teaspoonful every three or four hours.

The fumes of burning grindelia leaves are also efficacious in allaying the cough and relieving the dyspnea of asthma. A lotion consisting of an ounce (30.0 c.c.) of the fluid extract to a pint of water (0.5 L.) has been highly recommended in *rhus poisoning* and *erysipelas.* It should be applied on cloths and allowed to evaporate.

CREOSOTUM, U. S. P.

(Creasote.)

Creasote is a mixture of phenols, chiefly guaiacol and creosol, obtained during the distillation of wood-tar, preferably of that derived from the beech. It is a colorless or faintly yellow, oily liquid, having a penetrating, smoky odor and a burning, caustic taste. It is soluble in 150 parts of water, and in all proportions in alcohol, ether, chloroform, and oils. Much of the creasote of commerce is impure carbolic acid. Notable quantities of the latter, however, may be detected by one of the following tests: first, by the formation of a coagulum when equal quantities of the suspected liquid and collodion are mixed in a dry test-tube, and, secondly, by the absence of turbidity when one or more volumes of water are added to an intimate mixture of equal volumes of the liquid and glycerin. The dose of creasote is from 1–10 min. (0.06–0.6 c.c.), gradually increased to the limit of tolerance.

PREPARATION.   DOSE.
Aqua Creosoti, U. S. P. (about 1 per cent.)   1–4 fl. dr. (4.0–15.0 c.c.).

Physiologic Action.—The action of creasote closely resembles that of carbolic acid, but it is much less irritant and toxic. Very large doses, however, produce all the symptoms commonly observed in carbolic-acid poisoning—namely, burning pain, nausea and vomiting, vertigo, collapse, and almost simultaneous failure of the heart and respiration. As pointed out by Hare, soluble sulphates—Epsom or Glauber salts—are chemical antidotes.

Like all members of the phenol group, creasote is a powerful antiseptic. The drug is eliminated in large part by the kidneys as guaiacol sulphate and creosol sulphate of potassium, but a certain proportion escapes from the body through the lungs.

Therapeutics.—Creasote is employed chiefly as an expectorant, an antiseptic, and a local anesthetic.

Expectorant.—Creasote and its derivatives are useful expectorants in *chronic bronchitis* and *bronchiectasis* with copious purulent sputum. So long ago as Addison's time creasote was used as a remedy for *phthisis*. This use was revived by Bouchard and Gembert in 1877, and again by Sommerbrodt in 1887. While it has been shown conclusively that the drug has no specific influence on the tubercle bacilli in the lung, nevertheless the testimony of numerous observers is convincing that it has a positive value in allaying cough, lessening expectoration, and lowering temperature. It should be given in small doses gradually increased, and at once withdrawn if the stomach shows any intolerance. The best results are seen in phthisis attended with abundant sputum. Whether the drug has any other action than that of an expectorant and intestinal antiseptic is not known.

Antiseptic.—Creasote is a valuable internal antiseptic in *chronic gastric catarrh* with flatulence and in *simple dyspeptic diarrhea.* Creasote water has been recommended as a disinfectant lotion for *sloughing ulcers, uterine cancer*, etc., but for such a purpose it has no advantage over a dilute solution of carbolic acid.

Local Anesthetic.—Inhalations of creasote are often useful in allaying the cough of *laryngitis, bronchitis*, and *phthisis*. Ten min. (0.6 c.c.) of creasote to 10 ounces (300.0 c.c.) of boiling water makes a good inhalation when the catarrh is acute. A mixture of equal parts of creasote and spirit of chloroform, employed in a respirator, often serves to control the painful paroxysms of cough in phthisis. Doses of from 1–2 min. (0.06–0.1 c.c.) are sometimes efficacious in the *vomiting of pregnancy, peptic ulcer*, and *gastric cancer*. In *toothache* a drop or two on a pledget of cotton may be inserted in the cavity of the tooth. In the form of an ointment or lotion (10 min.–1 oz.—0.6–30.0 gm.) it has been successfully employed to allay itching in *chilblains*, *pruritus*, and *chronic eczema*, but its odor is objectionable.

Administration.—For internal use it is important to secure a preparation that is comparatively free from carbolic acid. The initial dose should be small (1–3 min.—0.06–0.2 c.c.), and the amount gradually increased to 10–15 min. (0.6–1.0 c.c.). In some cases of phthisis so much as 60 or even 100 min. (4.0–6.0 c.c.) three times daily are well borne, but such large doses are usually unnecessary. Creasote should be administered after meals, and withdrawn on the first appearance of indigestion. Small doses may be given in capsules, but large doses should be given well diluted with cod-liver oil, milk, or a weak wine. In phthisis the drug has been administered hypodermically, the injections consisting of 1 part of creasote to 2 of olive or almond oil. This method of administration, however, is painful, and has little if any advantage over that by the mouth. Intratracheal injections have also been employed, but guaiacol is preferable for this mode of administration. In diarrhea creasote may be prescribed in a powder of bismuth, as in the following formula :

> R   Creasoti,   ℥xij (0.75 c.c.);
>   Morphinæ sulphatis,   gr. j (0.065 gm.);
>   Bismuthi subnitratis,   ℥ss (15.0 gm.).—M.
> Fiant chartulæ No. xii.
> Sig. One every three hours.

Incompatibles.—Strong nitric and sulphuric acids act violently upon creasote. It reduces silver salts, and may explode when triturated with oxid of silver. It is also incompatible with ferric chlorid.

Creasote Carbonate (Creosotal).—This is an oily liquid containing a mixture of the carbonates of the various phenols found in creasote. Its chief ingredient is guaiacol carbonate. It is odorless, of a slightly bitter taste, insoluble in water, but freely so in alcohol, ether, and oils. It has pronounced advantages over creasote in having but little taste and in being more acceptable to the stomach. It may be employed in the same class of cases as creasote. Leyden and Cornet have highly recommended it in *pulmonary tuberculosis*. The dose is from 1–3 min. (0.06–0.2 c.c.), gradually increased to

10–20 min. (0.6–1.2 c.c.), thrice daily. It may be given in capsules, or in oil, milk, claret, or beef-tea.

Guaiacol (Methyl Pyrocatechin).—This phenol derivative is the chief constituent of creasote, from which it is obtained by fractional distillation. It is a colorless, oily liquid, having a rather unpleasant aromatic odor and taste. It is slightly soluble in water, but freely so in alcohol and ether. It readily unites with acid radicles to form crystalline compounds. The dose is from 1–3 min. (0.06–0.2 c.c.), gradually increased to 20 min. (1.2 c.c.). It should be given in the same manner as creasote.

The action of guaiacol resembles that of carbolic acid and creasote. In toxic doses it causes burning in the stomach, nausea and vomiting, brown-red urine, unconsciousness, and collapse. Wyss has reported a case of fatal poisoning in a girl of nine years from a dose of 80 min. (5.0 c.c.). It is absorbed and eliminated with great rapidity, undergoing partial oxidation in the body, and appearing in the urine as glycuronates and ether sulphates. When applied to the skin in sufficient quantity (20–30 min.—1.2–2.0 c.c.), evaporation being prevented, it causes in fever a marked fall of temperature. This antipyretic effect follows immediately upon its absorption, is attended with copious perspiration and depression, and is of comparatively short duration.

Guaiacol has been used internally as a substitute for creasote in *chronic bronchitis* and *pulmonary tuberculosis*, but it is now largely replaced by the more elegant carbonates of creasote and guaiacol. Intralaryngeal injections into the trachea of a mixture of guaiacol (1–2 per cent.) and olive oil are sometimes very useful in *allaying cough* in chronic pulmonary diseases. From 10–20 min. (0.6–1.2 c.c.) of the mixture may be injected once or twice a day. Guaiacol cannot be recommended as an antipyretic, since its external application in typhoid fever, tuberculosis, and other infections is sometimes followed by profound exhaustion and even by collapse. The drug has proved of some value as a local anesthetic. In *epididymitis* and *orchitis* the gentle inunction of an ointment of guaiacol and lanolin (1 : 5), followed by the application of a snugly fitting suspensory bandage, affords much relief. In *laryngeal tuberculosis* an oily spray of guaiacol (20–40 per cent.) is sometimes useful.

Guaiacol Carbonate (Duotal).—This compound is produced by the action of carbonyl chlorid on the sodium salt of guaiacol. It is a white, crystalline powder, neutral in reaction, odorless and tasteless, insoluble in water, and slightly so in alcohol, glycerin, and oils. The dose is from 3–10 gr. (0.2–0.65 gm.), gradually increased to 30 gr. (2.0 gm.), thrice daily.

According to Hesse, guaiacol carbonate is distinctly less poisonous than either guaiacol or creasote. On account of its freedom from odor and taste it is often much better borne than creasote. It is an excellent expectorant in *purulent bronchitis, bronchiectasis*, and *phthisis.* As it undergoes reduction in the intestine, it may be employed also as an *intestinal antiseptic.* It may be prescribed in pills, capsules, or powders.

<pre>
℞   Strychninæ sulphatis,   gr. ¼ (0.016 gm.) ;
    Codeinæ sulphatis,   gr. ij (0.13 gm.) ;
    Guaiacoli carbonatis,   gr. l (3.0 gm.) ;
    Terebeni,   ℳ l (3.0 c.c.).—M.
    Pone in capsulas No. xxiv.
    Sig.   Two capsules every four hours (subacute bronchitis).
</pre>

Potassium Guaiacol Sulphonate (Thiocol). — This compound occurs as a white, odorless powder, of a faintly bitter and then sweetish taste. It represents about 60 per cent. of guaiacol. Its chief advantage over the carbonate of guaiacol is its ready solubility in water. It may be given in doses of from 5–30 gr. (0.3–2.0 gm.), thrice daily, in some aromatic water.

Other preparations of guaiacol are sometimes employed, such as the benzoate (benzosol), phosphite (guaiacol-phosphal), cinnamate (styracol), and glycerol-ether (guaiamar), but they have no special advantages.

PIX LIQUIDA, U. S. P.

(Tar.)

Tar is an empyreumatic oleoresin obtained by the destructive distillation of *Pinus palustris* and other species of *Pinus.* It is a thick, blackish liquid, having a terebinthinate odor and a pungent taste. It is only sparingly soluble in water, but freely so in alcohol, oils, and alkaline solutions. Its active constituents are guaiacols and creosols. When subjected to distillation it yields a volatile oil—*oil of tar.* The dose is from 3–15 gr. (0.2–1.0 gm.).

PREPARATIONS.	DOSE.
Oleum Picis Liquidæ, U. S. P.	1–5 min. (0.06–0.3 c.c.).
Syrupus Picis Liquidæ, U. S. P.	1–4 fl. dr. (4.0–15.0 c.c.).
Unguentum Picis Liquidæ, U. S. P. (50 per cent.).	

A water (Aqua Picis Liquidæ) is also in use. It is made by shaking together 1 part of tar and 4 parts of water frequently for twenty-four hours, decanting and filtering. The dose is from 1–4 fl. oz. (30.0–120.0 c.c.).

Physiologic Action and Therapeutics.—Tar is a local stimulant and an antiseptic. It is chiefly employed in medicine as an expectorant, and as a local remedy in certain chronic skin diseases. In *chronic bronchitis with abundant sputum and hard, paroxysmal cough* it sometimes succeeds after other remedies have failed. It may be given in pills or capsules, or in the form of the syrup or water. Murrell has found inhalations of tar-water also useful in the same class of cases.

In *chronic eczema* and *psoriasis*, when the lesions are sluggish, an ointment of tar makes a useful application; the official ointment is generally too strong, $\frac{1}{2}$–2 dr. (2.0–8.0 gm.) to the ounce (32.0 gm.) of lard being quite sufficient in the majority of cases. It should always be employed tentatively, as in some persons the skin is exceedingly sensitive to the drug. It is absolutely contraindicated if the inflammation is at all active.

OTHER STIMULANT EXPECTORANTS.

Squill (see p. 218).—Syrup of squill, in doses of $\frac{1}{2}$–1 fl. dr. (2.0–4.0 c.c.), is an excellent expectorant in the *later stages of acute bronchitis.* The compound syrup of squill contains, in addition to squill, senega and tartar emetic, and is too depressing for ordinary use.

As syrup of squill contains free acetic acid, it is incompatible with ammonium carbonate and other alkaline carbonates.

Oil of Santal (see p. 230).—When pure, this drug often proves a useful expectorant in *subacute* and *chronic bronchitis*, with heavy, purulent sputum. Its chief drawback is its tendency to disturb the stomach. It should be given in doses of from 5–20 min. (0.3–1.2 c.c.), in capsules or in an emulsion. Terebene and oil of eucalyptus are efficient synergists.

Copaiba and Cubeb (see pp. 228 and 229).—The oil of copaiba and the oleoresin of cubeb have been employed as expectorants in *chronic bronchitis with profuse purulent sputum*, but they are neither so efficacious nor so elegant as terebene, oil of eucalyptus, and the preparations of guaiacol. Smoking cigarettes containing cubeb sometimes affords relief in the paroxysms of *asthma.* Lozenges of cubeb are useful in relieving hoarseness and fatigue of the larynx resulting for prolonged use of the voice.

Oil of Turpentine (see p. 230).—This drug, once much used as an expectorant in *chronic bronchitis*, has been largely replaced by terebene and terpin hydrate. Terebene (*Terebenum,* U. S. P.) is a liquid hydrocarbon made by oxidizing oil of turpentine with strong sulphuric acid. The dose is from 5–10 min. (0.3–0.6 c.c.). Terpin hydrate is a crystalline compound obtained by the interaction of oil of turpentine, alcohol, and nitric acid. The dose is from 2–15 gr. (0.13–1.0 gm.). Terebene is one of the most satisfactory remedies we possess in *bronchitis* with free expectoration. In *phthisis*, also, it is of service when the catarrhal symptoms are prominent. In moderate doses it is usually well received, but large doses may excite nausea or irritation of the kidneys and bladder. As it is insoluble in water, it should be given in capsules, on sugar, in an emulsion, or in an elixir. It may also be inhaled, either as a spray or from the sponge of a respirator.

Terpin hydrate has therapeutic properties similar to those of terebene, but it is less active. It may be given in pills, in capsules, or in an elixir. Terpinol, an oily body derived from terpin hydrate, has been employed as an expectorant in doses of from 3–15 min. (0.2–1.0 c.c.). It is not so efficient as terebene.

TONICS.

Tonics are drugs that impart strength or tone to the tissues. In one sense every drug that favorably influences disease is a tonic, since, by removing the cause of debility, it serves to restore normal vitality. The term, however, is customarily restricted to remedies that have a more or less general invigorating effect, without necessarily exerting a specific influence on any one organ. A more exact knowledge of the pharmacologic action of drugs and a clearer appreciation of the primary causes of general weakness and malnutrition have tended to diminish gradually, but steadily, the number of remedies grouped under this heading. The following drugs may still be classed conveniently as tonics:

Iron.	Calcium phosphate.
Phosphorus.	Calcium hypophosphite.
	Cod-liver oil.

Nux vomica, *cinchona*, and *arsenic* also exert a tonic influence, but as this is subordinate to their other actions, these drugs have been considered in other groups.

FERRUM, U. S. P.

(Iron; Fe.)

Iron is official in the form of fine, bright, non-elastic wire.

Physiologic Action.—As iron is a constituent of the

normal body, it may be regarded as a food as well as a medicine. It is especially abundant (0.4 per cent.) in hemoglobin, and its presence in this compound is intimately associated with the function of the red corpuscles to carry oxygen from the lungs to the tissues. When iron is taken in therapeutic doses it acts as an astringent, causing more or less constipation. Large doses have an irritant effect and disturb digestion. It is now definitely known that both inorganic and organic preparations of the metal are capable of absorption. Only a small portion, however, of the quantity ingested enters the blood, the larger portion being discharged from the bowel as a sulphid and an albuminate, the stools, owing to the presence of the former, acquiring a deep black color. Absorption takes place mainly in the duodenum, the iron entering the general circulation probably as an albuminate. According to Macallum, transportation is effected through the agency of the leukocytes, and more especially, of the blood-plasma. The amount that is absorbed varies with the nature of the preparation and the quantity that is administered. From the blood the iron is deposited temporarily in the spleen, liver, and bone-marrow, and in one of these organs, perhaps the liver, it is converted into hemoglobin. That which is absorbed in excess of the requirements of the blood is not eliminated in the secretions, but escapes from the body through the intestinal epithelium. Even when injected directly into a vein, the iron reappears chiefly in the intestines, and the amount of the metal in the urine is not materially increased.

In health the iron naturally contained in food is all-sufficient for the needs of the system, and the administration of medicinal iron is not followed by any notable increase in the number of red blood-cells or their hemoglobin value. In certain forms of anemia, however, especially chlorosis, chalybeate preparations increase the amount of hemoglobin and, to a less extent, the number of red blood-cells. Whether the drug acts simply by furnishing material for the corpuscles or by stimulating the functional activity of the blood-making organs is not known.

Untoward Effects.—The continued use of iron, especially of the ferric salts, frequently excites indigestion, constipation, and headache. Discoloration of the tongue and teeth also occasionally follows its use, and is thought to be due to the formation in the mouth of a black sulphid or tannate.

Therapeutics.—The most important indication for the use of iron is *anemia*. The best effects are seen in *chlorosis*, in which disease the drug has almost a specific action. In *secondary anemia* the results of its use are less decisive, but after the removal of the cause it serves to hasten the restoration of the blood to its normal condition. Iron plays indirectly the part of an emmenagogue in *amenorrhea* dependent upon anemia. Some of its salts are used as *astringents* and *styptics*.

Contraindications.—The chief contraindication is gastric irritation. It is generally badly borne, probably on account of impaired digestion, in febrile diseases and in gout.

Administration.—Ordinarily iron should be administered by the mouth and after meals. The least astringent preparations are the carbonate, citrate, ammonioferric citrate, ammonioferric tartrate, pyrophosphate, and reduced iron. The numerous organic compounds of iron with albumin have no great advantages over many of the older inorganic salts, but they are less liable to disturb digestion, and are, probably, somewhat more readily absorbed. In exceptional cases iron may be given hypodermically. For this purpose a freely soluble preparation, such as ammonioferric citrate, should be selected, and the solution should be freshly prepared and well diluted.

Incompatibles.—Ferrous and ferric salts are incompatible with all preparations containing tannic or gallic acid; with ammonia, alkaline carbonates, and mucilage of acacia.

FERRUM REDUCTUM, U. S. P.

(Reduced Iron; Iron by Hydrogen; Quevenne's Iron.)

Reduced iron is a fine, grayish-black, insoluble powder, without odor or taste. When pure it is quite free from irritating properties, and, being but feebly astringent, it has little tendency to cause constipation. It is employed only as a *hematinic.* It may be given in doses of from 1–5 gr. (0.06–0.3 gm.), in pills, capsules, or lozenges.

FERRI CARBONAS.

(Iron Carbonate; Green Ferrous Carbonate; $FeCO_3$.)

Iron carbonate is an unstable compound which is readily converted into ferric hydrate on exposure to air. It is not official.

Preparations.	Dose.
Ferri Carbonas Saccharatus, U. S. P. (contains 15 per cent. of freshly precipitated ferrous carbonate, protected from oxidation by sugar)	5–15 gr. (0.3–1.0 gm.).
Mistura Ferri Composita, U. S. P. (compound iron mixture; Griffith's Mixture: carbonate of iron in suspension with potassium sulphate, myrrh, and sugar)	½–1 fl. oz. (15.0–30.0 c.c.).

PREPARATION.	DOSE.

Massa Ferri Carbonatis, U. S. P. (Vallet's Mass: about 42 per cent. of ferrous carbonate with sugar and honey) . . . . . 3–5 gr. (0.2–0.3 gm.).

Pilulæ Ferri Carbonatis, U. S. P. (Blaud's Pills: each pill contains about 1 gr.—0.06 gm.—of ferrous carbonate, made with crystallized ferrous sulphate and potassium carbonate) . . . . 1–3 pills.

Therapeutics.—These preparations are comparatively free from astringency, and are convenient forms in which to administer iron. They are used solely in *anemia*. The saccharated carbonate and the pills of the carbonate (Blaud's pills) are the favorite preparations. The former should be given also in pills or else in capsules. Griffith's mixture is superfluous. Some practitioners prefer Blaud's pills made according to the original formula (equal parts of potassium carbonate and *dried* ferrous sulphate) to those made according to the official formula. The former contain a certain amount of free potassium carbonate.

FERRI CHLORIDUM, U. S. P.

(Ferric Chlorid; Sesquichlorid or Perchlorid of Iron; $Fe_2Cl_6 + 12\,H_2O$.)

Ferric chlorid occurs in deliquescent, orange-yellow crystalline pieces, having an astringent, chalybeate taste. It is not used internally, but sometimes externally as an astringent or a styptic.

PREPARATIONS.	DOSE.

Liquor Ferri Chloridi, U. S. P. (contains about 38 per cent. of the crystallized salt) . . . . 3–10 min. (0.2–0.6 c.c.).

Tinctura Ferri Chloridi, U. S. P. (contains about 13 per cent. of anhydrous ferric chlorid, with alcohol, water, and a trace of nitrous ether) . . . . 5–30 min. (0.3–2.0 c.c.).

Liquor Ferri et Ammonii Acetatis, U. S. P. (Basham's Mixture: Tincture of ferric chlorid, 2 parts; dilute acetic acid, 3 parts; solution of ammonium acetate, 20 parts; aromatic elixir, 10 parts; glycerin, 12 parts; water, to make 100 parts) . . 1–4 fl. dr. (4.0–15.0 c.c.).

Therapeutics.—The solution of ferric chlorid may be used locally as a styptic in controlling bleeding after the *extraction of teeth, removal of tonsils*, or *application of leeches.* In *postpartum hemorrhage* it should not be used except as a last resort. It may occasionally be of service in *hematemesis* and *entrorrhagia*, on account of its local action, but it cannot possibly do good in other internal hemorrhages.

External pedunculated piles have been successfully treated by injecting them, after they have been clamped, with a few minims of the tincture. The treatment of aneurysm by injections of ferric chlorid, as first advised by Pravaz, is dangerous, and fortunately obsolete. The tincture, in the strength of ½–1 fl. dr. (2.0–4.0 c.c.) to the ounce (30.0 c.c.) of water, may be used as an astringent application in *chronic pharyngitis* and *laryngitis*. Ferric chlorid in the form of the tincture or the solution is a valuable local remedy in *pharyngeal diphtheria*. It may be prescribed as follows:

℞	Tincturæ ferri chloridi,	f℥j–iij (4.0–11.0 c.c.);
	Glycerini,	f℥ss (15.0 c.c.);
	Aquæ,	q. s. ad f℥j (30.0 c.c.).—M.

Or as Löffler's solution:

℞	Menthol,	℥iiss (10 0 gm.);
	Toluol,	f℥x (36.0 c.c.);
	Alcoholis absoluti,	f℥ij (60.0 c.c.);
	Liquoris ferri chloridi,	f℥j (4.0 c.c.).—M.

The tincture is an efficient hematinic in *anemia*, but it is more liable than many of the other preparations of iron to injure the teeth, disturb digestion, and induce constipation. Basham's mixture is a favorite chalybeate diuretic in *chronic nephritis;* only small doses, however, should be given, and even these should be withheld if the drug causes constipation and headache. The tincture has long been regarded as a specific in *erysipelas.* This claim cannot be substantiated, although the drug is of benefit in some cases. It may be given in doses of from 15–20 min. (1.0–1.2 c.c.) every two or three hours.

Administration.—The tincture of ferric chlorid should be taken after meals, well diluted, through a tube.

Ferrum Dialysatum (*Dialyzed Iron*).—This is an unofficial preparation of the oxychlorid of iron from which acidulous matter has been removed by dialysis. It is a dark-red, tasteless, neutral liquid, quite free from astringency. It is a very unstable compound, and undergoes rapid decomposition in the stomach. It is an extremely feeble chalybeate, but a useful antidote in *arsenical poisoning*.

The dose as a hematinic is from 10–30 min. (0.6–2.0 c.c.); as an antidote, ½ fl. oz. (15.0 c.c.) frequently repeated.

FERRI OXIDUM HYDRATUM, U. S. P., AND FERRI OXIDUM HYDRATUM CUM MAGNESIA, U. S. P.

(Ferric Hydrate; $Fe_2(OH)_6$ and Ferric Hydrate with Magnesia.)

These preparations are used exclusively as antidotes in *arsenical poisoning*. The former may be precipitated from any liquid preparation of iron by adding to it an alkali (ammonia or sodium carbonate) ; the latter, by adding magnesia in excess. If a caustic alkali has been employed, such as ammonia, the excess should be removed by transferring the precipitate to a muslin filter, squeezing it and washing it with fresh water. Ferric hydrate with magnesia is preferable, since it needs no washing, and since the magnesia itself is antidotal. These preparations should be freshly made and given freely, while still moist, in doses of a tablespoonful or more every few minutes.

Ferric hydrate enters into iron plaster (*Emplastrum Ferri*, U. S. P.) and troches of iron (*Trochisci Ferri*, U. S. P.), but neither of these preparations is of much value.

FERRI SULPHAS, U. S. P.

(Ferrous Sulphate; Green Vitriol; $FeSO_4 + 7H_2O$.)

Ferrous sulphate occurs as pale, bluish-green, efflorescent prisms, having a saline, styptic taste. The dose is from 1–5 gr. (0.06–0.3 gm.).

PREPARATIONS.	DOSE.
Ferri Sulphas Exsiccatus, U. S. P.	1–3 gr. (0.06–0.2 gm.).
Ferri Sulphas Granulatus, U. S. P.	1–5 gr. (0.06–0.3 gm.).
Mistura Ferri Composita (Griffith's Mixture)	½–1 fl. oz. (15.0–30.0 c.c.).
Pilulæ Aloes et Ferri (contain about 1 gr. —0.06 gm.—of each)	1–3 pills.
Pilulæ Ferri Carbonatis (Blaud's Pills)	1–3 pills.

Therapeutics.—Sulphate of iron is an active astringent. It may be employed as a hematinic in *anemia* attended by relaxation of the bowels. The dried sulphate should be selected for pills.

Crude ferrous sulphate, or copperas, is used as a disinfectant and deodorizer for privies, cess-pools, etc., but its germicidal power is very feeble.

FERRI SUBSULPHAS.

(Ferric Subsulphate; Basic Ferric Sulphate; Monsel's Salt; $Fe_4O(SO_4)_5$.)

Ferric subsulphate is a yellow, hygroscopic, astringent powder, freely soluble in water.

PREPARATION.	DOSE.
Liquor Ferri Subsulphatis, U. S. P. (Monsel's Solution)	3–5 min. (0.2–0.3 c.c.).

Therapeutics.—Monsel's solution is a prompt and powerful styptic, which is somewhat less irritant in its action than the solution of ferric chlorid. It is employed chiefly as a *local hemostatic.* When used too freely it produces hard black clots, which may serve to conceal deep-seated hemorrhage. For this reason it should not be poured into the wound, but applied directly to the bleeding surface on a pledget of cotton. In *hematemesis* a few drops in water may prove efficient. It has been recommended in the form of a spray in *hemoptysis*, but it is of doubtful utility.

FERRI ET AMMONII SULPHAS, U. S. P.

(Ferric Ammonium Sulphate; Ammonioferric Alum.)

Ammonioferric alum occurs as pale-violet, efflorescent crystals, odorless, and of a sour, styptic taste. It is a more powerful astringent than ordinary alum (aluminum and potassium sulphate). It has been employed in saturated solution as a hemostatic in *capillary bleeding*, and in weak solutions as an astringent injection in *leukorrhea*.

FERRI IODIDUM.

(Ferrous Iodid; Protiodid of Iron; FeI_2.)

Ferrous iodid occurs in grayish-white, crystalline masses which are soluble in water with partial decomposition. It is rarely used in the pure form.

PREPARATIONS.	DOSE.
Ferri Iodidum Saccharatum, U. S. P. (contains 20 per cent. of pure ferrous iodid)	1–5 gr. (0.06–0.3 gm.).
Syrupus Ferri Iodidi, U. S. P. (contains 10 per cent. by weight of ferrous iodid)	5–60 min. (0.3–4.0 c.c.).
Pilulæ Ferri Iodidi, U. S. P. (contain about 1 gr.—0.06 gm.—of ferrous iodid)	1–3 pills.

Therapeutics.—Iodid of iron may be employed when it is desirable to combine an alterative with a chalybeate. It is a very useful preparation in *tuberculous adenitis, syphilitic cachexia*, and *rachitis.* It is usually well tolerated by children, to whom the syrup may be given in doses of from 3–20 min. (0.2–1.2 c.c.), or the saccharated iodid, in doses of from ½–1 gr. (0.03–0.06 gm.). J. C. Wilson has found large doses of the syrup efficacious in *articular rheumatism* and *other forms of acute arthritis* after failure with salicylic compounds.

As the syrup is injurious to the teeth, it should be taken well diluted, and the mouth should be thoroughly rinsed after its administration.

FERRI CITRAS, FERRI PHOSPHAS, AND FERRI TARTRAS.

These compounds are known as the scale preparations, because concentrated solutions of them are spread on plates of glass and allowed to evaporate, so that the salts may be obtained in the form of scales.   The following are official:

PREPARATIONS.	DOSE.
Ferri Citras, U. S. P.	2–5 gr. (0.13–0.3 gm.).
Ferri et Ammonii Citras, U. S. P. (contains 16 per cent. of metallic iron)	5–10 gr. (0.3–0.6 gm.).
Ferri et Quininæ Citras, U. S. P. (contains 11.5 per cent. of quinin and 14.5 per cent. of metallic iron)	5–10 gr. (0.3–0.6 gm.).
Ferri et Quininæ Citratis Solubilis, U. S. P. (same strength as preceding preparation, but more soluble)	5–10 gr. (0.3–0.6 gm.).
Ferri et Strychninæ Citras, U. S. P. (contains 1 per cent. of strychnin and 16 per cent. of metallic iron)	1–5 gr. (0.065–0.3 gm.).
Ferri et Ammonii Tartras, U. S. P. (contains 17 per cent. of metallic iron)	5–10 gr. (0.3–0.6 gm.).
Ferri et Potassii Tartras, U. S. P. (contains 15 per cent. of metallic iron)	5–10 gr. (0.3–0.6 gm.).
Ferri Phosphas Solubilis, U. S. P. (the solubility of this salt is dependent upon the presence of free sodium citrate)	3–10 gr. (0.2–0.6 gm.).
Ferri Pyrophosphas Solubilis, U. S. P. (the solubility of this salt is dependent upon the presence of free sodium citrate)	3–10 gr. (0.2–0.6 gm.).

Iron citrate is also official as *Liquor Ferri Citratis*, U. S. P., the dose of which is from 5–15 min. (0.3–1.0 c.c.).   From iron and quinin citrate, bitter wine of iron (*Vinum Ferri Amarum,* U. S. P.) is prepared, and from iron and ammonium acetate wine of ferric citrate (*Vinum Ferri Citratis*, U. S. P.) is prepared. The dose of either wine is from 1–3 fl. dr. (4.0–11.0 c.c.).   Iron phosphate enters into the syrup of the phosphates of iron, quinin, and strychnin (*Syrupus Ferri, Quininæ et Strychninæ Phosphatum*, U. S. P.), the dose of which is from 1–2 fl. dr. (4.0–8.0 c.c.).

Therapeutics.—The scale preparations of iron are mild and agreeable *hematinics*.   They are comparatively free from astringency, and are usually well borne by the stomach.   On account of their solubility they are well adapted for administration in liquid form.

FERRUM ALBUMINATUM.

Bunge first demonstrated that iron exists in the tissues and in various food-stuffs in the form of a nucleo-albuminate.   He subsequently succeeded in separating such a compound from

19

the yolk of eggs, and this he called *hematogen*. The latter differs from ordinary iron salts in being very resistant to the action of sulphids. According to Bunge, inorganic preparations of iron are not absorbed to any extent by the healthy mucous membrane of the alimentary canal, and are useful in anemia only by disposing of the sulphureted hydrogen in the intestine, and thus protecting the food-iron from reduction until its absorption has been accomplished. Recent investigations have proved conclusively that Bunge's view is fallacious, and that inorganic iron compounds are quite capable of absorption and assimilation. Albuminates and peptonates of iron, however, are often better borne by the stomach than the inorganic salts, are less injurious to the teeth, and are, perhaps, somewhat more readily absorbed. Many attempts have been made to produce artificially, and in a way that can be practically utilized, compounds having the characteristics of Bunge's hematogen. Schmiedeberg obtained from pig's liver a compound containing from 6–8 per cent. of iron, which he called *ferratin*. Later this preparation was obtained in larger quantities by the action of iron salts on egg-albumen. It is a fine, reddish powder, without odor or taste, insoluble in water, but soluble in dilute alkalis. The dose is from 3–10 gr. (0.2–0.6 gm.). It may be given in capsules, powders, or tablets, or with milk. Kobert has obtained two compounds—*hemogallol* and *hemol*—from ox-blood by reducing the hemoglobin with pyrogallol and zinc dust respectively. The dose of either compound is from 2–8 gr. (0.13–0.5 gm.). Grawitz has recently shown that hemorrhage into the intestinal canal or the administration of blood preparations by the mouth is speedily followed by granular degeneration of the red blood-corpuscles. He attributes this phenomenon to the action of poisonous substances derived from the blood during its passage through the digestive tract. Until further investigations prove Grawitz's observation to be erroneous the various derivatives of hemoglobin that have been recommended as hematinics should be regarded with considerable suspicion.

Ferri Lactas, U. S. P. (*Ferrous Lactate*).—This preparation occurs in pale, greenish-white, crystalline crusts, having a slight characteristic odor and a sweetish, ferruginous taste. It is slowly soluble in water. It may be given in doses of from 1–5 gr. (0.06–0.3 gm.). It is contained in the syrup of hypophosphites with iron (*Syrupus Hypophosphitum cum Ferro,* U. S. P.), the dose of which is from 1–4 fl. dr. (4.0–15.0 c.c.).

Ferri Oxalas (*Ferrous Oxalate*).—This salt occurs as a pale-yellow, odorless, crystalline powder, insoluble in water.

It has been lauded by Quincke and by Hayem as a hematinic in *chlorosis*, but it does not seem to have any special advantages. The dose is from 3–5 gr. (0.2–0.3 gm.).

Ferri Hypophosphis, U. S. P. (*Ferric Hypophosphite*).—This salt occurs as a white, odorless, and almost tasteless powder, slightly soluble in water. It has no special advantages. The dose is from 5–10 gr. (0.3–0.6 gm.).

Ferri Valerianas, U. S. P. (*Ferric Valerianate*).—This compound is a dark-red, amorphous powder, having the odor of valerian, and a mild styptic taste. It may be given with other valerianates when a combined chalybeate and anti-spasmodic action is desired. The dose is from 1–3 gr. (0.06–0.2 gm.).

Ferri Acetas (*Ferric Acetate*).—This salt is official in the form of a 31 per cent. aqueous solution—*Liquor Ferri Acetatis*, U. S. P. The dose of the latter is from 5–10 min. (0.3–0.6 c.c.). It is decidedly astringent.

Ferri Nitras (*Ferric Nitrate*).—This salt is official in the form of a 6.2 per cent. aqueous solution—*Liquor Ferri Nitratis*, U. S. P.—which has been used chiefly as an astringent.

PHOSPHORUS, U. S. P.

(P.)

Phosphorus is a yellowish, semitransparent, non-metallic element obtained from bones. It has a waxy consistence and luster, a peculiar, garlicky odor, and a disagreeable taste. By long keeping it turns red, and occasionally black. On exposure it emits white fumes, which are luminous in the dark, and on longer exposure it takes fire spontaneously. It is almost insoluble in water, soluble in 350 parts of absolute alcohol and in 50 parts of fatty oils, and readily soluble in chloroform. The dose is from $\frac{1}{100}$–$\frac{1}{30}$ gr. (0.00064–0.002 gm.).

PREPARATIONS.	DOSE.
Oleum Phosphoratum, U. S. P. (1 per cent. in oil of almond)	1–5 min. (0.06–0.3 c.c.).
Spiritus Phosphori, U. S. P. ($\frac{1}{5}$ per cent. in absolute alcohol, each fl. dr.—4.0 c.c.—contains about $\frac{1}{12}$ gr.—0.005 gm.—of phosphorus)	5–30 min. (0.3–2.0 c.c.).
Elixir Phosphori, U. S. P. (each fl. dr.—4.0 c.c.—contains about $\frac{1}{65}$ gr.—0.001 gm.—of phosphorus)	½–2 fl. dr. (2.0–8.0 c.c.)
Pilulæ Phosphori, U. S. P. (each pill contains $\frac{1}{100}$ gr.—0.0006 gm.)	1–3 pills.

Physiologic Action.—Wegner found that when phosphorus was given in small doses to growing animals it rendered the bones more dense, diminished the cancellous structure, and finally caused more or less obliteration of the marrow-cavity. That these changes were due to the stimulation of the bone-forming tissues, and not to the deposition of an excess of phosphates, was shown by the fact that in animals fed with phosphorus but deprived of phosphates the same hyperplasia in the bones resulted, but that the new tissue was soft and gelatinous instead of hard. Apart from these changes in the osseous system, and perhaps a slight increase in the number of red blood-cells, the effects of phosphorus in medicinal doses are not very obvious. There is considerable clinical testimony, however, to show that the drug improves the nutrition of other tissues, especially of the nervous system.

The manner in which phosphorus is absorbed has been a matter of some dispute; it probably enters the blood chiefly as phosphorus, absorption being facilitated by the solution of the drug in the fatty substances present in the bowel. A small part is volatilized in the alimentary canal, and probably enters the tissue in the form of vapor. Phosphorus is said to be eliminated in the urine partly as hypophosphoric acid.

Toxicology.—Toxic doses of phosphorus do not usually produce their effects for several hours. The earliest symptoms are intense abdominal pain, obstinate vomiting, thirst, a garlicky taste in the mouth, restlessness, and prostration. The ejected material contains mucus, bile, and occasionally disintegrated blood, and is luminous in the dark. At the end of from twenty-four to thirty-six hours the symptoms gradually subside and the patient feels comparatively comfortable, but soon the vomiting and pain return, jaundice develops, the liver becomes enlarged and painful, the pulse grows very feeble, and not infrequently ecchymoses appear in the skin and hemorrhages occur from the mucous membranes. The urine is scanty and contains albumin, tube-casts, bile, sarcolactic acid, an increased amount of ammonia, and sometimes leucin, tyrosin, and sugar. In fatal cases death occurs generally in from a few days to two weeks, and is often preceded by delirium, convulsions, stupor, and coma. Occasionally, phosphorus kills within a few hours, probably through its direct depressant influence on the heart. If the patient recovers, the symptoms abate very slowly and the convalescence is tedious.

The changes in the organs observed after death are attributable to grave disturbances of tissue-metabolism and not to the direct irritant action of the drug. The most characteristic feature is wide-spread fatty degeneration of the tissues. The latter is especially marked in the glands of the stomach and intestines, in the liver, kidneys, heart, and muscles. Ecchymoses are found in most of the organs. Areas of complete necrosis are often observed in the stomach and liver. In subacute poisoning the degenerated parenchyma may be replaced in part by newly formed connective tissue. Whether the fat found in the cells of the various organs is formed *in situ* or is carried to them from the subcutaneous tissues is a moot question.

Treatment.—If the phosphorus has been recently taken, the stomach should be washed out at once. Copper sulphate is probably the best emetic, since any excess may serve as an antidote by forming with the phosphorus insoluble phosphid of copper. A cathartic should be given to remove from the bowel any of the poison that may still be unabsorbed. All oily and fatty matters should be withheld, since they are active solvents of phosphorus. Various oxidizing agents have been employed as chemical antidotes. Potassium permanganate ($\frac{1}{2}$ per cent. aqueous solution) and the solution of peroxid of hydrogen are the most useful. Old French oil of turpentine, which is rich in ozone, has been repeatedly recommended, but it cannot be obtained in this country, and ordinary turpentine is useless.

Chronic Phosphorus-poisoning.—Workmen employed in match factories and who are exposed to the fumes of phosphorus develop a symptom-complex to which Magitot has applied the name *phosphorism.* This condition is characterized by anemia, loss of flesh, a garlicky odor of the breath, chronic diarrhea, albuminuria, fragility of the bones, and, in many instances, by necrosis of the maxillary bone. Since phosphorus necrosis occurs only in workmen who have carious teeth, it is usually attributed solely to the local action of the fumes, but there are good reasons for believing that systemic infection contributes largely to its development. Thus it rarely appears, despite the existence of carious teeth, until the worker has been some time, generally many years, at his trade; the sufferer nearly always manifests some of the symptoms of systemic saturation, and finally, Stubenrauch has demonstrated experimentally, that phosphorous fumes exert no especial action on exposed bone or periosteum.

Red or amorphous phosphorus being insoluble in ordinary liquids and non-volatile is free from toxic properties. Phosphoric acid has none of the qualities of phosphorus, and acts merely as a mild stomachic. Sodium phosphate is an agreeable laxative. The hypophosphites of calcium, sodium, and potassium are innoxious, and are employed as tonics. Sodium phosphite is extremely poisonous. Sodium pyrophosphate and sodium metaphosphate, when administered hypodermically, paralyze the central nervous system and the heart, and in slow poisoning produce fatty degeneration of the glandular organs and numerous hemorrhagic extravasations.

Therapeutics.—Phosphorus is employed as a tonic in certain diseases of the nervous system that are dependent upon exhaustion rather than upon organic changes. Thus, it is sometimes of service in *neurasthenia, neuralgia,* and *impotence* from sexual excesses. It has been employed in *rachitis,* according to Monti, since 1838, but its reputation in this disease is due largely to the very favorable reports of Kassowitz, in Europe, and of Jacobi, in this country. It is not unlikely that much of the benefit which has been observed in rachitic subjects while taking phosphorus has been due in reality to the cod-liver oil, simultaneously administered, and to the improvement in the diet and general hygiene. This view receives some support from the recent investigations of Zweifel, who found that phosphorus in cod-liver oil was soon converted, through oxidation, into phosphoric acid, and, therefore, lost its peculiar properties. Monti contends that phosphorus alone has not staid the progress of rachitis nor has it caused the slightest improvement.

Phosphorus has also been employed with asserted good results in other diseases of bones, such as *osteomalacia, caries,* and *delayed union of fractures.*

Administration.—As phosphorus undergoes oxidation on keeping in nearly all menstrua, only freshly made preparations should be prescribed. The official elixir is the most elegant form in which to administer the drug. Persons vary considerably in their susceptibility to phosphorus, and, therefore, the initial doses should be small, especially in children. Nebelthau has cited an instance in which death occurred in a well-developed child of two years after taking less than $\frac{1}{20}$ gr. (0.003 gm.) in the course of sixty hours. Even minute doses sometimes disturb the stomach and excite unpleasant eructations.

Zinci Phosphidum, U. S. P. (*Zinc Phosphid*).—This compound appears as a dark-gray, gritty powder, or as crystalline, metallic fragments having a faint odor and taste of phosphorus. It is insoluble in alcohol or water. The dose is from $\frac{1}{30}$–$\frac{1}{10}$ gr. (0.002–0.006 gm.), in pill. Its action closely resembles that of phosphorus, and it may be used for the same purposes as the latter.

Acidum Phosphoricum, U. S. P. (*Phosphoric Acid*).—

This is a colorless liquid, composed of 85 per cent. by weight of absolute orthophosphoric acid and 15 per cent. of water. It has a strongly acid taste. The dose is from 3–5 min. (0.2–0.3 c.c.), well diluted. Its action is somewhat like that of the dilute mineral acids, and bears no resemblance to that of phosphorus. It may be employed to allay *thirst in diabetes* and *febrile diseases.* It has also been used to *promote gastric digestion*, but for this purpose it is inferior to hydrochloric acid. It is usually prescribed as diluted phosphoric acid (*Acidum Phosphoricum Dilutum*, U. S. P.), the dose of which is from 10–60 min. (0.6–4.0 c.c.), in water.

CALCII PHOSPHAS.

(Calcium Orthophosphate ; $Ca_3(PO_4)_2$.)

Calcium phosphate is official as *Calcii Phosphas Præcipitatus*, a white, odorless, tasteless, amorphous powder, almost insoluble in water. The dose is from 5–30 gr. (0.3–2.0 gm.).

PREPARATION.	DOSE.
Syrupus Calcii Lactophosphatis, U. S. P. (contains about 3 per cent. of the double soluble salt)	1–4 fl. dr. (4.0–15.0 c.c.).

Physiologic Action and Therapeutics.—The insoluble salts of calcium, of which the phosphate may be taken as a type, have no very pronounced effects upon the system. When taken internally, the greater part resists absorption and reappears in the stools; a small portion, however, enters the circulation, and in case of any deficiency in the lime-salts of the food is appropriated by the tissues. In the absence of a deficiency in the natural supply, any excess of lime, if it does not fail altogether of absorption, is slowly excreted by the intestinal epithelium and kidneys. As food ordinarily contains more lime than the body requires, it is difficult to understand how the phosphate or hypophosphite of calcium can be of service when administered as a remedy. The difficulty, however, is no greater than that which is encountered in attempting to explain the undoubted efficacy of iron in chlorosis, a disease that is certainly not dependent upon a want of iron-salts in the food.

Calcium phosphate has been recommended, on theoretic grounds, in *rachitis* and *osteomalacia*, but it is of very doubtful value. The testimony to its favorable influence on general nutrition in such diseases as *tuberculosis, tertiary syphilis*, and *neurasthenia* is more convincing, but even here it is not unlikely that much of the good attributed to the phosphate is in reality due to the remedies usually associated with it.

Administration.—Calcium phosphate may be prescribed in.powders or capsules. The lactophosphate, being soluble, may be given in the form of the official syrup or in an emulsion of cod-liver oil.

Calcii Glycerophosphas.—The fact that phosphorus exists in the nervous system in the form of glycerophosphoric acid has led to the introduction of the salts of this acid as substitutes for phosphates and hypophosphites. They have been used chiefly in *exhaustion from overwork*, in *tardy convalescence from the specific fevers*, in *neurasthenia, phosphaturia*, and *exophthalmic goiter*. The dose of calcium glycerophosphate is from 2–5 gr. (0.1–0.3 gm.), in solution or gelatin-coated pills. Sodium glycerophosphate is employed in the same dose, and may be given by the mouth or hypodermically.

CALCII HYPOPHOSPHIS, U. S. P.

(Calcium Hypophosphite; $Ca_2(PH_2O_2)_2$.)

Calcium hypophosphite occurs in colorless, transparent crystals, or small, lustrous scales, of a nauseous, bitter taste. It is soluble in about 7 parts of water. The dose is from 5–30 gr. (0.3–2.0 gm.).

PREPARATION.	DOSE.
Syrupus Hypophosphitum, U. S. P. (calcium hypophosphite, 4.5 per cent.; sodium hypophosphite, 1.5 per cent.; potassium hypophosphite, 1.5 per cent.; dilute hypophosphorous acid, 0.2 per cent.; spirit of lemon, sugar, and water),	1–2 fl. dr. (4.0–8.0 c.c.).

Therapeutics.—Hypophosphites may be employed in the same class of cases as that in which calcium phosphate has been recommended.

OLEUM MORRHUÆ, U. S. P.

(Cod-liver Oil.)

Cod-liver oil is a fixed oil obtained from the livers of *Gadus Morrhua* and other species of *Gadus*. The *shore oil*, or *white oil*, the only variety at present used for medicinal purposes, is extracted by disintegrating the fresh livers with superheated steam, and subsequently straining the resultant pultaceous mass. It is a pale yellow, thin liquid, having a peculiar fishy odor and taste. In addition to olein, palmitin, stearin, and other fatty principles it contains traces of iodin, bromin, phosphorus, cholesterin, trimethylamin, and, possibly, of bile-salts. The dose is from 1–4 fl. dr. (4.0–15.0 c.c.).

Physiologic Action and Therapeutics.—Cod-liver oil is a food rather than a medicine. It is superior to other fats, because it is more digestible and assimilable. The amount of iodin and of phosphorus present is too small for these ingredients to exert any specific action. Attempts to isolate an active principle have not been attended with success. The ready absorption of cod-liver oil has been attributed by some authorities to the presence of biliary salts; by others, to the presence of free fatty acids, but no explanation that has yet been offered is perfectly satisfactory. In moderate doses cod-liver oil improves the general nutrition, increases the number of red blood-cells, and favors the accumulation of fat in the body. Large doses excite nausea, eructations, vomiting, and diarrhea.

Cod-liver oil, provided it is well borne by the stomach, may be employed in any disease in which malnutrition is a prominent feature; thus it is often of great service in *tuberculosis of the lungs, bones, or lymph-glands*, in *rickets, tertiary syphilis, chronic rheumatism, rheumatoid arthritis, chronic bronchitis, convalescence from the specific fevers*, and *secondary anemia*. In the early stage of phthisis it does more good than any other medicinal agent, but it must be given cautiously and withdrawn upon the first evidence of intolerance. When symptoms of secondary infection are prominent, such as fever, sweating, and rapid wasting, it is less likely to be well borne or to do good. In conjunction with general hygienic measures, the prolonged use of cod-liver oil is also valuable in preventing the development of tuberculosis in those who, through physique and heredity, have a marked predisposition to the disease.

Contraindications.—High fever and gastric or intestinal irritation are the chief contraindications to its use.

Administration.—Patients who at first have difficulty in taking cod-liver oil may ultimately acquire a tolerance for it if the dose be carefully adapted to their digestive powers. It is well, therefore, to begin with very small doses, even a few minims, and to increase the amount gradually as the stomach becomes accustomed to it. A single dose at bedtime may be well borne when the stomach will not tolerate it during the day. As the oil is digested in the intestine, and not in the stomach, its stay in the latter should be reduced to a minimum, and this can be accomplished by administering it two or three hours after meals. Although it is impossible to disguise completely the fishy odor and taste of cod-liver oil, there are many ways in which the unpleasantness of its administration may be lessened. It may be floated on the surface of whisky or porter, care being taken to prevent it from touching the glass, and swallowed quickly; it may be taken in ketchup or in tincture of orange peel, but perhaps the best way to prescribe it is in soft capsules holding from 10–60 min. (0.6–4.0 c.c.). or in a well-made emulsion. Emulsions containing 50 per cent. of oil may be made with yolk of egg, acacia, or malt extract, and flavored with oil of bitter almond, gaultheria, or cinnamon.

<pre>
℞   Olei morrhuæ.   f℥iv (118.0 c.c.);
    Olei amygdalæ amaræ,   ℳ x (0.6 c c);
    Acaciæ,   q. s.
    Syrupi calcii lactophosphatis,   f℥ij (60 0 c.c.) ;
    Aquæ,   q. s. ad f℥viij (235.0 c.c.).—M.
        Sig.   A tablespoonful twice daily, two hours after meals.
</pre>

As a rule, children take cod-liver oil readily, and even grow to like it; but if it cannot be taken by the mouth, owing to gastric irritability, it may be given by inunction. Just before retiring, and after a warm bath, 1 or 2 drams (4.0–8.0 c.c.) of the oil should be rubbed into the skin over the chest and abdomen, and the child put to bed in a flannel night-dress.

ALTERATIVES.

Alteratives are agents that favorably modify general morbid processes without exerting a demonstrable influence on any particular organ. It need scarcely be said that the manner of their action is quite unknown; if it were known, these remedies could be better apportioned among other classes. The term alterative is simply applied for convenience to a group of heterogeneous drugs that experience has shown to be more or less efficacious in certain constitutional diseases, the real nature of which is very imperfectly understood. The most important members of this group are:

Arsenic.

Iodids.

Iodin.

Iodoform.

Mercury.

Ichthyol.

Gold and sodium chlorid.

Colchicum.

Guaiac.

Sarsaparilla.

Jambul.

Mezereum.

Calx sulphurata.

Uranium nitrate.

Gland extracts (thyroid, supra- renal, thymus, and testicular extracts).

ARSENUM.

(Arsenic; As.)

Arsenic is not employed for medicinal purposes in the metallic form, but as arsenous anhydrid (arsenous acid), as a salt of the latter (an arsenite), or as a salt of arsenic acid (an arsenate).

PREPARATIONS.	DOSE.
Acidum Arsenosum, U. S. P.	$\frac{1}{60}$–$\frac{1}{20}$ gr. (0.001–0.003 gm.).
Liquor Acidi Arsenosi, U. S. P. (contains 1 per cent. of arsenous acid and 5 per cent. of diluted hydrochloric acid) . . .	1–5 min. (0.06–0.3 c.c.).
Liquor Potassii Arsenitis, U. S. P. (Fowler's Solution: contains the equivalent of 1 per cent. of arsenous acid and 3 per cent. of compound tincture of lavender),	1–5 min. (0.06–0 3 c.c.).
Sodii Arsenas, U. S. P.	$\frac{1}{60}$–$\frac{1}{20}$ gr. (0.001–0.003 gm.).
Liquor Sodii Arsenatis, U. S. P. (Pearson's Solution: contains 1 per cent. of sodium arsenate)	1–5 min. (0.06–0.3 c.c.).
Arseni Iodidum, U. S. P.	$\frac{1}{30}$–$\frac{1}{10}$ gr. (0.002–0.006 gm.).
Liquor Arseni et Hydrargyri Iodidi, U. S. P. (Donovan's Solution: contains 1 per cent. each of arsenic iodid and mercuric iodid)	1–5 min. (0.06–0.3 c.c.).

ACIDUM ARSENOSUM, U. S. P.

(Arsenous Acid; Arsenous Anhydrid; White Arsenic; As_2O_3)

Arsenous acid occurs either as an opaque, white, amorphous powder, or as heavy, amorphous masses of a translucent, glass-like appearance. It has neither odor nor taste. When thrown on hot coals, it volatilizes without melting, and emits a strong, garlicky odor. It is soluble in from 30–80 parts of cold water, slightly soluble in alcohol, and soluble in about 5 parts of glycerin. The dose is from $\frac{1}{60}$–$\frac{1}{20}$ gr. (0.001–0.003 gm.).

Physiologic Action.—Local Action.—On mucous membranes and denuded surfaces arsenic, in concentrated form, acts as a powerful and painful caustic. If it is applied over large surfaces a sufficient quantity may be absorbed to produce systemic poisoning.

Circulation.—In therapeutic doses, beyond increasing slightly the pulse-rate, arsenic exerts but little influence on the circulation. Toxic doses, however, lessen the frequency of the pulse and lower the blood-pressure. The fall of blood-pressure appears to be due to depression of the heart and of the vasomotor nerves, especially of those branches distributed to the blood-vessels of the abdominal organs.

Blood.—In health arsenic has little or no action on the blood, but in certain forms of anemia, especially pernicious anemia, it often increases decidedly the number of erythrocytes. The view that the blood improvement is due to stimulation of the bone-marrow receives some support from the experimental researches of Stockman and Grieg, who found that arsenic increased the vascularity of the marrow and led to a replacement of fat by red corpuscles. In some cases of poisoning there is marked disintegration of the red blood-cells.

Nervous System.—Clinical experience indicates that in medicinal doses arsenic exerts a stimulant influence upon the nervous system. In frogs toxic doses cause a rapid paralysis, which appears to be mainly of centric origin. In man nervous phenomena occur in both acute and chronic poisoning, but they are especially pronounced in the latter. The paralysis of chronic poisoning is, in the majority of instances, the result of a polyneuritis similar to that produced by other toxic agents; in a few cases, however, paralysis of spinal origin has been observed.

Alimentary Canal.—In minute doses arsenic stimulates the appetite and favors digestion. Large doses have an irritant effect, and excite colicky pains, vomiting, and diarrhea.

Effect on Metabolism.—It is generally conceded that arsenic in therapeutic doses favorably influences nutrition, but the manner in which this is accomplished is not understood. Toxic doses have a profound effect upon tissue-metabolism. They increase the nitrogenous output, suppress the glycogenic function of the liver, so that puncture of the floor of the fourth ventricle no longer causes glycosuria, and induce fatty changes in the epithelium of the alimentary canal, and, to a less extent, in the liver, heart, and kidneys.

Elimination.—Arsenic passes out of the body very slowly; most of it escapes in the urine, but traces may be detected in the other secretions.

Action on Lower Organisms.—While arsenic will readily kill insects and worms, it is not very destructive to the lowest forms of animal and plant life.

Toxicology.—*Acute arsenical poisoning* is characterized by severe abdominal pains, vomiting and purging of "rice-water" and, perhaps, bloody material, persistent thirst, oliguria, muscular cramps, dyspnea, cyanosis, and collapse. Death, which usually occurs in from one to three days, is often preceded by delirium, convulsions, and coma. If recovery follow, the symptoms of acute poisoning may be slowly replaced by those of chronic poisoning. Occasionally cases are encountered in which the symptoms depart somewhat from the usual type; thus there may be a rapid termination in collapse or coma without any preceding abdominal symptoms; sometimes a temporary remission in the symptoms occurs about the third day, and this is followed, as in the case of phosphorus poisoning, by jaundice and delirium; again, there may be, in addition to the abdominal symptoms, an extensive urticarial or vesiculo-papular rash.

It is difficult to fix upon the minimum fatal dose of arsenic, since much that is ingested often escapes absorption. According to Taylor, about 2 gr. (0.13 gm.) is the minimum fatal dose for an adult; but cases are on record in which very large doses, even 1–2 ounces (30.0–60.0 gm.), have been swallowed without destroying life.

After death from arsenic macroscopic changes are apparent in the stomach and upper bowel. The mucous membrane is swollen, congested, and more or less eroded. Microscopic examination of the organs reveals fatty degeneration in the epithelium of the alimentary canal, and in the liver, kidneys, and muscles. The erosion of the stomach cannot be the result of a local action, since it has occurred after the hypodermic injection of the poison and after its introduction through a wound, when only minute quantities could have reached the stomach. Moreover, arsenic lacks the power possessed by corrosive poisons of uniting with proteid matter to form an albuminate. The action of the drug in inducing cellular death appears to be a specific one, and in the present state of our knowledge no satisfactory explanation of it can be given.

Treatment of Acute Poisoning.—The stomach should be emptied by means of the stomach-pump or by an emetic. The best chemical antidote is the recently prepared ferric hydrate or ferric hydrate with magnesia (see p. 286), administered while still moist, in doses of a tablespoonful or more every ten or fifteen minutes. These compounds are in themselves harmless and act by forming insoluble arsenites. Dialyzed iron, as it liberates a certain amount of free ferric hydrate in the stomach, is also antidotal. As in poisoning from other irritants, diluents, demulcents, and opiates are usually indicated.

Chronic arsenical poisoning may be a sequel of acute poisoning, may follow the prolonged use of the drug for medicinal purposes, may result from the use of foods or liquors contaminated with arsenic, may occur from the constant inhalation of dust arising from wall-paper or other fabrics containing arsenical pigments, or may be acquired by workers in arsenic mines or in factories in which fumes of the metal are formed. It may be manifested by gastro-enteritis, conjunctivitis, and catarrh of the upper air-passages, anemia, peripheral neuritis, and various cutaneous lesions. Polyneuritis from the medicinal use of arsenic is by no means infrequent. It is most commonly observed in children, to whom the drug is being given for chorea. Railton has reported 4 cases in which the paralytic phenomena did not appear until from one to three weeks after the drug had been discontinued. The recent endemic of chronic arsenical poisoning occurring in Manchester, England, in which over 4000 beer-drinkers were affected, was traced to the presence of arsenic in the glucose used for brewing. The sulphuric acid employed in making the glucose contained from 1.4 per cent. to 2.6 per cent. of arsenous acid, while the beer contained from $\frac{1}{4}$ to 3 gr. (0.016–0.2 gm.) a gallon (4.0 L.). In this endemic gastro-intestinal symptoms were not marked, the chief phenomena being numbness and tingling in the hands and feet; a sense of burning in the feet, and painful flushing, resembling erythromelalgia; and certain skin-lesions, consisting of pigmentation resembling Addison's disease, herpes zoster, hyperidrosis, hyperkeratosis of the palms and soles, and pemphigoid eruptions.

Besides hyperkeratosis of the hands and feet, prolonged arsenical medication may result in other lesions of the skin indicative of increased cellular proliferation. Thus it may lead to corns, horny growths, multiple warts, and, perhaps, to epithelioma. Hartzell has collected 11 cases of epithelioma occurring in psoriasis, and apparently due to the prolonged use of arsenic.

Tolerance.—Under certain conditions, which are not well understood, the prolonged use of arsenic, instead of causing chronic poisoning, may result in the establishment of a tolerance for the drug. In some parts of Austria it is a custom among many of the peasants to take gradually increasing doses of arsenic, under the belief that it improves the complexion and increases the endurance. According to Knappe, so much as 5 or 6 grains (0.3–0.4 gm.) are sometimes taken at a single dose without harm. The practice, however, even among these peasants, is not altogether free from risk, and in other countries it is the exception, rather than the rule, for the continued administration of arsenic not to be followed by untoward effects.

Therapeutics.—Locally, arsenous acid is used successfully as a caustic in removing from the skin circumscribed, superficial new growths, such as *lupus* and *epithelioma*, especially the latter. For this purpose a paste made by adding equal parts of white arsenic and acacia to a saturated solution of cocain hydrochlorate is used. This should be spread over the diseased surface, and allowed to remain for from twenty-four to forty-eight hours, when a poultice may be applied to remove the slough. It is generally not advisable to spread the paste over a larger area than a square inch at one time.

Internally, arsenic is employed empirically in a number of quite diverse pathologic conditions. After iron, it is the most effective drug we have in *anemia*. In *pernicious anemia* Fowler's solution, although not curative, is our best remedy. When well borne, the dose should be gradually increased, so that at the end of three or four weeks the patient is taking from 10–15 min. (0.6–1.0 c.c.) or more, three times a day. Fowler's solution is also a valuable remedy in *Sydenham's chorea* or *St. Vitus's dance*. To be effective, however, it must be given in ascending doses until symptoms of saturation appear. *Neuralgia* that is dependent upon anemia is often benefited by arsenic. In *pulmonary tuberculosis* no general tonic, with the exception of cod-liver oil, is so serviceable. The drug has a long-standing reputation as a remedy in *diabetes mellitus*. While it is not nearly so efficacious as opium, it may be tried in the mild forms of the disease. It is a good alterative in *chronic rheumatism, chronic gout*, and *rheumatoid arthritis*. In *bronchial asthma* it is sometimes very efficient, although it is not so reliable as potassium iodid. According to Murray, it is especially useful in the asthma of children and of old emphysematous persons.

Both Bramwell and Balfour have spoken in the highest terms of the prolonged use of arsenic ($\frac{1}{100}-\frac{1}{50}$ gr.—0.0006–0.0012 gm.) in *myocardial degeneration* and *angina pectoris*. In these cases it may be combined advantageously with strychnin, and sometimes also with digitalis.

In *malaria* it is a useful adjuvant to quinin. It probably exerts no destructive influence upon the parasite, but when employed after the subsidence of the paroxysms, it proves a valuable blood-restorer.

Arsenic has been found efficacious in certain diseases of the digestive tract. It is often very useful in that obstinate affection know as " *geographic tongue*," in which whitish, circinate patches appear upon the tongue and creep from one part to another. In some cases of *gastralgia* it acts most happily. Ringer has derived great benefit from the use of Fowler's solution—a drop shortly before meals—in the *morning vomiting of drunkards*, and in cases of so-called *irritative dyspepsia*, in which the tongue is furred and its papillæ are red and prominent. The same treatment is also highly recommended in that form of indigestion which is characterized by an uncontrollable desire to evacuate the bowel immediately after taking food.

The administration of arsenic sometimes gives very gratifying results in certain chronic inflammatory skin diseases of a sluggish type. It is especially useful in *psoriasis, pemphigus,* and *chronic erythematous, squamous,* and *papular eczema.* The drug is contraindicated when the inflammatory process is of an acute or active type. It is a common practice to give Fowler's solution in combination with bromids to prevent the outbreak of acne. Ewald has confirmed the observation originally made by Mabille that arsenic obviates to a great extent the unpleasant symptoms excited by thyroid medication.

Administration.—As persons vary considerably in their susceptibility to the action of arsenic, it is always well to begin with small doses of the drug and to increase them gradually. Puffiness under the eyes, especially noticeable in the morning, and looseness of the bowels, with colicky pains, are the usual indications of saturation. For pills, arsenous acid is generally selected, and for solutions, the solution of potassium arsenite. When a very decided impression is desired, as in chorea and pernicious anemia, the best preparation to employ is Fowler's solution, and this should be prescribed by itself, so that changes in the doses can be readily made from day to day. When arsenic is prescribed for its constitutional effect, it should be given invariably after meals. When it is not tolerated by the stomach, it may be given subcutaneously in the form of sodium arsenate or sodium cacodylate (*vide*).

The following formulæ will illustrate the manner of prescribing the drug in combination:

R   Acidi arsenosi,   gr. ss (0.03 gm.);
  Quininæ sulphatis,   gr. l (3.2 gm.);
  Ferri reducti,   gr. xl (2.6 gm.);
  Pulveris capsici,   gr. x (0.65 gm.).—M.
 Fiant pilulæ No. xx.
Sig.   One after meals. (*Malaria after the arrest of the paroxysms.*)

R   Liquoris potassii arsenitis,   f ℥iss (5.5 c.c.);
  Sodii salicylatis,   ℥v (20.0 gm.);
  Glycerini,   f ℥j (30.0 c.c.);
  Aquæ menthæ piperitæ, q. s. ad. f ℥iv (118.0 c.c.).—M.
Sig.   A teaspoonful in water, gradually increased to a dessertspoonful, after meals. (*Diabetes mellitus.*)

Incompatibles.—Salts of iron, silver, copper, and ammonium, magnesia, lime, and tannic acid.

Arseni Iodidum, U. S. P.—Iodid of arsenic occurs in orange-red, glossy crystals, having the odor and taste of iodin. It is soluble in water and alcohol. The dose is from $\frac{1}{30}$–$\frac{1}{10}$ gr. (0.002–0.006 gm.). It has been found useful in *tuberculous adenitis* or *scrofula*. Saint-Phillippe has employed it with success in the troublesome *bronchitis* of strumous children. He recommends 5 min. (0.3 c.c.) of a 1 per cent. solution, gradually increased to 10–15 min. (0.6–1.0 c.c.), with meals. Aqueous solutions should be freshly prepared and kept in a cool place, since they are prone to decompose into arsenous and hydriodic acids.

Sodii Cacodylas.—Sodium cacodylate is an organic compound derived from cacodylic acid or dimethyl arsenic. It is a soluble, deliquescent, tasteless solid, containing about 50 per cent. of arsenic. Like other organic compounds of arsenic it is only feebly toxic—according to Renaut, 15 gr. (1.0 gm.) injected intravenously into a rabbit does not cause death. It may be employed medicinally in the same conditions in which the inorganic preparations of arsenic have been found useful. It is especially adapted for hypodermic administration, the dose being from $\frac{1}{3}$–$1\frac{1}{2}$ gr. (0.02–0.1 gm.), two or three times a day. The injections are not painful, and may be continued for a week or ten days, and then interrupted for a week. If administered by the mouth, it may become poisonous by meeting with reducing agents in the alimentary canal. Murrell has reported a case of poisoning in a tuberculous patient from the administration of the drug by the mouth in doses of 1 gr. (0.065 gm.) twice daily. The toxic symptoms appeared after the eleventh dose.

Liquor Arseni Bromidi.—This preparation, which is also known as Clemens' solution, is not, as it names implies, a solution of the bromid of arsenic, since the latter is decomposed in the presence of water, but rather a solution of arsenate and bromid of potassium. It is made by boiling 73 gr. (4.73 gm.) each of arsenous acid and of potassium bicarbonate in 2 fl. oz. (59.14 c.c.) of water until solution is affected; allowing to cool, adding 10 fl. oz. (295.73 c.c.) of water, then 117 gr. (7.581 gm.) of bromin, and finally enough water to make 16 fl. oz. (473.17 c.c.). The solution represents about 1 per cent. of arsenous acid. The dose is from 1–5 min. (0.06–0.3 c.c.), well diluted, after meals. It has been largely used in the treatment of *diabetes mellitus*, but with less success than opium, salicylic compounds, or antipyrin.

Cupri Arsenis.—Arsenite of copper is a yellowish-green, crystalline powder, slowly soluble in water. Under the name of Paris green or Scheele's green it is much used as an insect poison and as a pigment for paper, cotton fabrics, artificial flowers, etc. Cases of poisoning from this salt of copper have been of frequent occurrence. The antidotes are the same as those to arsenous acid. The therapeutic dose is from $\frac{1}{300}$–$\frac{1}{100}$ gr. (0.0002–0.00064 gm.), well diluted. It has been very highly recommended as an intestinal antiseptic in the *diarrhea of childhood*, when the stools are copious, green, and offensive.

POTASSII IODIDUM, U. S. P.

(Potassium Iodid; KI.)

Potassium iodid occurs as colorless, transparent or translucent crystals, or as a white granular powder, having a pungent, saline taste. It is soluble in 0.75 part of water, 2.5 parts of glycerin, and 18 parts of alcohol. The usual dose is from 5–10 gr. (0.3–0.65 gm.), well diluted, thrice daily; but in syphilis often three or four times this amount may be given with advantage.

PREPARATION.

Unguentum Potassii Iodidi, U. S. P. (contains 12 per cent. of potassium iodid and 1 per cent. of sodium hyposulphite).

Physiologic Action.—In health a single, moderate dose of potassium iodid produces no demonstrable effects beyond a slight increase in the secretion of urine and, perhaps, some disturbance of the stomach. The drug is rapidly absorbed from all parts of the digestive tract, and reappears in the secretion in less than fifteen minutes after its ingestion. The larger portion is eliminated through the kidneys, but small quantities escape in the saliva, tears, milk, and perspiration. It is likely that a small portion is retained for a time in the body, and that its continuous administration may result in accumulation. Large doses of potassium iodid cause burning in the stomach, nausea, vomiting, and diarrhea. The continuous use of the drug is usually followed, sooner or later, by a group of symptoms known as *iodism.* This condition has been attributed to the irritant effects of free iodin, into which a portion of the iodid appears to be converted. Its manifestations are most commonly associated with the mucous membrane of the respiratory tract and with the skin, and consist of frontal headache, lacrimation, running at the nose, sneezing, soreness of the throat, an increased flow of saliva, and, later, of a generalized acneiform eruption on the skin. More rarely the eruption is of an erythematous, purpuric, or bullous character. Intense

dyspnea from inflammatory edema of the larynx has occasionally been observed. In some instances symptoms suggesting
Graves' disease have developed, such as tremors, cardiac palpitation, sweating, and loss of weight. It is not improbable
that the last-named symptoms owe their origin to an influence
exerted by the iodid on the thyroid gland. Profound cachexia,
atrophy of the mammæ or testicles, paralysis, and blindness
have also been reported as occurring from the prolonged use
of iodids. The amount of the drug required to induce iodism
varies considerably in different subjects. Daily doses of from
200–300 gr. (13.0–20.0 gm.) are sometimes well borne, and, on
the other hand, doses of from 2–3 gr. (0.13–0.2 gm.) a day may
soon excite intense discomfort. Russell has reported a case in
which 5 doses of 4 min. (0.25 c.c.) of the syrup of ferrous
iodid with 2 gr. (0.13 gm.) of potassium iodid, over a period
of three days, caused acute iodism with fatal termination, in a
man of sixty-eight years, suffering from rheumatoid arthritis.
Patients with chronic Bright's disease are often extremely
intolerant. Syphilis does not always confer immunity.

Therapeutics.—Potassium iodid is of value in a variety
of morbid conditions, but the study of its action on healthy
subjects has thus far failed to throw much light upon the
manner in which its good effects are accomplished. Its best
effects are observed in *syphilis*, in which disease its efficacy is
equal to that of mercury. In the tertiary period it is more
valuable than mercury, and in the late secondary period it is
an excellent adjuvant. It should be given in ascending doses
until improvement follows or symptoms of iodism appear. To
secure permanent results, the treatment must be continued
for six months or a year. In *parasyphilitic affections*—locomotor ataxia, paretic dementia, and arteriosclerosis—the iodid
should always be tried, though, of course, no good can accrue
from its use if there has been much destruction of tissue.

Various opinions are held as to the value of potassium iodid
in chronic interstitial inflammations, like *cirrhosis of the liver*,
chronic interstitial nephritis, and *sclerosis of the spinal cord*,
when a syphilitic origin can be excluded. The author is of
the opinion that though the drug is of doubtful service, it
ought to be tried. Only small doses, however, should be given,
and these should be withdrawn on the slightest evidence of intolerance. In *subacute* and *chronic rheumatism* and in *chronic
gout* it sometimes proves beneficial. It is generally believed
that it aids in the absorption of the effusion in *serous pericarditis* and *pleurisy*. It is the most reliable remedy that we possess in *bronchial asthma* to prevent the return of the parox-

ysms. In *chronic bronchitis*, when the sputum is composed of thick, viscid mucus, it may be combined advantageously with expectorants.

In doses of from 10–20 gr. (0.65–1.3 gm.) thrice daily it is very beneficial in *aortic aneurysm;* even if it does not retard the progress of the disease, it nearly always lessens the pain. Whether it acts by lowering the blood-pressure, by promoting diuresis, and thus tending to inspissate the blood, or, as Balfour has suggested, by contracting the sac of the aneurysm, is not known. In *angina pectoris* no remedy with the exception of the nitrites is so useful in preventing recurrence of the attacks. To be effectual it should be given in daily doses of from 30–45 gr. (2.0–3.0 gm.) for several months. According to Osler, the patients who stand the treatment well are the robust, middle-aged men, in whom the angina is the sole symptom.

In *chronic metallic poisoning*, especially from lead and mercury, potassium iodid is undoubtedly efficacious. It aids in the elimination of the metal probably by forming with it in the tissues a double soluble salt. It has been employed internally, with complete success, as a curative remedy in a number of cases of *actinomycosis* occurring in man.

Administration.—Individual susceptibility to the iodids varies remarkably and idiosyncrasies are frequently encountered. Children generally bear the drug much better than adults. The initial dose should always be small, and the amount gradually increased as the tolerance of the patient permits. No drug is of much value in preventing iodism when the tendency to it is pronounced, but the tincture of belladonna (3–5 min.—0.2–0.3 c.c.) will sometimes relieve the coryza, and Fowler's solution (1–2 min.—0.06–0.12 c.c.), the acne and indigestion. Iodids should always be given after meals and well diluted. The unpleasant taste can be disguised with compound syrup of sarsaparilla. A convenient method of prescribing potassium iodid is in the form of a saturated aqueous solution, so prepared that 1 min. (0.06 c.c.) will contain 1 gr. (0.065 gm.) of the salt. As prescribing the iodid and water, ounce for ounce, results in a solution much weaker than the one indicated, Hynson suggests the following plan: Dissolve 480 gr. (31.1 gm.) of the salt in 5½ dr. (20.34 c.c.) of hot water, and then make up the solution to 8 dr. (29.57 c.c.) with water. This results in a solution containing 1 gr. (0.065 gm.) to a minim (0.06 c.c.). A drop from a medicine dropper will contain a little less than 1 gr. (0.065 gm.). If taken in milk, most patients do not find the solution disagreeable.

When not tolerated by the stomach, potassium iodid may be given as an enema in milk.

Incompatibles.—Mineral acids and salts, alkaloids, and spirit of nitrous ether. Potassium iodid is often added in excess to solutions of corrosive sublimate to form the soluble double iodid of mercury and potassium.

SODII IODIDUM, U. S. P.

(Sodium Iodid; NaI.)

Sodium iodid occurs in colorless, cubical crystals or white crystalline powder, of a bitter, saline taste. On exposure to moist air it deliquesces and becomes sodium carbonate and free iodin. It is soluble in 0.6 part of water and in 3 parts of alcohol. The dose is from 5–20 gr. (0.3–1.3 gm.) or more, well diluted. It has the same therapeutic value as the potassium salt, but it is sometimes better borne by the stomach.

AMMONII IODIDUM, U. S. P.

(Ammonium Iodid; NH$_4$I.)

Ammonium iodid occurs in minute colorless, cubical crystals or white granular powder, of sharp, saline taste. On exposure it attracts moisture from the air, and becomes yellowish-brown from the loss of ammonia and the liberation of iodin. It is soluble in 1 part of water and in 9 parts of alcohol. The dose is from 3–15 gr. (0.2–1.0 gm.) or more, well diluted. Excepting that it is somewhat more irritating to the stomach it has the same properties as potassium iodid.

STRONTII IODIDUM, U. S. P.

(Strontium Iodid; SrI$_2$ + 6H$_2$O.)

Strontium iodid occurs in colorless or faintly yellow, hexagonal plates, deliquescent, and of bitterish, saline taste. It is soluble in 0.6 part of water. The dose is from 5–20 gr. (0.3–1.3 gm.) or more, well diluted. While less irritating and less prone to induce iodism than potassium iodid, it does not seem to be so powerful as the latter salt.

PLUMBI IODIDUM, U. S. P.

(Lead Iodid; PbI$_2$.)

Lead iodid is a heavy, bright yellow powder, odorless and tasteless, and permanent in the air. It is almost insoluble in water. The dose is from ½–3 gr. (0.03–0.2 gm.).

PREPARATION.

Unguentum Plumbi Iodidi, U. S. P. (10 per cent.).

Iodid of lead is rarely used internally. Externally the ointment makes an excellent resolvent application for *tuberculous lymphatic glands* and other forms of *non-suppurative adenitis.* It should be applied very thoroughly and with gentle friction.

ACIDUM HYDRIODICUM.

(Hydriodic Acid; HI.)

Hydriodic acid is official in the form of the syrup of hydriodic acid (*Syrupus Acidi Hydriodici,* U. S. P.), which contains 1 per cent. by weight of the absolute acid. The dose is from $\frac{1}{2}$–2 fl. dr. (2.0–8.0 c.c.) or more. Its action is very feeble, but resembles that of the iodids. A fluidram (4.0 c.c.) represents about 1 gr. (0.06 gm.) of potassium iodid.

IODUM, U. S. P.

(Iodin; I.)

Iodin is a non-metallic element obtained chiefly from the ashes of seaweed. It occurs in heavy, bluish-black, friable crystals, having a metallic luster, a peculiar odor, and a sharp, acrid taste. On heating it emits a violet-colored vapor. It is soluble in 5000 parts of water, in 10 parts of alcohol, and freely soluble in ether, chloroform, and solutions of potassium iodid.

PREPARATIONS.　　　　　　　　　　　　　DOSE.

Tinctura Iodi, U. S. P. (7 per cent.) . . . . . 1–5 min. (0.06–0.3 c.c.).
Liquor Iodi Compositus, U. S. P. (Lugol's solu-
　tion : 5 per cent. of iodin in a 10 per cent.
　solution of potassium iodid) . . . . 1–10 min. (0 06–0.6 c.c.).
Unguentum Iodi, U. S. P. (4 per cent.).

Physiologic Action.—When applied to the skin, iodin stains it a yellowish-brown color and causes burning and itching. Strong solutions produce vesication. Mucous membranes are especially sensitive to its irritant action. Injections beneath the skin or into serous sacs excite intense pain and severe inflammatory reaction. When taken internally, it is rapidly absorbed in the form of iodids and soon reappears in all the secretions of the body. The bulk of it is eliminated in the urine, also in the form of iodids. In medicinal doses iodin exerts the same influence as potassium iodid, and if given continuously, induces all the phenomena of *iodism* (p. 306). Single large doses produce gastro-enteritis, respiratory failure, and collapse. Anuria and albuminuria have also been observed. Cases of fatal poisoning have been reported from its injection into large cysts and also from its too free use externally.

The *treatment of acute iodin-poisoning* consists in evacuating the stomach, administering starch or starchy foods (flour, arrow-root) as antidotes, and maintaining the respiration and circulation by hypodermic injections of alcohol, strychnin, digitalis, and ammonia.

Therapeutics.—Iodin is a useful counterirritant in conditions requiring a mild but persistent effect. The tincture may be applied externally in *pleurisy, bronchitis, laryngitis, synovitis, arthritis, neuritis, muscular rheumatism*, and similar inflammatory affections. It is best applied by means of a camel's-hair brush, one or two coats being painted over the part at intervals of a day or two. For children it should be diluted with 2 or 3 parts of alcohol. If the application proves too painful, the iodin should be removed with dilute alcohol, or, better, with a solution of potassium iodid, and the part dressed with starch jelly.

Preparations of iodin are very efficacious in bringing about resolution in various forms of *adenitis* provided they are applied early and before suppuration has commenced. A broad ring may be painted with the tincture around the swelling, or the ointment may be thoroughly rubbed into it. A better plan, however, and one less likely to produce painful dermatitis, is to rub in a mixture of equal parts of the ointments of belladonna and of lead iodid, and then to apply firm continuous pressure by means of a pad and a bandage.

Ganglion, hydrocele, housemaid's knee, and *spinal meningocele* are often successfully treated by evacuating the fluid and then injecting tincture of iodin into the sac. Iodin was formerly much used, both internally and externally, in the treatment of *simple goiter*, but thyroid extract has been found more efficacious. Injections of tincture of iodin are sometimes of value in reducing *cystic goiter*. From 3–5 min. (0.2–0.3 c.c.) should be injected into different parts of the tumor at intervals of three or four days. Local applications of iodin have long been held in high repute in certain chronic catarrhal processes attended by glandular hypertrophy, such as *rhinitis, pharyngitis*, and *endometritis*.

In *chilblain* an application of tincture of iodin and glycerin, equal parts, acts very favorably. Iodin has been used internally in the form of Lugol's solution in *syphilis, tuberculous adenitis*, and *goiter*, but on account of its irritant properties it is far less serviceable than the iodids.

Various attempts have been made to lessen the irritant effects of iodin by combining it with albumin or fat. The preparation known as *iodipin* represents, perhaps, the most successful of these attempts. It is an addition-product of iodin (10 per cent. or 25 per cent.) and oil of sesame, which, according to Winternitz, is not decomposed in the stomach, but in the intestine. The 10 per cent. preparation is administered by the mouth in doses of $\frac{1}{2}$–2 fl. dr. (2.0–8.0 c.c.). It is generally well borne, but its oily taste frequently proves objectionable. The 25 per cent. iodipin is used for subcutaneous injection, about $\frac{1}{2}$–1 fl. dr. (2.0–4.0 c.c.), slightly warmed, being injected daily between the skin and muscle of the gluteal region.

Incompatibles.—Alkaloids, mineral salts, ammonia, carbonates, starch, and mucilage of acacia. It acts violently upon turpentine and many other volatile oils. Tincture of iodin is also incompatible with aqueous preparations. In the so-called " colorless solutions of iodin " the iodin is replaced by iodids.

IODOFORMUM, U. S. P.

(Iodoform ; CHI_3.)

Iodoform, or tri-iodid of methane, is methyl with 3 atoms of hydrogen displaced by 3 of iodin. It is made by heating in a closed vessel iodin, alcohol, sodium hydrate, and water. It occurs in small, yellow, hexagonal crystals, having a peculiar penetrating odor, and a sweetish, iodin-like taste. It is freely soluble in ether, chloroform, and oils, soluble in about 52 parts of alcohol, and very feebly soluble in water. It contains more than 95 per cent. of iodin. The dose is from 1–3 gr. (0.06–0.2 gm.) in pill.

PREPARATION.

Unguentum Iodoformi, U. S. P. (10 per cent.).

Physiologic Action.—Upon mucous membrane and raw surfaces, especially the latter, iodoform acts as a mild anesthetic. It is readily absorbed from wounded surfaces, partly as iodin, which it liberates in the presence of moist organic matter, partly as albuminous compounds of iodin, and partly, probably, as iodoform. Shortly after absorption iodin appears in all the secretions; elimination, however, is mainly effected through the kidneys, which slowly excrete the drug in the form of iodids. When absorbed too freely it induces an intoxication, which not infrequently proves fatal. The symptoms of poisoning somewhat resemble those of cerebral meningitis, and include malaise, headache, mental depression, contraction of the pupils, the taste of iodoform in the mouth, nausea and vomiting, a very rapid pulse, delirium with hallucinations, stupor, and coma. In some cases there has been, in addition, high fever, and in others, a diffuse erythematous rash. Postmortem examination shows fatty degeneration of the organs and sometimes congestion of the cerebral meninges. As neither iodin nor iodids in overdoses cause cerebral symptoms, it is probable that the latter result from the action of iodoform itself. Old, anemic patients, especially sufferers from Bright's disease, are more readily poisoned than the young and robust.

Although the favorable action of iodoform on infected wounds can no longer be disputed, the manner of its operation is still obscure. It has little or no germicidal power, but it may act by retarding germ-growth through the iodin which it liberates, by inhibiting serous exudation and thus depriving the organisms of nourishment, or by neutralizing the toxins formed by bacteria.

In some persons the local application of iodoform proves irritating, and gives rise to a severe pustular or erythematous eruption.

Treatment of Poisoning.—The indications are to suspend the applications, to sustain the strength of the patient, and to favor elimination of the drug by means of subcutaneous injections of normal salt solution and the administration of alkaline diuretics.

Therapeutics.—Iodoform is extensively employed as a local application in *infected wounds*. In various *ulcers*—syphilitic, chancroidal, and tuberculous—it is very efficient. In the form of iodoform gauze it makes an excellent packing for deep *wounds, sinuses, fistulæ*, and the *rectal, vaginal*, and *nasal cavities. Cold abscesses* and *tuberculous joints* are often successfully treated by injecting into them, once a week, iodoform (10 per cent.) in sterile olive oil or glycerin or iodoform (5 per cent.) in ether, to the amount of $\frac{1}{2}$–3 fl. oz. (15.0–90.0 c.c.). The iodoform should always be sterilized by soaking it for several days in a solution of 1 : 1000 of corrosive sublimate and then thoroughly washing in sterilized water.

As a local analgesic it is particularly useful in relieving the pain and dysphagia of *tuberculous laryngitis*, in which affection it may be employed by insufflation, either in pure form (2–3 gr.—0.13–0.2 gm.) or mixed with morphin ($\frac{1}{16}$–$\frac{1}{12}$ gr.—0.004–0.005 gm.). Suppositories of iodoform (3 gr.—0.2 gm.) are of service in *painful hemorrhoids* and in *fissure of the anus*.

Semmola, Dreschfeld, Foxwell, and others have spoken very highly of the internal use of iodoform (6–12 gr.—0.4–0.8 gm. daily) in *pulmonary tuberculosis*, but the treatment has gone out of vogue. While many agents have been recommended for disguising the disagreeable odor of iodoform, none has proved very successful; the best, however, are the volatile oils, such as bergamot, fennel, anise, and cumarin, the odorous principle of Tonka bean.

IODOL.

(Tetra-iodo-pyrrol; C_4I_4NH.)

Iodol is prepared by acting upon pyrrol, a principle obtained from bone oil with iodin. It contains a little less than 90 per cent. of iodin, and appears as a yellowish, crystalline powder, free from odor and taste. It is almost insoluble in water, but it is freely soluble in alcohol, ether, and oils.

It is used as a substitute for iodoform. Though much more costly than the latter, it has a decided advantage in being odorless.

ARISTOL.

(Di-thymol-di-iodid; $C_{20}H_{24}O_2I_2$.)

Aristol is obtained by acting upon thymol in alkaline solution with iodin dissolved in potassium iodid. It contains about 45 per cent. of iodin, and appears as a brownish-red powder, tasteless and almost odorless. It is readily soluble in ether and oils, but it is insoluble in water. It is decomposed by heat, light, acids, alkalis, alcohol, and corrosive sublimate.

It is employed as a substitute for iodoform. While it has advantages in being odorless and less toxic than iodoform, it is more unstable, more costly, and less effective. It may be applied pure or dissolved in ether or oil.

EUROPHEN.

(Di-isobutyl-ortho-cresol-iodid; $C_{22}H_{29}O_2I$.)

Europhen is obtained by precipitating an alkaline solution of isobutyl-ortho-cresol with a solution of iodin in potassium iodid. It contains 28 per cent. of iodin, and appears as a very bulky, yellow, amorphous powder, of an aromatic odor. It is insoluble in water and glycerin, but freely soluble in alcohol, ether, and oils. It is a good substitute for iodoform.

NOSOPHEN.

(Tetra-iodo-phenol-phthalein; $(C_6H_2I_2OH)_2C_8H_4O_2$.)

Nosophen is obtained by the action of iodin on a solution of phenol-phthalein. It is a pale yellow, inodorous, and tasteless powder. It contains 60 per cent. of iodin. With bases it forms salts, the most important of which is the sodium salt (*antinosin*). It differs from iodoform in being an active antiseptic and in not yielding its iodin to the tissues.

HYDRARGYRUM.

(Mercury; Quicksilver; Hg.)

Mercury is a heavy, liquid metal, of a silvery luster, and without odor or taste.

Physiologic Action.—When small doses of an unirritating preparation of mercury are given continuously for a certain length of time, the first effects are observed in the mouth. There are increased flow of saliva, fetor of the breath, soreness of the teeth when the jaws are brought forcibly together, and redness of the gums near the insertion of the teeth. If the drug is not withdrawn, salivation becomes excessive, the gums become swollen and spongy, the teeth loosen and fall out, the tongue and parotid glands enlarge,—the former sometimes to such an extent that it protrudes from the mouth,—and finally the soft tissues become ulcerated and the bones necrosed. The term *ptyalism* is applied to this group of symptoms. At the same time the general health is more or less affected. The patient becomes pale and loses flesh. Fever, chilliness, thirst, anorexia, nausea, vomiting, and purging may appear.

The mercuric salts are more irritant and poisonous than the mercurous salts. The ingestion of a large dose of one of the former (corrosive sublimate) is speedily followed by intense burning in the esophagus, stomach, and abdomen, vomiting and purging of mucous and bloody material, anuria, or ischuria with albuminous urine, and profound collapse. Subsequently, if the patient survive the corrosive effects of the drug, ptyalism may develop from the absorption of the metal into the circulation. After death from acute mercury poisoning the mucous membrane of the alimentary canal is intensely inflamed and often deeply eroded. A diphtheric condition is sometimes observed. The kidneys are peculiarly affected. In addition to the evidences of acute parenchymatous degeneration, deposits of lime are found in the uriniferous tubules, especially the convoluted ones. These chalky deposits, while not pathognomonic of mercurial poisoning, are very suggestive of it.

Chronic mercurial poisoning is most frequently met with in workmen who handle the metal or who are exposed to its fumes. Thus it occurs in makers of thermometers, mirrors, and scientific instruments. Occasionally it is induced by the prolonged use of mercury as a medicine. Its chief manifestations are anemia, loss of flesh and strength, mental impairment, tremors similar to those of multiple sclerosis, motor palsies without atrophy of the muscles, gastro-intestinal disturbances, stomatitis, and salivation.

The form in which mercury is absorbed has not been definitely determined, but it is probable that the soluble preparations, at least, enter the circulation in the form of an albuminate. The metal is slowly eliminated through all the emunctories, but especially through the kidneys and bowel.

In medicinal doses the insoluble preparations of mercury act as cathartics (see p. 202), producing copious, loose stools without much griping. They also increase the flow of urine (see p. 226), probably by directly stimulating the renal epithelium. It is generally assumed that they stimulate also the hepatic cells, and thus increase the flow of bile, but recent investigations have shown that they have no such action. Mercurials, especially when locally applied, have a decided influence upon inflammatory exudations of a serous or fibrinous character, often aiding materially in their solution and reabsorption. That they have power to prevent or to lessen the outpouring of inflammatory material is very doubtful. There is reason for believing that the bichlorid in minute doses ($\frac{1}{100}$–$\frac{1}{50}$ gr.—0.0006–0.001 gm.) causes an increase in the number of red blood-corpuscles, when the latter have been reduced from overwork, acute disease, or hemorrhage. Soluble forms of mercury, owing to the avidity with which they unite with proteids, are very destructive to bacteria and other low forms of life. Finally, the drug exerts a specific influence in syphilis, but the manner of its action is unknown.

Untoward Effects.—Certain persons are exceedingly susceptible to the influence of mercury. A dose of less than a grain (0.065 gm.) of calomel has been known to excite severe stomatitis, persisting for several weeks. An erythematous or eczematous rash occasionally follows its administration by the mouth or its application to the skin. Calomel should not be applied to mucous membranes or be taken internally while the patient is under the influence of potassium iodid, since the latter is eliminated in all secretions and readily forms with mercurous compounds the irritant mercuric iodid.

Treatment of Poisoning.—Ptyalism.—The administration of mercury should be suspended as soon as the slightest tenderness of the gums manifests itself. The mouth should be frequently rinsed with a saturated solution of potassium chlorate. In severe cases the affected parts may be painted with slightly diluted sulphurous acid or with a saturated solution of iodoform in ether. To check the excessive flow of saliva, atropin ($\frac{1}{120}$ gr.—0.0005 gm.) may be given once or twice a day. Morphin may be required at night to relieve pain and to secure sleep. Potassium iodid is recommended to aid in the elimination of the mercury. Tonics may be needed to combat the anemia and exhaustion.

Acute Poisoning.—After evacuating the stomach, egg-albumen should be administered freely as an antidote. Opium is often required to allay the pain.

Chronic Poisoning.—The patient should be removed from the influence of the metal. Potassium iodid should be given to aid the elimination of the poison, and tonics to overcome the cachexia. Baths, massage, and electricity are useful adjuvants.

Therapeutics.—Mercury is used as an antisyphilitic, a germicide, a cathartic, a diuretic, and an absorbent.

Antisyphilitic.—Mercurials and iodids are the remedies relied upon to combat syphilis. Mercury should be given as soon as the diagnosis can be made with certainty, and continued throughout the secondary stage. In the tertiary stage iodids are generally more serviceable, but mercury is not without effect, and a combination of the two drugs is often employed with advantage. When very prompt effects are necessary, mercury should be given the preference. No matter which preparation of the metal is selected it should be given in ascending doses until the limit of tolerance is reached, when the dose should be cut down one-third or one-half, and this amount continued for a year or a year and a half. At the end of this time an iodid should be added, and the combination continued for about another year. The preparations of mercury most frequently prescribed in syphilis are the protiodid, biniodid, bichlorid, and mercury with chalk.

Germicide.—The soluble preparations of mercury, although among the most powerful germicides known, have certain drawbacks. Thus they are highly toxic, they are irritating to the tissues, they are destructive to metal instruments, and in the presence of albuminous matters they are readily converted into insoluble and inert albuminates. Notwithstanding these drawbacks they are the most popular germicides for general surgical work. Of the soluble salts, the bichlorid is usually chosen.

Cathartic.—Certain insoluble preparations of mercury—calomel, blue mass, mercury with chalk—have just enough irritant action on the bowel to induce catharsis. They are not suitable remedies for habitual constipation, but on account of their thorough and agreeable action they are well adapted for unloading the bowel in dyspeptic diarrhea and in the beginning of acute infectious diseases.

Diuretic.—Mercury in the form of calomel or blue mass is an active diuretic, and one that is well suited for carrying off dropsical effusions resulting from cardiac or hepatic disease.

Absorbent.—Ointments of mercury are very efficacious in promoting the absorption of the exudation thrown out in certain subacute and chronic inflammations. They are extensively employed for this purpose in synovitis, thecitis, arthritis, adenitis, and orchitis. It need scarcely be said that they are of no value in purulent inflammation.

Administration.—Mercurial preparations may be given by the mouth, by inunction, by subcutaneous injection, or by fumigation.

By the Mouth.—This is usually the best method of administration when mercury is employed for its constitutional effects. Blue mass, mercury with chalk, mercurous chlorid (calomel), mercuric chlorid (corrosive sublimate), mercurous iodid, and mercuric iodid, are the compounds most frequently chosen for exhibition by the mouth. The insoluble preparations are generally prescribed in pills or tablets. Corrosive sublimate may be given in solution or in pills or tablets. The biniodid is soluble in a solution of potassium iodid.

By Inunction.—This method of administering mercury is useful when the stomach is irritable or when it is deemed desirable to hasten mercurialization. The method has disadvantages in being uncleanly, troublesome, and uncertain as to dosage. About 1 dr. (4.0 gm.) of blue ointment may be rubbed into the groins, axillæ, or inner surface of the thighs and arms once a day. In order to avoid local irritation a different region should be selected each day. For infants an ointment may be used consisting of 1 part of mercurial ointment and from 6 to 8 parts of lard. A piece of this the size of a hickory-nut may be spread every day upon a flannel binder which is worn around the abdomen.

By Subcutaneous Injection.—Mercury may be given hypodermically in syphilis when a very rapid impression is necessary, or when other modes of administration have proved unsuccessful. This method is not suitable for ordinary cases. The injections are painful and are not unattended with danger. Local complications—induration, abscess, sloughing—are liable to occur, and grave ptyalism may develop, especially in patients with chronic nephritis. Soluble preparations only should be employed. Insoluble preparations may accumulate in the tissues for a time, and later give rise to severe constitutional symptoms. The best salts for hypodermic use are the bichlorid, benzoate, and salicylate. The last two are freely soluble in water containing sodium chlorid. From 20 to 30 daily injections should be made deep into the back or buttock. The initial dose of the bichlorid should not exceed $\frac{1}{8}$ gr. (0.008 gm.), and this amount should be gradually increased to a maximum of $\frac{1}{4}$ gr. (0.016 gm.). The injections should be given through a platino-iridium needle which has been previously sterilized.

By Fumigation.—This method is sometimes employed in syphilis instead of inunction, when the stomach is intolerant. It is especially useful when the cutaneous lesions prove obstinate. The mercurial salt, preferably calomel, of which about 20 gr. (1.3 gm.) are used, may be volatilized from a tin plate suspended over a spirit-lamp. The latter is placed under a stool or cane-seated chair, upon which is seated the patient, disrobed and surrounded by a blanket fastened to the neck. In about fifteen or twenty minutes the calomel is volatilized and deposited on the skin as a fine dust. The baths should be given just before retiring, and the sublimated mercury should be allowed to remain on the skin until the next morning. The treatment may be continued until slight tenderness of the gums develops.

Blue Mass (Massa Hydrargyri, U. S. P.).—This is a triturate of metallic mercury with honey of rose, glycerin, licorice, and althæa, containing 33 per cent. of the metal. The dose is from $\frac{1}{2}$–10 gr. (0.03–0.6 gm.). It is employed as a cathartic and diuretic. In the condition known as "*biliousness*" no treatment is so successful as the administration of a blue pill (5 gr.—0.3 gm.), followed in the morning by Epsom salts or a Seidlitz powder. In the beginning of *acute febrile diseases* and *dyspeptic diarrhea* blue mass is an excellent cathartic for unloading the bowel without inducing irritation. In combination with powdered digitalis and squill it is an efficient diuretic in the dropsy of *chronic heart* and *liver disease.*

Mercury with Chalk (Hydrargyrum cum Creta, U. S. P.). —This is a light-gray, damp powder, without odor, and of a sweetish taste. It contains 38 per cent. of mercury intimately mixed with chalk, honey, and water. The dose is from $\frac{1}{2}$–10 gr. (0.03–0.6 gm.). It is used in the same class of cases as blue mass. It is particularly serviceable in the *diarrhea of children* when the tongue is heavily coated, the breath fetid, and the stools are greenish or clay-colored. Hutchinson highly recommends mercury with chalk, in doses of from 1–4 gr. (0.065–0.26 gm.), thrice daily, as a convenient and efficient form in which to administer mercury in syphilis. In *congenital syphilis* from ¼–1 gr. (0.016–0.065 gm.) may be given three times a day.

Mercurial, or Blue, Ointment (Unguentum Hydrargyri, U. S. P.).—This is an ointment containing 50 per cent. of metallic mercury and 2 per cent. of oleate of mercury, 25 per cent. of lard, and 23 per cent. of suet. It is employed as an antisyphilitic, absorbent, and a parasiticide. In syphilis it is administered by inunction. In *synovitis, bursitis, rheumatic arthritis, glandular enlargements*, and *syphilitic nodes* blue ointment is a valuable absorbent. When the swelling is painful, an equal amount of belladonna ointment may be added for its sedative effect. In *subacute rheumatism* the local application to the affected joints of an ointment containing mercury and salicylic acid is often efficacious.

R   Acidi salicylici,   gr. xxx (2.0 gm.);
    Olei terebinthinæ,   f℥iiss (10.0 c.c.);
    Adipis lanæ hydrosi,   ℥ss (15.0 gm.);
    Unguenti hydrargyri,   ℥j (30.0 gm.).   M.
Sig.   Apply with gentle friction night and morning.

In *pediculosis pubis* the parasites are quickly destroyed by rubbing into the affected parts a small amount of blue ointment.

Mercurial Plaster (Emplastrum Hydrargyri, U. S. P.).—This is a plaster containing 30 per cent. of metallic mercury, 1.2 per cent. of oleate of mercury, and 68.8 per cent. of lead-plaster. It is employed as an absorbent in *chronic inflammatory swellings, glandular enlargements, syphilitic nodes*, etc., but is far less efficient than the ointment. Ammoniac plaster with mercury (Emplastrum Ammoniaci cum Hydrargyro, U. S. P.) is rarely used at the present time.

HYDRARGYRI CHLORIDUM CORROSIVUM, U. S. P.

(Mercuric Chlorid; Corrosive Sublimate; Bichlorid of Mercury; $HgCl_2$.)

Mercuric chlorid appears in the form of colorless, odorless crystals, having an acrid, metallic taste. It is soluble in 16 parts of water and in 3 parts of alcohol. The dose is from $\frac{1}{100}$–$\frac{1}{12}$ gr. (0.0006–0.005 gm.).

Therapeutics.—Corrosive sublimate is employed chiefly as a germicide, an antisyphilitic, and a tonic.

Germicide.—It is an energetic germicide, capable of destroying most bacteria even in solutions of 1 : 20,000, and their spores in solutions of 1 : 10,000. Some micro-organisms, however, like the bacillus of anthrax, are much less susceptible to its action. In the presence of hydrogen sulphid it is converted into an insoluble and inert sulphid of mercury; with albuminous matter it forms an impermeable albuminate, which prevents its further penetration; it has a corroding action on metal; it is irritating to the tissues, and when applied too freely to wounds or mucous membranes, it may be absorbed in sufficient quantity to induce poisoning. Notwithstanding these drawbacks, most surgeons accord corrosive sublimate the first rank among germicides for use upon the *skin of the patient and the hands of the operator*, and *for irrigating infected wounds and cavities*. On account of its irritant properties it should not be used on serous membranes. For the patient's skin and surgeon's hands a solution of from 1 : 1000 to 1 : 500 should be employed; for large wounds and cavities, 1 : 10,000 to 1 : 5000; for small wounds, 1 : 2000; for irrigating the bladder and vagina, 1 : 20,000 to 1 : 5000; for irrigating the urethra, 1 : 40,000 to 1 : 20,000; and for irrigating the conjunctiva, 1 : 5000. Tartaric or citric acid may be advantageously combined with bichlorid solutions to prevent the mercuric salt from forming an insoluble albuminate with the albuminous matter of the tissues. Compressed tablets, each containing $7\frac{1}{2}$ gr. (0.48 gm.) of the salt with tartaric acid, are in common use. One of these added to a pint of water makes a solution of 1 : 1000. Ordinary water, on account of the lime which it contains, partially precipitates corrosive sublimate in the form of an oxid of mercury. This precipitation, however, can be prevented by adding common salt to the water. Bichlorid solutions cannot be used for sterilizing metal instruments. As a household disinfectant mercuric chlorid has several disadvantages. It is very poisonous; it has a corroding action on metals; it becomes ineffective by contact with albuminous matters; and it renders indelible any stains of feces or blood that may be on clothing. Solutions of from 1 : 5000 to 1 : 1000, however, are serviceable for scrubbing floors, wood-work, and bare walls. It is not a suitable disinfectant for sputum, feces, or other albuminous matters.

As a parasiticide, corrosive sublimate is a useful remedy in *pediculosis pubis* and *ringworm*. In these affections it may be employed in the form of a lotion in the strength of from 2–4 gr. (0.13–0.26 gm.) to the ounce (30.0 c.c.) of water, or, better, tincture of benzoin.

The following solution, recommended by Jacobi, may be used as a local remedy in *diphtheria :*

℞	Hydrargyri chloridi corrosivi,	gr. $1\frac{1}{4}$ (0.08 gm.);
	Sodii chloridi,	gr. xl (2.6 gm.);
	Aquæ,	Oj (0.5 L.). M.

In nasal diphtheria this solution should be warmed and poured into the nares or from a nasal cup several times a day.

Before the introduction of antitoxin, mercury, in the form of calomel or of corrosive sublimate, was extensively used internally in the treatment of diphtheria. It seems to have some value in this affection, but the manner of its action is unknown. Children bear the drug remarkably well, and $\frac{1}{40}$ gr. (0.0015 gm.) of corrosive sublimate, gradually increased to $\frac{1}{20}$ gr. (0.003 gm.), may be given every three or four hours to a child of five years.

Compresses wrung out of bichlorid solutions (1 : 5000) are useful in *erysipelas* and in *small-pox.*

Antisyphilitic.—Corrosive sublimate, though somewhat more irritant than the protiodid, is a reliable salt of mercury for use in syphilis. It may be administered either in pill or in solution. One-twentieth of a grain (0.003 gm.), gradually increased to $\frac{1}{12}$ gr. (0.005 gm.), may be given three times a day after meals. The addition of a small amount of opium will usually serve to allay any gastric or intestinal irritation. In the late secondary period the bichlorid may be prescribed with potassium iodid, as in the following formula:

> R   Hydrargyri chloridi corrosivi,   gr. iss–ij (0.1–0.13 gm.);
> Potassii iodidi,   ℨiv–vj (15.0–23.0 gm.) ;
> Syrupi sarsaparillæ compositi,   f ℥iss (45.0 c.c.);
> Aquæ,   q. s. ad f ℥iij (90.0 c.c.).   M.
> Sig.   A teaspoonful in water after meals.

Corrosive sublimate is one of the best salts of mercury for hypodermic use. Sodium chlorid should be added to the solution to prevent the formation in the tissues of an insoluble albuminate.

> R   Hydrargyri chloridi corrosivi,   gr. vj (0.4 gm.) ;
> Sodii chloridi,   gr. xl (2.6 gm.) ;
> Aquæ destillatæ,   f ℥j (30.0 c.c.).   M.
> Sig.   Inject 10–20 min. (0.6–1.2 c.c.) daily.

Tonic.—Mercuric chlorid, in doses of $\frac{1}{100}$–$\frac{1}{60}$ gr. (0.0006–0.001 gm.), is sometimes useful as an adjuvant to iron in secondary anemia.

Incompatibles.—Corrosive sublimate has a wide range of incompatibilities. The most common substances precipitated by it are tannic acid, alkaline carbonates, albumin, iodids, silver nitrate, and solutions of lime. With potassium iodid it forms mercuric iodid, but if the potassium salt is present in excess, a colorless solution of the double iodid of mercury and potassium at once results, so that a combination of the two drugs is perfectly admissible.

Yellow wash (lotio flava) is made by adding 24 gr. (1.6 gm.) of mercuric chlorid to 16 ounces (474.0 c.c.) of lime-water. Yellow mercuric oxid is precipitated and calcium chlorid remains in solution. It is sometimes employed as a stimulating dressing in the treatment of *phagedenic venereal sores.*

HYDRARGYRI CHLORIDUM MITE, U. S. P.

(Mild Mercurous Chlorid; Calomel; Hg_2Cl_2.)

Calomel is a white, odorless, tasteless powder, insoluble in all ordinary menstrua. The dose is from $\frac{1}{10}$–10 gr. (0.0065–0.65 gm.).

PREPARATIONS.	DOSE.
Pilulæ Antimonii Compositæ, U. S. P. (contain $\frac{2}{3}$ gr. —0.04 gm.)	1–2 pills.
Pilulæ Catharticæ Compositæ, U. S. P. (contain 1 gr. —0.06 gm.)	1–3 pills.

Therapeutics.—Calomel is used internally as a cathartic (see p. 202), as a diuretic (see p. 226), as an antiphlogistic, and as an antisyphilitic. Externally it is employed as a stimulant, desiccant, and antiseptic.

External Use.—Zinc ointment to which calomel (5–15 gr. to the ounce—0.4–1.0 gm.) has been added makes an excellent application in *subacute and chronic eczema.* A dusting-powder composed of equal parts of calomel and zinc oxid is effective in *venereal warts.* Calomel is also efficacious in *corneal ulcers* and *phlyctenular conjunctivitis* when there is not much ciliary irritation. It should be flicked into the eye by gently tapping a camel's-hair brush loaded with the powder.

Antiphlogistic.—Some practitioners believe that calomel is of service in limiting fibrinous exudation in *inflammatory diseases of the serous membranes.* For this purpose it has been much used in pleurisy, meningitis, iritis, and pericarditis, but its beneficial effects are very doubtful.

Antisyphilitic.—Calomel has been used internally to combat syphilis, but the protiodid and the bichlorid are decidedly preferable, since they do not induce salivation so quickly. It is a suitable preparation, however, for volatilization when the patient is to be treated by fumigation. Some surgeons have recommended deep injections of oily mixtures of calomel in syphilis, but grave symptoms have occasionally resulted from its use in this way.

Incompatibles.—Calomel is incompatible with hydrochloric acid, chlorates, chlorids, iodids, bromids, and lime-water. With hydrocyanic acid and potassium cyanid it forms the highly poisonous bicyanid of mercury. Mixtures of calomel and iodoform turn red from the formation of mercuric iodid.

Black wash (lotio nigra) is made by adding 1 dr. (4.0 gm.) of calomel to 1 pint (0.5 L.) of lime-water. Black mercurous oxid is precipitated and calcium chlorid remains in solution. It is sometimes employed as a stimulating application in *venereal sores*. It is often very useful in *rhus poisoning* and *acute eczema*, when dabbed on the parts, allowed to dry, and followed by an application of zinc ointment.

HYDRARGYRI IODIDUM RUBRUM, U. S. P.

(Red Mercuric Iodid ; Biniodid of Mercury ; HgI_2.)

Red iodid of mercury is a bright-red, amorphous powder, free from odor and taste. It is almost insoluble in water, soluble in 130 parts of alcohol, and freely soluble in solutions of potassium iodid. The dose is from $\frac{1}{50}-\frac{1}{12}$ gr. (0.0012–0.005 gm.).

PREPARATION.	DOSE.
Liquor Arseni et Hydrargyri Iodidi, U. S. P. (Donovan's solution : 1 per cent. of each iodid)	1–5 min. (0.06–0.3 c.c.).

Therapeutics.—In its effect and strength the biniodid of mercury resembles the bichlorid. In the *late secondary stage of syphilis* it is often of value when given in a solution of potassium iodid, as in the following formula :

```
R   Hydrargyri iodidi rubri,            gr. j (0.065 gm.) ;
    Potassii iodidi,                    ℥ij–iv (8.0–15.0 gm.) ;
    Aquæ,                               f℥ij (60.0 c.c.) ;
    Syrupi sarsaparillæ compositi, q. s. ad f℥iv (120.0 c.c.).  M.
Sig.  A dessertspoonful in water after meals.
```

Donovan's solution is used internally as an alterative in *chronic rheumatism, tuberculous adenitis,* and *tertiary syphilis.*

HYDRARGYRI IODIDUM FLAVUM, U. S. P.

(Yellow Mercurous Iodid ; Protiodid of Mercury ; Green Iodid of Mercury ; Hg_2I_2.)

Protiodid of mercury is a yellow, amorphous, insoluble powder, free from odor and taste. The dose is from $\frac{1}{10}-\frac{1}{2}$ gr. (0.006–0.03 gm.).

Therapeutics.—The protiodid of mercury is far less irritant than the biniodid. Ordinarily, it is the best preparation for use in *syphilis.* It should be given in pills to which a little opium may be added if it excite colic or diarrhea.

HYDRARGYRI NITRAS.

(Mercuric Nitrate; $Hg(NO_3)_2$.)

Mercuric nitrate is official in two forms :

Liquor Hydrargyri Nitratis, U. S. P. (contains 60 per cent. of mercuric nitrate and 11 per cent. of free nitric acid).
Unguentum Hydrargyri Nitratis, U. S. P. (citrine ointment: contains 7 per cent. of mercuric nitrate).

Therapeutics.—The solution of mercuric nitrate is a powerful caustic. It is extensively employed for the cauterization of *mucous patches* and *sloughing venereal sores.* The danger of inducing salivation from using it too freely must be borne in mind. Citrine ointment, more or less diluted, may be used as a stimulant application in *indolent ulcers* and *chronic eczema.* Diluted with 8 parts of petrolatum it makes an efficient application for *granulating venereal ulcers.*

HYDRARGYRUM AMMONIATUM, U. S. P.

(Ammoniated Mercury; Mercuric Ammonium Chlorid; White Precipitate; NH_2HgCl.)

Ammoniated mercury is made by the action of ammonia on corrosive sublimate, and appears as a white, insoluble powder, free from odor and taste.

PREPARATION.

Unguentum Hydrargyri Ammoniati, U. S. P. (10 per cent.).

Therapeutics.—Ammoniated mercury is employed externally, in the form of the ointment, as a stimulant and as a parasiticide. It is often serviceable in *chronic eczema, psoriasis,* and *ringworm.* The official ointment, however, is too strong for ordinary use, a strength of from 20–30 gr. (1.3–2.0 gm.) to the ounce (30.0 gm.) being quite sufficient.

HYDRARGYRI CYANIDUM, U. S. P.

(Mercuric Cyanid; $Hg(CN)_2$.)

Mercuric cyanid occurs in the form of colorless, prismatic crystals, odorless, and of a bitter, metallic taste. It is freely soluble in water and alcohol. The dose is from $\frac{1}{40}-\frac{1}{10}$ gr. (0.0016–0.006 gm.).

Its action, though less irritant, resembles that of the bichlorid. It has been used as a substitute for the latter in surgical practice and also in the hypodermic treatment of syphilis.

HYDRARGYRI OXIDUM.

(Mercuric Oxid; HgO.)

Mercuric oxid occurs in two forms: Yellow oxid (*Hydrargyri Oxidum Flavum*, U. S. P.) and red oxid (*Hydrargyri Oxidum Rubrum*, U. S. P.). Both are heavy, permanent, insoluble powders, odorless, and of a somewhat metallic taste. The yellow oxid is an impalpable powder; the red oxid is more or less crystalline. They are not used internally.

PREPARATIONS.

Unguentum Hydrargyri Oxidi Flavi, U. S. P. (10 per cent.).
Unguentum Hydrargyri Oxidi Rubri, U. S. P. (10 per cent.).
Oleatum Hydrargyri, U. S. P. (20 per cent.).

Therapeutics.—The oxids of mercury are used externally for their stimulant and alterative effects. In certain *chronic inflammatory diseases of the eye*—phlyctenular conjunctivitis, keratitis, and blepharitis marginalis, an ointment of the yellow oxid is often very useful. In the last affection it is particularly efficacious when applied at night to the margins of the lids in the strength of 1 gr. of the oxid (0.06 gm.) to 1 dr. (4.0 gm.) of vaselin. In *chronic eczema* an ointment containing from 10–20 gr. (0.6–1.3 gm.) to the ounce (30.0 gm.) is sometimes serviceable. In the form of a dusting-powder the oxids of mercury have also been used in *chancroidal* and *syphilitic sores*. In *chronic adenitis* and other *indolent inflammatory indurations* one of the official ointments or the oleate may be employed for its sorbefacient effect.

The red oxid may be used in the same class of cases as the yellow oxid, but it is less satisfactory on account of its crystalline character.

HYDRARGYRI SUBSULPHAS FLAVUS, U. S. P.

(Yellow Mercuric Sulphate; Turpeth Mineral; $Hg(HgO)_2SO_4$.)

Yellow mercuric sulphate is a lemon-yellow, odorless, and tasteless powder, sparingly soluble in water. The dose is from 2–3 gr. (0.1–0.2 gm.), repeated once.

Therapeutics.—This preparation of mercury was formerly much used as an emetic in *croup*, but it has been largely replaced by less poisonous and less irritant drugs.

HYDRARGYRI BENZOAS.

(Mercuric Benzoate; $Hg(C_7H_5O_2)_2 + H_2O$.)

Mercuric benzoate is a white, crystalline powder, free from odor and taste. It is but slightly soluble in water, but freely so in water containing sodium chlorid. The dose is from $\frac{1}{30}$–$\frac{1}{10}$ gr. (0.002–0.006 gm.).

It is a satisfactory preparation for administering subcutaneously in *syphilis*.

HYDRARGYRI SALICYLAS.

(Mercuric Salicylate; $HgC_7H_4O_3$.)

Mercuric salicylate is a white, amorphous powder, without odor or taste. It is insoluble in water and alcohol, but freely soluble in water containing sodium chlorid. The dose, by the mouth, is from $\frac{1}{4}$–1 gr. (0.01–0.06 gm.).

It has been especially recommended for subcutaneous administration in *syphilis*, the claims for it being that it is more rapidly absorbed and is less irritant than the bichlorid.

MERCUROL.

Mercurol is a compound of mercury and yeast nuclein, containing about 10 per cent. of mercury. It is a light, brownish-white powder, insoluble in alcohol, but soluble in water. It has been used in the form of a 1 or 2 per cent. injection (5–10 gr. to the ounce—0.3–0.6 gm. to 30.0 c.c.), with asserted good results, in *acute gonorrhea*.

ICHTHYOL.

(Ammonium Sulpho-ichthyolate.)

Ichthyol is the ammonium salt of ichthyol sulphonic acid. The latter is the product of the action of sulphuric acid on an oily substance obtained by the destructive distillation of a bituminous mineral rich in fossil fish, found in the Tyrol. It is a thick, reddish-brown liquid, having a bituminous odor and taste. It is soluble in water and in a mixture of alcohol and ether, miscible with oils and glycerin in all proportions, and almost insoluble in strong alcohol or ether. It contains nearly 15 per cent. of sulphur, and to this ingredient its therapeutic properties, no doubt, are largely due. The dose is from 2–10 gr. (0.13–0.65 gm.), in capsules or pills.

Physiologic Action.—When applied to the skin in concentrated form, ichthyol produces slight redness and burning. Its absorption through the unbroken skin is readily effected if gentle friction be used in the application. The efficacy of the drug in many inflammatory diseases of the skin and mucous membranes has been attributed to a constricting action on the vessels and to an alterative influence on the deeper tissues, especially the glandular cells. Fessler, Abel, Neisser, and others have shown, also, that it possesses decided bactericidal properties. According to Neisser, even in solutions so dilute as 1 per cent. it is rapidly destructive to gonococci. Internally, large doses cause eructations, .nausea, vomiting, and diarrhea.

Therapeutics.—Ichthyol is very largely employed externally as an antiseptic and alterative. In the form of an ointment it is decidedly useful in reducing *inflammatory swelling in glands and joints.* The best vehicle is lanolin, and the strength of the application may vary from 25 to 50 per cent. It is of some value as a local remedy in *erysipelas,* and in this disease it may be combined with blue ointment, as in the following formula, so highly extolled by Roswell Park:

R	Ichthyol,	gr. xxx–xl (2.0–2.6 gm.);
	Resorcini,	ʒss (2.0 gm.);
	Unguenti hydrargyri,	ʒiv (15.5 gm.);
	Adipis lanæ hydrosi,	ʒv (20.0 gm.).—M.

It is said to be useful, either as a lotion or an ointment (10 to 20 per cent.), in certain diseases of the skin, especially *acne, vesicular and squamous eczema, pruritus,* and *urticaria.* It is sometimes of service in *bruises, sprains,* and *chilblains.* Tampons saturated with ichthyol and glycerin (1 to 20 or 1 to 10) are sometimes remarkably beneficial in *oöphoritis, perimetritis, endometritis, cervical catarrh,* and *gonorrheal vaginitis.* In *atrophic rhinitis* no remedy is so efficacious as ichthyol in relieving the disagreeable symptoms. After the nares have been thoroughly cleansed, pledgets of cotton soaked in an aqueous solution (20 to 50 per cent.) should be inserted and allowed to remain for a period of from fifteen to twenty minutes, or in bad cases the drug may be applied pure by means of a probe armed with cotton.

The chief drawback to the use of ichthyol externally is its unpleasant, bituminous odor; this can be disguised in a measure by the addition of oil of bergamot (1 : 40).

Ichthyol has been used internally in a variety of diseases, particularly *tuberculosis* and *rheumatism,* but the testimony to its efficacy is not convincing.

Incompatibles.—Acids, alkalis, and alkaloidal salts.

Ichthalbin.—This preparation is a combination of ichthyol and albumin, appearing as a brownish powder, odorless, and nearly tasteless. The dose is from 5–10 gr. (0.33–0.65 gm.). It has been thoroughly exploited as a substitute for ichthyol, both for internal and external use.

Ichthoform is a condensation product of ichthyol and formaldehyd. It has been recommended as a substitute for iodoform; but notwithstanding its freedom from odor it is less satisfactory than the older remedy.

Ichthargan is a compound of ichthyol and silver, containing 30 per cent. of metallic silver in organic chemical combination. It is a brown, amorphous powder, with a faint odor of chocolate. It is freely soluble in water, diluted alcohol, and glycerin. It has been used with asserted good results in *gonorrhea*, injections of a 1 : 10,000 to a 1 : 1000 solution being made three or four times daily.

Thiol.—This is a synthetic product obtained by the action of sulphur and sulphuric acid upon hydrocarbon as formed by the destructive distillation of peat. It was introduced as a substitute for ichthyol, but it has not proved a very formidable rival.

AURI ET SODII CHLORIDUM, U. S. P.

(Gold and Sodium Chlorid; $AuCl_3 + NaCl$.)

The gold and sodium chlorid of the Pharmacopeia is a mixture of equal parts, by weight, of dry gold chlorid and sodium chlorid. It is an orange-yellow powder, slightly deliquescent, of a saline and metallic taste. It is freely soluble in water. The dose is from $\frac{1}{20}-\frac{1}{4}$ gr. (0.003–0.016 gm.), in pill.

Physiologic Action and Therapeutics.—Gold and sodium chlorid is supposed to act as an alterative and a tonic. Large doses have an irritant action and excite gastro-enteritis. It has been recommended in a number of diseases, especially in *diabetes, hysteria, neurasthenia, tertiary syphilis, chronic alcoholism,* and *sclerosis of the spinal cord,* but it is of very doubtful value.

COLCHICUM.

(Meadow Saffron.)

Colchicum is the corm and seed of *Colchicum autumnale,* a bulbous perennial growing in Southern Europe and Northern Africa. The corm is official as *Colchici radix,* and the seed as *Colchici semen.* The active principle of the drug is the alkaloid, *colchicin,* which is a whitish, amorphous or crystalline powder, of a saffron odor and bitter taste, and readily soluble in water and alcohol. The dose of colchicin is from $\frac{1}{130}-\frac{1}{30}$ gr. (0.0005–0.002 gm.).

PREPARATIONS.	DOSE.
Extractum Colchici Radicis, U. S. P.	$\frac{1}{2}$–2 gr. (0.03–0.13 gm.).
Extractum Colchici Radicis Fluidum, U. S. P.	2–5 min. (0.1–0.3 c.c.).
Vinum Colchici Radicis, U. S. P.	5–15 min. (0.3–1.0 c.c.).
Extractum Colchici Seminis Fluidum, U. S. P.	2–5 min. (0.1–0.3 c.c.).
Tinctura Colchici Seminis, U. S. P.	10–30 min. (0.6–2.0 c.c.).
Vinum Colchici Seminis, U. S. P.	10–30 min. (0.6–2.0 c.c.).

Physiologic Action.—In warm-blooded animals large doses of colchicum or of its alkaloid excite severe abdominal pains, nausea, vomiting, and diarrhea. The discharges are at first serous, but later they may become mucous and even bloody. These symptoms are followed by progressive motor paralysis, enfeeblement of the circulation, collapse, and finally by death from asphyxia.

Postmortem examination usually reveals pronounced inflammatory lesions in the alimentary canal. In some instances, however, no morbid changes have been observed. Fatal poisoning has occurred from the ingestion of less than 3 dr. (11.0 c.c.) of the wine of the root, and of less than ½ gr. (0.03 gm.) of colchicin.

The gastro-intestinal features of the poisoning are probably due to the direct irritant action of the drug, although Jacobj attributes them to increased irritability of the motor nerves of the bowel, in consequence of which the muscular coat responds too vigorously to the ordinary stimuli. The paralysis, according to Rossbach, results from depression of the central nervous system. Upon the circulation colchicum appears to have no direct influence. Both the water and the solids of the urine are somewhat increased after moderate doses of the drug, but large doses may be followed by suppression.

Treatment of Poisoning.—The stomach should be evacuated as speedily as possible. Albumin and other demulcents are useful in allaying irritation and in preventing further injury to the mucous membrane. Tannic acid is of no value as an antidote. Morphin may be given hypodermically to relieve pain and to inhibit peristalsis. The usual measures will be required to combat the collapse.

Therapeutics.—The only disease in which colchicum is of value is *gout.* The good and bad effects of the drug in chronic joint affections seem to have been known as early as the sixth century of the Christian era. For many years it was neglected by regular practitioners, although it still served as the basis of many celebrated nostrums. Later the studies of Halford, Watson, and Garrod reestablished it in the confidence of the profession. The way in which gouty inflammation is affected by colchicum is not understood, and cannot be, until our knowledge of the pathology of the disease becomes more complete. The drug is most potent in acute gout, the pain and swelling of which it relieves as if by magic ; it is less efficacious in the chronic manifestations of the disease. In the different forms of rheumatism it is useless.

Administration.—As colchicum is powerful for harm as

well as for good, considerable care should be exercised in its administration. Only small doses should be employed, and these should be withdrawn or considerably reduced so soon as the pain has been relieved. Large doses, even if they do not excite irritation of the stomach or purging, may, by suppressing the local manifestations too abruptly, cause the grave visceral disturbances to which the term retrocedent gout has been applied. The wine of the root is the most reliable preparation; it should be taken, well diluted, after food. Alkalis are useful adjuvants, and may be combined with the colchicum, as in the following formula:

R   Potassii bicarbonatis,   ℨij (8.0 gm.);
    Vini colchici radicis,   f ℥iss (6.0 c.c.);
    Aquæ menthæ piperitæ,   q. s. ad f ℥iv (120.0 c.c.).   M.
  Sig.   A tablespoonful is a wineglassful of water thrice daily, after
  meals.

Colchicin is a convenient form for administering in pills or capsules.

GUAIACUM.

(Guaiac; Lignum Vitæ.)

Guaiac is official in two forms: *Guaiaci Lignum* and *Guaiaci Resina*. The former is the heart-wood of *Guaiacum officinale* and of *Guaiacum sanctum*, large trees growing in the West Indies; and the latter is the resin of *Guaiacum officinale*. The resin is the form in which the drug is chiefly employed medicinally. It appears as irregular masses, of a reddish-brown color, turning greenish-brown on exposure, and has an aromatic odor and an acrid taste. It is soluble in alcohol and in alkaline fluids, but is insoluble in water. Alcoholic solutions turn blue on the addition of oxidizing agents. It contains several resinous acids and aromatic oils and gum. The dose is from 5–30 gr. (0.3–2.0 gm.).

PREPARATIONS.	DOSE.
Tinctura Guaiaci, U. S. P. (20 per cent. of the resin)	½–1 fl. dr. (2.0–4.0 c.c.).
Tinctura Guaiaci Ammoniata, U. S. P. (20 per cent. of the resin in aromatic spirit of ammonia)	½–1 fl. dr. (2.0–4.0 c.c.).
Decoctum Sarsaparillæ Compositum, U. S. P. (2 per cent. of guaiacum wood, with sarsaparilla, sassafras, glycyrrhiza, and mezereum)	4–6 fl. oz. (120.0–180.0 c.c.).

Therapeutics.—Guaiac possesses considerable power, though less than colchicum, in relieving *gouty inflammation*. It is especially efficacious in the subacute and chronic forms of the disease. In acute gout it may be substituted advantageously for colchicum so soon as the pain has subsided. According to Garrod, when it is taken in the intervals of gouty attacks it is very effective in averting their recurrence. The drug is quite innocuous, and may be taken for indefinite periods without inducing untoward effects. It is also used in *tonsillitis* and *chronic rheumatism.*

Administration.—The tinctures are reliable preparations, but as they have a very disagreeable taste, they should be given in the form of an emulsion. The resin may be given in pills, but, like the tinctures, it is best given in an emulsion.

R   Guaiaci resinæ,   ℥iisss (10.0 gm.) ;
     Acaciæ,   q. s.
     Syrupi,   f ℥iv (15.0 c.c.);
     Aquæ cinnamomi,   q. s. ad f ℥iv (120.0 c.c.).
Misce et fiat emulsum.
Sig.   A dessertspoonful in water after meals.

Incompatibles.—Mineral acids and spirit of nitrous ether. Water is incompatible with the tinctures.

SARSAPARILLA, U. S. P.

Sarsaparilla is the root of *Smilax officinalis* and of other species of *Smilax,* large perennial climbers growing in swampy places in tropical America. It contains a volatile oil, resin, and several saponins.

PREPARATIONS.	DOSE.
Decoctum Sarsaparillæ Compositum, U. S. P. (sarsaparilla, 10 parts; sassafras, glycyrrhiza, and guaiacum wood, of each, 2 parts; mezereum, 1 part; water to make 100 parts)	1–4 fl. oz. (30.0–120.0 c.c.).
Extractum Sarsaparillæ Fluidum, U. S. P.	½–2 fl. dr. (2.0–8.0 c.c.).
Extractum Sarsaparillæ Fluidum Compositum, U. S. P. (sarsaparilla, 75 parts; glycyrrhiza, 12 parts; sassafras, 10 parts; mezereum, 3 parts; glycerin, 10 parts; alcohol and water, in the proportion of 1 to 2, to make 100 parts)	½–2 fl. dr. (2.0–8.0 c.c.).
Syrupus Sarsaparillæ Compositus, U. S. P. (fl. ext. sarsaparilla, 20 parts; fl. ext. glycyrrhiza, 1.5 parts; fl. ext. senna, 1.5 parts; sugar, 65 parts; oil of sassafras, oil of gaultheria, and oil of anise, of each, 0.01 part; water to make 100 parts)	1–4 fl. dr. (4.0–15.0 c.c.).

Therapeutics.—The action of sarsaparilla is very feeble. It has been used as an alterative in *syphilis* and *tuberculosis,* but it is without value. It is chiefly useful as a vehicle to disguise the taste of unpalatable drugs, particularly potassium iodid.

JAMBUL.

(Java Plum.)

Jambul is the root and seeds of *Eugenia jambolana*, a large tree growing in the East Indies. The active constituent of the drug has not been determined. The seeds are more powerful than the root, and may be given powdered in doses of from 5–20 gr. (0.3–1.3 gm.), in capsules.

PREPARATION.   DOSE.

Extractum Jambul Fluidum . . . . 10–30 min. (0.6–2.0 c.c.).

Therapeutics.—Jambul has been used solely as a remedy in *diabetes mellitus*, for which it was originally recommended by Banatvala, of Madras. Binz found that it materially lessened the excretion of sugar in phloridzin diabetes, but Minkowski found it absolutely useless in experimental pancreatic diabetes. Clinically, we have derived some benefit from it in a few cases of a mild type. Large doses sometimes lessen the glycosuria, but increase the quantity of urine. Von Noorden, in a study of 600 cases of diabetes, concludes that while jambul has no very marked action on the elimination of sugar, it is a good adjuvant to dietetic and hygienic procedures.

MEZEREUM, U. S. P.

(Mezereon.)

Mezereum is the bark of *Daphne Mezereum* and of other species of *Daphne*, small shrubs growing in mountainous districts in Europe and Asia. It contains a bitter glucosid and an irritant volatile oil.

PREPARATIONS.

Extractum Mezerei Fluidum, U. S. P.
Decoctum Sarsaparillæ Compositum, U. S. P. (1 per cent. of mezereum).
Extractum Sarsaparillæ Fluidum Compositum, U. S. P. (3 per cent. of mezereum).
Linimentum Sinapis Compositum, U. S. P. (20 per cent. of the fluid extract of mezereum).

Therapeutics.—Mezereon bark is an active irritant, and when applied to the skin, it causes vesication. Large doses of the fluid extract taken internally produce severe abdominal pain, followed by vomiting and purging. Mezereum has been recommended as an alterative in *chronic rheumatism, syphilis,* and *various skin diseases,* but its utility has never been satisfactorily demonstrated. Except in the form of the compound, fluid extract of sarsaparilla or the compound decoction of sarsaparilla is never prescribed internally. Externally the fluid extract has been used as an irritant application for *indolent ulcers.*

CALX SULPHURATA, U. S. P.

(Sulphurated Lime; Crude Calcium Sulphid.)

Sulphurated lime is a mixture containing at least 60 per cent. of calcium monosulphid, together with calcium sulphate and varying proportions of carbon. It is a pale gray powder, having a nauseous, alkaline taste and a faint odor of hydrogen sulphid. It is slightly soluble in water and insoluble in alcohol. On exposure to air it is gradually decomposed. The dose is from $\frac{1}{10}$–$\frac{1}{2}$ gr. (0.006–0.03 gm.), in pills, capsules, or powders.

Therapeutics.—As was first pointed out by Ringer, sulphurated lime possesses some power of preventing and arresting suppuration. It has been found especially useful in *pustular acne, boils,* and *carbuncles.* In *follicular tonsillitis* and in *quinsy,* doses of $\frac{1}{20}$ gr. (0.003 gm.) every hour sometimes yield excellent results. It is better to give small doses at short intervals than large doses infrequently, since the latter are more prone to derange digestion and to cause disagreeable eructations of sulphureted hydrogen. As the drug deteriorates on keeping, only fresh preparations should be used.

As a local remedy, in the form of Vleminckx's solution, it is often very efficacious in *acne rosacea.*

    R Calcis,   ℥ss (15.5 gm.);
         Sulphuris sublimati,   ℥j (31.1 gm.);
         Aquæ,   f℥x (296.0 c.c.).
    Coque ad f℥vj (177.5 c.c.), deinde filtra.
    Sig. Dilute with 10 parts of water and apply to the affected parts.

URANII NITRAS.

(Uranium Nitrate; Uranyl Nitrate; $UO_2(NO_3)_2 + 6H_2O$.)

Uranium nitrate occurs as light-yellow, rhombic prisms, soluble in water, alcohol, and ether. The dose is $\frac{1}{2}$ gr. (0.03 gm.), gradually increased to 5 gr. (0.3 gm.) or more, freely diluted with water, after meals.

Therapeutics.—In large doses uranium salts act as irritant poisons, producing gastro-enteritis and nephritis. Moreover, they affect the oxyhemoglobin of the blood in such a way that its oxygenating power is much reduced. This action on the blood may be responsible for the glycosuria, which, according to Leconte, results from the prolonged use of uranium compounds.

West, Morrison, Duncan, and others have observed improvement in cases of *diabetes mellitus* from the use of uranium nitrate in doses of from 5–20 gr. (0.3–1.3 gm.) two or three times a day. In general, however, the drug has proved disappointing. It certainly has no specific influence on the disease, and whatever action it may have in lessening the excretion of sugar probably depends, as Symonds suggests, on its effect in retarding the digestion of starches.

GLANDULA THYREOIDEA.

(Thyroid Gland.)

The profession is largely indebted to the observations of Kocher and to the experimental researches of Horsley for the treatment of myxedema and allied conditions by the administration of thyroid gland. Kocher found that complete removal of the thyroid gland was followed in many instances by the appearance of a peculiar cachexia, the symptoms of which were almost identical with those of myxedema. Horsley showed that thyroidectomy in monkeys induced a similar condition. Schiff proved that the bad effects of thyroidectomy in animals could be averted by transplanting the gland in the peritoneum or subcutaneous tissue, and thereupon Horsley suggested this treatment for myxedematous conditions in man. Murray, in 1891, found that transplantation of the gland was unnecessary, as the same results could be secured from hypodermic injections of thyroid juice, and a little later Mackenzie demonstrated that thyroid feeding was equally efficacious.

According to Baumann, the active constituent of the thyroid gland is *iodothyrin*, a non-proteid compound containing from 5 to 10 per cent. of iodin. This substance is localized in the colloid matter, and is probably secreted by the gland in association with an iodin proteid, the two forming the *thyreoglobulin* of Oswald.

Physiologic Action.—Large doses of thyroid extract frequently, but not invariably, produce a train of symptoms to which the term *thyroidism* has been applied. The most common manifestations of this intoxication are restlessness, insomnia, headache, palpitation of the heart, weakness of the circulation, anorexia, nausea, elevation of temperature, free perspiration, shortness of breath, tremors and twitchings of the limbs, prostration, and progressive loss of flesh. It will be observed that these symptoms are not unlike those of exophthalmic goiter, and to make the resemblance more complete, swelling of the thyroid gland and exophthalmos have occurred in a few instances. Coppez has reported 5 cases of optic neuritis from the continued use of large doses.

In man, the most constant effect of thyroid extract in medicinal doses is increased oxidation, in consequence of which a considerable reduction in the body-weight occurs. Both the proteids and the fats suffer disintegration, but, according to Vendelstadt, only one-sixth of the loss of weight can be attributed to the destruction of nitrogenous compounds, the rest being due to oxidation of fats and to increased excretion of water. The excessive catabolism of proteids is made evident by the increased excretion in the urine of nitrogen and of phosphorus; the more rapid combustion of fats, by the increased elimination of carbon dioxid and the greater demand for oxygen.

In excess, thyroid extract frequently increases the rate of the pulse and lowers the arterial pressure. The manner in which these effects on the circulation are produced is not known. The hurried respiration observed in thyroidism may be due to the increased demand for oxygen. Even in moderate doses the drug usually increases the quantity of urine, and in large doses it may induce albuminuria and glycosuria.

Elimination.—Iodothyrin must be decomposed in the body, at least to some extent, since after its administration iodin appears in the urine. Hutchinson found this element in the urine of a dog three hours after the administration of 15 gr. (1 gm.) of colloid matter. That a part of the active constituent resists destruction in the body and escapes through avenues other than the kidneys is evidenced by the fact that an infant may acquire thyroidism through the mother's milk.

Therapeutics.—The diseases in which thyroid extract is most efficacious are *cretinism* and *myxedema of the adult.* In cretinism the results are often truly remarkable, especially when the treatment is instituted early. "Within six weeks," as Osler writes, "a poor, feeble-minded, toad-like caricature of humanity may be restored to mental and bodily health." In both affections the remedy must be continued throughout life, otherwise relapses occur. After the symptoms have been relieved, a weekly or biweekly dose may be all that is necessary to maintain normal metabolism. Thyroid extract sometimes proves efficacious, also, in *infantilism,* and in certain other conditions on the borderland of myxedema, in which imperfect mental and physical development is a prominent feature.

Considerable success has attended the use of the drug in *simple goiter,* especially in that form met with in adolescents. No effect, of course, can be expected from the treatment in the old cystic goiters of adults. In exophthalmic goiter thyroid feeding is either useless or harmful.

The loss of flesh following the administration of thyroid extract suggested its use in *obesity*. Ebstein believes that the treatment is not a rational one, since the drug causes a waste of the body-proteids as well as of the fats. The nitrogen loss can be controlled in a measure, however, by increasing the amount of proteids in the food, so that it does not become an insurmountable objection to the treatment. In some cases of obesity the remedy proves entirely satisfactory and is not followed by any unpleasant consequences. It is most effective in the cases which bear a certain resemblance to myxedema, in which the skin is pale and the tissues are soft and flabby. Unfortunately, many patients rapidly acquire a tolerance for the drug, and for this reason relapses are common.

In certain skin diseases, such as *psoriasis* and *ichthyosis*, thyroid extract is occasionally serviceable, but in the majority of cases it proves disappointing. In *inoperable cancer* it is often useful as a palliative remedy in relieving pain, dispelling disagreeable odor, and retarding ulceration. The drug seems to have a special influence on the uterus. There is considerable testimony as to its efficacy in *metrorrhagia* from various causes, and some authorities go so far as to claim for it the power of reducing the size of *uterine fibroids*. Montgomery has found it useful in *sterility* dependent upon obesity.

The hope that thyroid extract might prove valuable in certain forms of insanity has not been realized. While some observers, notably Mabon and Babcock and Easterbrook, have found it useful in acute melancholia, acute mania, puerperal and climacteric insanities, and stuporous states, the testimony of most alienists has been unfavorable.

Administration.—As persons vary considerably in their susceptibility to thyroid preparations, it is always advisable to begin with small doses and gradually to increase them. As a rule, it is better to give small doses frequently than large doses at long intervals. Treatment should be suspended, at least temporarily, on the very first appearance of untoward symptoms. Mabille and Ewald have shown that the drug is much better received when combined with small doses of arsenic. Individuals with tuberculosis, Bright's disease, and cardiac insufficiency are generally intolerant to thyroid medication.

Thyroid gland may be given in the form of the fresh gland of the sheep, the dried gland (Thyroideum Siccum), the glycerin extract (Liquor Thyroidei, B. P.), or iodothyrin. The fresh gland may be given raw or slightly cooked in doses of from $\frac{1}{8}$–$\frac{1}{2}$ of a gland a day. The dried gland and the extract,

22

however, are far more convenient and are just as potent. Dry thyroid is a light, dull-brown powder, with a meat-like odor and taste. On exposure to air and moisture it is liable to decompose. The dose is from 1–5 gr. (0.06–0.3 gm.) three times a day. The glycerin extract is a pinkish, turbid fluid, without putrescent odor. One hundred min. (6.0 c.c.) equal one fresh gland. The dose is from 5–15 min. (0.3–1.0 c.c.). Tablets of iodothyrin triturated with milk-sugar are on the market, and may be given in doses of from 2–10 gr. (0.13–0.6 gm.) thrice daily.

GLANDULÆ SUPRARENALES.

(Suprarenal Glands.)

Physiologic Action.—In 1855 Addison first pointed out that the disease now known by his name was associated with lesions of the suprarenal glands. A little later Brown-Séquard demonstrated experimentally that ablation of these glands in animals proved rapidly fatal, and that the symptoms preceding death were very much like those of Addison's disease. Both of these observations have been repeatedly confirmed. Two explanations have been offered of the action of the adrenal bodies in maintaining the normal nutrition: first, that they furnish a secretion which neutralizes toxic substances formed elsewhere in the body, and secondly, that they manufacture a substance which exerts directly a tonic influence upon the tissues. In the light of recent researches the latter view appears to be the more plausible one.

Oliver and Schäfer published a paper in 1894 showing that an extract of the medulla of the gland, when injected intravenously, causes an enormous rise in the blood-pressure. As this phenomenon occurs after isolation of the heart and after the blood-vessels have been paralyzed by chloral, it must be due, at least in part, to stimulation of the heart itself or its contained motor ganglia. The fact, however, that the extract, when locally applied, powerfully constricts the vessels, supports the view that it raises blood-pressure partly, also, by directly acting upon the muscular coat of the vessels. Whether or not the vasomotor center shares in the stimulation has not been definitely determined. Intravenous injections first slow the pulse by stimulating the cardio-inhibitory center in the medulla, but later they increase the pulse-rate by stimulating the heart itself. As a result of centric stimulation the respiration-rate is also increased. After intravenous injection the effects on the circulation and respiration are prompt and powerful, but transient; after subcutaneous injections they are not nearly so pronounced unless very large doses be employed, and after administration by the mouth they are scarcely noticeable. The variations in the activity of suprarenal extract with different methods of administration are probably the result of its very rapid elimination. Blum, Zuelze, Richardson and Herter, and others have shown that the suprarenal capsules contain a substance which, when thrown into the blood of an animal, induces glycosuria; further, that the glycosuria is more intense after injections of this substance into the peritoneal cavity, and particularly after direct applications to the pancreas, and that it does not occur at all when the glands are given by the mouth. A postmortem examination on one of the animals revealed extensive necrosis of the pancreas, especially in the islands of Langerhans. While the exact cause of these effects has not been definitely determined, the evidence is sufficient to show that the suprarenal capsules manufacture a body capable of causing not merely glycosuria, but a true glycemia. When locally applied to mucous membranes or raw surfaces, preparations of the suprarenal gland produce intense ischemia by their direct action on the muscular coats of the blood-vessels.

Active Principle.—Ever since Oliver and Schäfer demonstrated that the suprarenal glands possessed physiologic activity many attempts have been made to separate from them a principle that would fully represent this activity. In 1896 Abel announced the separation of a body which he called *epinephrin;* the following year Von Furth isolated a compound which he named *suprarenin,* and in 1901 Takamine and Aldrich, working independently, obtained a substance in crystalline form which is now known as *adrenalin.* Adrenalin, the only one of these preparations to be produced on a commercial scale, represents more fully than the other two the active properties of the gland. According to Abel, however, it is not a pure chemical compound, but a mixture of native and reduced epinephrin. It is a white, microcrystalline substance, of a slightly bitter taste and a faintly alkaline reaction. It is with difficulty soluble in cold water, but it is easily soluble in acids, forming salts. In dry form it is perfectly stable, but its solutions absorb oxygen from the air, and in consequence turn first red and then brown. It is so active that one drop of a solution of 1 part to 50,000 parts of water blanches the conjunctiva within one minute.

Therapeutics.—There is considerable evidence to show that the extract of suprarenal capsule is a useful remedy in

Addison's disease. In most cases, however, the good effects have been only temporary, and this is what we should have expected, knowing the incurable nature of the local lesions. One of Osler's patients gained nineteen pounds, the asthenia disappeared, and he was alive two years subsequently, but was still pigmented. Kinnicutt has collected 48 cases treated with the extract. Of these, 6 were reported cured and 22 as improved. In estimating the value of these and similar statistics, due allowance should be made for the fact that favorable results are more likely to be published than those which are unfavorable. Intravenous injections of adrenalin have been recommended in *sudden heart failure, collapse during anesthesia,* and in *morphin poisoning.*

As a local vasoconstrictor, adrenalin has proved to be extremely valuable in *relieving congestion* and in *checking hemorrhage.* In operations on the eye, nose, and throat it not only prevents bleeding, but it renders anesthesia under cocain more complete and at the same time lessens the danger from absorption of the anesthetic. In severe epistaxis prompt relief is usually afforded by plugging the nostrils with cotton soaked in a solution of adrenalin, 1 : 1000. Weaker solutions are sometimes efficacious in *acute coryza* and *hay-fever.* In the eye it does not affect the pupil nor the muscles of accommodation. Its application is said to relieve pain in various *inflammatory diseases of the eye,* but Lemare urges against its use when there is a tendency to iritis, contending that it may prove harmful in driving the blood from the surface to the deeper anastomosing vessels of the iris and ciliary body.

Administration.—In Addison's disease the fresh gland, raw or partly cooked, may be employed, but the dried extract is more convenient. The latter may be given in doses of from 3–5 gr. (0.2–0.33 gm.) three times a day. For local purposes adrenalin chlorid or tartrate has completely supplanted all other preparations. It is best dissolved in normal salt solution, to which 0.5 per cent. of chloretone may be added as a preservative. The fresher the solution, the greater is its activity. For operations solutions of from 1 : 5000 to 1 : 1000 may be employed, and for local medication, solutions of from 1 : 10,000 to 1 : 5000.

GLANDULA THYMUS.

(Thymus Gland.)

The persistence of the thymus in the majority of cases of exophthalmic goiter suggested the possibility of an antagonism between this gland and the thyroid. Reasoning from this premise, many have employed thymus preparations in the treatment of *Graves's disease*. Kinnicutt, in 1897, collected 62 cases treated with the gland. Of these, 36 cases showed improvement; 25 were unimproved or showed aggravation of the symptoms. Of 20 cases treated by Mackenzie, 1 died, in 6 no improvement was observed, and in 13 there was some improvement. In none of the cases, however, was the effect so decided as to justify the conclusion that the thymus has any great therapeutic activity. Mendel, Mettenheimer, and others claim to have had good results from the use of thymus preparations in *rickets*. They base the treatment on Friedleben's observation that the thymus gland is often atrophied in rachitic children, and on Mendel's observation that removal of the thymus in dogs induces symptoms resembling rachitis. The dose of the dried gland is from 10–15 gr. (0.6–1.0 gm.) or more three or four times a day.

HYPOPHYSIS CEREBRI.

(Pituitary Body.)

Schaefer and Vincent have demonstrated the existence of two substances in the infundibular part of the pituitary body, one producing a rise and the other a fall of blood-pressure. They distinguish them as the pressor and depressor substances respectively, the former being insoluble in alcohol and ether, while the latter is soluble in both of these fluids. The pressor substance, the more active of the two, when given hypodermically, stimulates the heart and constricts the arterioles, in consequence of which it raises the blood-pressure, slows the pulse, and increases the quantity of urine.

The almost constant finding of lesions of the hypophysis in *acromegaly* suggested the use of pituitary extract in the treatment of this disease. Of 20 cases so treated collected by Hinsdale, 9 showed some improvement in the subjective symptoms. The dose of the dried extract is from 3–5 gr. (0.2–0.3 gm.) three times a day.

TESTIS AND OVARIUM.

In 1889 Brown-Séquard announced that he had personally experienced remarkable rejuvenating effects from injections of the extract of the testicle of rabbits. Emanating from such a high authority, this announcement led rapidly to the use of orchitic extracts, not only in senility, but in impotence, neurasthenia, hysteria, locomotor ataxia, and many other affections of the nervous system.   More or less favorable results of this treatment have been reported, but in no instance has the good achieved been so decided as to convince a judicious observer that it was not due to mental suggestion rather than to the physiologic activity of the drug itself.   So few have been convinced of the efficacy of the remedy that it has already been practically abandoned.

Upon the assumption that the ovary is a secreting gland, its substance has been recommended in a variety of conditions connected with menstruation, especially for the relief of the distressing symptoms attending the premature or artificial menopause.   The results of its use, however, have not been very satisfactory; moreover, there is no good reason for believing that the ovary has any other function than that of ovulation.

ANTIPYRETICS OR FEBRIFUGES.

The body derives its heat mainly from the oxidation of food-stuffs and from various mechanical movements, and loses it mainly by radiation and conduction from the skin and by the evaporation of water from the skin and lungs.   Normally, heat-production and heat-dissipation are so evenly balanced that a mean temperature of 98.6° F. in the axilla is maintained, notwithstanding considerable variations in the surrounding temperature.   The balance between income and expenditure is secured through the operation of a regulating mechanism located in the central nervous system.

Antipyretics are drugs that lower temperature when it is abnormally high.   They exert but little influence upon the normal temperature.   Some of them, like *acetanilid, phenacetin*, and *antipyrin*, act directly on the heat-regulating centers in such a way that the standard of temperature is brought to a lower level, where it is maintained for a limited period; some, like *alcohol* and *aconite*, through their influence on the circulation, increase the dissipation of heat; while others, like *quinin*, lessen the production of heat by retarding oxidation.

Diaphoretics lower temperature by increasing the loss of heat due to the evaporation of sweat.

Cold applications increase the heat-loss by conduction.

Indication.—Antipyretics are employed to lower temperature in febrile states. They are indicated, however, only when the temperature is sufficiently high as to be in itself a source of actual danger or of considerable discomfort to the patient. When used repeatedly at short intervals, they are all more or less depressing. On the other hand, cold bathing has a stimulant as well as an antipyretic effect, and for this reason, when it can be carried out satisfactorily, it is always preferable to the use of drugs.

The most important antipyretics are:

Acetanilid,
Antipyrin,
Phenacetin, } Coal-tar antipyretics.
Phenocoll,

Quinin.
Guaiacol.
Aconite.

Salicylic acid, carbolic acid, and *resorcin* also lower temperature in febrile states, but they are no longer used for this purpose.

ACETANILIDUM, U. S. P.

(Acetanilid; Antifebrin; $C_6H_5NH.C_2H_3O$.)

Acetanilid is a derivative of anilin, an atom of hydrogen in the latter being replaced by the acetic acid radical acetyl. It is a white, shining, crystalline powder, odorless, and of a faintly burning taste. It is soluble in 194 parts of water, in 5 parts of alcohol, and in 18 parts of ether. The dose is from 3–10 gr. (0.2–0.65 gm.).

Physiologic Action.—In health, single, moderate doses (5–10 gr.—0.3–0.65 gm.) of acetanilid produce no perceptible effect. In fever, however, they induce a marked fall in temperature, which usually begins within two or three hours, and lasts a variable time, generally four or five hours. The fall in temperature is often associated with free perspiration. The antipyretic action of the drug has not been satisfactorily explained, but it appears to be the result of a direct influence exerted on the heat-regulating centers, in consequence of which the heat-production is decreased and the heat-dissipation is increased.

Circulation.—In ordinary doses acetanilid has no pronounced action on the circulation. Very large doses, however, lower the arterial pressure by directly depressing the heart, and probably, also, the vasomotor centers.

Respiration.—The drug seems to have no direct action on respiration. In poisoning there is marked dyspnea, but this is probably due to the changes induced in the red blood-cells.

Nervous System.—Our knowledge of the action of acetanilid on the nervous system is very vague. The fact that the drug relieves headache and neuralgic pain without affecting movement or consciousness shows that it has a special affinity for the sensory neurons, but as in large doses it induces more or less somnolence and paresis, it must affect also, though to a less extent, other portions of the nervous apparatus than the sensorium. Locally, on mucous membranes and raw surfaces, it exerts an analgesic effect by depressing the peripheral sensory nerves. Bokai and others assert that in frogs it paralyzes the motor nerve-endings also when directly applied.

In animals, poisonous doses may excite convulsions, and these appear to be sometimes of spinal and sometimes of cerebral origin.

Blood.—After large doses the blood becomes of a chocolate color, owing to the production of methemoglobin. The corpuscles themselves, however, do not suffer disintegration unless the dose has been very large.

Metabolism.—The testimony as to the action of acetanilid upon nitrogenous elimination is so contradictory as to justify the conclusion that the drug has but little influence one way or the other upon tissue-metabolism.

Absorption and Elimination.—Acetanilid is absorbed and eliminated with great rapidity. It is largely oxidized in the organism, and appears in the urine as the sulphate and glycouronate of paramidophenol.

Action on Lower Organisms.—It has some power in inhibiting the growth of bacteria, but its germicidal powers are very feeble.

Toxicology.—The chief symptoms of acetanilid poisoning are marked cyanosis, feeble breathing, extreme weakness of the circulation, free perspiration, dilatation of the pupils, and collapse. *Treatment* consists in maintaining the temperature of the body by means of external heat and in supporting the respiration and circulation by the liberal use of such drugs as strychnin, ammonia, and atropin. Artificial respiration and oxygen inhalations are useful in combatting the cyanosis.

Untoward Effects.—Idiosyncrasies to acetanilid are not uncommon. Some persons are so susceptible to its action that cyanosis and collapse follow the administration of the drug even in small doses. Summers has reported an instance in which 4 grains, repeated in thirty minutes, caused cyanosis, partial loss of consciousness, and grave collapse in a healthy patient, who had many times previously taken the drug in much larger doses without ill effects.

Fortunately, in the vast majority of such cases recovery follows under appropriate treatment. Papular and erythematous rashes are occasionally induced by acetanilid, but not so frequently as by antipyrin.

Therapeutics.—Moderate doses of acetanilid (5 gr.—0.3 gm.) are sometimes useful in controlling the temperature in such diseases as *typhoid fever, scarlet fever, pneumonia,* and *erysipelas.* Nothing, however, is to be gained from the use of antipyretic drugs in these diseases when the temperature is not above 103° F. or 103.5° F., and is not causing much inconvenience to the patient. They should be employed only when the temperature itself is exciting considerable discomfort or when it is so high as to exert a baleful influence upon the nutrition of the vital organs; even then, hydrotherapy, on account of its stimulant effects, is nearly always preferable if it can be satisfactorily carried out and is not precluded by some special feature of the case.

In tuberculosis and other adynamic diseases coal-tar antipyretics, if used at all, must be used with the utmost caution.

Acetanilid and its congeners, judiciously employed, are of great value in relieving certain forms of pain, particularly *headache, neuralgia, migraine,* the *pains of influenza,* and even the *nerve-storms of locomotor ataxia,* but they have little or no influence upon the pain of acute inflammation, of traumatism, or of morbid growths.

Acetanilid has been recommended as a depressomotor in *epilepsy, chorea,* and *whooping-cough,* but for this purpose it has been employed much less extensively than antipyrin.

Externally, it has been frequently substituted for iodoform in the treatment of *wounds, burns, ulcers,* and *chancroidal sores.* Its advantages are its inexpensiveness and freedom from odor. It must be employed with some caution, as it may be absorbed in sufficient quantity to produce toxic symptoms. This use of the drug appears to be decidedly dangerous in young children, no less than 20 cases of poisoning from it in infants having been reported within recent years.

Administration.—It may be administered in powders, capsules, tablets, or pills. When a prompt effect is desired, it should be given in solution, alcohol being used as a solvent before the diluent is added.

Incompatibles.—With alkaline bromids and iodids, in aqueous solution, it forms insoluble compounds. When triturated with antipyrin, chloral, thymol, or resorcin, it forms a semiliquid mass. When added to spirit of nitrous ether, the solution after a time turns yellow and then red.

Exalgin or **methyl-acetanilid** is also a derivative of anilin. It has no advantages over acetanilid, which it resembles very closely in its action. It has been used more as an analgesic than as an antipyretic. The dose is from 3–10 gr. (0.2–0.6 gm.).

ANTIPYRINUM.

(Antipyrin; Phenazon; Phenyl-dimethyl-pyrazolon; $C_6H_5(CH_3)_2C_3HN_2O.$)

Antipyrin is a synthetic base obtained by acting on phenyl-hydrazin (an anilin derivative) with diacetic ether, and then methylating the resulting monoethyl compound. It occurs in white, crystalline scales, odorless, and of a somewhat bitter taste. It is freely soluble in water, alcohol, and chloroform. The dose is from 3–20 gr. (0.2–1.3 gm.).

Physiologic Action.—The action of antipyrin, so far as it is definitely known, closely resembles that of acetanilid. It is probably a little less toxic than acetanilid, although it is more prone than the latter to cause papular or erythematous skin eruptions. It is absorbed and eliminated very rapidly, most of it escaping undecomposed through the kidneys.

Therapeutics.—Antipyrin is employed internally for the same purposes as acetanilid. Both are used to *lower temperature in fever* and to *relieve pains of a neuralgic character.* While the two drugs are equally efficacious, one sometimes succeeds where the other fails.

Antipyrin has been used to a considerable extent as an antispasmodic. The diseases in which it is has acquired the most reputation in this respect are *epilepsy* and *whooping-cough.*

In epilepsy it sometimes renders excellent service as an adjuvant to the bromids. From 10–15 gr. (0.6–1.0 gm.) a day may be given for three or four weeks, and then discontinued for two or three weeks before being resumed. The drug should be withdrawn at once, however, if cyanosis appear. In whooping-cough no remedy is so generally useful in controlling the paroxysms as antipyrin. It may be given by itself or in combination with a bromid, as in the following formula:

<pre>
R   Antipyrini,   gr. 1 (3.2 gm.);
     Sodii bromidi,   ℥iss (6.0 gm.);
     Glycerini,   f ℥ss (15.0 c.c.);
     Aquæ menthæ piperitæ, q. s. ad f ℥iij (90.0 c.c.).   M.
Sig.   A teaspoonful in water every three hours for a child of three
       years.
</pre>

Antipyrin is of some value in *diabetes mellitus*, but the manner of its action is unknown. Locally it has been recommended as an analgesic and as a hemostatic. As an analgesic it has been used with more or less success in relieving the pain of *cancerous ulcers.* Deep injections of a 50 per cent. solution have been employed also in *lumbago* and in *sciatica*, but they are painful and liable to cause abscesses. As a hemostatic it sometimes is effective in arresting hemorrhage from small vessels; thus solutions of from 5 to 10 per cent. have been found satisfactory in checking *epistaxis* and in controlling the *capillary oozing after operations.* In some persons its application in any strength proves irritating.

Administration.—Antipyrin may be given by the mouth in the form of powders or capsules, but on account of its ready solubility it is perhaps best ordered in some aromatic water. In exceptional cases it may be administered hypodermically or by the rectum.

Incompatibles.—On account of its basic properties antipyrin has a wide range of incompatibility. It is incompatible with iron salts, calomel, corrosive sublimate, iodin, iodids, benzoates, carbolic acid, ammonia water, sodium bicarbonate, nitrites, and all preparations containing tannic acid. When triturated with chloral, butyl-chloral, sodium salicylate, orthoform, or beta-naphthol, it forms a semiliquid mass.

PHENACETINUM.

(Phenacetin; Para-acet-phenetidin; $C_6H_4.OC_2H_5.NHC_2H_3O.$)

Phenacetin is obtained by acting on paraphenetidin (an anilin derivative) with glacial acetic acid. In the reaction that takes place an atom of hydrogen in paraphenetidin is replaced by the acetic acid radicle, acetyl. It forms colorless, tasteless, inodorous scales, soluble in alcohol and in glycerin, but only sparingly so in cold water. The dose is from 5–10 gr. (0.3–0.65 gm.).

Physiologic Action and Therapeutics.—The studies that have been made upon the action of phenacetin indicate that it exerts much the same influence as acetanilid and antipyrin, save that it is less toxic than either of these drugs. On the whole, it is the most satisfactory of the coal-tar antipyretics, and when used in moderate doses, it is rarely followed by cyanosis, collapse, cutaneous eruptions, or other untoward effects. It may be employed with advantage as an antipyretic and analgesic in the same class of cases as that in which acetanilid has been found to be useful.

Administration.—It may be given in pills, capsules, powders, or tablets.

Incompatibles.—Salicylic acid, iodin, carbolic acid, chloral hydrate, and oxidizing agents.

Lactophenin.—This compound is closely allied to phenacetin in that it is para-phenetidin with an atom of hydrogen replaced by the lactic acid radicle, lactyl, instead of by the acetic acid radicle, acetyl. In addition to being an antipyretic and analgesic it has some power as a somnifacient. The dose is from 5–15 gr. (0.3–1.0 gm.).

Phenocoll Hydrochlorid.—This phenetidin derivative represents an attempt to produce a soluble phenacetin. Chemically, it is glycocoll para-phenetidin hydrochlorid. It is a white, crystalline powder, soluble in about 16 parts of water. It is readily decomposed by alkalis and alkaline carbonates. The dose is from 5–15 gr. (0.3–1.0 gm.). As an antipyretic and analgesic it appears to be a safe and an effective substitute for acetanilid and antipyrin. It has been used also in *rheumatism* and in *malarial fever*, but with very indifferent success.

Apolysin, Citrophen, and Kryofin.—These compounds, all phenetidin derivatives, have been introduced as rivals of phenacetin. Apolysin differs from phenacetin in containing a citric acid radicle instead of an acetic acid radicle. Citrophen differs from apolysin in containing 3 phenetidin groups instead of 1 to the molecule of citric acid. Kryofin differs from phenacetin in containing a methyl-glycollic acid radicle instead of an acetic acid radicle. The dose of any one of these compounds is from 5–15 gr. (0.3–1.0 gm.).

OTHER ANTIPYRETICS.

Quinin (see p. 410).—In health the bodily temperature is not appreciably lowered by quinin unless the dose be excessive, but in febrile states the drug usually exerts a marked antipyretic influence when given in doses of from 20 to 30 gr. (1.3–2.0 gm.). This effect on the temperature cannot be attributed to an action on the heat-regulating centers, since it occurs after section of the spinal cord; it appears to be due rather to an interference with metabolism, in consequence of which there is a decrease in the production of heat.

The best results are obtained by giving the drug in one large dose a few hours before a natural remission is expected to occur. As a matter of fact, however, quinin is rarely used as an antipyretic at the present time. When it is deemed necessary to lower the temperature by means of drugs, the coal-tar derivatives, on account of the certainty and promptness of their action, the ease with which they can be administered, and their power to relieve headache and other pains incident to fever, are preferable to quinin, which lacks these advantages, and which, moreover, often disturbs the stomach.

Guaiacol (see p. 279).—In febrile diseases guaiacol, applied to the skin, acts as a prompt and powerful antipyretic. To secure the desired effect about 30 min. (2.0 c.c.) should be slowly rubbed into the skin of the abdomen with a camel's-hair brush, and the part subsequently covered with a piece of waxed paper to prevent evaporation. The absorption of the drug, which usually occurs in a few minutes, is followed by a gradual fall of temperature, the lowest point being reached about three hours after the application. The reduction of temperature is associated with profuse perspiration, and is followed shortly by a rapid return of the fever, with marked chilliness. The applications, especially if made repeatedly, often cause considerable depression, and for this reason guaiacol cannot be recommended as an antipyretic, except for occasional use.

Aconite (see p. 59).—Compared with the coal-tar derivatives, aconite is but a feeble antipyretic. It lowers temperature probably by increasing the heat-dissipation. On account of its sedative influence on the circulation it is a useful febrifuge in *acute inflammatory conditions* and in the *febrile diseases of childhood* when the pulse is rapid and strong. It is contraindicated in asthenic fevers. The best results are obtained by giving the tincture in small doses at frequent intervals. Spirit of nitrous ether and the solution of ammonium acetate are useful synergists.

ASTRINGENTS.

Astringents are substances which, by their direct action, cause contraction or condensation of the tissues. Their influence is most marked upon raw surfaces and mucous membranes. Many of them have also the property of diminishing or arresting glandular secretion, and this they accomplish not so much by constricting the blood-vessels of the part as by affecting the secreting cells directly. While their action is probably due in part simply to shrinking of the protoplasm, it is undoubtedly chiefly due to precipitation or to coagulation of the albumin of the cells.

Astringents are divided into two classes, *vegetable* and *mineral.* The former owe their efficacy to the presence of tannic acid.   There are but few astringents that are not also irritants, especially if employed in concentrated form.   Some of the metallic salts are more irritant and caustic in their action than they are astringent.   This is due in one instance to the irritant properties of the acid liberated by the union of the metal with the albumin of the cells; in another instance to the intensely toxic nature of the metal itself; and in still another instance to the permeable texture of the coagulum that is formed by the first application of the salt.

Among the mineral astringents, the insoluble salts of bismuth, zinc oxid, and lead acetate have more of a sedative than an irritant action.

The chief vegetable astringents are:

Tannic acid.	Catechu.
Gallic acid.	Geranium.
Galls.	Hematoxylon.
Kino.	Sumac.
Krameria.	Oak bark.

Hamamelis.

The chief mineral astringents are:

Alum.	Zinc oxid.
Lead acetate.	Silver nitrate.
Copper sulphate.	Bismuth subnitrate.
Zinc sulphate.	Bismuth subcarbonate.

Calcium carbonate.

Iron salts (see p. 282) and weak solutions of mineral acids, especially sulphuric acid (see p. 441), also have an astringent action.

Indications.—The very mild astringents, like bismuth subnitrate and zinc oxid, are useful as protectives in acute superficial inflammations; the more active ones, like the tannic acid compounds and zinc sulphate, are employed to lessen the excessive secretion often remaining as a sequel of acute inflammation of mucous membranes.   Some astringents, like alum and tannic acid, are also used to check hemorrhage, their styptic properties depending upon their power to coagulate the albumin of the blood.

ACIDUM TANNICUM, U. S. P.

(Tannic Acid; Tannin; Gallotannic Acid; $HC_{14}H_9O_9$.)

Tannic acid is the active constituent of all the vegetable astringents.   It is derived from the nutgall, which contains from 30 to 60 per cent. of it. The pure acid is a light-yellow-ish, amorphous powder, almost odorless, and of a strongly astringent taste. It is soluble in about 1 part of water, in 0.6 part of alcohol, and in 1 part of glycerin. The dose is from 3–10 gr. (0.2–0.6 gm.).

PREPARATIONS.

Collodium Stypticum, U. S. P. (20 per cent.).
Glyceritum Acidi Tannici, U. S. P. (20 per cent.).
Unguentum Acidi Tannici, U. S. P. (20 per cent.).
Trochisci Acidi Tannici, U. S. P. (1 gr.—0.06 gm.—in each).

Physiologic Action.—When applied to raw surfaces tannic acid precipitates the albumin of the superficial cells and causes condensation of the tissues. At the same time it lessens the sensibility of the peripheral nerves. On mucous membranes it acts in a similar manner, but more decidedly, and in addition dries up secretion by combining with the proteids of the glandular cells. Whether or not it constricts the blood-vessels with which it comes in contact is a mooted question. It coagulates blood, and thus serves as a local styptic. It is said also to inhibit the diapedesis of the leukocytes. When taken internally, the acids of the gastric juice prevent it from forming a permanent union with any proteids that may be present, so that it is free to act as an astringent in both the stomach and bowel. In moderate doses it impairs digestion and causes constipation; in very large doses it acts as an irritant and excites vomiting and diarrhea.

In the intestine, tannic acid is transformed into gallic acid, and in this form a certain amount is absorbed and is subsequently eliminated in the urine.

Therapeutics.—Locally, tannic acid is much used to check excessive secretion and to impart tone to relaxed mucous membranes. In *subacute and chronic laryngitis and pharyngitis* a solution of from 1–5 gr. (0.06–0.3 gm.) to the ounce (30.0 c.c.) makes a useful spray. Some surgeons have found injections of the glycerite, more or less diluted, efficacious in *subacute and chronic urethritis.* The same preparation is frequently beneficial in *chronic vaginitis, leukorrhea, erosion of the uterine cervix,* and *chronic cervical endometritis.* Lotions and dusting-powders containing tannin sometimes act favorably in *hyperidrosis of the feet.*

When it can be brought in direct contact with a bleeding surface tannic acid is a reliable hemostatic. In *epistaxis* strips of lint spread with vaselin and tannic acid make an excellent tampon. In *hematemesis* or *enterorrhagia* the drug may be given in full doses by the mouth. As it is absorbed in the form of gallic acid, which has no styptic power, it is valueless as an internal remedy in hemorrhages outside of the alimentary canal.

As it forms more or less insoluble tannates with *tartar emetic* and the *vegetable alkaloids*, it may be employed as a chemical antidote in *poisoning* by one of these drugs. Cantani and others have spoken highly of enteroclysis with a hot tannic acid solution (2 per cent.) in *Asiatic cholera.*

Administration.—When its action is desired in the stomach, tannin should be given in powder; if the intestine is to be reached, it should be given in pill.

Incompatibles.—Alkaloids, gelatin, lime-water, tartar emetic, and the salts of iron, silver, lead, and copper. When tannic acid is triturated with potassium chlorate the mixture explodes with great violence; hence, when these two drugs are to be combined in solution, they should be dissolved separately before being brought together.

Tannalbin, Tannigen, and Tannoform.—In the last few years several attempts have been made to enhance the therapeutic value of tannic acid as an intestinal astringent by converting it into compounds that will pass through the stomach unchanged and will be decomposed in the intestine, setting free the acid. The most important of these new compounds are tannalbin, tannigen, and tannoform.

Tannalbin is a light-brown, odorless and tasteless powder, containing about 50 per cent. of tannin. It is prepared by subjecting tannin albuminate to dry heat for a considerable time. The dose for adults is from 15–20 gr. (1.0–1.3 gm.), in powders or cachets; for children, from 5–15 gr. (0.3–1.0 gm.), in some mucilaginous vehicle or in powders.

Tannigen is an acetic ester of tannic acid, and appears as a yellowish, odorless, almost tasteless, hygroscopic powder, insoluble in water. As in the case of tannalbin, its astringent properties are not manifested until it reaches the bowel, where its decomposition is effected. It may be given in the same dose as tannalbin.

Experience in the use of these two remedies warrants the opinion that they may be used interchangeably; that they are free from irritant properties, even when administered in large amounts; and that they have a definite, if but limited, field of usefulness in the treatment of *acute intestinal catarrh.* The chief indication for their employment is the continuation of profuse and watery discharges after the cause of the inflammation has been completely removed. In chronic diarrhea and in dysentery they usually prove disappointing, and in tuberculous enteritis, of course, no favorable results from them are to be expected.

Tannoform is a combination of tannic acid and formaldehyd. It appears as a pale-pink powder, insoluble in water. It escapes decomposition in the stomach, but in the intestine is slowly broken up, yielding free tannin and formaldehyd. The dose for an adult is from 3–5 gr. (0.2–0.3 gm.); for a child, 1–3 gr. (0.06–0.2 gm.). It has been recommended in *intestinal catarrh*, but it is most too irritating for internal use. As an external remedy, however, it is not without value. A mixture of tannoform 1 part and Venetian talc 2 parts often acts very favorably in *excessive sweating of the feet*, especially when there is more or less maceration of the tissues.

ACIDUM GALLICUM, U. S. P.

(Gallic acid; $HC_7H_5O_5 + H_2O$.)

Gallic acid is the hydrid of tannic acid, from which it is usually prepared by boiling with dilute sulphuric acid. It occurs in white or pale-fawn colored, silky needles, odorless, of an astringent or slightly acidulous taste, and permanent in the air. It is soluble in 100 parts of water, in 5 parts of alcohol, and in 12 parts of glycerin. It differs from tannic acid in not precipitating gelatin, albumin, or alkaloids. The dose is from 5–20 gr. (0.3–1.3 gm.) in powders or capsules.

Physiologic Action and Therapeutics.—Locally, gallic acid is a very feeble astringent, but as it does not coagulate blood, it cannot be recommended as a styptic. When taken internally, it is absorbed into the blood and is eliminated unchanged in the urine. It has an undeserved reputation as a remote astringent in checking hemorrhages from parts that can be reached only through the circulation, in controlling night-sweats, and in diminishing the excessive secretion of urine in diabetes insipidus.

Incompatibles.—Ferric salts, tartar emetic, lead acetate, silver nitrate, and spirit of nitrous ether.

GALLA, U. S. P.

(Nutgall; Gall.)

Galls are excrescences produced by the stings and deposited ova of an insect in the bark or leaves of a plant. The official galls are caused by a species of *cynips*, which deposits its eggs in the tender shoots of *Quercus lusitanica*, an oak growing in the countries bordering on the Mediterranean. Their therapeutic activities depend upon tannic acid, of which they contain from 30 to 60 per cent.

23

PREPARATIONS.	DOSE.
Tinctura Gallæ, U. S. P.	½–2 fl. dr. (2.0–8.0 c.c.).
Unguentum Gallæ, U. S. P. (20 per cent.).	

Therapeutics.—Neither galls themselves nor the tincture are at the present time used internally. Nutgall ointment with equal parts of stramonium ointment is a time-honored remedy in *painful hemorrhoids.*

KINO, U. S. P., KRAMERIA, U. S. P., AND CATECHU, U. S. P.

Kino is the inspissated juice of *Pterocarpus Marsupium,* a large tree growing in the East Indies. The dose of the powdered drug is from 5–30 gr. (0.3–2.0 gm.).

PREPARATION.	DOSE.
Tinctura Kino, U. S. P.	½–2 fl. dr. (2.0–8.0 c.c.).

Krameria, or rhatany, is the root of *Krameria triandra,* a low shrub growing on the mountains of Peru and Bolivia.

PREPARATIONS.	DOSE.
Tinctura Krameriæ, U. S. P.	½–2 fl. dr. (2.0–8.0 c.c.).
Syrupus Krameriæ, U. S. P. (contains 45 cent. by volume of the fluid extract) . .	1–2 fl. dr. (4.0–8.0 c.c.).
Extractum Krameriæ Fluidum, U. S. P. .	5–30 min. (0.3–2.0 c.c.).
Extractum Krameriæ, U. S. P.	5–10 gr. (0.3–0.6 gm.).
Trochisci Krameriæ, U. S. P. (each contains about 1 gr.—0.06 gm.—of the extract).	

Catechu is an extract prepared from the wood of *Acacia catechu,* a small tree abounding in India and Burmah.

PREPARATIONS.	DOSE.
Tinctura Catechu Composita, U. S. P. (10 per cent., with cassia cinnamon 5 per cent.)	½–2 fl. dr. (2.0–8.0 c.c.).
Trochisci catechu, U. S. P. (each contains about 1 gr.—0.06 gm.).	

Therapeutics.—All these drugs contain large amounts of tannic acid, to which their astringent properties are due. They are sometimes employed to check excessive secretion in *acute diarrhea.* They should never be given, however, until the bowel has been thoroughly cleared of irritant material. They are often prescribed with chalk and opium, as in the following formula:

℞	Cretæ præparatæ,	ʒij (8.0 gm.);
	Tincturæ opii deodorati,	℥xx (1.3 c.c.);
	Tincturæ krameriæ,	f ʒij (8.0 c.c.);
	Acaciæ,	q. s.;
	Aquæ cinnamomi,	q. s. ad f ℥iij (90.0 c.c.). M.

Sig. A teaspoonful every two or three hours for a child of two years.

Troches of krameria or of catechu are useful in *relaxed sore throat.*

GERANIUM, U. S. P.

(Cranesbill.)

Geranium is the rhizome of *Geranium maculatum*, a perennial herb growing in the woody places of North America. It contains from 10 to 28 per cent. of tannic acid.

PREPARATION.   DOSE.

Extractum Geranii Fluidum, U. S. P. . . ½–1 fl. dr. (2.0–4.0 c.c.).

Therapeutics.—Geranium is the therapeutic equivalent of kino, krameria, and catechu, for which it may be substituted with advantage on account of its less disagreeable taste.

HÆMATOXYLON, U. S. P.

(Hematoxylon; Logwood.)

Hematoxylon is the heart-wood of *Hæmatoxylon campechianum*, a small tree growing in Central America and the West Indies. Its chief constituents are tannin and a crystalline coloring principle, *hæmatoxylin.*

PREPARATION.   DOSE.

Extractum Hæmatoxyli, U. S. P. . . . . . 5–30 gr. (0.3–2.0 gm.).

Therapeutics.—Logwood is used chiefly as an astringent in the *diarrhea of young children.* It is less active than kino, krameria, or catechu, but more agreeable on account of its sweetish taste. It has a disadvantage in staining the diapers a blood-red color. *Hematoxylin* is not used medicinally, but it is extensively used for its tinctorial properties in preparing tissues for microscopic study.

RHUS GLABRA, U. S. P.

(Sumac.)

Sumac is the fruit of *Rhus glabra*, a shrub growing in waste places in North America. It contains, in addition to tannic acid, several acid mallates.

PREPARATION.

Extractum Rhois Glabræ Fluidum, U. S. P.

Therapeutics.—Sumac is never used internally, but the fluid extract, diluted with from 6 to 8 parts water, makes an excellent mouth-wash or gargle in *mercurial stomatitis* and in *acute pharyngitis*, especially if a small amount of potassium chlorate be added, as in the following formula :

R   Potassii chloratis,   ℥j (4.0 gm.);
    Extracti thois glabræ fluidi,   f ℥j (30.0 c.c.);
    Aquæ,   q. s. ad f ℥viij (250.0 c.c.).   M.
Sig.   Use as a gargle.

QUERCUS ALBA, U. S. P.

(White Oak.)

Quercus alba is the bark of a large tree of the same name growing in North America east of the Mississippi. It contains from 5 to 10 per cent. of tannin and a bitter principle, *quercin.*

Therapeutics.—In the form of a decoction (1 oz.: 1 pint —30.0 gm. : 0.5 L.) it is occasionally used as a gargle in *sore throat* or as an injection in *leukorrhea.*

HAMAMELIS, U. S. P.

(Witch-hazel.)

Hamamelis is the leaves of *Hamamelis virginiana*, a shrub widely distributed throughout North America. It contains a volatile oil, a small amount of tannin, and a bitter principle.

PREPARATION.   DOSE.

Extractum Hamamelidis Fluidum, U. S. P. . . . ½–1 fl. dr. (2.0–4.0. c.c.)

A distillate containing a little alcohol is more commonly used than the official preparation. It is made from the fresh twigs collected in autumn when the plant is flowering.

Physiologic Action and Therapeutics.—Although examinations of witch-hazel have not revealed the presence of any very active ingredient, the drug is credited with considerable power as an astringent, a hemostatic, and a sedative. Dujardin-Beaumetz was of the opinion that it exerted a tonic influence upon the muscular coats of the blood-vessels, but the investigations of Guy and of Wood indicate that it has no such action. According to Ringer, large doses sometimes cause severe headache.

The distillate has a popular reputation as a topical remedy for *sprains, bruises,* and *small wounds.* As an injection or lotion it makes a very soothing application in *painful hemorrhoids.* Diluted with water, 1 part to 3, it is used with benefit as a spray in *acute coryza.* Both the distillate and the fluid extract have been used internally in various *hemorrhages* with asserted good results.

ALUMEN, U. S. P.

(Alum; Potassium Alum; Aluminum and Potassium Sulphate; $Al_2K_2(SO_4)_4 + 24H_2O$.)

The alum of the United States Pharmacopeia is a double sulphate of aluminum and potassium. In the British Pharmacopeia both this salt and the double sulphate of aluminum and ammonium are official. These compounds are of equal value. Potassium alum occurs in large, colorless, octahedral crystals, odorless, and of a sweetish and strongly astringent taste. It is freely soluble in water and in warm glycerin, but it is insoluble in alcohol. The usual dose is from 5–15 gr. (0.3–1.0 gm.), but as an emetic from 1–2 dr. (4.0–8.0 gm.) may be given.

PREPARATION.

Alumen Exsiccatum, U. S. P. (dried or burnt alum).

Physiologic Action.—When applied to the broken skin or to mucous membranes, alum acts as a powerful astringent, precipitating the albumin of the superficial cells, coagulating the fluids, and contracting the tissues. When used too freely or in the form of dried alum, it acts as an irritant. It forms a firm coagulum with blood, and thus tends to arrest hemorrhage. Taken internally in small doses it has an astringent effect and causes constipation. Large doses usually excite vomiting, but if retained, they induce gastro-enteritis.

Therapeutics.—Alum may be used as a local styptic in arresting *hemorrhages from small wounds*. A solution of from ½–1 dr. (2.0–4.0 gm.) to the pint (0.5 L.) is sometimes efficacious in *leukorrhea*. Insufflations of dried alum (1 : 20 of starch) or sprays of a weak solution—5–10 gr. (0.3–0.6 gm.) to the ounce (30.0 c.c.)—are sometimes of service in *subacute and chronic pharyngitis* and *laryngitis*, especially when there is much mucous secretion. Its prolonged use in the mouth is contraindicated on account of its destructive action on the teeth. Lotions of alum and diluted alcohol are sometimes employed in *hyperidrosis*. Dried alum has long been used as a mild caustic for destroying *exuberant granulations*.

Internally alum is no longer in use as an astringent. It is, however, a safe but somewhat uncertain emetic, and may be given to children in doses of a teaspoonful of the powdered drug in syrup, repeated once or twice if vomiting does not follow.

Incompatibles.—Alkalis and their carbonates, lead acetate, mercury, iron salts, and tannic acid.

PLUMBUM.

(Lead; Pb.)

Metallic lead is obtained from a native sulphid and is not official; the following preparations, however, are recognized by the United States Pharmacopeia: Acetate, subacetate, carbonate, oxid, nitrate, and iodid.

Physiologic Action.—Upon the skin soluble lead-salts have little or no effect, but when applied to denuded surfaces or to mucous membranes, they act as astringents, combining with the albumin of the cells and fluids to form a delicate but impervious coagulum. Unless applied in concentrated form they exert, with the exception of the nitrate, a sedative rather than a corrosive action.

Taken internally, single, moderate doses of lead acetate have no action outside of the alimentary canal. They leave a sweetish, metallic taste in the mouth, with a feeling of dryness, and, owing to their astringent effects, they tend to cause constipation. Most of the drug escapes absorption and is discharged in the stools in the form of a sulphid. In very large doses lead acetate acts as an irritant poison and excites pain in the stomach, nausea and vomiting, great thirst, diarrhea, or, more rarely, constipation, and collapse. Death is often preceded by coma and convulsions. The form in which lead is absorbed is not definitely known, but it is presumed that it circulates as an albuminate. It is eliminated in the urine, bile, intestinal secretions, saliva, and milk, and probably, also, in the sweat. Ordinarily its excretion is effected rather slowly, so that its continuous absorption, even in small quantities, is liable to lead to an accumulation of the metal in the tissues.

Chronic Lead-poisoning or Plumbism.—This condition may be brought about by the too prolonged use of the salts of lead for medicinal purposes, but it is much more frequently induced in workmen who are exposed to the fumes or dust of lead, or who handle the metal or paints containing it. It may follow, also, the accidental introduction of lead into the system through drinking-water, articles of food, hair-dyes, and cosmetics.

Hard waters may be stored in cisterns lined with lead or may be conveyed through lead pipes without harm resulting, since they soon deposit on the metal an insoluble coating of lead sulphate; on the other hand, soft waters, especially those containing considerable quantities of carbon dioxid, have a dangerous lead-dissolving action.

Symptoms of Chronic Poisoning.—The most common

symptom is *colic (colica pictonum)*. This condition consists in paroxysms of severe abdominal pain, with retraction and rigidity of the abdominal muscles and obstinate constipation, although rarely there may be diarrhea. During the attacks there is often a marked increase in the arterial tension and slowing of the pulse. A characteristic feature of plumbism is the *blue line* on the edge of the gums near the insertion of the teeth. This discoloration is due to the deposit of lead sulphid formed from the sulphureted hydrogen evolved in the decomposition of particles of food. Perfect cleanliness of the teeth may prevent its appearance. *Lead palsy* is of common occurrence. It usually attacks the muscles supplied by the musculospiral nerve,—the extensors of the fingers and of the wrist,—causing the so-called " wrist drop." The affected muscles ultimately atrophy and yield the reactions of degeneration. Occasionally other muscles are involved, such as the extensors of the legs, the recti of the eye, and the adductors of the larynx. Lead-paralysis usually results from peripheral neuritis ; in some instances, however, there appears to be also degeneration of the ganglionic cells of the spinal cord, but whether this is primary or is secondary to the neuritis has not been determined.

Anemia is rarely absent. Both the hemoglobin and the number of red cells are decreased, and, moreover, as Grawitz has shown, the red cells usually contain numerous basophilic granules.

Cerebral symptoms (encephalopathia saturnina) occasionally occur. They include persistent headache, epileptiform convulsions, delirium, optic neuritis, stupor, and coma. A symptom-complex resembling paretic dementia has also been described. In some instances the cerebral symptoms are simply expressions of uremia, but in others they are the consequence, without doubt, of the toxic action of the lead itself. McCarthy has observed in dogs poisoned by lead degeneration of the ganglion-cells, increase in the gliar elements, and marked endarteritis. Finally, workers in lead are especially prone to *chronic nephritis with arteriosclerosis* and to *gouty arthritis*.

Treatment of Lead-Poisoning.—*Acute Poisoning.*—The stomach should be emptied by means of the stomach-pump, unless vomiting has rendered this procedure unnecessary. A soluble sulphate (Epsom or Glauber's salt) is a chemical antidote, forming with the lead an insoluble sulphate ; it should be given in excess, so that a purgative effect may also be secured. The resulting gastro-enteritis should be treated by the application of warm fomentations and by the administration of opium and demulcent drinks.

Chronic Poisoning.—Much may be done to prevent plumbism in lead-work establishments. Absolute cleanliness, especially of the hands and nails, is of the utmost importance. Fans and suitable means of ventilation should be provided wherever dust is generated. Respirators should be worn by persons employed in dry processes. No food should be eaten in any part of the works.

Curative treatment consists in eliminating the poison and in relieving the immediate symptoms. Potassium iodid, in doses of from 5–15 gr. (0.3–1.0 gm.) thrice daily, is employed to meet the first indication. It favors excretion by uniting with the lead in the tissues to form a double lead and potassium iodid, which is soluble. Sulphur baths are also recommended. They are prepared by mixing in a wooden tub 3 or 4 oz. (90.0–125.0 gm.) of potassium sulphuret with about 20 gallons (75.0 L.) of water. Constipation should be relieved by saline cathartics, preferably by Epsom salt. Colic will require hot applications and hypodermic injections of morphin and atropin. For the paralysis massage, electricity, and strychnin should be used.

PLUMBI ACETAS, U. S. P.

(Lead Acetate; Sugar of Lead; Pb $(C_2H_3O_2)_2 + 3H_2O$.)

Lead acetate occurs in heavy, colorless, efflorescent, prismatic crystals or crystalline masses, having a sweetish, metallic taste, and a faintly acetous odor. It is soluble in 2.3 parts of water and in 21 parts of alcohol. The dose is from 1–4 gr. (0.065–0.26 gm.) in pills.

Therapeutics.—Internally, pills of lead acetate and opium are sometimes useful in controlling *subacute* and *chronic diarrhea.* Solutions of from 1–5 gr. (0.06–0.3 gm.) to the ounce (30.0 c.c.), employed as injections, are of service in the stationary stage of *gonorrhea.*

The well-known "lead-water and laudanum" is usually made of the subacetate of lead, but the acetate may be substituted with advantage, as in the following formula: ·

℞	Plumbi acetatis,	ʒj (4.0 gm.);
	Tincturæ opii,	f ʒj (30.0 c.c.);
	Aquæ,	q. s. ad f ℥viij (236.0 c.c.).—M.

Although the ingredients in this combination are chemically incompatible, the mixture makes an excellent sedative application in *bruises, sprains, superficial inflammations,* and *erysipelas.*

Incompatibles.—Acids, alkalis, sulphates, carbonates, chlorids, and tannin.

PLUMBI SUBACETAS.

(Lead Subacetate; Pb_2O $(C_2H_3O_2)_2$.)

The subacetate of lead is so unstable that it is employed only in solution.

PREPARATIONS.

Liquor Plumbi Subacetatis, U. S. P. (Goulard's extract: contains 25 per cent. of lead subacetate).

Liquor Plumbi Subacetatis Dilutus, U. S. P. (lead-water: contains 3 per cent. by volume of the solution of lead subacetate).

Ceratum Plumbi Subacetatis, U. S. P. (Goulard's cerate: contains 20 parts of the solution of lead subacetate and 80 parts of camphor cerate).

Therapeutics.—The solution of the subacetate of lead, diluted with 3 or 4 parts of water, is employed as a sedative lotion in *acute eczema, rhus poisoning,* and *erysipelas.* The official diluted solution is too weak to be of much service, although it is often used in preparing "lead-water and laudanum" (2 parts of lead-water to 1 part of laudanum). The cerate is occasionally used as an application for *chapped hands.*

PLUMBI CARBONAS, U. S. P.

(Lead Carbonate; White Lead; $(PbCO_3)_2Pb(OH)_2$.)

Lead carbonate is a heavy, white, opaque powder, odorless and tasteless, and insoluble in ordinary menstrua.

PREPARATION.

Unguentum Plumbi Carbonatis, U. S. P. (10 per cent. of lead carbonate in benzoinated lard).

Therapeutics.—The ointment of lead carbonate is sometimes used as a sedative and protective dressing for *burns* and *scalds,* but on account of the danger of absorption other remedies have been largely substituted.

PLUMBI OXIDUM, U. S. P.

(Lead Oxid; Litharge; PbO.)

Lead oxid is a heavy, reddish-yellow powder, odorless and tasteless, and insoluble in ordinary menstrua.

Lead oxid is rarely employed except in preparing the solution of lead subacetate (*Liquor Plumbi Subacetatis,* U. S. P.) and diachylon or lead-plaster (*Emplastrum Plumbi,* U. S. P.). The latter is really oleate of lead, and is made by boiling together in suitable proportions lead oxid, olive oil, and water. It makes a useful protective dressing for *superficial ulcers* and *bed-sores*. An ointment composed of equal parts of lead-plaster and of vaselin makes an efficacious application in *subacute eczema* and *hyperidrosis*. In the latter affection, after the parts have been cleaned and dried, the ointment should be applied on strips of muslin, and renewed twice daily for two or three weeks, instructions being given to avoid washing the feet in water during the progress of the treatment. Lead-plaster also enters into diachylon ointment (*Unguentum Dia-chylon*, U. S. P.) and into all but three of the official plasters.

PLUMBI NITRAS, U. S. P.

(Lead Nitrate; $Pb(NO_3)_2$.)

Lead nitrate occurs in white, translucent, octahedral crystals, odorless, and of a sweetish, astringent, and metallic taste. It is soluble in two parts of water.

Therapeutics.—At the present time lead nitrate is employed only as a caustic in the treatment of *onychia*. It is usually applied as a powder.

PLUMBI IODIDUM, U. S. P.

(Lead Iodid; PbI_2.)

Lead iodid is a heavy, bright yellow powder, odorless and sparingly soluble in water and in alcohol. It owes its activity more to the iodin element than to the lead, and is used only as a local absorbent (see p. 309).

CUPRI SULPHAS, U. S. P.

(Copper Sulphate; Blue Vitriol; $CuSO_4 + 5H_2O$.)

Copper sulphate occurs in large, transparent, deep-blue crystals, odorless, and of a nauseous, metallic taste. It is soluble in 0.5 part of water, and almost insoluble in alcohol. The dose as an astringent is from $\frac{1}{4}$–1 gr. (0.016–0.06 gm.); as an emetic, 5–10 gr. (0.3–0.6 gm.).

Physiologic Action.—Upon mucous membranes and raw surfaces copper sulphate in dilute form acts as an astringent; in concentrated forms it acts as a mild caustic. Taken internally in large doses it causes emesis by its direct irritant action on the stomach.

Toxicology.—Acute copper-poisoning is characterized by severe abdominal pain, a metallic taste in the mouth, and violent vomiting and purging, the ejecta often being mucous and bloody. Death may be preceded by delirium, convulsions, and coma. The bluish or greenish color of the vomit may serve to distinguish it from poisoning by other irritants. After death the gastro-intestinal tract is found to be intensely inflamed, and sometimes ulcerated. Unless death occurs quite promptly, there may be found also fatty changes in the liver and kidneys.

Treatment.—The antidotes are potassium ferrocyanid, magnesia, sodium carbonate, and soap. Demulcents like milk and eggs should be given freely. Opium will be required to relieve the pain.

Chronic copper-poisoning is of doubtful occurrence; it is said to occasion gastro-intestinal disturbances, cachexia, a green line on the gums, and a greenish discoloration of the hair.

Therapeutics.—Copper sulphate is a prompt and powerful emetic, but it is most too irritant for ordinary use. When administered once without effect it is best not to repeat the dose. In *phosphorus poisoning* it is useful not only as an emetic, but also as an antidote, since it forms on the phosphorus an insoluble coating.

In *indolent ulcers, ulcerative stomatitis, chronic granular conjunctivitis (trachoma)*, light applications of the solid crystal are often very useful for their stimulant effects. In *gonorrhea*, after the acute symptoms have subsided, an injection containing 2 gr. (0.13 gm.) to the ounce (30.0 c.c.), gradually increased in strength, is quite efficacious.

Internally, in pill form combined with opium, it is sometimes of service in *obstinate chronic diarrhea*.

Cuprol.—This is a combination of copper (6 per cent.) and nucleinic acid. In the form of a 10 per cent. solution it is recommended in *chronic conjunctivitis* as being less painful and less irritant than copper sulphate.

ZINCUM, U. S. P.

(Zinc; Zn.)

Metallic zinc is official in the form of thin sheets, or in irregular granulated pieces, or molded into thin pencils, or in a state of fine powder. The metal itself is not used medicinally.

ZINCI SULPHAS, U. S. P.

(Zinc Sulphate; White Vitriol; $ZnSO_4 + 7H_2O$.)

Zinc sulphate occurs in colorless, transparent, rhombic crystals, odorless, and of an astringent metallic taste. It is soluble in 0.6 part of water and in 3 parts of glycerin, and is insoluble in alcohol.   The dose as an astringent is from $\frac{1}{2}$–3 gr. (0.03–0.2 gm.); as an emetic, 10–30 gr. (0.6–2.0 gm.).

Physiologic Action.—In weak solution zinc sulphate exerts an astringent effect; in concentrated solution, an irritant or caustic effect.   Administered by the mouth, large doses (20 gr.—1.3 gm.) excite emesis through their action on the stomach. Toxic doses induce severe gastro-enteritis.

The **treatment of poisoning** consists in the free exhibition of alkalis or their carbonates, and of demulcents like eggs and milk.

Therapeutics.—A solution containing $\frac{1}{2}$ gr. (0.03 gm.) gradually increased to 5 or 6 gr. (0.3–0.4 gm.) to the ounce (30.0 c.c.) makes an excellent injection in *gonorrhea* after the subsidence of acute symptoms.   Weak solutions are sometimes of service in leukorrhea.

> R   Zinci sulphatis,
>   Aluminis,   aa ʒiss (5.8 gm.);
>   Glycerini,   f ℥ vj (180.0 c.c.).—M.
> Sig.   Add a tablespoonful to a quart of hot water, and use as an
>   injection.

A solution of from 1–2 gr. (0.06–0.1 gm.) to the ounce (30.0 c.c.) is a favorite collyrium in the later stages of *simple conjunctivitis*.

Internally, zinc sulphate is used as an emetic (see p. 160), and very rarely as an astringent in *chronic diarrhea*.

Incompatibles.—Alkalis and their carbonates, vegetable astringents, lead acetate, silver nitrate, and lime-water.

ZINCI ACETAS, U. S. P.

(Zinc Acetate; $Zn\,(C_2H_3O_2)_2 + 2H_2O$.)

Zinc acetate occurs in thin, colorless, six-sided plates, of a pearly luster, an acetous odor, and an astringent, metallic taste. It is soluble in 2.7 parts of water and in 36 parts of alcohol. It is used as an astringent for the same purposes as zinc sulphate.

ZINCI OXIDUM, U. S. P.

(Zinc Oxid; ZnO.)

Zinc oxid is an amorphous white powder, free from odor and taste, and insoluble in water and in alcohol.   The dose is from 1–5 gr. (0.06–0.3 gm.) in pill.

PREPARATION.

Unguentum Zinci Oxidi, U. S. P. (contains 20 per cent. of zinc oxid).

Therapeutics.—The ointment of zinc oxid is extensively used as a slightly astringent and protective dressing for *burns, acute ulcers,* and *acute inflammatory skin diseases.* Dusting-powders containing zinc oxid, starch, and Venetian talc, in various proportions, are very serviceable in *vesicular eczema* and in *erythema intertrigo.*

R Zinci oxidi,
   Pulveris talci Veneti,   aa ʒj (4.0 gm.);
   Pulveris amyli,   ʒij (8.0 gm.).—M.

Internally, zinc oxid has been employed as an antispasmodic and an antihydrotic, but it is useless for either of these purposes.

ZINCI CARBONAS.

(Zinc Carbonate; $2ZnCO_3 . 3Zn(OH)_2$.)

Zinc carbonate is official in the form of precipitated zinc carbonate (*Zinci Carbonas Præcipitatus*, U. S. P.), which is an impalpable white powder, of variable composition, odorless and tasteless, and insoluble in ordinary menstrua.

An impure precipitated carbonate of zinc, known as *calamin,* was formerly official (1860).

Therapeutics.—Zinc carbonate resembles zinc oxid in appearance and in therapeutic properties. It is chiefly employed as a sedative and protective application in *acute inflammatory* affections of the skin, such as *eczema, erythema intertrigo,* and *dermatitis venenata.* It is often combined with zinc oxid, as in the following formula :

R Zinci carbonatis præcipitati,
   Zinci oxidi,   aa ʒiiss (10.0 gm.);
   Glycerini,   fʒj (4.0 c.c.);
   Liquoris calcis,   f℥ij (60.0 c.c.);
   Aquæ rosæ,   q. s. ad f℥vj (178.0 c.c.).—M.

ZINCI CHLORIDUM, U. S. P.

(Zinc Chlorid; $ZnCl_2$.)

Zinc chlorid occurs in white, granular powder or porcelain-like masses, or molded pencils, very deliquescent, odorless, and of a caustic metallic taste. It is freely soluble in water and in alcohol.

PREPARATION.

Liquor Zinci Chloridi, U. S. P. (50 per cent. by weight of zinc chlorid).

Therapeutics.—Zinc chlorid is the least astringent and the most caustic of the zinc salts. At present it is rarely employed except as an escharotic in removing *superficial epitheliomata.* Its action is rather slow, and is attended with severe pain. It is best applied in the form of a paste, which may be made by mixing 1 part of zinc chlorid and 3 parts of flour with a saturated solution of cocain hydrochlorate.

The solution of zinc chlorid has an unmerited reputation as a disinfectant.

ARGENTUM.

(Silver; Ag.)

The following preparations of silver are official: Silver cyanid (*Argenti Cyanidum,* U. S. P.), silver iodid (*Argenti Iodidum,* U. S. P.), silver nitrate (*Argenti Nitras,* U. S. P.), and silver oxid (*Argenti Oxidum,* U. S. P.). Of these, the only one of any importance is silver nitrate.

ARGENTI NITRAS, U. S. P.

(Silver Nitrate; Lunar Caustic; AgNO₃.)

Silver nitrate occurs in colorless, transparent, tabular, rhombic crystals, odorless, of a caustic metallic taste, and soluble in 0.6 part of water and in 26 parts of alcohol. It turns dark on exposure to light. The dose is from $\frac{1}{6}$–$\frac{1}{2}$ gr. (0.01–0.03 gm.).

PREPARATIONS.

Argenti Nitras Fusus, U. S. P. (silver nitrate molded into hard, white cones or pencils).

Argenti Nitras Dilutus, U. S. P. (mitigated caustic: molded cones or pencils of silver nitrate and potassium nitrate, containing about 33 per cent. of the former).

Physiologic Action.—When applied undiluted to the skin, silver nitrate acts as a superficial escharotic, producing a white slough which subsequently turns black on exposure to light. Upon mucous membranes and raw surfaces it acts in dilute form as an unirritating astringent, precipitating the albumin of the cells with which it comes in contact, and contracting the blood-vessels; in concentrated form it acts as a caustic, coating the part with a white pellicle of silver albuminate. Its corrosive action, however, never extends very deeply on account of the impenetrable nature of the coagulum that is at once formed.

Silver nitrate is an active germicide, solutions of 1 : 2500 destroying the organisms of diphtheria, cholera, glanders, and typhoid fever in two hours.

When taken internally in medicinal doses, it exerts no other influence than that of an astringent and an antiseptic. Being so readily precipitated by chlorids, proteids, and acids, its astringent action is chiefly expended on the mucous membrane of the stomach. The bulk of it escapes absorption, but that a small percentage enters the circulation is evident from the fact that the prolonged use of the drug is followed by a deposit of silver in the skin and internal organs. Very little is known concerning its elimination; it is supposed to be excreted slowly and imperfectly in the urine.

Toxicology.—In large doses silver nitrate acts as an irritant poison, producing intense abdominal pain, persistent vomiting and purging, and collapse. White or blackish patches may be present on the lips and mucous membrane of the mouth. Death is sometimes preceded by delirium, convulsions, and coma.

Treatment.—Common salt is the best antidote. It forms with the poison an insoluble and inert chlorid of silver. Demulcents should also be used freely.

Argyria.—The long-continued use of silver nitrate, either internally or locally, may result in the permanent deposition of dark granules of metallic silver in various tissues of the body. This pigmentation is known as argyria. Clinically, its chief manifestation is a peculiar, bluish-gray discoloration of the skin and mucous membranes. In the majority of cases there is no impairment of the general health. It is incurable.

Therapeutics.—At the present time silver nitrate is chiefly used for its action on *inflamed mucous membranes* and *ulcerated surfaces.* In *chronic ulcers* a solution of from 10 to 40 grains to the ounce (0.6–2.6 gm. to 30.0 c.c.) or the solid stick may be employed as a stimulant application. Silver nitrate is perhaps the best caustic for removing *exuberant* or *superfluous granulations.* Its employment in *poisoned wounds,* especially in the punctured variety, and in those resulting from the bites of rabid animals, is to be condemned on account of its superficial action and the premature closure of the orifice. A strong solution (30 gr. to 1 oz.—2.0 gm. to 30.0 c.c.) will sometimes abort a *felon* if applied early. In *acute epididymitis* counter-irritation may be secured by drawing a pencil of lunar caustic several times over the posterior surface of the scrotum, but this treatment is less efficacious than the application of guaiacol (p. 279). The treatment of *erysipelas* by local applications of silver nitrate, so highly recommended by Higginbottom, has few advocates at the present time.

In *simple conjunctivitis,* when the discharge has become mucopurulent, the membrane may be painted with a solution containing from 3 to 5 grains to the ounce (0.2–0.3 to 30.0 c.c.).

In *purulent ophthalmia* and *ophthalmia neonatorum* the conjunctiva, after being thoroughly cleansed, may be touched with a 10-grain solution (0.65 gm. to 30.0 c.c.). Credé's prophylactic treatment, which consists in the instillation of 2 drops of a 2 per cent. solution into the eyes of the new-born child when gonorrhea in the mother is suspected, has given excellent results. It must be remembered that the long-continued use of collyria containing silver nitrate may be followed by permanent discoloration of the conjunctiva.

Copious irrigation of the urethra with hot solutions of silver nitrate (1 : 10,000) has been found serviceable in both *acute* and *subacute gonorrhea.* Excellent results are obtained in *chronic cystitis* by washing out the bladder first with distilled water and then with a solution of silver nitrate (1 : 8000).

Solutions varying in strength from 10 to 60 grains to the ounce (0.6–4.0 gm. to 30.0 c.c.) are extensively employed in the treatment of *chronic stomatitis, chronic pharyngitis,* and *chronic laryngitis.* Light touches of the solid stick act very favorably upon *mucous patches.*

In *chronic dysentery* great benefit often follows the use of copious injections into the bowel of solutions of silver nitrate (10–30 grains to the pint—0.65–2.0 gm. to 0.5 L.). These injections should be given two or three times a week, the fluid being introduced very slowly through a tube passed well up into the bowel. At first the amount injected should not exceed 2 pints (1.0 L.), but later 4 or 5 pints (2.0–2.5 L.) may be employed with advantage.

Internally, silver nitrate is employed chiefly for its local action on the gastro-intestinal tract. In *chronic gastric catarrh* and in *ulcer of the stomach* no remedy, with the exception of bismuth subnitrate, is so generally useful. In obstinate catarrh of the stomach, especially when there is supersecretion or marked hyperesthesia of the mucous membrane, douching the stomach once or twice a week, first with a solution of silver nitrate (1 : 5000 to 1 : 2000) and then with plain water, is a valuable method of treatment. About a pint (0.5 L.) of the solution should be introduced through the douche at each operation. Silver nitrate is also used in *acute* and *chronic enteritis,* but with less benefit than in inflammatory diseases of the stomach.

The treatment of *epilepsy, locomotor ataxia,* and *chorea* by the administration of the salts of silver possesses little more than a historic interest.

Administration.—In affections of the stomach silver nitrate should be given in pill form, half an hour before meals.

Powdered opium or extract of hyoscyamus may be used as an excipient. When intended to act in the bowel, it should be given in keratin-coated pills. The administration of silver should not be continued for a longer period than six or eight weeks without interrupting the treatment for a like period.

Incompatibles.—Organic matter, bromids, chlorids, iodids, cyanids, sulphids, carbonates, phosphates, arsenites, and hydrochloric acid. With creasote it is explosive.

Organic Silver Salts and Soluble Silver.—Many attempts have been made recently to produce compounds of silver that will not precipitate albumins and chlorids, and yet will retain the germicidal properties of the metal. The most important compounds of this class thus far introduced are argentamin, silver lactate (actol), silver citrate (itrol), protargol, argyrol, and largin. In addition to these, Créde has also suggested a soluble allotropic form of silver (argentum solubile). The therapeutic value of these preparations has not yet been definitely determined.

Argentamin.—This is a solution of silver phosphate (10 per cent.) in an aqueous solution of ethylendiamin (10 per cent.), the purpose of the ethylendiamin being to prevent the precipitation of the salt by albumins and chlorids. Like other silver preparations, it is decomposed by light. It has been used chiefly in *gonorrhea* and in *conjunctivitis;* in the former, in solutions of from 1 : 5000 to 1 : 1000, and in the latter, in solutions of from 3 to 5 per cent. In the eye its action is somewhat irritating.

Silver Lactate (Actol).—This is a white powder, free from odor and taste, and soluble in about 20 parts of water. It is a powerful germicide, solutions of 1 : 1000 destroying most microorganisms within five minutes. Solutions of from 1 : 1000 to 1 : 500 have been employed for injection in *gonorrhea*, but they have proved very irritating.

Silver Citrate (Itrol).—This salt resembles the lactate in appearance, but as it is almost insoluble in water, it is much less irritating. It has been used as an antiseptic dusting-powder for wounds, and in solution (1 : 5000) as a disinfectant for instruments and the skin.

Protargol.—This is a proteid compound of silver containing about 8 per cent. of metallic silver. It is a fawn-colored powder, freely soluble in water. It is an energetic germicide and comparatively free from irritant properties. It has been used quite extensively and with good results in *purulent conjunctivitis* (1 to 10 per cent. solution), in *gonorrhea* (0.25 per cent. solution gradually increased to 2 per cent.), and in *sup-purative middle-ear disease* (3 to 5 per cent. solutions).

Argyrol (Silver Vitellin).—This is a compound of silver and a vegetable proteid containing 30 per cent. of metallic silver. Its high percentage of silver (nearly four times that of protargol), extreme solubility, and comparative freedom from irritant properties make it one of the most serviceable of the organic silver compounds. In solutions of from 5 to 20 per cent. it has been used with considerable success in both *acute* and *chronic gonorrheal infections.*

Largin.—This is also a proteid compound of silver. It is a grayish powder, freely soluble in water, containing about 11 per cent. of metallic silver. It has been used with asserted good results in *gonorrhea* (0.5 to 1.5 per cent. solution) and in *purulent ophthalmia* (1 to 10 per cent. solutions).

Soluble or Colloidal Silver (Argentum Solubile).—This is an allotropic form of silver, occurring in heavy, greenish-black particles, readily soluble in water, forming a reddish-brown solution. It is most commonly used in the form of a 15 per cent. ointment, known as Credé's ointment, which has been found of service as a local remedy in such conditions as *erysipelas, lymphadenitis, septic phlebitis,* and *gonorrheal arthritis.* With the view of producing a constitutional impression, daily inunctions of the ointment (45 gr.—3.0 gm.) have also been employed in various septic processes, but this method of using the drug is not very promising.

BISMUTHUM.

(Bismuth ; Bi.)

The following preparations of bismuth are official : Bismuth subnitrate (*Bismuthi Subnitras,* U. S. P.), bismuth subcarbonate (*Bismuthi Subcarbonas,* U. S. P.), bismuth citrate (*Bismuthi Citras,* U. S. P.), and bismuth and ammonium citrate (*Bismuthi et Ammonii Citras,* U. S. P.). Bismuth citrate itself is not used medicinally ; it is employed only in preparing bismuth and ammonium citrate. Among the compounds of bismuth which are in use, but which are not official, may be mentioned the subgallate, salicylate, benzoate, beta-naphtholate, and tribrom-phenolate. The last four of these compounds will be consid-ered in the section devoted to germicides and antiseptics (p. 379).

BISMUTHI SUBNITRAS, U. S. P.

(Bismuth Subnitrate; $BiONO_3 + H_2O$ (?).)

Bismuth subnitrate is a heavy white powder, of a somewhat varying composition, odorless, almost tasteless, and permanent in the air. It is insoluble in water and in alcohol. The dose is from 5–40 gr. (0.3–2.6 gm.).

Physiologic Action.—When taken internally, even in very large doses, the effects of bismuth subnitrate are confined to the mucous membrane of the alimentary canal, its action being that of a feeble astringent and protective. Since traces of the metal have been found in the urine there must be some absorption, but the amount of the drug entering the blood is undoubtedly too small to exert any special influence. Several cases of fatal poisoning have been reported as resulting from large doses, but these were due to the contamination of the bismuth with arsenic, an impurity no longer to be feared. It is unlikely that harm could result, save mechanically through impaction, even from enormous doses of a pure preparation. The characteristic black stools following the administration of the drug are attributed to its partial conversion into a sulphid in the intestine. The breath of patients taking bismuth sometimes acquires a peculiar garlicky odor; this is due, as Reisert has demonstrated, to the minute tellurium impurities often present in commercial bismuth compounds.

Upon the unbroken skin bismuth subnitrate acts simply as a protective, but on raw surfaces it acts also as an astringent and antiseptic. Absorption takes place rather readily from denuded parts, and poisoning may occur if the drug be applied too freely. The symptoms of bismuth poisoning are salivation, stomatitis, dysphagia, a black discoloration of the mucous membrane of the mouth, ulceration of the throat, diarrhea, and albuminuria.

Therapeutics.—No drug is so generally useful as bismuth subnitrate in allaying *gastric irritation* from various causes. It is a standard remedy in both *acute* and *chronic gastric catarrh* and in *gastric ulcer*. In the latter affection excellent results are often obtained by giving a single daily dose of from 1–2 dr. (4.0–8.0 gm.) in the morning, on an empty stomach, and then having the patient assume various positions—on the back, on the right, and on the left side, remaining in each position for five or ten minutes. In many cases of ulcer an alkali, such as sodium bicarbonate or magnesia, may be added with advantage. Even in *gastric cancer* the drug sometimes affords temporary relief. It is of much service in *gastralgia* dependent upon excessive acidity of the stomach. *Vomiting* the result of gastric irritation generally yields to it. It is especially efficacious as an antiemetic when conjoined with diluted hydrocyanic acid.

In *catarrhal diarrhea*, after the intestine has been thoroughly emptied, bismuth subnitrate is invaluable as a protective and an astringent. In these cases it may be combined advantageously with morphin and an antiseptic, as in the following formula:

R   Morphinæ sulphatis,   gr. j (0.06 gm.);
     Salol,   gr. xxx (2.0 gm.);
     Bismuthi subnitratis,   ℥ss (15.5 gm.).—M.
Fiant chartulæ No. xv.
Sig.   One every three hours.

Topically, bismuth subnitrate, alone or with starch, may be used as a dusting-powder in *acute erythema* and in *intertrigo.* In the form of a paste made with glycerin and water it has been employed rather extensively also as a protective dressing for *burns;* in a number of instances, however, such applications have been followed by poisoning. In the *second stage of gonorrhea* injections of bismuth subnitrate, 15 gr. (1.0 gm.) to the ounce (30.0 c.c.) of glycerin and water, have a favorable action.

In *acute coryza* a snuff composed of bismuth subnitrate, gum acacia, and a little morphin sometimes affords relief.

Administration.—It is usually prescribed in powders or in capsules. Very large doses, however, may be given conveniently in water, the mixture being well shaken before each administration. Alkaline bicarbonates and sodium hyposulphite are incompatible with such mixtures. In affections of the stomach the drug should be taken half an hour before meals.

BISMUTHI SUBCARBONAS, U. S. P.

(Bismuth Subcarbonate; $(BiO)_2CO_3 + H_2O$ (?).)

Bismuth subcarbonate closely resembles bismuth subnitrate in its physical properties and action. Its uses and dose are exactly the same as those of the subnitrate.

BISMUTHI SUBGALLAS.

(Bismuth Subgallate; Dermatol; $C_6H_2(OH)_3CO_2Bi(OH)_2$.)

Bismuth subgallate is a fine, yellowish-white powder, odorless and tasteless, permanent, and insoluble in ordinary solvents. It is said to contain 55 per cent. of bismuth oxid. The dose is from 5–30 gr. (0.3–2.0 gm.).

Therapeutics.—Bismuth subgallate has been employed somewhat extensively as a protective and an astringent applications for *burns* and *moist eczema.* It does not appear, however, to have been less harmful than other bismuth compounds when used externally. Internally it has been used instead of the subnitrate in *catarrhal affections of the stomach and intestine,* but it has no advantages over the official preparation.

CALCII CARBONAS.

(Calcium Carbonate; Chalk; $CaCO_3$.)

Chalk is official in two forms: Precipitated calcium carbonate (*Calcii Carbonas Præcipitatus*, U. S. P.) and prepared chalk (*Creta Præparata*, U. S. P.). The dose of either preparation is from 10–40 gr. (0.65–2.3 gm.) or more.

PREPARATIONS.	DOSE.
Pulvis Cretæ Compositus, U. S. P. (prepared chalk, 30; acacia, 20; sugar, 50) . . .	5–40 gr. (0.3–2.3 gm.).
Mistura Cretæ, U. S. P. (compound chalk powder, 20; cinnamon water, 40; water, q. s. 100)	1–4 fl. dr. (4.0–15.0 c.c.).
Trochisci Cretæ, U. S. P. (prepared chalk, 25; acacia, 7; sugar, 40; spirits nutmeg, 3).	
Hydrargyrum cum Creta, U. S. P. (57 per per cent. of chalk)	½–10 gr. (0.03–0.6 gm.).

Therapeutics.—Chalk is a mild astringent and antacid, free from irritant properties. It may be employed in *acute inflammatory diarrhea* in the same manner as bismuth subnitrate. It is an excellent antacid when acidity of the stomach is associated with relaxation of the bowels. It may be prescribed as an *antidote* in poisoning by a *mineral acid* or by *oxalic acid*.

Externally, it is used as a dusting-powder in *erythematous eczema* and *intertrigo*. It is particularly useful in the *chafing of the genitalia and buttocks* of young children from irritation by urine. It enters into the composition of most *tooth-powders*.

HEMOSTATICS.

Hemostatics are agents that arrest hemorrhage. They all act in accordance with the methods pursued by nature in spontaneously closing a bleeding vessel. If an artery be divided, it shrinks within its sheath, the contiguous structures fall in upon the bleeding orifice, and if the hemorrhage be copious, the force of the circulation diminishes and the coagulability of the blood increases. The choice of the hemostatic will largely depend upon the seat of the hemorrhage—that is, whether it is in an accessible or in an inaccessible region, and upon its source—that is, whether it is of capillary, of venous, or of arterial origin. Hemorrhage from a large vessel in an accessible region should always be controlled by *mechanical* means—that is, by ligature, torsion, pressure, acupressure, or cauterization.   Many parts formerly considered inaccessible are no longer so regarded, and to-day surgical aid is sometimes wisely invoked to arrest excessive bleeding from such organs as the stomach and bowel, and even the brain.

Small external hemorrhages, especially capillary oozing, are often satisfactorily controlled by the direct application of agents that contract the vessels or that coagulate the albumin of the blood.   The most potent local vasoconstrictors are *cold*, in the form of water or of ice, and *adrenalin ;* and the most powerful styptics are *ferric subsulphate, ferric chlorid, alum,* and *tannic acid*.

Three classes of drugs have been recommended to control hemorrhage in organs that must be reached through the circulation.   The first includes drugs like tannic acid and certain mineral salts, which have been found useful as local styptics or astringents; the second, drugs like *ergot*, which cause vasoconstriction by stimulating the vasomotor centers or by acting directly on the muscular walls of the vessels ; and the third, drugs like *gelatin* and *calcium chlorid*, which are thought to increase the natural coagulability of the blood.

There is no reason to believe that tannic acid exerts any influence as a hemostatic outside of the alimentary canal, since it enters the blood in such a form (*gallic acid*) that it loses its styptic and astringent properties.   As regards the mineral astringents, it is exceedingly doubtful whether any one of them can enter the circulation in sufficient quantity to exert a favorable influence upon a hemorrhage in a territory remote from the point of absorption.   Owing to its power of causing contraction of the unstriped muscle of the uterus, ergot is unquestionably of value in controlling certain forms of metrorrhagia, but the testimony to its efficacy in other internal hemorrhages, such as hemoptysis and hematuria, is not convincing.   Granting that it has the power of causing vasoconstriction through a central influence, it is unreasonable to suppose that its effects are limited to the vessel concerned in the bleeding, and if such be not the case, it should tend to do harm rather than good, since a universal narrowing of the blood-paths and a consequent rise in the blood-pressure must favor the escape of blood and also hinder the formation of an occluding thrombus.   Like ergot, *hydrastinin* and *cotarnin* also act as vasoconstrictors. While clinical experience has shown them to be of value in some forms of metrorrhagia, it is doubtful whether they are of any avail in bleeding from other organs.   Gelatin and calcium chlorid have recently been recommended as internal hemostatics on the ground that they increase the normal coagulability of the blood. The former appears to be more reliable than the latter, but sufficient time has not yet elapsed to allow of any decided opinion of its efficacy.

In the treatment of internal hemorrhage absolute rest of body and mind is essential. An ice-bag may be applied over the seat of the bleeding. *Opium* is of great value ; it not only exerts a calmative effect, but it also serves to control involuntary movements, such as cough and peristalsis. For the purpose of lowering the blood-pressure a ligature may be applied to a leg or an arm. For the profound anemia which accompanies or follows profuse hemorrhage normal salt solution may be given subcutaneously.

The only drugs employed as hemostatics which have not been considered in other sections are hydrastis, cotarnin, gelatin, and calcium chlorid.

HYDRASTIS, U. S. P.

(Golden Seal.)

Hydrastis is the rhizome and root of *hydrastis canadensis*, a perennial herb growing in the woods of North America east of the Mississippi. It contains *hydrastin, berberin,* and *canadin.* By oxidizing hydrastin with potassium permanganate the artificial alkaloid *hydrastinin* is obtained, which is official as hydrastininæ hydrochloras. The dose of crystalline hydrastin is from $\frac{1}{4}$–$\frac{1}{2}$ gr. (0.016–0.03 gm.) ; of hydrastinine hydrochlorate, from $\frac{1}{4}$–1 gr. (0.016–0.06 gm.).

PREPARATIONS.	DOSE.
Extractum Hydrastis Fluidum, U. S. P.	$\frac{1}{4}$–1 fl. dr. (2.0–4.0 c.c.).
Tinctura Hydrastis, U. S. P.	$\frac{1}{2}$–2 fl. dr. (2.0–8.0 c.c.).
Glyceritum Hydrastis, U. S. P.	$\frac{1}{2}$–2 fl. dr. (2.0–8.0 c.c.).

Physiologic Action.—Hydrastis owes its activities largely to the presence of the alkaloid hydrastin. The latter is first an excitant and then a paralyzant of the spinal cord, consequently toxic doses are followed first by tetanic convulsions and later by motor paralysis. Large doses primarily raise the blood-pressure by stimulating the vasomotor centers and thus constricting the arterioles ; secondarily, they lower the blood-pressure by depressing the vasomotor centers and the heart. Hydrastin probably increases peristalsis and induces uterine contractions, although its power to do the latter has been denied by some authorities. Finally, the drug, when locally applied, seems to exert a favorable influence on chronic catarrhal inflammations, but the manner of its action is not understood.

Hydrastinin in very large doses is a powerful paralyzant to the motor centers of the brain and spinal cord. The paralysis is apparently not preceded, as in the case of hydrastin, by a stage of motor excitation. Like hydrastin, it raises the arterial pressure, but its action in this respect differs from that of hydrastin in being more pronounced and more lasting and in not being followed secondarily by circulatory depression. The rise in the arterial pressure is due in part to stimulation of the vasomotor centers, and in part, also, according to some authors, to stimulation of the heart and of the muscular coats of the arterioles. Whether or not hydrastinin can induce uterine contractions is a moot question, but the trend of experimental evidence favors the view that it does possess ecbolic properties and is capable, in sufficient dose, of causing abortion.

Therapeutics.—Hydrastis is a useful local remedy in certain catarrhal affections. In *leukorrhea*, injections of the fluid extract, 2 fluidrams (8.0 c.c.) to the pint (0.5 L.) of hot water, are often beneficial. In the catarrhal stage of *gonorrhea* injections of hydrastin hydrochlorate, 1–5 gr. (0.06–0.3 gm.) to the ounce (30.0 c.c.), have been used with excellent results. The same preparation, in doses of from $\frac{1}{12}$–$\frac{1}{6}$ gr. (0.005–0.01 gm.), has also been highly recommended as an internal remedy in *chronic gastric catarrh*, especially when there is muscular atony.

Hydrastinin is chiefly used as a uterine hemostatic. It is particularly efficacious in *menorrhagia* or *metrorrhagia* dependent upon constitutional disorders, endometritis, or disease of the adnexa.

COTARNINÆ HYDROCHLORAS.
(Stypticin.)

Cotarnin is an artificial alkaloid closely allied to hydrastinin in its chemical composition, and obtained by the oxidation of the opium alkaloid, narcotin. It appears in the form of yellow crystals of a bitter taste, and freely soluble in water. The dose is from $\frac{1}{2}$–3 gr. (0.3–0.2 gm.), three or four times a day. It may be administered by the mouth or subcutaneously. Its action resembles that of hydrastinin, but is less rapid. It has been found useful as a hemostatic in *uterine hemorrhages* not accompanied by gross lesion of the uterine mucous membrane.

GELATINUM.
(Gelatin.)

Gelatin is the air-dried product of the action of boiling water on gelatinous animal tissues, such as skin, tendons, ligaments, and bones. It occurs in brittle, transparent sheets or shreds, without odor or taste. It is insoluble in cold water, but freely so in hot water, and if the solution contain more than 2 per cent., it solidifies on cooling. Solutions heated above a temperature of 230° F. (110° C.), however, remain permanently liquid.

Physiologic Action.—In 1896 Dastre and Floresco demonstrated that blood drawn from a dog into which a 5 per cent. solution of gelatin had been injected intravenously solidified almost immediately, and further that the same result could be secured by adding gelatin to the blood outside of the body. That the solidification was due to clotting and not merely to jellifying was evident from the fact that it occurred at a temperature (38° C.) at which a 5 per cent. solution of gelatin will not jellify, and, moreover, that it occurred also with solutions containing a percentage of gelatin much under that which is necessary for jellying—that is, under 2 per cent. These observers also showed that gelatin could offset the anticoagulant action of peptones, but not of concentrated saline solutions or of substances which decalcify the blood, like oxalic acid. In 1898 Lancereaux and Paulesco found that subcutaneous injections of gelatin were quite as efficient in increasing the coagulability of the blood as intravenous injections. Whether or not its power is affected by digestion has not been definitely determined, although a number of practitioners have attributed good results in hemorrhage to its administration by the mouth or rectum. No satisfactory explanation has yet been offered of the action of gelatin in promoting the coagulability of the blood.

Therapeutics.—Carnot appears to have been the first to bring gelatin prominently forward as a hemostatic; although, according to Miwa, it has been used for this purpose for many centuries in Oriental countries. Considerable testimony has already accumulated as to its efficacy in hemorrhages from both external and internal parts. As a local remedy it has been used with success in *epistaxis, metrorrhagia, bleeding from hemorrhoids,* and in *oozing from wounds.* As an internal hemostatic it has apparently rendered service in *hemoptysis, hematuria,* and *enterorrhagia.* It has been found efficacious also in *hemophilia, purpura hæmorrhagica,* and *melæna neonatorum.* For bleeding in accessible parts a 10 per cent. solution may be applied on tampons. The solution should be sterilized and should have added to it a small quantity of carbolic acid to prevent decomposition. The dose for subcutaneous administration is from 15–45 gr. (1.0–3.0 gm.), in the form of a 5 or 10 per cent. solution, once or twice a day. The best vehicle is normal salt solution. A 10 per cent. solution may be prepared by dissolving 1½ ounces (50.0 gm.) of gelatin and 35 gr. (2.5 gm.) of salt in a pint (0.5 L.) of hot distilled water. After the mixture has been allowed to cool, the white of an egg should be stirred in, and the whole brought quickly to the boiling-point. The mixture should next be filtered first through gauze and then through paper. Finally, it should be poured into small flasks and sterilized in a steam sterilizer, fifteen or twenty minutes, once daily, for three days. Prepared in this way the mixture will keep for several weeks. When required, it should be liquefied by standing the flask in hot water, and then injected to the amount of from 3 to 5 fluidrams (10.0–20.0 c.c.) into the thigh or under the breast, using for the purpose a perfectly sterile syringe. When the fluid is free from turbidity, the injections are not painful. It is exceedingly important that the solution should be absolutely sterile, since commercial gelatin not infrequently contains tetanus bacilli. Several cases of tetanus developing after gelatin injections have recently been reported. Acute nephritis is regarded by some authorities as a contraindication to the employment of gelatin subcutaneously.

When administered by the mouth, it may be given in doses of from 1–8 dr. (4.0–30.0 gm.) or more daily.

The treatment of *aortic aneurysms* by subcutaneous injections of gelatin, first recommended by Lancereaux, although it seemed promising, has not proved very satisfactory. It is followed, however, in a certain proportion of cases by a temporary amelioration of the symptoms. It is credited with having caused embolism in one or two instances.

Gelatin makes a good protective dressing for *indolent leg ulcers*, especially when they are accompanied by chronic eczema. It may be employed in the form of Unna's dressing:

> ℞ Zinci oxidi,
>   Gelatini,   āā ʒj (4.0 gm.);
>   Glycerini,
>   Aquæ,   āā f℥iv (120.0 c.c.). M.
> Sig. After the ulcer has been thoroughly cleansed and disinfected, coat the eczematous regions with Lassar's paste (starch and zinc oxid, of each, 2 parts; vaselin, 4 parts), dust the ulcer with iodoform, and cover it with cotton. Then paint the entire limb with Unna's dressing melted and applied with a brush. Bandage evenly and firmly, first with gauze soaked in hot water, then with dry gauze, and finally with a cotton roller bandage. The dressing should be changed every two, four, or six days, according to the amount of discharge.

Incompatibles.—Tannic acid and formaldehyd.

CALCII CHLORIDUM, U. S. P.

(Calcium Chlorid; CaCl$_2$.)

Calcium chlorid occurs in white, slightly translucent, hard fragments, very deliquescent, odorless, and of a sharp saline taste. It is soluble in 1.5 parts of water and in 8 parts of alcohol. The dose is from 5–20 gr. (0.3–1.3 gm.) three times a day.

Physiologic Action and Therapeutics.—In 1893 Wright, of Netley, England, first called attention to the action of calcium chlorid in increasing the coagulability of the blood. According to Dastre and Floresco, it restores coagulability that has been destroyed by decalcifying agents, like potassium oxalate, but that unlike gelatin it does not antagonize the anti-coagulant action of peptone. It has been used with some success as a hemostatic in *purpura, hemophilia,* and *small persistent hemorrhages.* Wright has found that the prolonged administration of the drug actually decreases the coagulability of the blood, and that the best results are obtained by giving it in doses of from 3–10 gr. (0.2–0.6 gm.) thrice daily, for a period of two or three days, and then omitting it for a like period. As it is decidedly irritating to the stomach, it should be given after meals, well diluted. It has been recommended also in *urticaria* on the supposition that this disease depends upon diminished coagulability of the blood, but the results have been disappointing. Savill states that he has used it successfully in a large number of cases of *pruritus* from various causes.

GERMICIDES, ANTISEPTICS, AND DEODORIZERS.

Germicides or **disinfectants** are agents that destroy micro-organisms and their spores. They may act by exerting some specific influence on the micro-organisms, by coagulating their albuminous constituents, or by entering into chemical combination with the microprotein. The chief germicides are:

Mercuric chlorid.

Phenol (carbolic acid),
Cresols,
Creosol,
Guaiacol,
Resorcin,
Naphthalin,
Naphthol,
Salicylic acid,
Benzoic acid,

AROMATIC SERIES.

Benzene derivatives or coal-tar germicides.

Formaldehyd.	Sulphurous acid.
Urotropin.	Hydrogen dioxid.
Silver compounds.	Potassium permanganate.
Chlorin.	Lime.
Bromin.	

The various preparations of *quinin* also possess a moderate amount of germicidal power; they are especially destructive to certain protozoa, such as the hematozoa of malaria and the amœbæ of dysentery.

Heat is the most certain disinfectant. All known pathogenic bacteria and their spores are destroyed by boiling water in from five to ten minutes, and by steam (212° F.—100°C.), at ordinary pressure, within an hour. Steam under pressure is a still more energetic disinfectant. Dry heat, on account of slight penetrating power, is much less effective than moist heat. According to Koch, sporeless bacteria are destroyed in one and one-half hours by hot air at a temperature slightly above 212° F. (100° C.), and spores of bacilli in three hours by hot air at a temperature of 316° F. (140° C.).

Antiseptics are agents that prevent or hinder the growth of micro-organisms without necessarily destroying them. Whether a substance act as a germicide or as an antiseptic depends largely upon the degree of concentration in which it is employed. By dilution germicides may be reduced to the rank of antiseptics. The following substances may be classed as antiseptics, since in the solutions ordinarily used they arrest the development of bacteria, but do not destroy them:

Boric acid.
Picric acid.
Ferrous sulphate.

Many volatile oils.
- Eucalyptus.
- Thyme.
- Sandalwood.
- Copaiba.
- Cubeb.

Many anilin derivatives.
- Methyl-blue.
- Methylene-blue.
- Acetanilid.

Iodoform (see p. 312), although it is practically without antiseptic power, nevertheless makes a valuable dressing for infected wounds. It probably acts indirectly by absorbing moisture, by liberating iodin, which is antiseptic, and, perhaps, by neutralizing the toxins formed by pyogenic bacteria.

Deodorizers are agents that destroy offensive odors. They do not, of necessity, possess either germicidal or antiseptic power. They may act by oxidizing or deoxidizing fetid compounds or by abstracting hydrogen from them. The most powerful deodorizers are:

Chlorin.	Sulphurous acid.
Formaldehyd.	Potassium permanganate.
Lime.	Hydrogen dioxid.
	Charcoal.

Pulverized dry earth is a useful deodorant of fecal matter, but it is not to be considered as a disinfectant.

General Surgical Antisepsis.—Mercuric chlorid (see p. 320) is the most popular of the germicides for general surgical work. It is energetic, soluble, and cheap. On the other hand, it has three disadvantages: it is very poisonous, it is destructive to metal instruments, and it is readily converted into an inert compound in the presence of albuminous matters. This last drawback may be offset by adding to its solutions a weak acid (tartaric acid). For the patient's skin and the surgeon's hands solutions of from 1:1000 to 1:500 should be employed; for large wounds and cavities, 1:10,000 to 1:5000; and for small wounds, 1:2000. On account of its irritant properties it should not be used on serous membranes.

Carbolic acid is not nearly so reliable as corrosive sublimate, but it is quite destructive to pus-organisms. It is readily soluble, is not affected by albuminous matters, and does not seriously injure metal instruments, although it dulls them. On the other hand, it is decidedly toxic, and it has a benumbing effect on the hands of the operator. A solution of 1:20 is often employed for cleansing suppurating wounds, sinuses, and abscess cavities. A solution of 1:40 is sometimes used as a bath for instruments.

Formaldehyd is a very powerful germicide, but it is too irritating to be generally useful in surgical work. One or 2 per cent. solutions of formalin (40 per cent. solution of the gas in water), however, are sometimes employed to disinfect wounds and to irrigate sinuses and suppurating cavities. Instruments may be disinfected by immersion in a 2 per cent. solution of formalin, or, better, by subjecting them to the action of the gas in a closed chamber (see p. 399).

Disinfection of the Hands.—Fürbringer's method, or some modification of it, is usually followed: after removing all dirt from around and beneath the nails, the forearms, hands, and nails are thoroughly scrubbed with soap and hot water. The hands are then soaked for at least a minute in alcohol (95 per cent.), and before the alcohol has evaporated they are plunged in a hot solution of corrosive sublimate, 1 : 500. Finally, they are rinsed in sterile water and dried. The alcohol removes the soap and grease from the furrows and pores, and favors the penetration of the germicide. Kelly, after scrubbing the hands, dips them in a warm, saturated solution of potassium permanganate, then in a warm, saturated solution of oxalic acid, and, finally, in distilled water. Weir's method has been found very satisfactory. It is applied as follows: Scrub the hands and forearms in running hot water, using a brush and green soap; clean under and around the nails with a bit of soft wood; place in the palm a tablespoonful of chlorinated lime and an equal quantity of washing soda (not sodium bicarbonate); add enough sterile water to make a creamy mass, and rub this cream over the hands and forearms until the harsh soda granules can no longer be felt. This rubbing should occupy about five minutes. Push the paste under and around the nails by means of a piece of sterile orange-wood, and then wash in hot, sterile water.

Antisepsis of Mucous Membranes.—The most important bacterial poisons for use on mucous membranes are mercuric chlorid, silver nitrate, organic silver compounds, boric acid, potassium permanganate, and formalin.

Gastro-intestinal Antisepsis.—Internally, antiseptics are of service in controlling fermentation and putrefaction in the contents of the stomach and bowel. They cannot affect, of course, bacteria like the typhoid bacillus and the tubercle bacillus, which multiply in the intestinal walls. The chief gastro-intestinal antiseptics are carbolic acid, creasote, guaiacol, naphthalin derivatives, salicylic compounds, sulphites, resorcin, and thymol. Silver nitrate and the mercurial preparations also owe, no doubt, some of their efficacy in affections of the digestive tract to their antiseptic properties.

Urinary Antiseptics.—It is quite possible to inhibit the growth of micro-organisms in the urine by administering antiseptic agents by the mouth. Urotropin, salicylic compounds, benzoic acid, and several volatile oils (copaiba, cubeb, sandalwood) are used for this purpose.

Disinfection of Stools and Sputum.—Stools should be thoroughly mixed with twice their volume of a 1 per cent. chlorinated lime solution, of a 5 per cent. carbolic acid solution, or of a 2 per cent. formalin solution, allowed to stand for two or three hours, and then buried or discharged into the closet.

Infected sputum should be received into cups containing a 1 per cent. chlorinated lime solution or a 5 per cent. carbolic acid solution; or the patient may expectorate into moist rags or into impermeable paste-board cups, which should be burnt before the sputum has had time to dry.

Disinfection of Rooms and their Contents.—Articles of little value should be burnt. Bedding, clothing, carpets, etc., should be disinfected by steam. Towels, napkins, and sheeting should be soaked in a 5 per cent. carbolic acid solution and then boiled. Woodwork, floors, and plain furniture should be washed with a chlorinated lime solution (1 per cent.) or with a corrosive sublimate solution (1 : 5000). Finally, the room should be fumigated with formaldehyd gas (see p. 398) or with sulphur dioxid (see p. 402), preferably with the former.

ACIDUM CARBOLICUM, U. S. P.

(Carbolic Acid; Phenol; C_6H_5OH.)

Carbolic acid is obtained from the distillation of coal-tar, and when pure occurs in the form of colorless, needle-shaped crystals of a characteristic odor and of an acrid, burning taste. In the light it acquires a reddish tint, and on exposure to the air it deliquesces. It is soluble in 15 parts of water, and freely in alcohol, glycerin, chloroform, ether, and oils. Although it combines with salifiable bases, it is chemically not an acid, but an alcohol of the benzene group. The dose is from $\frac{1}{2}$–2 gr. (0.03–0.13 gm.).

Crude carbolic acid (*Acidum Carbolicum Crudum*, U. S. P.) is a nearly colorless or reddish-brown liquid, of a somewhat uncertain composition, but consisting chiefly of cresol and phenol. It is employed only as a disinfectant.

PREPARATIONS.

Unguentum Acidi Carbolici, U. S. P. (5 per cent.).
Glyceritum Acidi Carbolici, U. S. P. (20 per cent.).

Physiologic Action.—When applied to the skin, carbolic acid blanches the surface and causes a burning sensation, which is soon followed by numbness. Later the part becomes red and then brown, and the epidermis desquamates. On account of the albuminous coagulum that it forms its caustic action does not extend very deeply into the tissues. Its prolonged application, however, even in dilute solution (5 per cent.), has been followed by gangrene in a number of instances. Absorption readily occurs from cutaneous surfaces, mucous membranes, and wounds.

No appreciable effects follow the internal administration of carbolic acid unless a toxic dose is given.

Circulatory System.—In poisoning there is a marked fall in the blood-pressure from depression of the heart and the vasomotor centers.

Nervous System.—Toxic doses are quickly followed by stupor and coma, the result of cerebral depression. In man, symptoms referable to the action of the drug on the spinal cord are rarely noted, but in the lower animals tetanic seizures often develop and are followed by paralysis, indicating that the spinal centers are at first excited and then depressed. Locally, the drug acts as an anesthetic by paralyzing the peripheral sensory nerve filaments.

Respiratory System.—The respirations are first quickened and then slowed, in consequence, probably, of the primary stimulation and the secondary paralysis of the respiratory center. In fatal poisoning death usually results from asphyxia.

Temperature.—Large doses lower the bodily temperature, probably by decreasing heat-production and increasing heat-dissipation.

Elimination.—Carbolic acid is eliminated by all the emunctories, but chiefly by the kidneys. In the tissues a part combines with sulphuric and glycuronic acids, and is eliminated as the double sulphate and glycuronate of phenol; a part is oxidized into hydroquinon and pyrocatechin, which leave the body also largely as double sulphates and glycuronates. When the amount of phenol ingested has been very large, a part is excreted unchanged. After large doses the urine acquires a smoky or greenish-black color, which is due, probably, to oxidation products of hydroquinon and pyrocatechin. In poisoning ischuria and albuminuria are common symptoms.

Action on Lower Organisms.—Carbolic acid is fatal to most pathogenic bacteria, but it is not very effective against spores. A solution of 1 : 1000 prevents the development of most bacteria; a solution of 1 : 100, under favorable conditions, destroys pyogenic cocci, tubercle bacilli, cholera bacilli, and many other non-spore-forming bacteria. Anthrax bacilli in the spore stage resist the action of a 5 : 100 solution for many days. Alcohol and ether as solvents lessen its germicidal power, and oils practically destroy it. Heat increases its destructive action.

Toxicology.—Poisoning may result either from the ingestion of the drug or from its external application. After the ingestion of a very large dose (1 oz.—30.0 c.c.) the usual symptoms are unconsciousness, contraction of the pupils, stertorous breathing, a rapid, feeble pulse, and collapse. If the dose has not been so large, these symptoms may be preceded by those of gastro-enteritis. The characteristic phenomena are the odor on the breath, the white, corrugated patches on the buccal mucous membrane, and the smoky urine. In poisoning from the use of the acid externally the initial symptoms generally are discoloration of the urine, headache, vertigo, pallor, and muscular weakness.

Treatment.—Two drugs are of special value when the poison has been taken by the mouth: Alcohol, which neutralizes its caustic action, and a soluble sulphate (Epsom salt), which forms with it an innocuous sulphocarbolate. Two or three ounces (60.0–90.0 c.c.) of diluted alcohol should be poured into the stomach through a tube, and then lavage practised with water containing Epsom salt. Demulcents are useful in allaying the irritation. The application of external heat and the subcutaneous administration of diffusible stimulants are also indicated.

Therapeutics.—As an antiseptic for general surgical purposes carbolic acid no longer holds a prominent place. Solutions that are really effective are too irritant and too toxic to be employed. It is rarely used at present even for disinfecting instruments, because it dulls them and at the same time benumbs the surgeon's fingers. Its prolonged application in the form of a moist dressing, even in weak solution, is dangerous. Harrington has collected 132 cases of gangrene of the fingers or toes from its external application. In most of these cases the strength of solution was less than 5 per cent. A solution of 1 : 20 is still employed to some extent for cleansing *suppurating wounds* and *abscess cavities*. *Carbuncles* are sometimes aborted by injecting into them in the early stage from 5–10 min. (0.3–0.6 c.c.) of a 10 per cent. solution. Excellent results have been obtained in *anthrax* from the injection of from $\frac{1}{2}$–1 dr. (2.0–4.0 gm.) of the pure acid into and around the eschar in the course of a day. Bacelli speaks very favorably of carbolic acid injections in *tetanus*. He recommends that from $\frac{1}{3}$–$\frac{2}{3}$ gr. (0.02–0.04 gm.) should be given hypodermically every two or three hours until a daily amount of from 8–15 gr. (0.5–1.0 gm.) has been reached. Pure carbolic acid is sometimes employed to destroy *chancroids* and *venereal warts* and to purify *sloughing wounds*.

A 5 per cent. solution may be used for disinfecting *soiled clothing* and *various excretions*, like sputum, feces, and vomited matters. In the form of a 2 per cent. spray it may be employed as an inhalation to destroy the fetor of the breath in such affections as *bronchiectasis* and *gangrene of the lung*.

25

Carbolic is very useful as a local sedative and antipruritic. Carbolized oil, 10 gr. (0.6 gm.) to the ounce (30.0 c.c.), although it has no antiseptic power, makes a soothing dressing for *superficial burns.* In the form of a lotion it is the most valuable remedy we possess to allay itching in *eczema, urticaria, jaundice,* and *pruritus.* It may be employed in the strength of from 2–3 dr. (8.0–12.0 gm.) to the pint (0.5 L.), as in the following formula :

<pre>
R   Acidi carbolici,   ℥ij–iij (8.0–12.0 gm.) ;
    Acidi borici,   ℥iv (15.0 gm.) ;
    Alcoholis,   f ℥j (30.0 c.c.) ;
    Glycerini,   f ℥ss (15.0 c.c.) ;
    Aquæ,   q. s. ad Oj (0.5 L.).   M.
</pre>

Internally, carbolic acid is used as an antiemetic and as an antiseptic. Doses of from 1–2 min. (0.06–0.12 c.c.) are sometimes efficacious in checking *vomiting* the result of gastric irritability. In the same doses, combined with bismuth, it is of service as an antiseptic in *flatulent dyspepsia* and in *acute diarrhea.*

Administration.—It may be prescribed in pills, in capsules, in powders of bismuth subnitrate, or dissolved in some aromatic water.

Incompatibles.—Alkalis, metallic salts, soluble sulphates, collodion.

Sodium Sulphocarbolate (*Sodii Sulphocarbolas,* U. S. P.). —This salt occurs in colorless, rhombic crystals, odorless, of a saline taste, and freely soluble in water. The dose is from 5–20 gr. (0.3–1.3 gm.). It is much less poisonous than carbolic acid. It is sometimes of service as an antiseptic in *flatulent dyspepsia* and in *diarrhea.* When intended to act in the stomach, it may be prescribed in solution; when intended to act in the intestine, in keratin-coated pills.

Zinc Sulphocarbolate (*Zinci Sulphocarbolas*).—This salt is somewhat more astringent than sodium sulphocarbolate. A solution of 5 gr. (0.3 gm.) to the ounce (30.0 c.c.) is sometimes used as a spray in *catarrhal affections of the throat* and as an injection in the second stage of *gonorrhea.* Internally it often acts favorably in *diarrhea.* As an intestinal antiseptic the dose is from 2–3 gr. (0.1–0.2 gm.) in keratin-coated pills.

Sozoiodol (*Sozoiodolic Acid*).—This compound is paraphenol-sulphonic acid with 2 atoms of hydrogen displaced by 2 of iodin. Paraphenol-sulphonic acid is prepared by acting on phenol with sulphuric acid in the presence of heat. Sozoiodol occurs in needle shaped crystals, soluble in water and in alcohol. It has been used as a substitute for iodoform. Its salts—mercury, potassium, sodium, and zinc—have also been employed to some extent as antiseptics.

Nosophen (*Tetra-iodo-phenol-phthalein*).—This compound is prepared by the action of iodin on a solution of phenolphthalein. It is a pale, yellow, inodorous, and tasteless powder. With bases it forms salts, the most important of which is the sodium salt (*antinosin*). It is recommended as an antiseptic dusting-powder.

Xeroform (*Tribromphenol-bismuth*).—This is a yellowish, odorless, tasteless, and insoluble powder, representing about 50 per cent. of bismuth oxid. It has been used in place of iodoform as an antiseptic application for infected wounds, burns, etc.

CRESOLS, CREOSOL, AND GUAIACOL.

Cresols, of which there are three,—metacresol, orthocresol, and paracresol,—are homologues of carbolic acid. They are obtained from coal-tar by fractional distillation, and are present in large quantity in crude carbolic acid. They closely resemble carbolic acid in their action, and, like the latter, they possess distinct germicidal properties. Their slight solubility in water is an important disadvantage. They are used chiefly for disinfecting specific excreta.

Tricresol is a refined mixture of the three cresols. It is a colorless liquid, of a creasote-like odor, and soluble in water to the extent of 2.5 per cent. Its germicidal power is nearly three times greater than that of carbolic acid. A 1 per cent. solution may be employed in surgical work for the same purposes as carbolic acid. De Schweinitz recommends a 1 : 1000 solution as a solvent for atropin, cocain, and eserin in ophthalmic practice. Such solutions remain free from bacteria and are not irritant to the eye.

Creolin.—This is an emulsion of cresols prepared by means of resin soap. It is a brownish, syrupy liquid, which, diluted with water, forms an opaque mixture. Aqueous mixtures containing from 3 to 5 per cent. of creolin are employed for disinfecting excreta. Mixtures of from 1 to 2 per cent. have been used rather extensively in obstetric practice. Injections containing from 0.5 to 1 per cent. have been efficacious in chronic cystitis. It has been recommended as a disinfectant for instruments, but its opacity in aqueous mixtures is a serious disadvantage.

Lysol is a preparation of cresols obtained by dissolving in fat, and subsequently saponifying with alcohol, that portion of tar-oil which boils between 374° and 392° F. (190° and 200° C.). It is a brown, oily liquid, with the odor of creasote. It contains about 50 per cent. of cresols, and mixes with water to form a clear, saponaceous, frothy liquid. It may be used for the same purposes as creolin.

While the cresol preparations are comparatively safe, they are not altogether harmless. Poisoning from their absorption has not been very rare. There are at least 20 cases of poisoning by lysol on record.

Creosol and **guaiacol** are the chief constituents of creasote (see pp. 276 and 279).

THYMOL, U. S. P.

(Thymic Acid; $C_{10}H_{14}O$.)

Thymol is a homologue of phenol obtained from the oil of thyme (*Thymus vulgaris*) and certain other volatile oils. It appears in the form of large, colorless crystals having the odor of thyme and an aromatic pungent taste. It is only sparingly soluble in water, but freely so in alcohol, ether, and oils. The dose is from 1–10 gr. (0.06–0.6 gm.).

Physiologic Action and Therapeutics.—Thymol resembles carbolic acid in its action, although it is decidedly less toxic and less irritant. It has been used as an antiseptic in the dressing of *wounds*, but its aromatic odor soon becomes disagreeable to the patient, and, moreover, is apt to attract flies. A solution of a grain (0.06 gm.) to the ounce of water (30.0 c.c.), a little alcohol being used as a solvent, makes a good mouth-wash in the infectious forms of *stomatitis* and *pharyngitis*. An ointment of thymol, 5–30 gr. (0.3–2.0 gm.) to the ounce (30.0 gm.), has been used with some success by Crocker and others in chronic *eczema* and *psoriasis*. The drug makes an excellent lotion in *senile pruritus*.

Internally, thymol is sometimes used as an antiseptic in *subacute diarrhea*. In large doses (10–20 gr.—0.6–1.3 gm.) it has been used to some extent as a teniacide against *tape-worm* and *anchylostomum duodenale*. It may be administered in wafers, capsules, or pills.

Aristol.—This compound, which is chemically dithymoldi-iodid, is prepared by acting upon thymol in alkaline solution with iodin dissolved in potassium iodid. It is an unstable preparation, containing about 45 per cent. of iodin, and appears as a brownish-red powder, tasteless and almost odorless. It is employed as a substitute for iodoform. Although it has advantages over the latter in being odorless and less toxic, it is generally not so efficacious.

RESORCINUM, U. S. P.

(Resorcin; Metadioxybenzol; $C_6H_4(OH)_2$.)

Resorcin is a diatomic phenol, occurring in the form of colorless or faintly reddish prisms or needles having a slightly urinous odor and a sweetish, pungent taste. It is readily soluble in water, alcohol, ether, or glycerin. The dose is from 1–5 gr. (0.06–0.3 gm.).

Physiologic Action and Therapeutics.—Resorcin acts much like carbolic acid, but it is less poisonous and less irritant. Toxic doses produce vertigo, ringing in the ears, tremors, epileptiform convulsions, quickening of the pulse and respiration, unconsciousness, and collapse. The urine becomes of an olive-green color. In health it exerts but little effect on the temperature, but in the febrile state it causes a marked fall, which is accompanied by free perspiration.

At the present time resorcin is chiefly employed externally except in certain diseases of the skin. It is a valuable remedy in *seborrhea*, especially in dandruff, in which disease it may be employed in the form of a 3 to 5 per cent. lotion, as in the following formula:

℞	Resorcini,	℥iss (6.0 gm.);
	Olei ricini,	ℳxx--xxx (1.2–2.0 c.c.);
	Spiritus myrciæ,	q. s. ad f℥iv (120.0 c.c.). M.

A lotion containing from 5 to 20 gr. (0.3–1.3 gm.) to the ounce (30.0 c.c.) is often serviceable in *subacute* and *chronic eczema*, especially of the vesicular variety. It is also a useful antipruritic remedy. According to Hartzell, its power to relieve itching is increased considerably by the addition of 0.5 per cent. of sodium chlorid to the solution, as in the following formula:

℞	Resorcini,	gr. xv–xxx (1.0–2.0 gm.);
	Sodii chloridi,	gr. xv (1.0 gm.);
	Glycerini,	f℥ij (8.0 c.c.);
	Liquoris calcis,	q. s. ad f℥iv (120.0 c.c.). M.

Internally, resorcin is occasionally used as an antiseptic in *chronic diseases of the stomach* and in *diarrhea*. It should be given in pill or capsule.

ACIDUM PICRICUM.

(Picric Acid; Carbazotic Acid; Trinitrophenol; $C_6H_2(NO_2)_3$. OH.)

Picric acid is the product of the action of nitric acid on phenol-sulphonic acid, which is obtained by dissolving crystallized carbolic acid in sulphuric acid. It occurs in yellow, flat crystals, odorless, and of an intensely bitter taste. It is sparingly soluble in water, but freely so in alcohol and in ether.

Physiologic Action and Therapeutics.—Locally, picric acid in pure form acts as a caustic. Poisonous doses produce vomiting, diarrhea, yellowness of the skin, mucous membranes, and urine, convulsions, and collapse. It is a feeble bactericide.

Gauze wet with a 1 per cent. solution of picric acid, covered with cotton, makes an excellent dressing for *burns* of the first and second degree. For fear of poisoning, however, the drug should not be applied over large or deep burns. A 1 per cent. lotion has also been found serviceable in *acute eczema* and in *herpes zoster*. Occasionally local applications are followed by an erythematous rash. Injections of picric acid, 1 : 200 to 1 : 100, are recommended by some surgeons in the treatment of the second stage of *gonorrhea*. A saturated solution makes a delicate test for albumin in the urine.

Internally, ammonium picrate was used for a time as an antimalarial and anthelmintic, but it soon proved unsuitable for either purpose.

NAPHTALINUM, U. S. P.

(Naphtalin ; Naphtalen ; Tar Camphor; $C_{10}H_8$.)

Naphtalin is a hydrocarbon obtained by distilling coal-tar between 180°–250° C. (356°–482° F.). It occurs in shiny, white scales, having a strong odor of coal-tar and a burning taste. It is insoluble in water, soluble in 15 parts of alcohol, and in all proportions in ether and chloroform. The dose is from 2–10 gr. (0.13–0.6 gm.).

Therapeutics.—Naphtalin was at one time used as an intestinal antiseptic in *diarrhea*, but it has been almost completely displaced by naphtol.

NAPHTOL, U. S. P.

(Beta-naphtol; $C_{10}H_7OH$.)

Naphtol is a phenol occurring in coal-tar, but usually prepared artificially from naphtalin. It occurs in white, shining scales or as a yellowish-white crystalline powder, having a faint, phenol-like odor and a pungent taste. It is soluble in about 1000 parts of water, and freely soluble in alcohol, ether, and chloroform. The dose is from 2–10 gr. (0.13–0.6 gm.).

Physiologic Action and Therapeutics.—Naphtol resembles carbolic acid in its action, but it is less poisonous. Large doses, however, not infrequently give rise to considerable irritation of the genito-urinary tract. It is eliminated in the urine in combination with glycuronic acid. Locally, in concentrated form, it is irritating to mucous membranes and raw surfaces. Absorption readily follows its local application. Its germicidal power is much greater than that of carbolic acid.

Internally, naphtol is useful as an antiseptic in *chronic diseases of the stomach* and in *diarrhea.* Bouchard and others have recommended it also in *typhoid fever ;* the course of the disease, however, is not affected in the least by it, although it is sometimes of service in lessening flatulence and diarrhea. It is best administered in capsules.

Externally, it is efficacious as a parasiticide and stimulant application in certain diseases of the skin. In *scabies* and in *ringworm of the scalp or body* it may be employed as an ointment in the strength of a dram (4.0 gm.) to the ounce (30.0 gm.). Ointments containing from ½–1 dr. (2.0–4.0 gm.) to the ounce (30.0 gm.) have been used also with some benefit in psoriasis, but, as a rule, the drug is less useful in this disease than either chrysarobin or tar.

Benzonaphtol.—This compound is prepared by acting on beta-naphtol with benzoyl chlorid. It is said to escape the action of the gastric juice, and to be split up into its components in the intestine. In doses of from 5–10 gr. (0.3–0.6 gm.) it is a good intestinal antiseptic, having at least one advantage over beta-naphtol in being tasteless.

Betanaphtol-bismuth (*Orphol*).—This preparation is a combination of bismuth oxid and beta-naphtol. It is a light-brown, insoluble powder, odorless and tasteless. As its decomposition is mainly effected after it leaves the stomach, it is found efficacious in the various forms of intestinal catarrh. The dose is from 10–30 gr. (0.6–2.0 gm.).

ACIDUM SALICYLICUM, U. S. P.

(Salicylic Acid; $HC_7H_5O_3$.)

Salicylic acid is an organic acid contained in wintergreen, birch, and various other plants, but chiefly prepared artificially by acting on phenol with caustic soda and carbon dioxid. It occurs in fine white needles or as a crystalline powder, odorless, and of a sweetish, acrid taste. It is soluble in 450 parts of water, 2.4 parts of alcohol, 2 parts of ether, and 60 parts of glycerin. The dose is from 5–20 gr. (0.3–1.3 gm.).

Physiologic Action.—Single, moderate doses of salicylic acid are not followed, as a rule, by any appreciable symptoms. Large doses, or even moderate doses in susceptible subjects, however, give rise to a feeling of fulness in the head and ringing in the ears, and, perhaps, to deafness, dimness of vision, flushing of the surface, and perspiration. Toxic doses, in addition to the above symptoms, produce rapid, deep breathing, followed by shallow respiration, subnormal temperature, a slow, feeble pulse, olive-green urine, albuminuria, delirium, ocular palsies, unconsciousness, paralysis of the sphincters, and collapse. Death usually results from asphyxia. In some cases the delirium has been associated with delusions and hallucinations; in others, there has been an erythematous, urticarial, or ecchymotic eruption.

Circulatory System.—Moderate doses sometimes raise the blood-pressure, probably by stimulating the vasomotor centers. Large doses lower the blood-pressure by directly depressing the heart. Its use generally results in a slight leukocytosis. In poisoning the red cells suffer disintegration.

Nervous System.—Salicylic acid has comparatively little influence on the nervous system. The fulness in the head and the disorders of hearing are probably the result of some disturbance of the circulation; the amblyopia has been attributed to the same cause, but there is a suspicion that it is produced by changes in the ganglion-cells of the macula.

Respiratory System.—Toxic doses appear first to stimulate and then to paralyze the respiratory center.

Alimentary Canal.—Its presence in the alimentary canal, even in small quantities, tends to retard the action of the digestive ferments. It is a distinct irritant to the mucous membrane of the stomach, and, in consequence, nausea and vomiting are frequently induced by its administration. It appears to have some power to increase the secretion of bile.

Metabolism.—There is considerable testimony to show that salicylic acid increases the output of urea, uric acid, and the sulphur compounds, but how this is brought about has not been determined.

Temperature.—In health salicylic compounds do not exert much influence on the bodily temperature; in fever, however, they have a marked antipyretic effect, which, according to Hare, probably results from an increase in the heat-dissipation and a decrease in the heat-production.

Absorption and Elimination.—Salicylic acid enters the blood as an alkaline salicylate. It is eliminated from the body in most of the secretions, but chiefly in the urine, in which it appears for the most part as salicyluric acid, a compound of salicylic acid and glycocoll, although some of the drug is excreted unchanged. The greenish discoloration of the urine is said to be due to the formation of indican or of pyrocatechin. Its excretion is attended by some increase in the flow of urine, owing, no doubt, to stimulation of the kidneys. Large doses may excite nephritis.

Action on Micro-organisms.—Its germicidal properties are no less pronounced than those of carbolic acid.

Therapeutics.—The chief use of salicylic acid and its allied compounds is in *acute rheumatism*, on which disease they exert a specific influence, mitigating the pain, lowering the temperature, lessening the liability to cardiac complications, and shortening the duration of the attack. The manner in which these drugs produce their favorable influence is not understood, but it is possible that they may act by neutralizing the toxins of the infective agent. From 10 to 15 gr. (0.6–1.0 gm.) of the acid or of one of its salts should be given every two or three hours until tinnitus develops or the articular symptoms are favorably affected, when the interval between the doses may be gradually increased. Salicylic compounds are also efficacious in certain affections which seem to be in some way related to rheumatism, such as *tonsillitis, lumbago, sciatica, erythema nodosum*, and *neuritis* due to cold. In chronic rheumatism they are of little value, and in rheumatoid arthritis and gout they are useless.

In *diabetes mellitus*, for which they were originally recommended by Ebstein, they are frequently of service in lessening the excretion of urine and in relieving pruritus.

Salicylic acid is no longer in use as a topical antiseptic, but some of its preparations are employed with advantage in checking *fermentation in the alimentary canal.*

Locally, the acid is an excellent remedy in certain diseases of the skin. The following dusting-powder often affords relief in *hyperidrosis*:

℞	Acidi salicylici,	gr. xxx (2.0 gm.);
	Acidi borici,	ℨj (4.0 gm.);
	Amyli,	℥ss (15.0 gm.). M.

Ointments or pastes containing from 20–40 gr. (1.3–2.6 gm.) to the ounce (30.0 gm.) are of service in many cases of *subacute* and *chronic eczema.* Dissolved in collodion, a dram (4.0 gm.) to the ounce (30.0 c.c.), it is employed with success in removing *corns* and *warts.*

Contraindications.—Salicylic compounds must be withheld or used with great care when there is renal irritation or active disease of the middle ear.

Administration.—Salicylic acid is more irritating to the stomach than its salts, and for this reason the latter are generally selected for internal use. The acid obtained from oil of wintergreen is preferable to the much cheaper synthetic acid, since the latter is not infrequently contaminated with poisonous by-products. Salicylic compounds should be given after meals, in some aromatic water, well diluted. When in rheumatism they are not tolerated in any form by the stomach, the acid may be applied directly to the joints, as in the following formula:

R   Acidi salicylici,   ℥iss (6.0 gm.);
    Olei terebinthinæ,   f ℥j (4.0 c.c.);
    Adipis lanæ hydrosis,   ℥ss (15.0 gm.);
    Adipis benzoinati,   q. s. ad ℥ij (60.0 gm.).   M.
   Sig.   Spread on muslin and keep in place by means of a flannel
          bandage.

SODII SALICYLAS, U. S. P.

(Sodium Salicylate; $NaC_7H_5O_3$.)

Sodium salicylate is a white amorphous powder, odorless, and of a sweetish, saline taste. It is soluble in 0.9 part of water and in 6 parts of alcohol. The dose is from 5–20 gr. (0.3–1.3 gm.).

Therapeutics.—It is the most commonly used of all the salts of salicylic acid. Its action is the same as that of the acid, but it has advantages in being more soluble in water and less irritant to the stomach. Like all salicylic compounds, however, it should be given after meals, well diluted.

LITHII SALICYLAS, U. S. P.

(Lithium Salicylate; $LiC_7H_5O_3$.)

Lithium salicylate is a white, deliquescent powder of a sweetish taste. It is freely soluble in water and in alcohol. It is somewhat less irritating to the stomach than sodium salicylate, for which it may be substituted. The dose is from 5–20 gr. (0.3–1.3 gm.).

AMMONII SALICYLAS.

(Ammonium Salicylate; $NH_4C_7H_5O_3$.)

Ammonium salicylate occurs in clear, colorless prisms, freely soluble in water. Wood recommends it as being more agreeable to the taste and less depressing and less nauseating than the sodium salt. The dose is from 5–20 gr. (0.3–1.3 gm.).

STRONTII SALICYLAS.

(Strontium Salicylate; $Sr(C_7H_5O_3)_2$.)

Strontium salicylate occurs in colorless, octahedral crystals, freely soluble in water. It is much less irritating to the stomach than the sodium salt.   The dose is from 5–20 gr. (0.3–1.3 gm.).

METHYL SALICYLAS, U. S. P.

(Methyl Salicylate; Artificial Oil of Wintergreen; $CH_3C_7H_5O_3$.)

Methyl salicylate is prepared synthetically by distilling methyl-alcohol with salicylic and sulphuric acids.   It is a colorless or slightly yellowish oily liquid, having the characteristic odor and taste of wintergreen.   It appears to be almost identical in its properties and actions with the natural oil of wintergreen (*Oleum Gaultheriæ*, U. S. P.) and the oil of sweet birch (*Oleum Betulæ Volatile*, U. S. P.), which are composed almost entirely of pure methyl salicylate.   The dose of either the artificial product or of the natural oils is from 5–20 min. (0.3–1.2 c.c.).

Therapeutics.—These preparations may be used internally in *rheumatism* interchangeably with salicylic acid and the mineral salicylates.   They are more likely, however, to disturb digestion than either ammonium or strontium salicylate, and, moreover, their strong flavor often turns the patients against them.   Externally, they serve an excellent purpose as local applications to the affected joints.   For internal use they may be given on sugar, in capsules, or in an emulsion.

BISMUTHI SALICYLAS.

(Bismuth Salicylate; $Bi(C_7H_5O_3)_3Bi_2O_3$.)

Bismuth salicylate is a white, odorless, tasteless, and insoluble powder, containing about 64 per cent. of bismuth trioxid. It is used almost entirely as an intestinal antiseptic and astringent in *diarrhea*.   The dose is from 5–20 gr. (0.3–1.3 gm.).

SALOL, U. S. P.

(Phenyl Salicylate; $C_6H_5C_7H_6O_3$.)

Salol is a white, crystalline powder, of a faintly aromatic odor, and almost tasteless.   It is nearly insoluble in water, but soluble in 10 parts of alcohol, 0.3 part of ether, and readily in oils.   The dose is from 5–20 gr. (0.3–1.3 gm.).

Physiologic Action and Therapeutics.—Salol resists to a great extent the action of the gastric juice, but is slowly broken up by the alkaline fluids of the intestine into salicylic acid (60 parts) and carbolic acid (40 parts).   After very large doses poisoning results from carbolic rather than from salicylic acid.

Like other salicylic compounds, it is useful in the various

manifestations of *rheumatism*, but in the more severe forms of the disease its action is less prompt and powerful than that of sodium or ammonium salicylate. As its dissociation does not occur until it has escaped from the stomach, it may be employed advantageously as an antiseptic in all forms of *intestinal catarrh* associated with abnormal fermentation. In *cystitis* and *urethritis* it is often combined with such drugs as oil of sandal-wood and balsam of copaiba for the purpose of sterilizing the urine.

It should always be used with considerable caution when there is evidence of active nephritis.

It may be prescribed in powders, capsules, pills, or in an emulsion. Tablets and hard pills may fail of solution and escape in the feces.

Salophen (*Acetyl-paramido-phenyl Salicylate*).—This is a compound ether of salicylic acid, occurring in tasteless and in-odorous white scales, insoluble in water, but soluble in alcohol and ether. It contains about 50 per cent. of salicylic acid. It was introduced as a substitute for salol, on account of the ill effects thought to result from the liberation of phenol from the latter. It is decomposed in the intestine into salicylic acid and the comparatively innocuous acetyl-paramido-phenyl. It is much less active in *rheumatism* than sodium salicylate, but being tasteless and unirritating to the stomach, it is a valuable remedy in the *milder manifestations of the disease.* The dose of salophen is from 5–30 gr. (0.3–2.0 gm.).

Betol differs from salol in containing a naphtyl instead of a phenyl group. It is decomposed in the intestine into beta-naphtol and salicylic acid. It is used in the same doses and for the same purposes as salol.

SALICINUM, U. S. P.

(Salicin; $C_{13}H_{18}O_7$.)

Salicin is a neutral principle obtained from several species of willow (*salix*) and poplar (*populus*). It occurs in white, silky needles, odorless, and of a very bitter taste. It is soluble in 28 parts of water and in 30 of alcohol. The dose is from 10–40 gr. (0.6–2.6 gm.). Salicin was the first of the salicylic preparations to be used in rheumatism, Maclagen, of Dundee, having prescribed it in 1874, a year before the introduction of salicylic acid by Buss, of Basle. It is partially converted in the body into salicylic acid, and rapidly appears in the urine partly as salicin and partly as saligenin, salicyluric acid, and salicylic acid. It is well borne by the stomach, but it is far less active and reliable than salicylic acid or the salicylates.

ACIDUM BENZOICUM, U. S. P.

(Benzoic Acid; $HC_7H_5O_2$.)

Benzoic acid (see p. 270) is an organic acid contained largely in benzoin, balsam of Peru, and balsam of tolu.   It is obtained from benzoin, from the urine of herbivorous animals, and artificially from toluene, a coal-tar derivative.   It occurs in white or slightly yellow scales or needles, of an aromatic odor and taste.   It is sparingly soluble in water, but freely so in alcohol, ether, and chloroform.   The dose is from 5–20 gr. (0.3–1.3 gm.).

Therapeutics.—The germicidal power of benzoic acid is about equal to that of salicylic acid.   Internally, it is very useful as a urinary antiseptic in *cystitis with ammoniacal urine.* The acid itself and its salts have also been employed to some extent as intestinal antiseptics in *catarrhal enteritis.*   Senator and others have recommended the benzoates in *rheumatism,* but they have been found much less efficacious than the salicylates.

Externally, in the form of balsam of Peru (see p. 272), benzoic acid is much used as a parasiticide in *scabies.*

FORMALDEHYD.

(Formic Aldehyd; Formol; Oxymethylene; CH_2O.)

Formaldehyd is a colorless gas obtained by oxidizing methyl-alcohol (wood-alcohol).   A 40 per cent. aqueous solution of the gas is known commercially as *formalin.*   The latter is a clear, colorless, neutral liquid, of a very pungent odor and a caustic taste.   In solution the gas readily polymerizes, forming *paraform* (trioxymethylene, (CH_2O_3)), a white crystalline powder, which yields formaldehyd on heating.

Physiologic Action and Therapeutics.—Formaldehyd is an intensely irritating gas, causing, when inhaled, severe congestion and even inflammation of the mucous membrane of the entire respiratory tract.   On cutaneous surfaces concentrated solutions or even dilute solutions if the contact be prolonged, not infrequently excite an erythematous or eczematous eruption.   Concentrated solutions coagulate albuminous matters, but weak solutions do not.   To higher animals it is comparatively harmless, but prolonged inhalation may cause fatal pneumonitis.   In a case cited by Zorn, an ounce (30.0 c.c.) of formalin produced the following symptoms: burning in the throat and stomach, nausea, vomiting, diarrhea, a weak, feeble pulse, dyspnea, cyanosis, vertigo, and anuria lasting for several hours.   The urine passed later contained albumin and casts.

Large doses, as in Klüber's case, may cause also stupor and coma. The antidote for formalin poisoning is ammonia in the form of the water or spirit, which combines with formaldehyd to form the harmless hexamethylene-tetramin (urotropin). The fate of formaldehyd in the body is not definitely known, but a portion, at least, appears to be eliminated in the urine unchanged.

Germicidal Power.—The germicidal power of formaldehyd is little, if at all, inferior to that of mercuric chlorid. A 1 per cent. solution of formalin kills pure cultures of pathogenic bacteria in an hour, and, according to Harrington, bacteria of the highest resistance, when freely exposed to an atmosphere produced by vaporizing 110 c.c. of formalin in each 1000 cubic feet, are killed within two and a half hours. Though very diffusible, formaldehyd has little penetrating power; it cannot be considered, therefore, except under the most favorable conditions, more than a surface disinfectant. As it unites readily with hydrogen sulphid, mercaptan, ammonia, and fetid ammonia bases to form inodorous compounds, it is a powerful deodorizer.

As a *surface disinfectant for rooms* it has decided advantages in being comparatively harmless to the higher forms of animal life, and in having no injurious effects on delicate fabrics or metals. On account of its weak penetrating power, however, it is not a suitable disinfectant for upholstered furniture, carpets, bedding, books, etc. As formalin when vaporized from an open vessel loses most of its germicidal power through the polymerization of the formaldehyd into paraform, some special device has been found necessary for securing an adequate supply of the gas. Lamps devised for generating formaldehyd directly from methyl-alcohol were first employed, but they did not prove reliable. A sufficient quantity of gas may be obtained by vaporizing paraform tablets in a suitable lamp or by subjecting a fine stream of formalin by means of a special contrivance to a high degree of heat. If the paraform lamp be used, it must be placed inside the room to be disinfected, but if the second form of apparatus be employed, it may be placed outside, the gas being conducted into the room by means of a tube passed through the keyhole. At least 60 paraform tablets or a quart (0.1 L.) of formalin should be vaporized for every 1000 cubic feet of room to be disinfected. To secure the best results the room should be sealed as hermetically as possible before the operation, and should be kept closed after it for at least ten hours. Subsequently ammonia water may be evaporated in the room to destroy the formaldehyd.

An excellent method of sterilizing *small instruments* without dulling or tarnishing them is to expose them for ten minutes in an air-tight tin box to the vapor evolved from a 5 gr. (o.3 gm.) paraform tablet.

A 5 per cent. solution of formalin (1 : 20 parts of water) in excess is a reliable disinfectant for *stools* and *sputa*.

The irritant properties of formaldehyd detract from its usefulness in general surgical work, but a 1 or 2 per cent. solution of formalin may be employed to cleanse *wounds* and to irrigate *sinuses* and *suppurating cavities*. A solution of the same strength containing glycerin makes a useful *wash for the nose and throat* in infectious diseases. Solutions of from 1 : 1000 to 1 : 500 have been found serviceable in *chronic otorrhea*. A solution of 1 : 2000 has been used successfully as a collyrium, especially in *ulcerative keratitis*. A lotion containing from 2 to 4 per cent. of formalin is sometimes very efficacious in *bromidrosis* and *hyperidrosis*.

There is considerable evidence to show that formaldehyd added to milk as a preservative lessens its nutritive value and interferes with digestion; it is but fair to state, however, that Tunnicliffe and Rosenheim conclude from a recent study of the subject that in the proportion ordinarily used the drug has no appreciable influence on perfectly healthy children.

Incompatibles.—Ammonia, alkalis, gelatin, tannin, and mineral salts.

Glutol.—This is a combination of formaldehyd and gelatin, appearing as a gray, odorless, and tasteless powder. It has been well spoken of as a substitute for iodoform in the treatment of *infected wounds* and *ulcers*.

Tannoform (see p. 353) is a combination of formaldehyd and tannic acid. It is chiefly useful as a dusting-powder in *hyperidrosis*.

UROTROPIN.

(Hexamethylene-tetramin; Formin; $(CH_2)_6N_4$.)

Urotropin is the product of the action of formaldehyd on ammonia. It occurs in colorless, rhombic crystals, odorless, and of a sweetish taste. It is soluble in 1.3 parts of water, less soluble in alcohol, and insoluble in ether. It is readily decomposed by acids and by heat. The dose is from 3–10 gr. (o.2–o.6 gm.), dissolved in water, thrice daily between meals.

Physiologic Action and Therapeutics.—When taken internally in moderate doses, urotropin produces no special symptoms, but is rapidly absorbed and rapidly eliminated in the urine, partly unchanged and partly, according to Casper, as formaldehyd. Urine passed shortly after the administration of urotropin remains sterile for a long period, even when exposed to the air. It does not materially increase the quantity of urine nor affect the elimination of solids, but it has more power than any other internal remedy of ridding that secretion of bacteria. Large doses, especially if long continued, sometimes cause frequent micturition, burning in the bladder, and even hematuria, but these symptoms speedily disappear upon the withdrawal of the drug. Baum and Belfield each report a single instance in which urotropin appears to have excited palpitation and weakness of the heart's action in a patient without cardiac disease. As a urinary antiseptic urotropin is most efficacious in *typhoid cystitis*, in *chronic cystitis* accompanying *prostatic enlargement* or *urethral stricture*, and in *simple bacteriuria*. Its power to arrest ammoniacal decomposition of the urine is sometimes truly remarkable. To secure a lasting effect, however, it should be given for a considerable period after the urine has become apparently sterile. If the urine be alkaline when secreted by the kidneys, formaldehyd may not be liberated ; in such cases urotropin to be effectual should be preceded by a course of benzoic acid.

When bacteria, instead of being free in the urine, are chiefly in the tissues of the bladder, as they are in tuberculous and gonorrheal cystitis, urotropin is of little value.

As the urine of typhoid patients frequently contains typhoid bacilli long after the establishment of convalescence, and as this secretion is frequently the means, no doubt, of spreading the infection, Horton-Smith, Richardson, Gwyn, and others have advocated as a routine prophylactic measure the administration of urotropin during the last weeks of the disease.

Keyes, Otis, and others have found the drug of value as a prophylactic remedy in preventing infection when administered for several days before and several days after *operations on the genito-urinary tract*. It is important, however, that it should not be used too freely, since when eliminated in large quantities it tends to retard healing.

CHLORUM.

(Chlorin ; *Cl.*)

Chlorin is a heavy, yellowish-green gas, of a suffocating odor and a caustic taste. It may be prepared by heating together sodium chlorid, sulphuric acid, and manganese dioxid or by acting upon chlorinated lime with an acid.

<table>
<tr><td align="center">PREPARATION.</td><td align="center">DOSE.</td></tr>
</table>

Aqua Chlori, U. S. P. (contains at least 0.4
    per cent. of the gas) . . . . 1–2 fl. dr. (4.0–8.0 c.c.).

Physiologic Action and Therapeutics.—In concentrated form chlorin acts as a powerful irritant, even upon cutaneons surfaces. When inhaled it excites pain in the chest, cough, dyspnea, spasm of the vocal cords, and, ultimately, mucopurulent or fibrinous bronchitis with inflammation and edema of the lungs. Owing to its affinity for hydrogen and its power to separate nascent oxygen from water, it is an energetic germicide, especially in the presence of moisture. For the same reasons it is also an active deodorizer.

As a gaseous disinfectant for rooms, however, it has serious disadvantages in being exceedingly irritant and poisonous, in being destructive to wall-paper and other fabrics, and in being so heavy that it diffuses with difficulty. Chlorin water is a very unstable preparation; for disinfecting excreta it is, therefore, less reliable than chlorinated lime. Largely diluted, it has been used to some extent as a wash for *fetid sores* and as a *spray for the nose and throat* in infectious diseases. Internally, it has been recommended as an intestinal antiseptic in *typhoid fever*, but it is of very doubtful utility.

CALX CHLORATA, U. S. P.

(Chlorinated Lime.)

Chlorinated lime is a preparation containing not less than 35 per cent. of available chlorin, and is obtained by acting on slaked lime with chlorin gas. It consists chiefly of the hypochlorite and the chlorid of calcium, and appears as a grayish-white powder, having a strong odor of chlorin and a disagreeable saline taste. It is partially soluble in water and in alcohol.

Action and Uses.—A solution of 0.5–1 per cent. of freshly prepared chlorinated lime kills most bacteria within ten minutes. A 0.5 per cent. solution makes a reliable disinfectant wash for *bare walls* and *woodwork*. A solution of the same strength is also useful for disinfecting *white clothes*. A 1 per cent. solution may be employed to sterilize *feces and sputa*.

Liquor Sodæ Chloratæ, U. S. P. (Solution of Chlorinated Soda or Labarraque's Solution).—This is a solution of several chlorin compounds of sodium containing at least 2.6 per cent. of available chlorin. It is used for the same purposes as chlorinated lime.

ACIDUM SULPHUROSUM, U. S. P.

(Sulphurous Acid; H_2SO_3.)

Sulphurous acid gas, or sulphur dioxid (SO_2), when required in large amounts, is usually obtained by burning sulphur. In the official preparation, however, which is a 6.4 per cent. aqueous solution of sulphur dioxid, the gas is obtained by reducing sulphuric acid with charcoal.

Physiologic Action and Therapeutics.—Sulphur dioxid is an intensely irritating, suffocating gas, quite capable, if inhaled in sufficient quantity, of destroying life. It has considerable germicidal power, especially in the presence of moisture, and before the introduction of formaldehyd it was the most popular gaseous disinfectant for rooms, hospital wards, and ships. Formaldehyd, however, being much more reliable and being without injurious effects on colored fabrics, has largely displaced it. To be at all effective, at least 4 pounds of sulphur should be burned for every 1000 cubic feet of airspace in the apartment. The sulphur, in the form of small fragments, should be put in a pan, and the latter should be placed inside of a tub partly filled with water. Before being ignited the fragments should be well saturated with alcohol. To secure the required amount of moisture steam may be generated at the same time or the walls and contents of the room may be previously sprayed with water. Key-holes and other openings should be carefully sealed, and the gas should be allowed to act for from ten to twelve hours.

A more convenient method of sulphurous acid disinfection, but one not always available, is to use the gas that has been liquefied under pressure and has been stored in metal cylinders.

Official sulphurous acid, though somewhat irritant, is a reliable parasiticide in *ringworm of the body.* It may be applied in full strength or diluted. As light and air gradually convert sulphurous into sulphuric acid, only fresh preparations should be employed.

AQUA HYDROGENII DIOXIDI, U. S. P.

(Solution of Hydrogen Dioxid; Solution of Hydrogen Peroxid.)

Hydrogen dioxid (H_2O_2) is a very unstable compound, prepared by the action of mineral acids or barium dioxid. It is employed in medicine only in the form of aqueous solutions. The official solution, when freshly prepared, contains about 3 per cent., by weight, of the pure dioxid, an amount corresponding to about 10 volumes of available oxygen. This solution is a colorless liquid, odorless, of a slightly acidulous taste, and producing a foam in the mouth. It contains a small amount of free acid as a preservative. Sunlight, protracted agitation, heat, and many metallic substances serve to decompose it into oxygen and water.

Physiologic Action and Therapeutics.—The therapeutic value of hydrogen dioxid depends upon the readiness with which it parts with oxygen when it is brought in contact with the tissues and fluids of the body. When it is applied to a suppurating wound, effervescence follows from the liberation of oxygen, the pus is discharged, and the surface is left perfectly clean and protected by a delicate coagulum. When taken internally it causes no special symptoms, its decomposition into oxygen and water being speedily effected in the stomach. According to Egbert, it has no inhibitive influence on the unorganized ferments, such as ptyalin and pepsin. When injected intravenously or subcutaneously, however, it may kill suddenly by forming gaseous emboli in the blood. The same accident has also resulted from its introduction into one of the large serous sacs. Owing to its oxidizing power also it is an active germicide and deodorizer. A 20 per cent. solution of the official preparation quickly destroys pyogenic cocci and other non-spore-bearing bacteria. As an antiseptic for general surgical purposes its drawbacks are its proneness to deteriorate and the short duration of its action; its advantages are its freedom from odor and from irritant and toxic properties. Its chief use is in the preliminary treatment of *septic wounds, abscess cavities*, and *fistulous tracts;* its efficacy in these cases being due in part to its antiseptic action and in part to its mechanical action in expelling pus, blood-clots, and detritus. It is generally applied diluted with from 1 to 3 parts of water. It should not be injected into deep abscesses, unless there is a free exit for the gas and pus, as otherwise serious harm may result from the increased tension within the cavity; neither should it be introduced, under any circumstances, into a large serous sac, like the pleura or peritoneum.

A 25–50 per cent. solution of the official preparation makes an excellent wash or spray for the nose and throat in *diphtheria*. If an atomizer be used, the tube and nozzle should be of hard rubber or of glass, as contact with metal favors the decomposition of the dioxid. A solution of the same strength is also serviceable in *mercurial stomatitis*, in *noma*, and in *follicular tonsillitis*. Harris has found copious rectal injections of a dilute solution (1 : 4 or 8) of some benefit in *amebic dysentery*.

The bleaching properties of hydrogen dioxid render it useful, according to Bulkley, in *hiding superflous black hair* upon the face of women. Its physical action also makes it useful in removing the particles of carbon in gunpowder burns.

With the hope that it might impart some of its oxygen to the blood it has been recommended internally in a number of *diseases*, but in none has it gained any reputation.

POTASSII PERMANGANAS, U. S. P.

(Potassium Permanganate; $KMnO_4$.)

Potassium permanganate occurs in the form of slender, dark-purple prisms, odorless, and of a sweetish, astringent taste. It is soluble in 16 parts of water, and is decomposed by alcohol. The dose is from 1–3 gr. (0.06–0.2 gm.) in pill form, after meals.

Physiologic Action and Therapeutics.—In concentrated solution potassium permanganate acts as an irritant or mild caustic. By the mouth, if the dose be sufficiently large, it produces the symptoms of a corrosive poison. In the presence of organic matter it quickly yields its oxygen; hence it is a prompt germicide and deodorant. Unfortunately, its usefulness is considerably restricted by the facility with which it is rendered inert by deoxidation. Solutions of potassium permanganate, however, make valuable deodorizing and detergent washes in many diseases attended with offensive discharges; thus, solutions of from 1–3 gr. (0.06–0.2 gm.) to the ounce (30.0 c.c.) may be used with advantage in putrid *sore throat, cancer of the tongue, ozena, cancer of the uterus, foul ulcers*, and *bromidrosis*. A warm saturated solution makes a good *disinfectant for the hands*, provided it be followed by a saturated solution of oxalic acid to reinforce the permanganate and to remove the stain. A solution of from 1 : 6000 gradually increased to 1 : 1000, used by irrigation or by injection, is quite efficacious in *gonorrhea*.

Internally, potassium permanganate is used as an *emmenagogue* (see p. 242) and as an antidote to *morphin poisoning* (see p. 88).

Incompatibles.—Potassium permanganate has an exceedingly wide range of incompatibilities. With oxidizable substances, especially organic ones, like alcohol and glycerin, it is explosive.

CALX, U. S. P.

(Lime; Quicklime; Burned Lime; CaO.)

Lime is obtained by burning the purest natural varieties of calcium carbonate. It occurs in hard, white lumps, odorless, and of a sharp, caustic taste. It is soluble in 750 parts of water and is insoluble in alcohol. In the presence of water it evolves heat, and is gradually converted into calcium hydrate or slaked lime. Mixed with 3 or 4 parts of water, it forms a magna which is known as milk of lime.

PREPARATIONS.	DOSE.

Liquor Calcis, U. S. P. . . . . $\frac{1}{2}$–2 fl. oz. (15.0–60.0 c.c.).
Syrupus Calcis, U. S. P. (lime, 6.5 parts;
   sugar, 40; water, to make 100 parts) . . $\frac{1}{2}$–1 fl. dr. (2.0–4.0 c.c.).
Linimentum Calcis, U. S. P. (Carron oil).
Potassa cum Calce, U. S. P. (Vienna paste;
   50 per cent. of each).

Physiologic Action and Therapeutics.—Locally, lime is a quickly acting but superficial escharotic. It is never employed by itself as a caustic, but in the form of potassa cum calce it is occasionally used to destroy *small epitheliomata.* When thoroughly mixed with putrefying matter quicklime favors disintegration, dispels offensive odors, and acts directly to a limited extent as a disinfectant. Freshly prepared milk of lime is sometimes employed to disinfect fecal discharges. Calcium hydrate, in the form of lime-water (see p. 165) and the syrup of lime, is used as an antacid and a mild astringent.

SODII HYPOSULPHIS, U. S. P.

(Sodium Hyposulphite; Sodium Thiosulphate; $Na_2S_2O_3 . 5H_2O$.)

Sodium hyposulphite occurs in white, transparent prisms, odorless, and of a cooling, somewhat bitter taste. It is freely soluble in water. The dose is from 5–20 gr. (0.3–1.3 gm.).

Therapeutics.—It is an excellent unirritating parasiticide in ringworm—*tinea circinata* and *tinea sycosis*—and in *tinea versicolor.* It may be applied either as a lotion or as an ointment in the strength of a dram (4.0 gm.) to the ounce (30.0 c.c. or gm.). A solution of 30 gr. (2.0 gm.) to the ounce (30.0 c.c.) makes a useful mouth-wash in *thrush.* Internally, it has been found of service as an antiseptic in *gastrectasis,* when sarcinæ and yeast are present in large quantities in the stomach-contents.

Sodii Sulphis, U. S. P. (Sodium Sulphite; $Na_2SO_3 . 7H_2O$) and **Sodii Bisulphis,** U. S. P. (Sodium Bisulphite; $NaHSO_3$). —These are salts sometimes employed as substitutes for sodium hyposulphite.

ACIDUM BORICUM, U. S. P.

(Boric Acid; Boracic Acid; H_3BO_3.)

Boric acid occurs in colorless, transparent, pearly scales or crystals, odorless, and of a bitterish taste. It is soluble in

25.6 parts of water, 15 of alcohol, and 10 of glycerin. The dose is from 5–20 gr. (0.3–1.3 gm.).

PREPARATION.

Glyceritum Boroglycerini, U. S. P. (contains 31 per cent. boric acid in glycerin).

Physiologic Action.—In single moderate doses boric acid produces no appreciable symptoms. It is rapidly absorbed and rapidly eliminated, the bulk of it escaping in the urine within twenty-four hours after its administration. In very large doses it acts as an irritant poison. Its continued use in doses of a dram (4.0 gm.) a day not infrequently leads to a train of symptoms to which the term *borism* has been applied. The most important features of this condition are digestive disturbances, marked dryness of the skin and mucous membranes, a tendency to alopecia, erythematous or eczematous eruptions, areas of local edema, albuminuria, and cachexia. The same phenomena have also been observed after repeated external applications and rectal injections of boric acid. Although Annett found that kittens could not live longer than four weeks if fed upon milk containing small quantities of boric acid, it is the opinion of most experimenters (Rideal and Foulerton, Chittenden and Gies, Tunnicliffe and Rosenheim, Vaughan and Veenboer) that the drug when used as a food preservative does not, at least in adults, unfavorably affect nutrition.

According to Sternberg, boric acid has pronounced antiseptic properties, but is inefficient as a germicide.

Therapeutics.—As an unirritating, antiseptic wash, a solution of boric acid (10 gr. : 1 fl. oz.—0.6 gm. : 30.0 c.c.) has a wide range of usefulness in *inflammatory diseases of the nose and throat.* A solution of from 5–15 gr. (0.3–1.0 gm.) to the ounce (30.0 c.c.) may be employed with advantage as an eyewash in *simple conjunctivitis* and during *operations on the eye.* Insufflations of boric acid are very valuable in *chronic otorrhea;* in acute inflammation of the middle ear, however, especially when the discharge is profuse and contains much mucus, the drug may prove dangerous by clogging the outflow. Daily injections into the bladder of a warm solution (5 : 10 gr. to 1 fl. oz.—0.3–0.6 gm. to 30.0 c.c.) are often useful in *cystitis.* Lotions, ointments, and dusting-powders containing boric acid are extensively used in many acute inflammatory skin diseases, such as *erythema intertrigo, eczema, superficial burns,* and *miliaria.* In *pruritus* a lotion of boric and carbolic acids frequently affords relief.

> R Acidi carbolici,   ℞xx (1.2 c.c.);
>   Acidi borici,   gr. xl (2.6 gm.);
>   Glycerini,   f ʒiss (6.0 c.c.);
>   Aquæ,   q. s. ad f ʒiv (120.0 c.c.). M.

Dusting-powders containing boric acid are sometimes of service in *bromidrosis:*

> R Acidi borici,   ʒiss (6.0 gm.);
>   Acidi salicylici,   gr. xx (1.3 gm.);
>   Pulv. amyli,   ʒiv (16.0 gm.). M.

Internally, boric acid is of some value in *cystitis with alkaline urine*, but, as a rule, it is less efficacious than urotropin or benzoic acid.

Incompatibles.—Carbonates and bicarbonates.

Sodii Boras, U. S. P. (Sodium Borate; Borax; $Na_2B_4O_7 . 10H_2O$).—This salt occurs in colorless, transparent prisms or as a white powder, inodorous, and of a sweetish, alkaline taste. It is soluble in 16 parts of water and in 1 part of glycerin, and is insoluble in alcohol. The dose is from 5–20 gr. (0.3–1.3 gm.). Its action is very similar to that of boric acid, for which it is often substituted. Internally, it has been used with some success in *epilepsy*, but it is much less useful than the bromids, and, moreover, in the large doses required it is often productive of grave constitutional symptoms. It is incompatible with acids, metallic salts, and alkaloids. Glycerin slowly converts it into boric acid.

CARBO.

(Carbon.)

Carbon is official in the following forms:

Animal charcoal (*Carbo Animalis*, U. S. P.), *purified animal charcoal* (*Carbo Animalis Purificatus*, U. S. P.), and *charcoal* (*Carbo Ligni*, U. S. P.). Animal charcoal, or bone-black, is prepared by burning bones in closed iron cylinders. It is composed of carbon and calcium carbonate and phosphate. Purified animal charcoal is bone-black from which the earthy salts have been removed by hydrochloric acid. Wood charcoal is prepared from soft wood and is finely powdered. The dose is from ½–2 dr. (2.0–4.0 gm.).

Charcoal has the property of absorbing many times its own volume of gases or vapors. Owing to the oxygen condensed within its pores it has considerable oxidizing power, which may be utilized to destroy offensive gases, like hydrogen sulphid, and to hasten the decomposition of organic matter. Thorough wetting destroys its activity. As it is not absorbed when taken internally, it exerts no specific action on the body.

Charcoal is employed chiefly as an absorbent and a deodorant. In the form of a poultice it was at one time a favorite application for *foul ulcers*, but it has been largely displaced by more cleanly dressings. While it is a satisfactory agent for *deodorizing fecal discharges*, it has less disintegrating action than dry earth.

Internally, it is sometimes useful as an absorbent in *flatulent dyspepsia*. In may be given as a powder or in lozenges.

Animal charcoal is not used for medicinal purposes, but it is largely employed by chemists for removing coloring-matter from alkaloids. It was formerly regarded as an excellent filtering medium for drinking-water, but it has been shown to be unsafe for this purpose, since by adding phosphates and nitrates to the water it actually favors the development of bacteria.

OTHER GERMICIDES, ANTISEPTICS, AND DEODORIZERS.

Silver Compounds (see pp. 366–370).—The salts of silver are powerful germicides. Many of the organic salts are not so irritant as the nitrate, and, moreover, are not affected by albuminous matters or chlorids. They are used especially on mucous membranes.

Bromin (see p. 447).—This substance is an energetic disinfectant, but it is rarely used on account of its destructive and intensely irritant properties. A weak solution (1 : 400), however, is occasionally employed as a deodorizer for cisterns, trenches, slaughter-houses, etc.

Ferrous Sulphate (see p. 287).—This salt is a very feeble disinfectant. It is sometimes used as a deodorizer for middens, cesspools, etc.

Volatile Oils.—Many of the volatile oils, while they are not actively germicidal, have considerable power as antiseptics. The oils of sandalwood, copaiba, and cubeb (see pp. 228–230) are largely used as genito-urinary antiseptics.

Methyl-blue (Pyoktanin).—This substance, like many of the anilin dyes, has some power as an antiseptic, but very little as a germicide. A solution of 1 : 1000 has been used with some success in ophthalmic surgery.

Methylene-blue (see p. 415).—This compound, in doses of from 1–3 gr. (0.06–0.2 gm.), has been used as a urinary antiseptic in gonorrhea. It is also of value as an antimalarial, though it is decidedly inferior to quinin.

Acetanilid (see p. 343).—This compound has been used to some extent as a substitute for iodoform in the treatment of burns and ulcers.

ANTIMALARIALS.

Antimalarials are drugs that exert a curative influence in malaria by acting destructively upon the specific parasites present in the blood. The *alkaloids of cinchona*, especially *quinin*, are by far the most important members of this class. *Methylene-blue* and *Warburg's tincture*, while they have some virtue, do not approach quinin in efficacy. *Arsenic*, though useful in correcting the anemia caused by the parasites, does not appear to have any special action. Many other drugs have been recommended as antimalarials, but they have not been found trustworthy.

CINCHONA, U. S. P.

(Peruvian Bark.)

Cinchona is the bark of *Cinchona Calisaya* and other species of *cinchona*, tall evergreen trees indigenous in South America, and at present largely cultivated in India, Java, and Jamaica. Its activities depend upon a number of alkaloids, the most important of which are *quinin, quinidin, cinchonin,* and *cinchonidin.* To be up to the official standard the bark should contain at least 2.5 per cent. of quinin and not less than 5 per cent. of total alkaloids.

In addition to its alkaloids, cinchona contains quinic acid, quinovic acid, and quinotannic acid.

PREPARATIONS.	DOSE.
Extractum Cinchonæ, U. S. P.	5–20 gr. (0.65–1.3 gm.).
Extractum Cinchonæ Fluidum, U. S. P.	½–1 fl. dr. (2.0–4.0 c.c.).
Infusum Cinchonæ, U. S. P.	½–2 fl. oz. (15.0–60.0 c.c.).
Tinctura Cinchonæ, U. S. P.	1–2 fl. dr. (4.0–8.0 c.c.).
Tinctura Cinchonæ Compositæ, U. S. P. (Huxham's Tincture: red cinchona, 10; bitter orange-peel, 8; serpentaria, 2; glycerin, 7.5; water, 7.5; alcohol, to make 100)	1–4 fl. dr. (4.0–15.0 c.c.).

The following alkaloids and their salts are also employed medicinally:

Quinina, U. S. P. . . . . ⎫
Quininæ Sulphas, U. S. P. . . . . . ⎪
Quininæ Bisulphas, U. S. P. . . . . ⎬ 1–30 gr. (0.06–2.0 gm.).
Quininæ Hydrochloras, U. S. P. . . . ⎪
Quininæ Hydrobromas, U. S. P. . . . ⎭
Quininæ Valerianas, U. S. P. . . . . . . 1–10 gr. (0.06–0.6 gm.).
Cinchonina, U. S. P. . . . . ⎫
Cinchoninæ Sulphas, U. S. P. . . . . ⎪
Quinidinæ Sulphas, U. S. P. . . . . . ⎬ 1–30 gr. (0.06–2.0 gm.).
Cinchonidinæ Sulphas, U. S. P. . . . ⎪
Quininæ Dihydrochloras . . . . . . ⎪
Quininæ Tannas . . . . ⎭
Quininæ   Dihydrochloras   Carbamidata
   (quinin and urea hydrochlorate) . . . . 1–20 gr. (0.06–1.3 gm.).
Cinchonidinæ Salicylas . . . . 1–10 gr. (0.06–0.6 gm.).
Euquinin (an ethyl-carbonic ester of quinin) 5–60 gr. (0.3–4.0 gm.).

Quinin also enters into iron and quinin citrate, bitter wine of iron, syrup of the phosphates of iron, quinin, and strychnin, and Warburg's tincture.

Quinin is the most important alkaloid of cinchona, and represents very largely its active properties.

QUININA, U. S. P.

(Quinin.)

Quinin occurs in the form of a white, flaky, amorphous or crystalline powder, odorless, and of an intensely bitter taste. It is almost insoluble in water, but is readily soluble in acidulated water, alcohol, and ether.

Physiologic Action.—In large doses (20.0 gr.—1.3 gm.) quinin causes a feeling of fulness in the head, ringing in the ears, deafness, headache, and sometimes · dimness of vision. The term *cinchonism* is applied to this group of symptoms. Toxic doses may cause in addition circulatory weakness, dyspnea, delirium, stupor, convulsions, and coma.

Although alarming symptoms have been produced by overdoses of quinin, it is very doubtful whether in a human being death has ever resulted directly from the use of the drug.

Circulatory System.—Small doses do not materially affect the circulation, but very large doses decrease the force and frequency of the pulse by directly depressing the heart or its contained ganglia.

Nervous System.—Toxic doses congest the brain, and probably first stimulate and then depress the cerebral and spinal centers. The tinnitus aurium and deafness occurring in cinchonism are doubtless the result of an intense congestion of the middle ear and labyrinth. In rare instances, apparently owing to chronic inflammation of the auditory passages or to hemorrhage into the labyrinth, the deafness has been permanent. Recent studies tend to show that quinin amblyopia results directly from the toxic effects of the drug on the ganglion-cells of the retina, and not, as was formerly supposed, from constriction of the retinal arteries and secondary changes in the optic nerves. Permanent blindness from quinin is exceedingly rare.

Respiratory System.—Respiration is not affected by quinin except in poisoning, and then it may be slow and labored from depression of the center in the medulla.

Alimentary Canal.—In small doses it whets the appetite and stimulates gastric peristalsis. Large doses have an irritant effect and sometimes excite nausea and vomiting.

Uterus.—Quinin will sometimes intensify labor pains when they are inefficient owing to weakness or fatigue. The contractions induced by it are intermittent, never tetanic. Whatever ecbolic power the drug has it probably owes to its general tonic effects and not to any influence which its exerts directly upon the uterus itself. It does not seem capable, even in large doses, of originating labor pains, and thereby acting as an abortifacient.

Blood.—As Binz originally noted, quinin when added to freshly drawn blood arrests the ameboid movement of the white cells. Moreover, when applied in very dilute solution to the exposed mesentery of a frog, it suspends, almost immediately, the migration of the leukocytes. This cessation of diapedesis was believed by Binz to be due to poisoning of the corpuscles, but Metschnikoff and his followers attribute it rather to the repelling influence (negative chemotaxis) which quinin in common with many other substances exerts upon motile cells. Drawn blood when mixed with quinin loses in oxidizing power, as shown by its failure to strike a blue color with guaiac in the presence of turpentine and to decolorize indigo by transforming it into isatin.

Metabolism.—Under the influence of quinin there is a considerable falling off in the nitrogenous excretion, and as this continues for some time after the withdrawal of the drug, it would seem to be due to a retardative effect upon metabolism.

Temperature.—In health the bodily temperature is not appreciably influenced by quinin; in febrile states, however, large doses exert a pronounced antipyretic effect, which is thought to be due to an inhibition of metabolism, in consequence of which heat-production is diminished.

Absorption and Elimination.—Quinin is absorbed chiefly from the stomach, and is eliminated for the most part by the kidneys. Under favorable conditions it enters the blood very quickly, and traces of it may be found in the urine within twenty minutes after its ingestion. Its elimination, however, does not keep pace with its absorption, and after large doses, several days may elapse before all of it has left the body. It is excreted very largely unchanged, but a small part appears to be converted in the tissues into dihydroxyl-quinin.

Action on Lower Organisms.—Quinin is not an active germicide, although it is very destructive to certain protozoa, like the ameba and the hematozoon of malaria.

It is, however, an energetic antiseptic, and, in the proportion even of 1 : 800, it inhibits the growth of bacteria in fluids containing considerable organic matter.

Untoward Effects.—Idiosyncrasies to quinin are not infrequently encountered. In some individuals a dose of from 2–3 gr. (0.1–0.2 gm.) will cause intense cinchonism. Impairment of vision is fortunately rare, and has generally been produced by very large doses. Quinin rashes are not uncommon; of 60 cases analyzed by Morrow, 38 were erythematous, 12 urticarial, 5 purpuric, and 2 vesicular and bullous. Irritability of the bladder and urethra is occasionally noted.

Therapeutics.—Owing to its destructive action on the parasites of *malaria,* quinin is to be regarded as a specific in this disease. The quartan and tertian hematozoa readily succumb to it, but the estivo-autumnal hematozoa are more resistant, especially when in the crescentic or ovoid form. In ordinary *intermittent fever* doses of from 15–30 gr. (1.0–2.0 gm.) a day will generally arrest the paroxysms within three or four days. The drug is most effective when its administration is so timed that the maximum amount shall be in the blood during the paroxysm—that is, while sporulation is in progress. Thus, the daily dose may be divided into smaller doses of 5 gr. (0.3 gm.) each, the first being given about eight hours and the last not less than three hours before the expected chill. When, however, the patient is not seen until within a short time of the expected paroxysm, a single large dose should be given at once. The remedy should be continued in full doses until the paroxysms fail to appear, and then gradually withdrawn over a period of several weeks. The administration of a laxative dose of calomel as a preliminary measure increases the efficacy of the quinin, probably by facilitating its absorption. During convalescence arsenic may be advantageously associated with the quinin.

In *estivo-autumnal fever* large doses—30–40 gr. (2.0–2.6 gm.) a day—are usually required. In *pernicious malarial fever* the

patient should be cinchonized as quickly as possible by injecting at once into the tissues of the thigh or buttock about 30
gr. (2.0 gm.) of a soluble salt of quinin like the dihydrochlorate.

Whether the administration of quinin is at times responsible
for the occurrence of hemoglobinuria in malarial subjects, as
was first suggested by Verétas in 1858, is still a mooted question. Even if the affirmative be true, there does not seem
to be any good reason for withholding the drug in cases
of hemoglobinuria when malarial parasites can be detected in
the blood. Bastianelli has summed up the matter in the following rules: If hemoglobinuria occurs during the paroxysms and parasites are found, use quinin; if parasites are
not found, do not use quinin; if quinin has been used before
the hemoglobinuria begins and there are no parasites, discon-
tinue the quinin.

Quinin is not only a curative remedy in malaria, but also a
valuable prophylactic agent. Daily doses of from 4–8 gr.
(0.26–0.5 gm.) will generally prevent the incidence of the disease in persons living in malarial regions. The drug does not
afford immunity from the disease, but prevents the development of the parasites.

Quinin, in doses of from 1–3 gr. (0.06–0.2 gm.), has long
been used as a general tonic in *states of lowered vitality* following acute disease or brought on by overwork. In such cases
it may be combined conveniently with iron and strychnin.
Small doses are thought to be of some value also in *acute infections*, like septicemia, diphtheria, and influenza. It is possible that the drug owes its efficacy in these diseases, at least in
part, to its retarding influence on the tissue changes. Warm
rectal injections of the alkaloid (1 : 5000 to 1 : 2000) have been
found of service in *amœbic dysentery*. In *whooping-cough*
moderately large doses are sometimes useful in lessening the
severity and frequency of the paroxysms, but it is not known
how the drug acts. Quinin has been recommended in *neuralgia* and in *headache*, when the attacks are more or less
periodic, but unless these affections can be traced to a malarial
origin, the drug is not likely to do good.

The use of quinin as an *antipyretic* (see p. 348) is almost
obsolete. In doses of from 20–30 gr. (1.3–2.0 gm.) it undoubtedly has a pronounced effect upon high temperature,
especially if it be given a few hours before a natural remission
is expected to occur; but when drugs must be used at all as
antipyretics, the preference should generally be given to the
coal-tar derivatives, on account of the promptness and certainty

of their action and their comparative freedom from disagreeable by-effects.

Quinin (10–15 gr.—0.6–1.0 gm.) is sometimes of value as an ecbolic (see p. 250) in the first stage of labor when the pains are infrequent and inefficient, owing to *simple uterine inertia.*

Contraindications.—The chief contraindications are meningitis and acute middle-ear disease. It should be administered cautiously when acute inflammation of the genito-urinary tract exists. There is some evidence to show that the drug sometimes acts unfavorably in epileptics in increasing the number of seizures. Of course, quinin should be withheld from patients in whom, owing to idiosyncrasy, it has previously caused grave symptoms.

Administration.—Under ordinary circumstances quinin should be given by the mouth. It may be prescribed in capsules, cachets, freshly made pills, or in solution. The last method of prescribing it, however, is objectionable on account of its intensely bitter taste. Old pills should be avoided, since they are liable to escape from the stomach before the alkaloid has been liberated and absorbed. The sulphate, though it is most commonly selected, is not so eligible as the more soluble bisulphate (10 parts of water) and hydrochlorate (34 parts of water). If the sulphate be used, it should be rendered soluble by associating with it a few drops of one of the mineral acids.

To children the drug may be given suspended in syrup of yerba santa, syrup of chocolate, or elixir of licorice. The tannate, while it has less than half of the alkaloidal strength of the sulphate, is only slightly bitter, and may, therefore, be given to children in the form of chocolates. Euquinin is a tasteless and insoluble preparation, somewhat more active than the tannate.

Quinin is absorbed very imperfectly from the bowel, but for young children it may be prescribed in the form of suppositories, each containing 2 or 3 gr. (0.13–0.2 gm.) of a soluble salt like the dihydrochlorate.

For hypodermic use only the most soluble salts should be used, such as the dihydrochlorate, carbamidated dihydrochlorate, or bisulphate. If the bisulphate be chosen, the solution should be slightly acidulated with tartaric acid to prevent the precipitation of the salt by the alkaline juices of the tissues. The injections should be intramuscular rather than subcutaneous, and should be given with the most thorough antiseptic precautions.

Bacelli has reported excellent results in pernicious malarial

fever from the employment of intravenous injections of quinin.
He recommends the following solution:

℞ Quininæ hydrochloratis,   gr. xv (1.0 gm.);
    Sodii chloridi,   gr. xij (0.8 gm.);
    Aquæ destillatæ,   f ℥iiss (9.2 c.c.).   M.
  Sig.   Boil and filter; inject while still warm.

Of the fluid preparations of cinchona, the compound tincture is the most effective. It is rarely used, however, except as a restorative.

Incompatibles.—Alkalis, tannic acid, iodids, and spirit of nitrous ether.

Quinidin, Cinchonin, and Cinchonidin.—The action of these alkaloids is very similar to that of quinin, but it is much less powerful.

METHYLENE-BLUE.

(Tetramethyl-thionin Hydrochlorate; $C_{16}H_{18}N_3SCl$.)

Methylene-blue is a complex anilin derivative, occurring in dark-blue crystals or as a bronze-like powder, slightly soluble in water and in alcohol. The dose is from 1–4 gr. (0.065–0.26 gm.). Its action has not been carefully studied. When taken by the mouth or injected subcutaneously, it enters the blood and soon reappears in the secretions, especially in the urine, to which it imparts an intensely blue or greenish-blue color. Large doses irritate the stomach and also excite frequent and painful micturition. Achard and Castaigne, having found that in cases of renal disease the elimination of methylene-blue when injected intramuscularly is retarded as compared with the elimination in a normal subject, suggested that the degree of permeability of the kidneys might be determined by using as a criterion the time intervening between the injection of the drug and its reappearance in the urine. Subsequent studies, however, showed that this method, while of some value in determining the presence of interstitial nephritis, could not be depended upon in cases of parenchymatous nephritis.

The fact that methylene-blue is one of the best stains for the hematozoa of *malaria* suggested its use as a remedy in this disease. While it undoubtedly has some virtue, its efficacy is distinctly inferior to that of quinin. It may be used with advantage, however, when, owing to an idiosyncrasy, quinin cannot be taken. Berthier has found it useful in *dysentery ;* he recommends warm injections containing from 1½–3 gr. (0.1–0.2 gm.) to a pint or quart (0.5–1.0 L.). Flint found it of service in cases of *filariasis*, but in a case studied by Henry it was without effect. The drug appears to be of some value as an antiseptic in inflammatory diseases of the genito-urinary tract, especially in *gonorrhea*.

The strong affinity shown by Ehrlich to exist between the axis-cylinders of nerves and methylene-blue prompted the use of the drug as an analgesic and a sedative. It has been employed as an analgesic by Lemoine, Klemperer, and others with asserted good results in *neuralgia, sciatica*, and *migraine*, and as a sedative by Bodoni and others in various forms of *insanity characterized by excitement.*

Methylene-blue may be administered by the mouth or sub-cutaneously. The former is the better method, as the injections are quite painful. It should be prescribed in pills or capsules, combined with half its weight of powdered nutmeg, the latter serving to prevent gastric disturbances and irritation of the bladder. If it be given hypodermically, the solution should first be sterilized by boiling, otherwise abscess is liable to occur. Patients should always be warned of the discoloration of the urine caused by the drug.

WARBURG'S TINCTURE.

(Antiperiodic Tincture.)

Warburg's tincture was for a time a proprietary preparation, but in 1875 the originator himself made known its composition. As a number of the ingredients recommended by Warburg are no longer obtainable, the remedy is at the present day prepared after a somewhat simpler formula than the original one. Each fluidounce (30.0 c.c.) contains: Quinin, 10 gr. (0.65 gm.); rhubarb and angelica seed, of each, 3½ gr. (0.2 gm.); elecampane, saffron, fennel, and extract of aloes, of each, 1⅓ gr. (0.1 gm.); gentian, zedoary, cubeb, myrrh, white agaric, and camphor, of each, ⅞ gr. (0.05 gm.). The value of this tincture as an antimalarial, especially in the *severer forms of estivo-autumnal fever* which prevail in tropical countries, appears to be well established. To be effective it is necessary that it should produce copious diaphoresis, and to this end it should be given in doses of half an ounce (15.0 c.c.), undiluted, all drink being withheld from the patient. Playfair has found it very efficacious as an antipyretic in protracted cases of *puerperal sepsis.*

ANTHELMINTICS

Anthelmintics or vermifuges are drugs which destroy or expel intestinal worms. The principal members of this class are:

For tape-worms

Aspidium.
Pomegranate.
Kousso.
Chloroform.

Pumpkin-seed.
Kamala.
Oil of turpentine.
Thymol.

For round-worms:

Santonin.

Spigelia.

Chenopodium.

For thread-worms or pin-worms

Santonin or spigelia by the mouth and one of the following drugs by rectal injection:

Quassia.
Lime-water.

Vinegar.
Sodium chlorid.

Tannin.

For ankylostoma

Thymol.

Anthelmintics are most effective when preceded and fol lowed by purges. In the case of the tape-worm it is advisable that the patient should fast, or, at least, be restricted to very small quantities of liquid food for twenty-four hours before taking the anthelmintic.

ASPIDIUM, U. S. P.

(Male Fern; Filix-mas.)

Aspidium is the rhizome of *Dryopteris Filix-mas* and of *Dryopteris marginalis*, ferns growing in North America, Europe, and Asia. It contains a fixed oil, a volatile oil, resin, *filicic acid*, and a number of neutral bodies the chief of which is *aspidin*. Of these, filicic acid and aspidin are probably the most active constituents.

PREPARATION.	DOSE.
Oleoresina Aspidii, U. S. P.	. ½–1 fl. dr. (2.0–4.0 c.c.).

vomiting and purging, vertigo, headache, increased reflex activity, tonic spasms, collapse, and coma. Temporary or permanent blindness has also been present in many of the cases. Of 78 cases of poisoning by male-fern collected by Sidler-Huguenin, in 12 death occurred, and in 18 there was lasting impairment of sight in one or both eyes. According to Okamoto, microscopic examination of the eyes of poisoned dogs shows degenerative changes in the optic nerves.

Aspidium is perhaps the most generally useful remedy for *tape-worm.* A dram (4.0 c.c.) of the oleoresin may be given in emulsion or in capsules at bedtime, and followed in the morning by a saline purgative. Castor oil should not be used, as it is believed to increase the toxicity of aspidium by favoring the absorption of its active constituents.

GRANATUM, U. S. P.

(Pomegranate.)

Pomegranate is the bark of the stem and root of *Punica Granatum,* a small tree indigenous in Southwestern Asia, and cultivated in most subtropical countries. It contains, in addition to a large quantity of tannin, a number of alkaloids, the chief of which is *pelletierin.* The latter represents very largely the anthelmintic properties of the bark, and occurs as a colorless, oily liquid, capable of uniting with acids to form crystalline salts. Of the salts, the tannate is generally preferred, as it is less soluble than the sulphate or hydrochlorate. It may be given in doses of from 5–8 gr. (0.3–0.5 gm.).

Physiologic Action and Therapeutics.—Locally, pomegranate is astringent. Internally, large doses of the bark or of the alkaloid may cause headache, vertigo, dimness of vision, nausea, vomiting, and extreme muscular weakness. The last symptom is said to result from paralysis of the peripheral ends of the motor nerves.

Pomegranate is a reliable remedy for *tape-worm,* ranking next to aspidium in efficiency. A decoction of the bark (1 oz.–31.0 gm.) is sometimes used, but it is very unpalatable and less certain in its action than the alkaloid. Adjuvant treatment is necessary as in the case of the other teniacides.

CUSSO, U. S. P.

(Kousso; Brayera.)

Cusso is the female inflorescence of *Hagenia abyssinica,* an ornamental tree growing in the mountainous districts of Abyssinia. A neutral body, *kosotoxin,* is probably the active principle. The dose of the powdered drug is from 6–8 dr. (24.0–31.0 gm.). *Koussein*, which is probably an impure preparation of kosotoxin, may be given in doses of from 15–30 gr. (1.0–2.0 gm.). The crystalline body known as *kosin* is inactive.

<table>
<tr><td>PREPARATION.</td><td>DOSE.</td></tr>
<tr><td>Extractum Cusso Fluidum, U. S. P. . . . .</td><td>2–6 fl. dr. (8.0–24.0 c.c.).</td></tr>
</table>

Therapeutics.—Cusso, provided it be fresh, is a powerful teniacide. The fluid extract is less efficacious than an infusion in which the powdered flowers have been allowed to remain; both preparations, however, are unpalatable and are apt to excite nausea and vomiting. Koussein is an active preparation, but costly. It is readily taken, especially if dispensed in cachets. Cusso does not usually require the assistance of a purgative, but one should be given if the bowels do not move within six or eight hours.

PEPO, U. S. P.

(Pumpkin-seed.)

Pepo is the seed of the common pumpkin, *Cucurbita Pepo.* It contains a fixed oil and a resin, the latter probably being the active principle. The dose is from 1–3 ounces (31.0 gm.–93.0 gm.).

Pumpkin-seed is a perfectly safe, but somewhat uncertain, remedy for *tape-worm.* It is usually prescribed in the form of an emulsion, made by beating the decorticated seeds into a paste, adding sugar, and diluting with water or milk. It should be taken in the morning on an empty stomach, and followed in two or three hours by castor oil.

KAMALA, U. S. P.

(Rottlera.)

Kamala is a brownish-red, tasteless powder, consisting of the minute glands and hairs obtained from the surface of the capsules of *Mallotus philippinensis,* a small tree growing in western Asia and the neighboring islands. It contains several resinous principles, one of which occurs in yellow needles and is known as *rottlerin.* The dose of the powder is from 1–2 dr. (4.0–8.0 gm.).

Therapeutics.—Kamala has been highly spoken of as a teniacide by East Indian surgeons. It is rarely employed in this country. It is best given suspended in syrup or honey. A purgative is seldom required after it, as the drug itself causes considerable intestinal irritation and diarrhea.

SANTONINUM, U. S. P.

(Santonin.)

Santonin is a neutral principle obtained from santonica, or Levant wormseed (*Artemisia pauciflora*), a perennial shrub growing in Turkestan. It occurs in white, shining prisms, odorless, and of a slightly bitter taste. It is nearly insoluble in water, but readily so in alkaline solutions. The dose for a child is ¼–1 gr. (0.016–0.065 gm.); for an adult, 1–5 gr. (0.065–0.3 gm.).

PREPARATION.	DOSE.
Trochisci Santonini, U. S. P. (each troche contains ½ gr.—0.03 gm.).	1–5 troches.

Physiologic Action and Therapeutics.—In therapeutic doses santonin produces no special symptoms except an intensely yellow coloration of the urine and sometimes yellow vision, or xanthopsia. The last symptom is probably the result of a specific action of the drug on the nerve-centers or retina, although some investigators believe it to be due simply to staining of the humors of the eye. The fate of santonin in the body is not definitely known. A part is undoubtedly absorbed, probably as a santoninate, is largely oxidized in the tissues, and finally excreted in the form of oxysantonins.

Santonin poisoning is characterized by yellow vision, mydriasis, vertigo, tremors, unconsciousness, and violent convulsions, first epileptiform and then tetanic. Nausea and vomiting have been observed in some instances. When death results it is usually through asphyxia. The convulsions are due to stimulation of the motor centers—first the cerebral and then the spinal.

Santonin is used almost exclusively as a remedy for the *round-worm*. It may be prescribed in lozenges or in powders, mixed with a few grains of sugar. A good plan is to give the drug morning and evening and to follow it next day by a brisk cathartic like castor oil or calomel. For a child under two years the dose should never be more than ½ gr. (0.03 gm.); and for a child under five years, never more than 1 gr. (0.065 gm.). Santonin has been recommended in certain other affections, notably in amaurosis, amenorrhea, incontinence of urine, and epilepsy, but its claims to confidence are very doubtful.

SPIGELIA, U. S. P.

(Pinkroot.)

Spigelia is the rhizome and roots of *Spigelia marilandica*, a perennial herb growing in the Southern United States. It contains a volatile oil, tannin, a bitter principle, and an alkaloid known as *spigelin.*

PREPARATION.   DOSE.

Extractum Spigeliæ Fluidum, U. S. P. . . . ½–2 fl. dr. (2.0–8.0 c.c.).

Therapeutics.—Spigelia is a very efficient, and, if reasonable care be used in its administration, a perfectly safe remedy for the *round-worm.* Toxic doses produce excitement, flushing of the face, swelling of the eyelids, dilatation of the pupils, dimness of vision, and, finally, stupor. It should always be administered in association with a purge.

CHENOPODIUM, U. S. P.

(American Wormseed.)

Chenopodium is the fruit of *Chenopodium ambrosioides*, a perennial herb indigenous in Central and South America, and naturalized in the United States. The active principle is a volatile oil (*Oleum Chenopodii*, U. S. P.), the dose of which is from 3–10 min. (0.2–0.6 c.c.).

Therapeutics.—Chenopodium is a reliable vermicide for *round-worms*—less pleasant to take, however, than santonin or spigelia. The oil, which is almost exclusively used, should be given on sugar or in an emulsion three times a day, and followed by a brisk purgative every other day.

OTHER ANTHELMINTICS.

Oil of Turpentine (see p. 230).—This drug was formerly much used as an anthelmintic for *tape-worm,* but it has been almost entirely supplanted by remedies that are more agreeable and quite as efficacious. If it be selected as a teniacide, it should be given in doses of half an ounce (15.0 c.c.) combined with an equal amount of castor oil.

Chloroform (see p. 109).—This drug, in doses of ½–1 fl. dr. (2.0–4.0 c.c.), has been used to a limited extent as a teniacide, but it is unreliable.

Quassia (see p. 168).—Rectal injections containing quassia are very efficacious in *oxyures,* or *pin-worms.* To secure the best results the lower bowel should first be thoroughly emptied by means of a soap-and-water enema, after which 2 ounces (60.0 c.c.) of a cold infusion of quassia (1 oz. to 1 pint —31.0 gm.–0.5 L.) should be slowly injected. As the parasites often occupy the cecum and small intestine as well as the rectum, it may be necessary, in some instances, in order to secure permanent relief, to give also by the mouth an anthelmintic (spigelia or santonin) with a cathartic.

Lime-water, Vinegar, Sodium Chlorid, and Tannin. —These drugs are sometimes used in the form of rectal injections to destroy *pin-worms.* None is quite so efficacious as quassia. They may be used in the following proportions: Lime-water, undiluted; vinegar, 1 to 3; sodium chlorid, 1 dr. (4.0 gm.) to a pint (0.5 L.); tannin, ½ dr. (2.0 gm.) to a pint (0.5 L.).

Thymol (see p. 388).—This drug, in doses of 30 gr. (2.0 gm.), repeated in two hours and followed by a purgative, is almost a specific in *ankylostomiasis.* It has been used also with some success as an anthelmintic against the *tape-worm,* especially the *bothriocephalus latus.*

ANTITOXINS AND TOXINS.

Our knowledge of the subject of serum-therapy is the outgrowth of the studies upon immunity.

Immunity.—The inborn insusceptibility of an individual to a disease to which others are commonly susceptible constitutes *natural immunity.* Immunity the result of changes which have taken place in the body during the life-time of the individual is termed *acquired immunity.* The latter may be the result of disease naturally contracted (*unintentional immunity*); or it may be the result of injections of specific micro-organisms or their spores in an attenuated form, or of minute doses of virulent, specific micro-organisms, or of the dead bacteria with their contained toxins, or of the serum of an animal that has been previously protected from the disease by one of these methods (*unintentional* or *artificial immunity*).

As Ehrlich, Calmette, and others have shown, the power to confer immunity is not confined solely to bacterial toxins, but is shared also by other organic poisons, like ricin (from castor-oil beans), abrin (from jequirity seeds), and the venom of poisonous serpents. The study of the mechanism of immunity has scarcely reached beyond the hypothetical stage. Pasteur reasoned that a second infection was impossible in many diseases, because in the first infection the bacteria had used up substances that were necessary to their growth. This theory at once became untenable when it was shown that the metabolic products of bacteria were as powerful in conferring immunity as the bacteria themselves. Chauveau conceived the idea that the bacteria left material in the body which rendered the tissues unsuitable for subsequent infection. A serious objection, however, to this theory is the fact that the blood-serum of the animal artificially protected is in many cases a good culture-medium for the special bacteria concerned. Metschnikoff believes that immunity depends entirely upon cellular activity (phagocytosis). He originally taught that the main factor in the process was the capacity of the leukocytes and of certain other cells to ingest and destroy the bacteria. The discovery of the immunizing power of specific sera necessitated a modification of this view, and this distinguished observer now teaches that the injection of soluble protective substances exerts an "educational" or stimulating effect upon the phagocytes, in consequence of which their activity is so intensified that they ultimately become invincible. He further asserts that any bactericidal power that the body-fluids may possess is to be attributed to substances derived from the stimulated leukocytes. That phagocytosis is a factor in the establishment of immunity is generally admitted; that it is the chief factor, however, is highly improbable. Pfeiffer demonstrated that cholera vibrios injected into the peritoneal cavity of the artificially immunized guinea-pig were destroyed without the aid of the phagocytes. The humoral theory evolved from the studies of Nuttal, Flügge, Behring, and others is the most plausible of all theories yet offered. This theory attributes immunity to the power of the extra-cellular fluids, especially of the blood-serum, to neutralize the poisons made by bacteria or to destroy the bacteria themselves. These antitoxic or bactericidal powers are believed to be due, however, to substances derived from the cells. Those substances which neutralize toxins are known as *antitoxins*, and those which disintegrate and dissolve the bacteria are known as *bacteriolysins*. In certain infections, like diphtheria and tetanus, immunity is chiefly antitoxic, while in others, like cholera and typhoid fever, it is mainly antibacterial.

The manner in which these protective bodies act has not yet been clearly explained, but the " lateral-chain " theory of Ehrlich is highly suggestive. This theory presupposes that the protoplasm of the cells contains complex molecules having a comparatively stable central group of atoms, to which are attached much less stable lateral chains of atoms. Poisons entering the body unite with the lateral chains for which they have an affinity, and in so doing stimulate the cells to form new lateral chains even in excess of those previously existing. These extra lateral chains, separated from the cells and free in the blood, are the antitoxins which unite with the toxins, thus neutralizing them and preventing their union with the lateral chains still attached to the cells. In the case of anti-bacterial immunity, Ehrlich holds that the extra lateral chains serve as an *immunizing body*, without which another body normally present in the blood, which he calls the *complement* (alexin of Buchner), cannot exert any bactericidal influence.

The subject of artificial immunity is closely related to that of serum-therapy, inasmuch as it has been well established that a susceptible animal which has been rendered artificially immune against a certain infection is capable of furnishing a serum which, injected into other susceptible animals, will pro-tect them from a like infection, and even arrest such an in-fection if it is just beginning. Many pathogenic bacteria furnish both an antitoxin and a bacteriolysin, but usually one in greater proportion than the other. Thus far, decided thera-peutic results have been obtained only in those infections the bacteria of which furnish serum especially rich in antitoxin bodies. As antitoxins only neutralize toxins and have no power to undo the damage that the latter have already wrought, it is evident that the earlier they are administered in any infection, the more effective they will be.

DIPHTHERIA ANTITOXIN.

Diphtheria antitoxin is obtained from the horse, the animal having been rendered artificially immune by repeated injec-tions extending over a period of several months, of gradually increasing quantities of the strongest diphtheria toxin. As the bacilli themselves are not injected, the horse does not become infected with diphtheria, but he gradually acquires a tolerance for the toxins of the disease and develops in his blood a sub-stance (antitoxin) which has the power to neutralize those toxins. At the proper time, when it is thought that his blood has acquired the requisite degree of potency, the animal is bled, and the serum—the part of the blood containing the antitoxin—is carefully separated from the clot, filtered, and standardized. The last procedure is accomplished by deter-mining the quantity of antitoxin serum required to offset the effects of the minimum quantity of toxin necessary to kill a guinea-pig in a definite time. The strength of the antitoxin is measured in units, a unit containing the amount of antitoxin required to save the life of a guinea-pig which has been injected with 100 fatal doses of toxin. Irrespective of the toxin in-jected and of the duration of the immunizing treatment, the serum yielded by different horses varies considerably in antitoxin strength. Only a very small percentage of horses give more than 1000 units in a cubic centimeter.

Therapeutics.—Diphtheria antitoxin has both preventive and curative power. The immunity afforded by it is but temporary, rarely lasting more than two or three weeks. The prophylactic dose is from 250–500 units. That it is a powerful curative agent is evident from the fact that the mortality of diphtheria has been reduced at least 50 per cent. since the introduction of serum treatment. Baginsky claims that the mortality has fallen from 41 per cent. to 8 or 9 per cent. In 5576 cases (moribund cases excluded) collected by the American Pediatric Society, mostly from the private practice of American physicians, the death-rate was 8.8 per cent.

In the laryngeal cases the figures have been reversed; instead of a mortality of 70 per cent. or more, the recoveries now number 70 per cent. or more. The chances of recovery in all cases vary directly with the time of administration. The statistics of the American Pediatric Society show that if antitoxin is injected upon the first day the death-rate is 4.9 per cent.; if upon the second day, 8.6 per cent.; if upon the third day, 12.7 per cent.; if upon the fourth day, 22.9 per cent.; and if after the fourth day, 38 per cent. In pharyngeal cases, if the patient is seen early, the dose should be from 1500 to 2000 units; when treatment is inaugurated rather late and the case is severe, the dose should be not less than 3000 units. Laryngeal cases require 3000 to 4000 units. Unless decided improvement follows the initial dose within from four to six hours, a second and much larger dose should be given. As antitoxin is practically harmless, the danger lies not in using too much, but in using too little. No allowance should be made for age, as children are more susceptible to the disease than adults. Antitoxin deteriorates with age, and therefore only fresh preparations should be used. Whether the serum acts indirectly through the cells (vital reaction) or directly by combining with the toxin (chemical reaction) is not known, though the weight of evidence is in favor of the latter view. Antitoxin is probably eliminated through all the emunctories. It has been found in the urine and milk.

Administration.—As antitoxin is absorbed very slowly and imperfectly from the stomach, it should always be given subcutaneously. The needle may be inserted into the loose subcutaneous tissue of the pectoral region, side of the abdomen, or interscapular space. Strict antiseptic precautions should be taken in the operation. Manipulation of the swelling to hasten the absorption of the serum is undesirable.

Untoward Effects.—The most common of these is an urticarial or erythematous rash. A rise of temperature of short duration occurs in about 20 per cent. of the cases. Joint pains, and even slight effusion into the joints, occur in about 6 per cent. of the cases. These effects are probably not produced by the antitoxin itself, but by other substances in the serum, since they may occur on injecting normal horse's serum. The increased frequency of complications like paralysis and albuminuria since antitoxin has been used is to be attributed to the fact that there are more recoveries than formerly.

OTHER ANTITOXINS AND TOXINS.

Tetanus Antitoxin.—Tetanus antitoxin is obtained in the same manner as diphtheria antitoxin—that is, by inoculating the horse with increasing doses of tetanus toxin. The strength of the serum is based upon the amount required to protect 1 gram of animal, like the mouse, from a dose of toxin sufficient to kill it within four days. The dose of the serum as a prophylactic is from 5–10 c.c.; as an antidote, from 20–50 c.c.

Unfortunately, the curative effects of tetanus antitoxin are not comparable to those of diphtheria antitoxin. The chief reasons for this are, first, the great virulence of the toxin produced in many of the cases of tetanus, and, secondly, the advanced stage that the disease must reach before it is recognizable. In the case of diphtheria the diagnosis is usually made before much toxin has been absorbed into the system; in the case of tetanus, however, diagnosis is impossible before the nerve-centers are fully under the influence of the toxin. Serum-therapy does not appear to have materially lowered the mortality in grave cases of tetanus—that is, in cases in which the period of incubation was less than seven or eight days; on the other hand, it seems to have been of some service in the more chronic cases of the disease, cases in which the period of incubation ranged between eight and twenty-four days.

In order that the antitoxin might be brought in direct contact with cells of the central nervous system, for which the toxin of tetanus has a great affinity, Roux and Borrell recommended the intracerebral injection of the serum, but the results from this method of administration have not been very encouraging. Recently (1902) Jacob has obtained promising results from subarachnoidal injections through a lumbar puncture. Despite the poor showing of the antitoxin treatment, we are of the opinion that it should be used early and energetically in every case of tetanus, but not to the exclusion of other measures known to be of benefit. As the serum, even in small doses, has considerable immunizing power, there seems to be no good reason for not using it as a prophylactic remedy in all suspicious wounds, especially in those which have become contaminated with garden earth, street dirt, or stable refuse.

Antistreptococcus Serum.—A serum obtained from horses after they have received repeated injections of streptococcus cultures has been offered as a remedy in various diseases due to streptococcus infection. These diseases include erysipelas, puerperal fever, septicemia, ulcerative endocarditis, and infective lymphangitis. Numerous cases are on record in which good results are attributed to the serum, but the statistics collectively do not indicate that it has had a very decided influence in lowering the mortality of septic diseases. However, as the serum when sterile and pure is practically harmless, there seems to be no good reason for not giving it a thorough trial in cases of infection demonstrated by bacteriologic examination to be due to the streptococcus. As yet there are no accurate data upon which the dosage may be computed, but from 20 to 30 c.c. of a reliable serum may be given two or three times a day so long as the fever remains high. As the potency of the serum is notably impaired by keeping, preparations more than two or three weeks old should not be used.

Injections of the *toxins* produced by streptococci have been advocated in *inoperable malignant tumors*, but the treatment is not without danger and has failed to secure the indorsement of most surgeons. It must be admitted, however, that it has been apparently successful in a few instances in arresting the growth of sarcomatous tumors and even in effecting a cure.

Antipneumococcic Serum.—It has been clearly demonstrated that the blood-serum of animals artificially immunized against pneumococcic infection possesses definite antitoxic properties. A serum obtained from horses which have been repeatedly inoculated with cultures of pneumococci of great virulence has been applied to the treatment of pneumonia in human beings. It is too soon to express a positive opinion upon the merits of this treatment, but the reports thus far presented are to an impartial observer certainly not very encouraging. The mortality in 36 cases of pneumonia in which J. C. Wilson employed antipneumococcic serum was 27.7 per cent.

Tuberculin.—In 1890 Koch introduced as a specific remedy for tuberculosis a preparation to which the name *tuber-*

culin was applied. This was subsequently shown to be a glycerin extract of the metabolic products of tubercle bacilli preserved in glycerin. Koch did not claim for tuberculin bactericidal or even antitoxic properties ; he contended, however, that by reinforcing the toxin already in the body it would increase the irritation at the seat of infection, attract a certain number of leukocytes, and ultimately lead to the complete encapsulation of the bacilli. Unfortunately, in many cases the reaction induced by the injections proved to be too severe, and, in consequence, softening of quiescent foci followed, resulting in a wider dissemination of the bacilli. The preparation is now rarely used as a therapeutic agent, although it is still retained for diagnostic purposes.

A substance known as *new tuberculin*, which is really a watery extract of the soluble portions of tubercle bacilli, has also been tried as a remedy in tuberculous affections, but with no better results than were achieved with the older product.

Bubonic Plague Antitoxins and Toxins.—In 1895 Yersin demonstrated that animals could be successfully immunized against bubonic plague by subcutaneous injections of sterile cultures of plague bacilli, and, further, that the blood-serum of animals thus artificially immunized possessed protective and curative properties. Serum obtained after the method of Yersin has been used with asserted good results in Canton, Amoy, Annam, Oporto, and elsewhere. Immunity conferred by this serum is said to last about a fortnight. Haffkine has prepared a prophylactic fluid which appears to be more reliable than Yersin's serum. It is a fluid culture of pest bacilli sterilized by heating to 70° C. for one hour. Of 11,362 individuals from the plague-infected regions near Bombay who were inoculated with Haffkine's fluid, 12 died, 3 of whom were sick when inoculated, and 33 developed the disease and recovered. The immunity afforded by this vaccine is said to last from four to six months.

Lustig recommends as a prophylactic a nucleoproteid obtained from dead plague bacilli, and as a remedy a serum obtained by treating horses with this nucleoproteid. It is stated that the rate of recovery under this serum in 475 cases was 39.36 per cent., against 20.6 per cent. of recoveries in 5962 cases treated during the same time without serum.

Cholera Toxin.—It has been shown by Pfeiffer and others that the blood-serum of human beings who have recently recovered from attacks of cholera and of animals which have been artificially immunized has the property of disintegrating and dissolving cholera vibrios. Unfortunately, this bacteriolytic serum has thus far proved unavailing when applied to the treatment of cholera in human beings. Haffkine, however, appears to have been measurably successful in producing artificial immunity in human beings by means of subcutaneous injections of attenuated cholera cultures.

Typhoid Toxin.—It has been shown that the blood-serum of animals artificially immunized by gradually increasing doses of living or dead typhoid bacilli acquires specific bacteriolytic and antitoxic properties. But the attempt to use this serum in the treatment of typhoid fever has resulted in a complete failure. Protective vaccination, however, by means of sterile cultures of the typhoid bacillus, as practised by the British military surgeons in South Africa, appears to have been attended with considerable success.

Antivenene.—The studies of Sewall, Phisalix and Bertrand, Calmette, Frazer, and others have established the fact that animals can be successfully immunized against the venom of poisonous snakes by being inoculated with increasing doses of venom, and, further, that the blood-serum of animals thus treated possesses both protective and antidotal properties. To this serum the name antivenene has been applied. Calmette asserts that a dose of 10 c.c. of his serum will be more than sufficient to prevent intoxication and death, provided the serum be injected two, or even four, hours after the snake's bite. The practical tests of antivenene already reported lend considerable support to Calmette's claims.

McFarland summarizes the treatment of snake-bite as follows: Stop immediately the circulation in the bitten part, so as to prevent absorption of the poison. Incise and enlarge the fang-wound freely, and suck forcibly to extract the poison, the poison being harmless when swallowed. Inject hypodermically from 3 to 6 drops of a fresh 10 per cent. aqueous solution of calcium chlorid into about a dozen areas around the wound. Give strychnin to stimulate the respiratory center. *Immediately* inject from 10 to 20 c.c. of antivenomous serum, and repeat these injections frequently. Persons living in or going into regions where there is danger of snake-bites should carry a bottle of antivenene with them.

IRRITANTS AND COUNTERIRRITANTS.

Irritants are substances which, when applied to the surface of the body, cause active hyperemia or inflammation. When they are applied not simply for their local action, but to influence for the better neighboring or remote morbid processes, they are termed *counterirritants.* We have yet no clear understanding of the manner in which counterirritants do good; the most plausible theory, however, is that they call forth reflexly from the central nervous system centrifugal influences, vasomotor or trophic or both, which act favorably upon local perturbations.

Counterirritants which produce merely an acute hyperemia of the skin are known as *rubefacients;* those which act more severely and lead to the formation of blisters are termed *vesicants* or *epispastics*; and those which act especially on the ducts of the sudoriferous glands, producing a crop of pustules, are called *pustulants.*

Counterirritation may be carried to a still greater degree of intensity by producing ulceration, and for this purpose it is now customary to use the *actual cautery.* This may be applied by means of irons of various shapes, heated to redness, or, better still, by means of the Paquelin cautery, in which heated platinum-sponge is brought to incandescence by contact with the vapors of naphtha or rhigolene.

Light applications of the thermocautery are often very efficacious in certain deep-seated affections attended with severe pain, such as sciatica, chronic meningitis, arthritis, and locomotor ataxia.

Still another method of producing a counterirritant effect, applicable to certain painful affections, is *acupuncture*, which consists in thrusting fine needles deeply into the tissues. In sciatica and lumbago this treatment sometimes gives brilliant results. The needles should be four or five inches long and about the thickness of a bonnet-pin. Having been sterilized, they should be inserted into the most painful spots to a depth of two or three inches, and allowed to remain for from ten to fifteen minutes.

The most important rubefacients are:

Mustard.	Capsicum.
Iodin.	Chloroform.
Oil of turpentine.	Ammonia.
Arnica.	Camphor.
Pitch.	Aconite.
Veratrin.	Volatile oils.

Rubefacients may be applied with advantage in acute congestion of the internal organs, in inflammatory affections of a mild type, and in functional pains. Thus they are useful in acute congestion of the lung, acute bronchitis, acute catarrhal enteritis, gastralgia, intestinal colic, muscular rheumatism, and neuralgia of the superficial nerves. When applied over large surfaces they serve also as general stimulants, and as such may be employed to arouse the system in collapse, shock, and narcotic poisoning. They may be applied in the form of plasters (mustard), stupes (oil of turpentine), liniments (chloroform, arnica), pigments (iodin), or ointments (veratrin).

The chief vesicant is cantharides.

A number of the drugs classed as rubefacients (ammonia, mustard, chloroform, iodin), if applied in concentrated form, will also produce vesication, but they are never employed for that purpose.

Vesicants exert a more profound and a more lasting effect than rubefacients. They are often of service in inflammatory diseases of a severe type, such as pleurisy, pericarditis, gastritis, iritis, arthritis, neuritis, etc., and in obstinate forms of neuralgia.

Vesicants are usually applied directly over the affected part, though some authorities hold that they should be placed some little distance from it. In neuralgia they are often most beneficial, as Anstie pointed out, when applied to a posterior branch of the spinal nerve-trunk from which the painful nerve issues. In trifacial neuralgia it is customary to apply the blister behind the ear.

The only pustulant in ordinary use is croton oil.

A mixture of croton oil (see p. 204) and olive oil (1 to 3), or the liniment of croton oil of the British Pharmacopœia (croton oil, 1 part ; cajuput oil and rectified spirit, of each, $3\frac{1}{2}$ parts), is sometimes of benefit as a counterirritant in bronchitis, pleurodynia, and dry pleurisy.

A large number of drugs which produce irritation when applied locally are used for their stimulant effects upon the superficial tissues themselves in various forms of chronic inflammation, such as subcutaneous indurations, indolent ulcers, and certain chronic skin diseases. Only those, however, which are used almost exclusively for this purpose will be considered here :

Chrysarobin.	Chaulmoogra oil.
Oil of cade.	Gurjun balsam.

Jequirity.

CANTHARIS, U. S. P.

(Cantharides; Spanish Flies.)

Spanish fly is the dried beetle, *Cantharis vesicatoria,* secured chiefly in southern Europe. The active principle is *cantharidin,* the anhydrid of cantharidic acid. The latter does not exist uncombined, but its salts are obtained by acting on cantharidin with alkalis. Cantharidin occurs in colorless prisms, readily soluble in alcohol, ether, chloroform, or oils, but only sparingly soluble in water.

PREPARATIONS.	DOSE.
Tinctura Cantharidis, U. S. P.	1–5 min. (0.06–0.3 c.c.).
Ceratum Cantharidis, U. S. P. (cantharides, 32; oil of turpentine, 15; yellow wax, 18; resin, 18, and lard, 22).	
Collodium Cantharidatum, U. S. P. (cantharidal collodion : contains 60 per cent. of cantharides).	
Emplastrum Picis Cantharidatum, U. S. P. (cantharidal pitch plaster or warming plaster : 8 per cent. of the cerate in Burgundy pitch).	

Physiologic Action.—When applied to the skin, preparations of cantharides produce redness and burning, and, later, vesication. If the action of the drug be allowed to continue, it may ultimately lead to pustulation, ulceration, and sloughing. A cerate of good quality requires from six to ten hours to produce a blister, the time varying with the part to which it is applied and the condition of the skin. To mucous membranes cantharides is also highly irritating, that of the genito-urinary tract being especially susceptible to its influence.

Internally doses of from 5–10 min. (0.3–0.6 c.c.) increase to some extent the excretion of urine, and often cause frequent desire to micturate, with some dysuria. The active principle of the drug is rapidly eliminated by the kidneys. Toxic doses are followed by intense irritation of the alimentary canal and of the genito-urinary tract. The symptoms of poisoning are burning pain in the mouth, throat, and stomach, dysphagia, great thirst, ptyalism, vomiting and purging of mucous and bloody material, and extreme prostration. Later, when a sufficient quantity of the poison has been absorbed, symptoms referable to the genito-urinary tract appear, such as aching pains in the loins, a constant desire to urinate, severe vesical tenesmus, with the passage of merely a few drops of bloody, albuminous urine, and, in males, priapism. Erotic excitement sometimes occurs, and in women abortion may follow the powerful irritant effects of the drug upon the pelvic viscera. In fatal cases death may be preceded by coma and convulsions. If the cantharides has been taken in the form of powder, the characteristic shining green parts of the insect may be recognized in the vomitus. Section after death shows gastro-enteritis and glomerulonephritis.

Cantharides is absorbed readily from the skin, and a number of cases are on record in which severe and even fatal nephritis has resulted from too free vesication. Owing to unusual susceptibility, the use even of very small blisters in some persons is followed by strangury.

Treatment of Poisoning.—This consists in evacuating the stomach, and in relieving the local irritation by the free administration of demulcents. Opium may be required for the pain. Fatty substances should be withheld, as they dissolve cantharidin and so favor its absorption. The treatment of cantharidal nephritis does not differ from that usually adopted in other forms of the disease. The treatment of strangury consists in applying hot fomentations to the lower part of the abdomen, in giving freely diluent drinks, and in administering rectal injections of laudanum.

Therapeutics.—The most important use of cantharides is as a counterirritant. As a vesicant it is often of decided value in acute inflammation of serous membranes—*pleurisy, pericarditis, synovitis*, etc. In the early stage of these diseases it tends to relieve the pain, and, later, it aids in the absorption of the effusion. In severe forms of *neuritis* small blisters may be applied with advantage over the course of the affected nerve. In *facial neuralgia* blistering behind the ear may afford speedy relief. In *acute rheumatism* the application of blisters around the inflamed joint will often be found to relieve the pain and to reduce the swelling. In *pneumonia* of the usual type blisters are of doubtful value, but they are sometimes efficacious when used repeatedly in cases of *delayed resolution*. In the early stage of *phthisis*, when the cough is troublesome, counterirritation with cantharides often proves very beneficial. In *oöphoritis* the application of a blister over the affected gland is very advantageous. *Obstinate vomiting* due to acute irritation of the stomach sometimes yields promptly to a vesicant applied to the epigastrium.

The tincture of cantharides, well diluted, has an established reputation as a stimulant lotion in the treatment of *premature alopecia* and *alopecia areata*.

Internally cantharides has been used to some extent as a stimulant to the genito-urinary tract, in *incontinence of urine*

28

from atony of the bladder, and in *chronic pyelitis*, *cystitis*, and *urethritis*, but it is neither so safe nor so efficacious as many other remedies in common use. Ringer and others have recommended small doses of the tincture in the treatment of *acute nephritis* when all symptoms of active inflammation have subsided and there is a tendency for the disease to assume a chronic type, but we would urge the utmost caution in using the drug in such cases. In very large doses, probably by irritating the bladder and urethra, cantharides sometimes excites the sexual appetite, hence it has been employed as an aphrodisiac; for this purpose, however, it is without merit.

Administration.—The tincture is the only form in which the drug is used internally. It should be taken after meals, well diluted. For blistering it is customary to use a plaster made of the cerate. The market is well supplied with good ready-made plasters which may be cut the desired size and shape. To obtain the best results the skin should be washed, and shaved if necessary, and then thoroughly dried; before the plaster is applied it should be moistened with vinegar or a few drops of oil. Vesication is generally induced in from six to ten hours; even if it is not, it is advisable to remove the plaster at the end of that time and to complete the operation by applying a poultice. In delicate subjects the poultice should be used at the end of three or four hours. When the bleb has fully developed, it should be opened with a large needle, and then dressed with a pledget of dry cotton. In many cases excellent results are obtained from a succession of blisters applied to different parts of the affected region and allowed to remain only long enough to produce a rubefacient effect (*flying blisters*). Cantharidal collodion may be used as a blistering agent in cases in which there is difficulty in controlling the patient. Cantharidal pitch plaster may be employed when it is desired to produce vigorous rubefaction, but not vesication.

Contraindications.—Vesicants must be used with the greatest caution in very young, old, or debilitated subjects, as they may occasion sloughing. For the same reason they should be avoided in patients with diabetes. When there is active nephritis, some other irritant than cantharides should be selected.

SINAPIS ALBA, U. S. P., AND SINAPIS NIGRA, U. S. P.

(White Mustard and Black Mustard.)

White mustard is the seed of *Brassica alba*, and black mustard is the seed of *Brassica nigra*. Both of these herbs are largely cultivated in Europe and America. The *mustard* sold as a condiment is a mixture of powdered white and black mustard, often more or less adulterated.

White mustard contains a ferment, *myrosin*, and a glucosid, *sinalbin*. In the presence of water the ferment acts upon the glucosid and separates from it an acrid fixed oil—*acrinyl sulphocyanate*. Black mustard also contains myrosin, and this ferment, in the presence of water, acts upon another constituent, *sinigrin*, separating from it an intensely irritant volatile oil —*allyl isosulphocyanate*.

PREPARATIONS.

Charta Sinapis, U. S. P. (paper spread with a mixture of powdered and purified black mustard and an India-rubber solution).

Oleum Sinapis Volatile, U. S. P. (distilled from a mixture of powdered black mustard and water).

Linimentum Sinapis Compositum, U. S. P. (contains volatile oil, 3 ; fl. ext. mezereum, 20 ; camphor, 6 ; castor oil, 15 ; alcohol, to make 100).

Physiologic Action and Therapeutics.—Mustard made into a paste with water and applied to the skin causes redness, heat, and burning pain, and, if the contact be prolonged, vesication. As the blisters produced by mustard are healed with difficulty, the drug is used externally only as a rubefacient. Applied in the form of a plaster (*sinapism*) or of the official paper, it often affords much relief in various painful affections, such as *bronchitis, cerebral congestion, headache, intestinal colic, gastritis, muscular rheumatism,* and *neuralgia.* A plaster may be made by spreading between two layers of thin muslin a paste made by mixing ordinary mustard and flour, equal parts, with warm water. Hot water should not be used, since high temperature destroys the ferment required to evolve the irritant oil from the glucosid. Such a plaster should be left on from twenty to thirty minutes, or until the skin is quite red. A very mild rubefacient effect may be secured by sprinkling a little mustard on the surface of a flaxseed poultice.

Internally, mustard acts as an irritant emetic, evacuating the stomach promptly and thoroughly. It may be employed in cases of *narcotic poisoning,* the dose being a tablespoonful stirred up in warm water, and repeated, if necessary, in fifteen minutes.

Thiosinamin (*Allyl-sulphourea*).—This is a compound produced by acting on the volatile oil of mustard with ammonia. It occurs in colorless, soluble crystals, of bitterish taste and a garlicky odor. The reputed action ascribed to it is the softening of scar tissue. Hebra, Unna, Juliusberg, Crocker and

Pernet, and others have recommended it in *keloid*, *post-lupus scarring*, *scleroderma*, and *cicatricial strictures* and *deformities*. It may be given hypodermically every other day, in doses of from 1–2 gr. (0.065–0.13 gm.), dissolved in glycerin and water, or by the mouth in daily doses of 3 gr. (0.2 gm.). Another method of application is by the local use of thiosinamin plasters (10 to 30 per cent.).

Hebra has warned against the use of the drug in all cases of partially healed tuberculous foci.

ARNICA.

Arnica is official as the flowers (*Arnicæ Flores*, U. S. P.) and as the root (*Arnicæ Radix*, U. S. P.) of *Arnica montana*, a perennial herb growing in the temperate regions of Europe, Asia, and America. It contains a volatile oil, tannin, and a glucosid, *arnicin*.

PREPARATIONS.	DOSE.
Tinctura Arnicæ Florum, U. S. P.	10–30 min. (0.6–2.0 c.c.).
Tinctura Arnicæ Radicis, U. S. P.	5–20 min. (0.3–1.2 c.c.).
Extractum Arnicæ Radicis Fluidum, U. S. P.	5–10 min. (0.3–0.6 c.c.).
Extractum Arnicæ Radicis, U. S. P.	2–5 gr. (0.13–0.3 gm.).
Emplastrum Arnicæ, U. S. P. (contains 33 per cent. of the extract).	

Physiologic Action and Therapeutics.—Applied to the skin, arnica causes redness and burning, and occasionally very severe erysipelatous inflammation. Internally, according to Hare, moderate doses slow the pulse and raise the blood-pressure. Toxic doses cause gastro-intestinal irritation, a feeble pulse, profound prostration, and, sometimes, delirium and stupor.

Arnica, in the form of the tincture of the flowers, has been used very largely as a stimulant application in *sprains* and *bruises*. Internally, it has been recommended in a variety of diseases, but there is little evidence to show that it is of value in any one of them.

PIX BURGUNDICA, U. S. P.

(Burgundy Pitch.)

Burgundy pitch is a resinous exudation obtained from *Abies excelsa*, or Norway spruce, a lofty conifer growing in Europe and Northern Asia. It contains a volatile oil and resin.

Emplastrum Picis Burgundicæ, U. S. P. (Burgundy pitch, 80; olive oil, 5; yellow wax, 15).

Emplastrum Picis Cantharidatum, U. S. P. (Burgundy pitch, 92; cerate of cantharides, 8).

Emplastrum Ferri, U. S. P. (14 per cent.).

Emplastrum Opii, U. S. P. (18 per cent.).

Therapeutics.—Applied to the skin, Burgundy pitch is slightly irritating. In the form of a plaster it is sometimes used for its gentle but persistent rubefacient effect in such conditions as *chronic bronchitis, lumbago, pleurodynia,* and *chronic articular rheumatism.*

VERATRINA, U. S. P.

(Veratrin.)

Veratrin is an alkaloid obtained from the seed of *Asagræa officinalis,* a bulbous herb growing in Mexico and Central America. The official preparation is a mixture of pure veratrin and a number of other less active alkaloids. It is a white, amorphous, or semicrystalline powder, odorless and intensely acrid. It is readily soluble in alcohol, but very slightly so in water.

PREPARATIONS.

Unguentum Veratrinæ, U. S. P. (4 per cent.).

Oleatum Veratrinæ, U. S. P. (2 per cent.).

Physiologic Action and Therapeutics.—Upon cutaneous surfaces veratrin acts first as an irritant and subsequently as a depressant, causing tingling and burning, and then a feeling of numbness and coldness. Inhaled, it produces violent sneezing and cough, and applied to the conjunctiva severe inflammation. Internally large doses cause burning in the throat and stomach, salivation, vomiting and purging, a weak and irregular pulse, muscular tremors, convulsions, fall of temperature, collapse, and, finally, death from asphyxia. Upon striated muscles the drug has a peculiar effect, which consists in a more intense response to stimuli and a remarkable prolongation of the period of contraction. Finally, after toxic doses, the muscles become paralyzed. The cardiac inhibitory center and the vasomotor center are primarily stimulated, in consequence of which the pulse is slowed and the blood-pressure is raised; later, however, the medullary centers, and probably the heart also, are depressed, so that the pulse becomes rapid and weak. The fall of temperature is apparently due to failure of the circulation, and the convulsions to stimulation of the centers in the spinal cord.

Veratrin is not used internally. In the form of the ointment or oleate it is rarely employed as a local remedy in *neuralgia*. It should never be applied over large surfaces or where there are any abrasions. In employing it about the face great care must be exercised to keep it out of the eye.

CHRYSAROBINUM, U. S. P.

(Chrysarobin.)

Chrysarobin is a neutral principle, more or less impure, obtained from *goa powder*, the medullary matter of *Andira Araroba*, a large tree growing in the forests of Brazil. It is a yellow, crystalline powder, odorless and tasteless, and very slightly soluble in water, alcohol, chloroform, or ether. On exposure to air it turns brownish, owing to partial oxidation into chrysophanic acid.

PREPARATION.

Unguentum Chrysarobini, U. S. P. (5 per cent.).

Physiologic Action and Therapeutics.—Chrysarobin is a local irritant and a parasiticide. Internally, large doses cause gastro-enteritis and nephritis. It is eliminated chiefly by the kidneys, partly as chrysophanic acid and partly unchanged.

Although it stains the skin temporarily and the clothing permanently, and, in some persons, causes severe dermatitis, chrysarobin is probably the most efficient remedy we have in *psoriasis*. It is adapted to cases in which the patches are comparatively few and large, or to the large patches in extensive cases (Stelwagon). If used too freely, it may be absorbed and give rise to untoward symptoms. As the conjunctiva is exceedingly sensitive to its irritant effect, it should not be employed about the face. It may be prescribed in an ointment, 10–60 gr. (0.65–4.0 gm.) to the ounce (30.0 c.c.); suspended in chloroform, 1–2 dr. (4.0–8.0 gm.) to the ounce (30.0 c.c.); or suspended in collodion, 30–60 gr. (2.0–4.0 gm.) to the ounce (30.0 c.c.). When the infected areas are small, chrysarobin may be applied with advantage also in *ringworm*, either of the body or of the scalp.

OLEUM CADINUM, U. S. P.

(Oil of Cade ; Juniper Tar Oil.)

Oil of cade is a volatile oil distilled from the wood of *Juniperus Oxycedrus*, a shrub resembling the common juniper and growing in Southern Europe. It is a thick, dark-brown liquid having a tarry odor and taste. It is completely soluble in ether or chloroform and partially so in alcohol.

Therapeutics.—The action of oil of cade is very similar to that of the oil of tar. The drug is chiefly used as a stimulant application in *psoriasis* and *chronic eczema.* It may be prescribed in ointment, 1–3 fl. dr. (4.0–11.0 c.c.) to the ounce (30.0 gm.), or as a pigment made by diluting the oil with one or two parts of alcohol.

OLEUM GYNOCARDIÆ.

(Oil of Gynocardia; Chaulmoogra Oil.)

Chaulmoogra oil is an acrid, whitish fat obtained from the seed of *Gynocardia odorata*, a native of the East Indies. Its activity is believed to be due to *gynocardic acid.* It has been used with asserted good results, both internally and externally, in *leprosy.* The dose is from 10–30 min. (0.6–2.0 c.c.) in capsules or in emulsion. Externally, it is prescribed by inunction, a 50 per cent. ointment being thoroughly rubbed into the affected parts for an hour or two daily. It has been employed also to some extent as a local stimulant in *sprains* and *bruises.*

BALSAMUM DIPTEROCARPI.

(Balsam of Dipterocarpus; Gurjun Balsam; Wood Oil.)

Gurjun balsam is an oleoresin obtained by incising *Dipterocarpus turbinatus* and other species of *Dipterocarpus*, large trees growing in India and the East Indies. It contains a volatile oil, gurjunic acid, and resin. It has been used largely to adulterate the oleoresin of copaiba, which it resembles very closely in its properties. It is said to be of value, both as an internal remedy and as a local application in *leprosy.* The dose is from 10–40 min. (0.6–2.5 c.c.), in emulsion. Externally it is applied by inunction, diluted with from 1 to 3 parts of lime-water or olive oil.

OTHER COUNTERIRRITANTS.

Oil of Turpentine (see p. 230).—In the form of a stupe, oil of turpentine makes an excellent rubefacient application in a large number of affections attended with pain or irritation, such as *muscular rheumatism, intestinal colic, gastralgia, bronchitis*, and *pulmonary congestion.* Turpentine liniment applied with friction is useful in *muscular rheumatism, chronic articular rheumatism*, and *chilblain.*

Iodin (see p. 310).—As a counterirritant, iodin is well

adapted to conditions requiring a mild but persistent effect. The tincture applied as a pigment is extensively employed in such affections as *laryngitis, pleurisy, synovitis, arthritis, neuritis, periostitis, bubo,* and *chilblain.*

Capsicum (see p. 179).—Tincture of capsicum, more or less diluted, may be applied as a lotion in *acute torticollis* and *chilblain.* Capsicum plaster is sometimes serviceable in relieving *lumbago* and other *muscular pains.*

Chloroform (see p. 109), **Aconite** (see p. 59), **Ammonia** (see p. 40), **Camphor** (see p. 115).—These drugs are extensively employed as rubefacients in the form of liniments. They are often combined with advantage, as in the following formula :

    ℞   Tincturæ aconiti,
        Chloroformi,
        Olei terebinthinæ,   āā f ℥ss;
        Linimenti saponis,   f ℥iij. M.

JEQUIRITY.

(Abrus; Prayer Beads.)

Jequirity is the seed of *Abrus precatorius,* a plant growing in India and in Brazil. It contains a toxalbumin known as *abrin.* In the form of a freshly prepared infusion (3 per cent.) it has been used as a local remedy in the treatment of *chronic granular conjunctivitis* and of the *corneal pannus* associated with this disease. Applied to the eye, it produces a severe membranous conjunctivitis which must be controlled by cold compresses. As the treatment is somewhat hazardous, it is rarely employed at the present day except in long-standing cases of trachoma with very dense pannus.

ESCHAROTICS OR CAUSTICS.

Escharotics or caustics are agents that corrode or disorganize the tissues. Some, like arsensic, have a specific poisonous action upon the cells; some, like the strong mineral acids, extract water from the tissues and precipitate the proteids; some, like the alkaline hydrates, not only extract water, but dissolve the proteids and form with them soluble compounds; some, like the nitrate of lead, form with the proteids insoluble albuminates, and in so doing also liberate an acid; some, like the nitrate of mercury, poison the cells directly, form albuminates with the proteids, and also set free an acid;

while others still, like bromin, are powerful oxidizers of organic matter. The most important escharotics are:

Sulphuric acid.	Osmic acid.
Nitric acid.	Bromin.
Carbolic acid.	Potassium hydrate.
Arsenous acid.	Sodium hydrate.
Acetic acid.	Sodium ethylate.
Trichloracetic acid.	Zinc chlorid.
Chromic acid.	Mercuric nitrate.
Lactic acid.	Silver nitrate.
Pyrogallic acid.	Lead nitrate.

The *actual cautery* (hot iron; Paquelin cautery; galvano-cautery) also offers a prompt and powerful means of destroying tissue.

Escharotics are used to destroy exuberant granulations, the tissue around poisoned wounds, and small growths, such as warts, epitheliomata, and lupus infiltrations, and also to modify the specific character of phagedenic ulcers and sloughing wounds.

Depilatories are agents employed to remove superfluous hair. The most important are *barium sulphid* and *calcium sulphydrate.* Permanent removal of hair can be effected only by *electrolysis.*

ACIDUM SULPHURICUM, U. S. P.

(Sulphuric Acid; Oil of Vitriol; H_2SO_4.)

Official sulphuric acid is a colorless, heavy, oily liquid, odorless, and of intensely sour taste and acid reaction. It has a marked affinity for water, with which it mixes with the evolution of heat. It rapidly chars and destroys organic matter.

PREPARATIONS.	DOSE.
Acidum Sulphuricum Dilutum, U. S. P. (contains 10 per cent. of absolute sulphuric acid)	10–20 min. (0.6–1.2 c.c.).
Acidum Sulphuricum Aromaticum, U. S. P. (an alcoholic solution containing about 20 per cent. of the official acid, with ginger and cinnamon)	5–20 min. (0.3–1.2 c.c.).

Physiologic Action.—When applied to the skin in concentrated form, sulphuric acid causes intense pain and a rapid destruction of tissue, the eschar being first white and then brown or black. Upon mucous membranes it has a still more irritant and corrosive effect. The local corrosion results from the extraction of water, the neutralization of the alkalis, and the precipitation of albumins. Internally, large doses produce burning in the throat and gullet, violent pain in the abdomen, constant vomiting of dark matter, often mixed with blood and mucus, intense thirst, difficult breathing, and collapse. Death usually occurs within twenty-four hours, and if the acid has reached the larynx, a fatal termination may result almost immediately from suffocation. Blackish stains on the clothing may serve as a clue to the recognition of the poison.

Even in very dilute form sulphuric and other mineral acids, in large doses, may prove fatal to herbivorous animals by overcoming the alkalinity of the blood, thereby rendering it incapable of removing carbonic acid from the tissues. Such an untoward effect, however, can scarcely occur in carnivorous animals or in man, since in them the tissues under the influence of the acid yield large quantities of ammonia, which unites with acid, thus protecting the alkalis of the blood from neutralization. Sulphuric acid circulates in the blood in the form of its salts, and is rapidly eliminated in the urine chiefly as acid salts.

Treatment of Poisoning.—This consists in administering alkalis,—chalk, magnesia, soap, white-wash, or alkaline carbonates,—together with demulcent drinks, such as albumin-water, milk, or barley-water. Opium is always required to relieve the severe pain. The stomach-pump should be avoided on account of the risk of piercing the softened esophagus.

Therapeutics.—Sulphuric acid, owing to the severity of its action, is less useful as an escharotic than some of the other acids. It is occasionally employed, however, as a caustic in *phagedenic chancroid*, being applied in the form of a paste made by mixing the acid with some indifferent substance, such as charcoal or asbestos. It has been used also to a limited extent in *caries*, to hasten the removal of the diseased bone.

Internally, in the form of aromatic sulphuric acid, it is sometimes of service as an antihydrotic in the *night-sweats of phthisis*. Combined with opium, it has been recommended in *serous diarrhea*. Sulphuric acid lemonade has been used by workers in lead as a prophylactic against *plumbism*.

Administration.—Sulphuric acid should be given well diluted, precautions being taken to prevent its action on the teeth.

Incompatibles.—Alkalis, alkaline carbonates, iodids, and salts of lead and of lime. It is explosive with sugar and oil of turpentine. The undiluted acid carbonizes syrups.

ACIDUM NITRICUM, U. S. P.

(Nitric Acid; HNO$_3$.)

The official nitric acid is a colorless, intensely acid, fuming liquid composed of 62 per cent. of absolute nitric acid and 32 per cent. of water.

PREPARATIONS.	DOSE.
Acidum Nitricum Dilutum, U. S. P. (10 per cent. by weight of absolute nitric acid)	5–20 min. (0.3–1.2 c.c.).
Acidum Nitrohydrochloricum, U. S. P. (aqua regia: nitric acid, 18; hydrochloric acid, 82)	2–4 min. (0.1–0.25 c.c.).
Acidum Nitrohydrochloricum Dilutum, U. S. P. (nitric acid, 4; hydrochloric acid, 18; water, 78)	5–20 min. (0.3–1.2 c.c.).

Physiologic Action and Therapeutics.—Locally, pure nitric acid is a powerful caustic, but somewhat less painful and less penetrant than sulphuric acid. It stains the skin yellow. In overdoses it produces symptoms and lesions (except the yellow coloration) similar to those produced by sulphuric acid, and in poisoning the same treatment is applicable.

As an escharotic, nitric acid is generally preferred to other mineral acids, as its action is more readily controlled. It has been found reliable in *phagedenic ulcers*, *chancroids*, and *chancres*, and in *hospital gangrene*. It has been used very successfully also, when milder measures have failed, as a means of producing inflammatory contractile tissue in *prolapse of the rectum* occurring in children.

Internally, dilute nitric acid has been recommended by Ringer and others in *indigestion*, especially when eructation of offensive gas is a prominent symptom, or when there is a tendency to persistent aphthous stomatitis. It should be taken well diluted through a glass tube.

Incompatibles.—Alkalis, carbonates, oxids, lead acetate, and iron sulphate. The strong acid forms explosive compounds with readily oxidizable substances, such as glycerin, alcohol, phenol, resins, volatile oils, etc.

Nitrohydrochloric Acid.—This is a yellow, fuming, corrosive liquid produced by mixing nitric acid with hydrochloric acid. Both acids undergo decomposition with the formation of nitrosyl chlorid (NOCl) and chlorin. Nitrohydrochloric acid is believed to have a special action on the liver; whether this be true or not, the drug is often of value in that form of *indigestion* which culminates at frequent intervals in so-called *bilious attacks.* It has a well-deserved reputation also in *mal-assimilation* manifested by gaseous eructations, great mental depression, and persistent oxaluria. The best results are obtained by using the strong acid, freshly prepared, though considerable care must be exercised in handling it. It should be taken after meals, well diluted, through a tube. It should not be prescribed undiluted with tinctures or other alcoholic preparations, as such mixtures are liable to explode.

Nitrohydrochloric-acid baths were formerly employed rather extensively in various affections of the liver, but it is doubtful whether they exerted any influence beyond a beneficial stimulation of the skin. Nitrohydrochloric acid is not used as an escharotic.

ACIDUM ACETICUM.

(Acetic Acid; $HC_2H_3O_2$.)

Acetic acid is official in the following forms:

Acidum Aceticum, U. S. P. (36 per cent. by weight of absolute acetic acid).
Acidum Aceticum Dilutum, U. S. P. (6 per cent. by weight of absolute acetic acid).
Acidum Aceticum Glaciale, U. S. P. (nearly absolute acetic acid).

Vinegar (*acetum*) is not official, but has the same therapeutic value as diluted acetic acid.

Physiologic Action and Therapeutics.—Locally applied, concentrated acetic acid produces redness, vesication, and ultimately superficial sloughing. Taken internally, it causes severe gastritis. Even small doses of acetic acid, if long continned, cause chronic gastric catarrh. It is eliminated as carbonates, so it does not increase the acidity of the urine.

Glacial acetic acid is sometimes employed to remove *warts* and *corns*. Inhalations of diluted acetic acid or of vinegar are often useful in checking vomiting occurring after general anesthesia. Vinegar is of service as a local hemostatic in *epistaxis*, *leech-bites*, and *small wounds*, as well as in *postpartum hemorrhage*. In the form of an enema it is occasionally employed to destroy *oxyures*. Internally, it may be used as an antidote in *poisoning by alkalis*.

ACIDUM TRICHLORACETICUM.

(Trichloracetic Acid; $C_2HCl_3O_2$.)

Trichloracetic acid is formed by acting on glacial acetic acid with chlorin or on chloral with fuming nitric acid. It occurs

in colorless, hygroscopic crystals, of pungent odor and strong
acid reaction.

Therapeutics.—Pure or in concentrated solution, trichloracetic acid has been used to some extent as a caustic for removing small growths, such as *venereal* or *cutaneous warts, nevi,* and
corns. A solution of from 5 to 10 per cent. makes an excellent stimulant application in *chancroids, mucous patches, tuberculous ulcers,* and *fistulous tracts.* In this strength it causes
but little pain or inflammatory reaction.

ACIDUM CHROMICUM, U. S. P.

(Chromic Acid; Chromic Anhydrid; CrO_3.)

Chromic acid occurs in crimson, deliquescent crystals, odorless, and readily soluble in water. With organic matter, alcohol, glycerin, ether, tannin, sugar, etc., it is explosive.

Physiologic Action and Therapeutics.—Owing to its
strong oxidizing power chromic acid is an energetic caustic,
slower and somewhat less painful in its action, however, than
nitric acid, potassa, or silver nitrate. Internally, it is an active
poison, producing severe abdominal pain, vomiting, purging,
albuminuria, and collapse. The vomited matters are often of
a green or bluish-green color. The postmortem lesions consist in inflammation and erosion of the gastro-intestinal mucous membrane, nephritis, and fatty degeneration of the tissues,
especially of the liver. Alkalis neutralize the acid, but as they
form poisonous salts, the stomach should be thoroughly
emptied after their administration. The local application or inhalation of chromium compounds may also result in poisoning.
Workers in these preparations not infrequently suffer from ulcerations of the skin and mucous membranes, chronic bronchitis, and interstitial nephritis.

Chromic acid may be used to destroy small growths, like
common warts, venereal warts, and *corns.* For this purpose it
may be applied in substance or in strong solution—2–3 dr. (8.0–
12.0 gm.) to the ounce (30.0 c.c.). The fused acid on a probe
may also be employed successfully in removing *soft hyper-
trophies* of the nasopharyngeal mucous membrane and in closing small *salivary fistulæ.* A solution of from 20–30 gr. (1.3–
2.0 gm.) to the ounce (30.0 c.c.) makes an excellent stimulant
application in *mucous patches.* A solution of 40 gr. (2.6 gm.) to
the ounce (30.0 c.c.), applied once a week, has been well recommended in *hyperidrosis* and *bromidrosis of the feet.*

Potassium Bichromate.—This salt, in doses of from
$\frac{1}{12}-\frac{1}{10}$ gr. (0.005–0.0065 gm.), has been used with good results

by Frazer, Bradbury, White, and others in *chronic gastric catarrh* and *gastric ulcer.* It should be given in capsules or in pills made with kaolin, thrice daily, on an empty stomach.

ACIDUM LACTICUM, U. S. P.

(Lactic Acid; $HC_3H_5O_3$.)

Lactic acid is an organic acid usually obtained by subjecting milk-sugar or grape-sugar to lactic fermentation. The official preparation is composed of 75 per cent. by weight of absolute lactic acid and 25 per cent. of water.

Therapeutics.—Lactic acid, in solutions of from 20 to 100 per cent., is a valuable application in *tuberculous* and *lupous ulcerations of mucous membranes.* In the proportion of 1 part of the acid to from 5–20 parts of water it sometimes proves efficacious in *freckles* and *chloasma.*

PYROGALLOL, U. S. P.

(Pyrogallic Acid; $C_6H_3(OH_3)$.)

Pyrogallic acid is produced by the action of heat on gallic acid. It occurs in white, lustrous needles or scales, odorless, and of a bitter taste. It is readily soluble in water, alcohol, or ether.

Physiologic Action and Therapeutics.—Locally, pyrogallol is an irritant or caustic, according as it is applied in dilute or in concentrated form. It is also an active parasiticide. It stains the skin a brownish color. As a caustic its action is slow and attended with but little pain. Owing to its power to destroy the red blood-cells and to convert hemoglobin into methemoglobin, grave, or even fatal, toxic effects may result from the application of the drug over too large a surface.

Pyrogallol is an efficient caustic in *lupus* and *epithelioma*, when the lesions are small. It is generally applied in the form of an ointment, from 20–40 per cent. in strength. Stelwagon recommends the following formula:

℞	Pyrogallol,	ʒij–iij (8.0–12.0 gm.);
	Cerati resinæ,	
	Petrolati,	aa q. s. ad ℥j (31.0 gm.). M.

Such an ointment should be applied on lint, and renewed twice a day for two or three weeks, the superficial sloughs being removed every few days by poulticing or gentle scraping.

An ointment containing from 20–60 gr. (1.3–4.0 gm.) of pyrogallol to the ounce (31.0 gm.) is sometimes useful in *psoriasis.*

ACIDUM OSMICUM.

(Osmic Acid; Osmium Tetroxid; OsO_4.)

Osmic acid occurs in yellow, crystalline needles, having a pungent odor and a burning taste. It is readily soluble in water, alcohol, or ether.

Therapeutics.—Even in weak solution, it is exceedingly irritant and caustic. Stekoulis, Schapiro, Bennett, and others claim excellent results from injections of osmic acid in *neuralgia.* The nerve having been exposed by a small incision, a 1.5 per cent. solution, freshly prepared, is injected into it at several points, to the amount of from 5–10 min. (0.3–0.6 c.c.). The blood and tissues are stained intensely black by the acid, but healing of the wound is said to proceed rapidly without suppuration.

BROMUM, U. S. P.

(Bromin; Br.)

Bromin is a non-metallic element obtained from sea-water. It is a dark, reddish-brown liquid, evolving, even at ordinary temperatures, suffocating and irritating fumes. It is soluble in 30 parts of water and readily soluble in alcohol or ether.

Physiologic Action and Therapeutics.—Bromin is an intensely active and painful escharotic. It is also a disinfectant and deodorant. When ingested, it causes inflammation and extensive necrosis of the gastro-intestinal mucosa. Inhaled, it acts like chlorin, producing mucopurulent bronchitis, edema of the lungs, and foci of catarrhal pneumonia.

Bromin has been used as an escharotic in *phagedenic* and *cancerous ulcerations*, but, owing to the difficulty in handling it and the great pain caused by its application, it is rarely employed at the present time. It is still occasionally used in very weak solutions as a *deodorant* (see p. 408).

Bromipin.—This preparation is an addition-product of sesame oil and bromin (10 per cent.). It is a yellowish, oily liquid, free from the caustic properties of bromin. In daily doses of from 1–3 fl. dr. (4.0–11.0 c.c.) it has been used with asserted good results as a substitute for the bromids in *epilepsy.*

POTASSA, U. S. P.

(Potassium Hydrate; Caustic Potash; KOH.)

Caustic potash occurs in hard white pencils or fused masses, odorless, deliquescent, strongly alkaline, and corrosive. It is soluble in 0.5 part of water and in 2 parts of alcohol.

| Preparations. | Dose. |

Liquor Potassæ, U. S. P. (5 per cent.) . . 5–30 min. (0.3–2.0 c.c.).
Potassa cum Calce, U. S. P. (Vienna paste :
  50 per cent. each of potassa and lime).

Physiologic Action and Therapeutics.—In concentrated form, potassium hydrate is a diffusive and deeply penetrating escharotic. Its action is rapid and painful. It produces a grayish, soft, and pultaceous slough, which separates in a few days, leaving an ulcer. Its caustic properties are due to the power of its hydroxyl molecule to abstract water from the tissues, to soften and dissolve them, and to form with the proteids a soluble alkaline albuminate. When swallowed in strong solution it produces all the symptoms of a violent corrosive poison—burning in the throat and gullet, intense abdominal pain, vomiting and purging of mucous and bloody matter, dysphagia, hoarseness, and collapse. If recovery follows, stricture of the esophagus or cicatricial contractions of the stomach may result from the extensive ulceration.

Small doses, in dilute solution, exert an influence similar to that of the alkaline carbonates (see Gastric Antacids). The drug is eliminated as a carbonate, chiefly in the urine.

Treatment of Poisoning.—This consists in neutralizing the alkali with a weak acid, such as vinegar or lemon-juice, and in allaying the irritation and pain with demulcents and opium.

As an escharotic, caustic potash may be used to remove *common warts* and small *cutaneous cancers*. For this purpose the solid stick should be used, the surrounding parts being protected by a coating of oil. When the desired effect has been attained, further action may be prevented by the application of vinegar. Caustic potash is also an efficient agent with which to cauterize the *bites of rabid animals*. In *ingrowing toe-nail*, the insertion under the nail of a pledget of cotton soaked in liquor potassa (1 part to 4 parts of water) affords an excellent means of removing the redundant tissue, thus facilitating the raising of the nail. The diffusive action of potassa renders it unsuitable for use on mucous membranes.

Vienna paste, being less deliquescent than pure potassa, is more manageable than the latter, but slower in its action.

Internally, solution of potassa, well diluted, is rarely used as an *antacid*.

Incompatibles.—Acids, acid salts, metallic salts, and alkaloids.

SODA, U. S. P.

(Sodium Hydrate; Caustic Soda; NaOH.)

Caustic soda occurs in white, translucent pencils or fused masses, odorless, of an acrid, caustic taste and an intensely alkaline reaction. It is readily soluble in water and in alcohol.

PREPARATION.	DOSE.
Liquor Sodæ, U. S. P.	5–30 min. (0.3–2.0 c.c.).

Therapeutics.—Caustic soda may be used for the same purposes as caustic potash, the action of the two drugs being similar.

SODII ETHYLAS.

(Sodium Ethylate; Sodium Alcohol; NaC_2H_6O.)

A solution of sodium ethylate, made by dissolving 1 part of metallic sodium in 20 parts of absolute alcohol, has been used, on the recommendation of B. W. Richardson, as a caustic to remove *vascular nevi* and *warts*. It should be applied with a glass rod, and the superficial dry crust formed allowed to come off spontaneously. It is said to cause less scarring than other caustics. It is decomposed by water.

BARII SULPHIDUM.

(Barium Sulphid; BaS.)

Barium sulphid is an amorphous, pale yellow, phosphorescent powder, soluble in water. Diluted with from 1 to 3 parts of some inert powder, it is used as a *depilatory*. Duhring recommends:

℞   Barii sulphidi,   ʒj–ij (4.0–8.0 gm.);
     Amyli,
     Zinci oxidi,   aa ʒiss (6.0 gm.). M.

At the time of application sufficient water is added to make a paste, which is thickly spread over the affected part and allowed to remain for a minute or two or until it excites slight burning.

Barium sulphid deteriorates with age and on exposure to air.

CALCII SULPHIDUM HYDRATUM.

(Hydrated Calcium Sulphid; Calcium Sulphohydrate; $CaS + H_2O$.)

Hydrated calcium sulphid is a whitish or pale-pink powder, made by heating together at a high temperature calcium sulphate and granulated wood-charcoal. In the form of a paste it has been recommended by Kaposi, Brayton, and others as a harmless *depilatory.* When wetted, it gives off hydrogen sulphid, which is sometimes a drawback to its use.

OTHER ESCHAROTICS.

Arsenous Acid (see p. 299).—Arsenic is a slowly acting but deeply penetrating escharotic. It causes severe pain and considerable inflammatory swelling, and, if applied over too large a surface, may prove dangerous through absorption. On the other hand, its action, while very thorough, is largely limited to the diseased tissue. Its *modus operandi* is not definitely known, but as it does not combine chemically with the tissues, it is supposed to act directly upon the cells in a specifically poisonous manner. The employment of arsenic as an escharotic is practically confined to the removal of *small cutaneous cancers* and *lupus infiltrations* (see p..302) and the destruction of the *pulp in carious teeth.*

Carbolic Acid (see p. 383).—Carbolic acid is a comparatively superficial escharotic, the deeper tissues being protected by an albuminous coagulum. Owing to the property which it possesses of paralyzing the peripheral sensory nerves, its application is not followed by very severe or persistent pain. Absolute alcohol, as Phelps and Powell have shown, destroys its caustic action. After thorough curetment of the parts pure liquid carbolic acid may be applied with advantage in *inflammatory gangrene, irritable ulcers, carbuncles,* and *fistulæ.* Local injections of the pure acid are also efficacious in *anthrax* (see p. 385). Von Bruns claims excellent results in the treatment of *infected wounds* from the application of pure carbolic acid in small quantities and for one minute, followed immediately by irrigation with absolute alcohol.

Mercuric Nitrate (see p. 325).—This salt is an exceedingly irritant and powerful escharotic. Its action depends upon several factors: (1) The union of the metal with the proteids to form a soluble albuminate; (2) the liberation of nitric acid, which is also caustic; and (3) the toxic effects upon the cells of the mercury itself. As an escharotic it has been extensively employed in the form of the solution of mercuric nitrate in *phagedenic venereal sores.* If used too freely, it is liable to be absorbed and to induce salivation.

Zinc Chlorid (see p. 365).—This salt is an energetic and very painful caustic. It produces a dry, whitish slough, the separation of which usually requires from one to three weeks.

Unlike arsenic, it does not spare the healthy tissue. In the form of a paste to which cocain has been added, it is sometimes used to remove *cutaneous epitheliomata.*

Silver Nitrate (see p. 366).—The insoluble and impenetrable pellicle of silver albuminate which lunar caustic immediately forms when brought in contact with the tissues confines its destructive action to the superficial cells. In the form of the solid stick it is an excellent caustic for removing such formations as *small warts, mucous patches,* and *exuberant granulations.*

Lead nitrate (see p. 362).—The corrosive action of this salt of lead is due largely to the nitric acid which is liberated when it is brought in contact with proteid matter. It is sometimes used in the form of a powder as a caustic in *onychia.*

PROTECTIVES.

Protectives are agents that serve to protect inflamed or injured surfaces from external irritation. Their action is chiefly mechanical. Those of a mucoid or colloid nature, which exert a soothing effect upon inflamed parts, are known as **demulcents.** The most important members of this subdivision are:

Flaxseed.	Marshmallow.
Acacia.	Licorice root.
Tragacanth.	Almond.
Sassafras pith.	Starch.
Slippery elm.	White and yolk of egg.

Internally, demulcents are especially useful in protecting the walls of the stomach from the action of corrosive poisons. Made up into lozenges and allowed to dissolve slowly in the mouth they tend to subdue pain and cough resulting from irritation of the throat. In the form of enemata, they are used to relieve the tenesmus of acute dysentery. Externally they are rarely employed except in the form of poultices, and even in this form they are used more as a means of applying heat and moisture than as protectives. Hot poultices are very efficacious in relaxing local spasms, such as occur in asthma and the various forms of colic. In deep-seated inflammations— pneumonia, nephritis, cystitis, etc.—they usually afford considerable comfort, and no doubt exert a reflex influence on the morbid process itself (see Counterirritants). In superficial inflammations hot antiseptic fomentations have largely supplanted ordinary poultices, because the latter afford such excellent media for the growth of bacteria. When suppuration has already occurred or is obviously inevitable, poultices are wholly inadmissible.

Protectives of a fatty character, intended to shield the skin from external irritants and at the same time to soften and relax it, are known as **emollients.** The most important are:

Lard.	Almond oil.
Suet.	Linseed oil.
Wool-fat.	Petrolatum.
Spermaceti.	Glycerin.
Olive oil.	Wax.

Emollients are used for their protective and soothing properties in burns, in fissuring of the lips and hands, and in acute inflammatory diseases of the skin. They are also extensively employed as bases for special ointments.

The so-called **dusting-powders** constitute another group of protectives. The requisites of a good dusting-powder are dryness, insolubility, inertness, and absolute freedom from grittiness. Those in common use are:

Starch.	Zinc oxid.
Talc.	Chalk.
Kaolin.	Magnesium carbonate.
Lycopodium.	

Dusting-powders are prescribed in various forms of erythema and in acute eczema to protect the parts from air, moisture, and friction.

Finally, certain protectives which have strong adhesive properties and can be kept in place for several days without change are employed as **fixed dressings.** In this class are:

Collodion.	Solution of gutta-percha.

Lead plaster (*Emplastrum Plumbi,* U. S. P.), isinglass or court-plaster (*Emplastrum Ichthyocollæ,* U. S. P.), and soap-plaster (*Emplastrum Saponis,* U. S. P.) also belong in this group.

Fixed dressings are frequently of service in the treatment of abrasions, small wounds, fissures, and bed-sores.

LINUM, U. S. P.

(Linseed; Flaxseed.)

Linseed is the seed of *Linum usitatissimum,* or common flax, an annual cultivated in all temperate countries. From a therapeutic standpoint its chief ingredients are a mucilaginou. principle and a fixed oil.

PREPARATIONS.

Oleum Lini, U. S. P.
Linimentum Calcis, U. S. P. (Carron-oil : equal parts of linseed oil and lime-
　water).
Sapo Mollis, U. S. P. (soft or green soap).
Linimentum Saponis Mollis, U. S. P. (tincture of green soap: soft soap,
　65; oil of lavender, 2; alcohol, 30; water to make 100).

Therapeutics.—An infusion or tea of flaxseed serves as an excellent *demulcent drink.* Ground flaxseed is the material most commonly selected for making ordinary *poultices.* Soft soap and liniment of soft soap are used for detergent purposes and also for removing *crusts* and *scales* in sluggish diseases of the skin like seborrhea and psoriasis.

ACACIA, U. S. P.

(Gum Arabic.)

Acacia is a gummy exudation obtained from *Acacia Senegal,* a small tree growing in Senegal, Kordofan, and Abyssinia. It contains *arabic acid* in combination with calcium, magnesium, and potassium.

PREPARATIONS.	DOSE.
Mucilago Acaciæ, U. S. P.	Indefinite.
Syrupus Acaciæ, U. S. P.	Indefinite.
Pulvis Cretæ Compositus, U. S. P.	5–60 gr. (0.3–4.0 gm.).

Acacia also enters into chalk mixture, compound mixture of glycyrrhiza, emulsion of almond, and several troches and pills.

Therapeutics.—In therapeutics acacia is used chiefly as a *demulcent;* in pharmacy it is used for holding together the active ingredients in pills and lozenges and for suspending insoluble substances in water. It is precipitated from its solution by alcohol, subacetate of lead, ferric salts, and borax.

TRAGACANTHA, U. S. P.

(Tragacanth.)

Tragacanth is a gummy exudation from *Astragalus gum-mifer,* a shrub growing in western Asia. It contains *tragan-thin* or *bassorin* and the *calcium salt of gummic acid.* It swells up in water into a gelatinous mass, but, unlike acacia, does not dissolve in it.

PREPARATION.
Mucilago Tragacanthæ, U. S. P.

Tragacanth also enters into emulsion of chloroform and most of the official troches.

Therapeutics.—On account of its insolubility, it is rarely used as a demulcent. It is employed in pharmacy for suspending resins, oils, and heavy powders in water.

SASSAFRAS MEDULLA, U. S. P.
(Sassafras Pith.)

Sassafras pith is obtained from the branches of *Sassafras variifolium*, a tree growing in the woods of North America.

PREPARATION.
Mucilago Sassafras Medullæ, U. S. P.

Therapeutics.—The mucilage has demulcent properties, but it is rarely used except in domestic medicine.

ULMUS, U. S. P.
(Slippery Elm.)

Slippery elm is the inner bark of *Ulmus fulva,* a large tree growing in the Eastern States of North America. It contains a large quantity of mucilaginous matter.

PREPARATION.
Mucilago Ulmi, U. S. P.

Therapeutics.—The powdered bark is sometimes used instead of flaxseed meal for making small *poultices.* In the form of lozenges it is sometimes employed for its soothing properties in *pharyngitis.*

ALTHÆA, U. S. P.
(Marshmallow.)

Marshmallow is the root of *Althæa officinalis*, a perennial herb growing in most temperate countries. It contains a large amount of mucilage and of starch.

PREPARATION.
Syrupus Althææ, U. S. P.

Althea is an ingredient also of blue-mass, pills of phosphorus, and pills of carbonate of iron.

Therapeutics.—Although it is an agreeable demulcent, marshmallow is rarely prescribed except as a vehicle in the form of the syrup.

GLYCYRRHIZA, U. S. P.

(Licorice Root.)

Glycyrrhiza is the root of *Glycyrrhiza glabra*, a perennial herb growing in southern Europe and western Asia and cultivated in England and the United States. It contains a glucosid, *glycyrrhizin*, to which it owes its sweet taste.

PREPARATIONS.	DOSE.
Extractum Glycyrrhizæ, U. S. P.	Indefinite.
Extractum Glycyrrhizæ Furum, U. S. P.	Indefinite.
Glycyrrhizinum Ammoniatum, U. S. P.	5–15 gr. (0.3–1.0 gm.).
Extractum Glycyrrhizæ Fluidum, U. S. P.	½–2 fl. dr. (2.0–8.0 c.c.).
Mistura Glycyrrhizæ Composita, U. S. P. (brown mixture: Pure extract of licorice, 3; syrup, 5; mucilage of acacia, 10; paregoric, 12; wine of antimony, 6; spirit of nitrous ether, 3; water, to make 100)	1–4 fl. dr. (4.0–15.0 c.c.).
Pulvis Glycyrrhizæ Compositus, U. S. P. (glycyrrhiza, 23.6; senna, 18; washed sulphur, 8; fennel oil, 0.4; sugar, 50)	½–2 dr. (2.0–8.0 gm.).
Trochisci Glycyrrhizæ et Opii, U. S. P. (each contains extract of glycyrrhiza, 2⅓ gr.—0.15 gm.—and powdered opium, $\frac{1}{12}$ gr.—0.005 gm.).	
Trochisci Ammonii Chloridi, U. S. P. (each contains extract of glycyrrhiza, about 4 gr.—0.25 gm.—and ammonium chlorid, 1½ gr.—0.1 gm.).	

Therapeutics.—Licorice is a popular remedy for *cough* resulting from irritation of upper respiratory tract. It is also largely employed to cover the disagreeable taste of other medicines.

AMYGDALA.

(Almond.)

Almonds are official in two forms: bitter almonds (*Amygdala Amara*, U. S. P.) and sweet almonds (*Amygdala Dulcis*, U. S. P.). Both varieties contain a *fixed oil* and a ferment, *emulsin*. Bitter almonds contain also a glucosid, *amygdalin*, which, in the presence of water, is acted upon by the emulsin and broken up into hydrocyanic acid, benzaldehyd, and glucose. Oil of bitter almond (*Oleum Amygdalæ Amaræ*,

U. S. P.) is a loose combination of hydrocyanic acid (1.5–4.0 per cent.) and benzaldehyd, obtained by distillation. As sweet almonds do not contain amygdalin, they do not yield hydrocyanic acid when triturated with water.

PREPARATIONS.	DOSE.
Oleum Amygdalæ Amaræ, U. S. P.	$\frac{1}{4}$–1 min. (0.016–0.06 c.c.).
Oleum Amygdalæ Expressum, U. S. P.	
Aqua Amygdalæ Amaræ, U. S. P. ($\frac{1}{10}$ per cent. of oil of bitter almond)	1–4 fl. dr. (4.0–15.0 c.c.).
Spiritus Amygdalæ Amaræ, U. S. P. (1 per cent. of oil of bitter almond).	5–20 min. (0.3–1.2 c.c.).
Syrupus Amygdalæ, U. S. P. (4 per cent. of bitter almond and 14 per cent. of sweet almond)	1–4 fl. dr. (4.0–15.0 c.c.).
Emulsum Amygdalæ, U. S. P. (6 per cent. of sweet almond)	2–4 fl. dr. (8.0–15.0 c.c.).
Unguentum Aquæ Rosæ, U. S. P. (cold cream: spermaceti, 12.5; white wax, 12; expressed oil of almond, 60; stronger rose-water, 19; sodium borate, 0.5).	

Therapeutics.—Almond is nutritive and demulcent. As almond flour contains no starch, it may be used, after the sugar and gum have been extracted from it, as a *diabetic food.* Expressed oil of almond, ointment of rose-water, and emulsion of almond are useful *emollients.* Syrup of almond is employed chiefly as a pleasant vehicle. Oil of bitter almond is an active poison. It was for a time used as a substitute for hydrocyanic acid, but its uncertain composition proved to be a great disadvantage. It is now rarely employed except as a flavoring agent.

AMYLUM, U. S. P.

(Starch.)

Official starch is the fecula of the seed of *Zea mays*, or Indian corn. Wheat-starch and rice-starch are also used in therapeutics.

PREPARATION.

Glyceritum Amyli.

Therapeutics.—Finely powdered starch is much used as an absorbent and protective application in *erythema intertrigo,* or *chafing.* As it readily decomposes in the presence of heat and moisture, boric acid should be added to it as a preservative. The starch poultice, made by mixing starch with cold water and then adding boiling water until the mass is converted into a gelatinous paste, is sometimes employed to remove *crusts* in

chronic inflammatory diseases of the skin. Starch-and-lauda-
num enemata, made by adding laudanum to thin starch mucilage, are often efficacious in *acute dysentery* and other *inflammatory diseases of the rectum.* Starch is an antidote to *iodin-
poisoning.*

ADEPS LANÆ HYDROSUS, U. S. P.

(Hydrous Wool-fat; Lanolin.)

Hydrous wool-fat is the purified cholesterin fat of sheep's
wool, mixed with not more than 30 per cent. of water. It is
a yellowish-white, unctuous substance, having a faint, peculiar
odor.

Therapeutics.—It was introduced by Liebreich, under the
name of lanolin, as a vehicle for external medicaments. As
an ointment base it is bland and unirritating, does not become
rancid, and is miscible with twice its weight of water without
losing its ointment-like character; but the claim that it has
greater penetrating power than all other fats is not borne out
by experience. Singly, it is an unsatisfactory base, but it sometimes makes a valuable addition to other bases, especially when
aqueous substances are to be incorporated.

CETACEUM, U. S. P.

(Spermaceti.)

Spermaceti is a concrete fatty substance obtained from the
head of the sperm-whale (*Physeter macrocephalus*). It occurs
in pearly-white, translucent masses, odorless, and of a bland,
mild taste. It is employed chiefly to give proper consistence
to cerates and ointments.

PREPARATIONS.

Ceratum Cetacei, U. S. P. (spermaceti, 10; white wax, 35; olive oil, 55).
Unguentum Aquæ Rosæ, U. S. P. (12.5 per cent.).

OLEUM OLIVÆ, U. S. P.

(Olive Oil; Sweet Oil.)

Olive oil is a fixed oil expressed from the ripe fruit of *Olea
europæa.* It is an ingredient of several official ointments, liniments, and plasters.

Therapeutics.—Externally, olive oil is a useful emollient.
With carbolic acid (5 per cent.) it makes a soothing, protective
application for *superficial burns.* In *scarlatina* and *measles*
daily inunctions of olive oil are of service in allaying irritation

of the skin and in preventing diffusion of the scales. It may be employed to soften *adherent crusts* in such diseases as eczema and seborrhea. Enemata of olive oil (6 to 8 oz.—175.0–235.0 c.c.) are very efficacious in removing *fecal concretions* from the rectum. Internally, its effects are nutritive, demulcent, and feebly laxative. Leube and others have employed, with good results, daily subcutaneous injections of olive oil (2–6 oz.—60.0–175.0 c.c.) as a means of nourishing patients with stricture of the esophagus, but the treatment is not without danger. In one instance, at least, it was the cause of fatal oil-embolism. By some authorities olive oil is held in high repute as a remedy for *gall-stones,* but its value is somewhat doubtful. Five or six ounces (150.0–175.0 c.c.) are given in the attacks of colic, and smaller doses in the intervals. If the drug possesses any efficacy in cholelithiasis, this may depend upon an antispasmodic influence · exerted upon the common bile-duct, as Cohnheim has found large doses of the oil very beneficial in *gastrectasis* depending upon *spasm of the pylorus* from ulcer or fissure, and Weill, Duplant, and others have spoken highly of olive oil in *lead-colic.*

PETROLATUM.

(Paraffin; Cosmolin; Vaselin.)

Petrolatum is a mixture of hydrocarbons, chiefly of the marsh-gas series, obtained by distilling off the more volatile portions of petroleum and purifying the residue when it has the desired melting-point. Three forms are official: Liquid petrolatum (*Petrolatum Liquidum,* U. S. P.); soft petrolatum (*Petrolatum Molle,* U. S. P.); and hard petrolatum (*Petrolatum Spissum,* U. S. P.).

Therapeutics.—Petrolatum is largely used as an emollient and protective dressing and as a substitute for animal and vegetable fats in ointments. It is bland and unirritating, and has little tendency to change. It probably has somewhat less penetrating power than lard.

Subcutaneous and submucous injections of paraffin, as recommended by Gersuny in 1900, have been employed with gratifying results for the correction of various defects and for cosmetic purposes. They have been used successfully in preventing rectal and vaginal prolapse, in relieving incontinence of urine and of feces, in filling depressions caused by the removal of tissue, in correcting nasal deformities, etc. The paraffin should have a melting-point of from 100° to 104° F. (38°–40° C.), should be sterilized by boiling, and should be injected while warm and semisolid. The paraffin is not absorbed, but remains where injected and finally becomes encapsulated.

GLYCERINUM, U. S. P.

(Glycerin; $C_3H_5(OH)_3$.)

Glycerin is a liquid obtained from fats and fixed oils by saponification with alkalis or by the action of superheated steam. It is colorless, transparent, hygroscopic, without odor, and of a sweet, warm taste and a syrupy consistence. It is freely soluble in water and in alcohol, but insoluble in ether, chloroform, and oils. The dose is from 1–4 fl. dr. (4.0–15.0 c.c.).

PREPARATIONS.

Glyceritum Acidi Carbolici, U. S. P. (20 per cent. of carbolic acid).
Glyceritum Acidi Tannici, U. S. P. (20 per cent. of tannic acid).
Glyceritum Amyli, U. S. P. (Plasma: starch, 10; water, 10; glycerin, 80).
Glyceritum Boroglycerini, U. S. P. (50 per cent. of boroglycerin).
Glyceritum Hydrastis, U. S. P. (each cubic centimeter contains 1 gm. of drug).
Glyceritum Vitelli, U. S. P. (45 per cent. of fresh yolk of egg).
Suppositoria Glycerini, U. S. P. (each suppository contains 92 gr.— 6.0 gm.—of glycerin).

Physiologic Action.—Owing to its avidity for water, pure glycerin, when applied to sensitive skin, produces considerable burning and irritation. Internally, in large doses (1 oz.—30 c.c.), it sometimes acts as a cathartic, probably by abstracting water from the intestinal vessels and also by stimulating peristalsis. In moderate doses glycerin has some nutritive value, inasmuch as it is largely oxidized in the body and protects from combustion fats and carbohydrates. When large amounts are injected directly into the circulation of an animal they cause muscular weakness, thirst, vomiting, hemoglobinuria, a fall in temperature, a rapid, weak pulse, tetanic convulsions, coma, and, finally, death from asphyxia.

Glycerin possesses some power in inhibiting the growth of micro-organisms.

Therapeutics.—Externally, glycerin is employed in various forms as an emollient. In the form of an ointment or lotion it is useful in *chapped hands, fissured nipples,* etc. It is highly recommended as a preventive against *bed-sores.* The parts should be bathed twice daily with warm water, carefully dried, and gently rubbed with the glycerin. A paste made by mixing glycerin and kaolin (hydrated aluminum silicate), and applied hot, is sometimes used as a *substitute for poultices.* In *inflammatory disease of the mouth, nose, and throat* it makes an excellent vehicle for other drugs. Being hygroscopic, it not only tends to deplete the turgid capillaries, but it also serves to spread the medicament over the entire surface. In *endometritis, uterine congestion,* and *subinvolution* tampons holding glycerin and tannin afford an efficient means of securing local depletion. In the form of a suppository or of an injection (2–3 fl. dr.—8.0–11.0 c.c.—with 1–2 fl. oz.—30.0–60.0 c.c.—of water) it is often employed with advantage to unload the bowel in *simple constipation.*

Glycerin was at one time recommended as a substitute for cod-liver oil in *phthisis* and other *wasting diseases,* but it proved to be of little value. In *diabetes* it may be employed as a sweetening agent instead of sugar, but, as a rule, patients soon tire of it.

In pharmacy glycerin is extensively used as a solvent, an excipient, and a preservative.

Incompatibles.—Glycerin is explosive with powerful oxidizing agents like chromic acid and potassium permanganate. Undiluted mixtures of potassium chlorate, tincture of ferric chlorid, and glycerin are liable to explode if warmed.

TALCUM.

(Talc; Venetian Talc; Magnesium Silicate.)

Talc occurs in grayish-green masses having a waxy luster and a greasy feel. It is insoluble in acids and other liquids. Finely powdered, it is employed as a dusting-powder in inflammatory diseases of the skin, especially in *erythema intertrigo* and *acute erythematous eczema.* It is usually prescribed with an antiseptic and an absorbent, as in the following formula:

℞	Pulveris talci,	ℨiv (15.5 gm.);
	Pulveris acidi borici,	ℨj (4.0 gm.);
	Pulveris zinci oxidi,	ℨiij (11.5 gm.). M.

KAOLIN.

(Fuller's Earth; Hydrated Aluminum Silicate.)

Kaolin is a white, impalpable powder, unctuous when moist. Like talc, it is inert and unalterable. It is used as a protectant in *inflammatory skin diseases,* sometimes as a dusting-powder, but more often in the form of a paste, when ointments and lotions are found to be irritating. Mixed with glycerin and applied hot, it is employed as a *substitute for poultices.* In pharmacy it is used as a *pill-basis* for substances readily decomposed, like silver nitrate and potassium permanganate.

LYCOPODIUM, U. S. P.

Lycopodium is the spores of *Lycopodium clavatum* and of other species of *Lycopodium*, mosses growing in the dry woods of nearly all temperate countries. It is a fine, pale-yellowish powder, mobile, inodorous, and tasteless. It is employed in therapeutics chiefly as a protective and absorbent application to *excoriated surfaces.* In pharmacy it is much used to prevent the adhesion of pills, suppositories, etc.

COLLODIUM, U. S. P.

(Collodion.)

Collodion as a solution of pyroxylin, or gun-cotton (3), in ether (75) and alcohol (25). It is a colorless, syrupy, highly inflammable liquid, having a strong ethereal odor.

PREPARATIONS.

Collodium Flexile, U. S. P. (collodion, 92; Canada turpentine, 5; castor oil, 3).
Collodium Stypticum, U. S. P. (tannic acid, 20; alcohol, 5; ether, 25; collodion, to make 100).
Collodium Cantharidatum, U. S. P. (Blistering Collodion: Cantharides, 60; flexible collodion, 85; chloroform, to make 100).

Therapeutics.—Collodion was introduced into surgery by Schoenbein, in 1846. When applied to an exposed part it quickly dries and forms a thin film, which in shrinking exerts a constricting and compressing effect. It makes a useful protective for *abrasions, fissures,* and *aseptic punctures.* It also affords an efficient means of securing antiseptic dressings on small wounds, especially those about the face and scalp. Flexible collodion makes less pressure and cracks less readily than plain collodion. With the addition of a little benzoin it makes an excellent application for *fissured nipples, chapped hands,* etc. A mixture of plain collodion and flexible collodion is often employed as a vehicle for such drugs as chrysarobin, pyrogallol, resorcin, etc., in the treatment of certain skin diseases, especially psoriasis, chronic eczema, lupus erythematosus, and ringworm.

Styptic collodion is sometimes applied to small wounds to check capillary oozing. Cantharidal collodion is employed to produce vesication (see p. 432).

LIQUOR GUTTA-PERCHÆ.

(Solution of Gutta-percha; Traumaticin.)

Solution of gutta-percha is a 9 per cent. solution of gutta-percha in chloroform. It has been used as a substitute for collodion, but it possesses no advantages over the latter.

FLAVORING AGENTS.

LIMON.

(Lemon.)

Lemon is the ripe fruit of *Citrus Limonum,* a tree cultivated in most subtropical countries.   The rind contains a *volatile oil,* and the juice, *citric acid.*

PREPARATIONS.	DOSE.
Limonis Cortex, U. S. P.	
Limonis Succus, U. S. P.	1–4 fl. oz. (30.0–118.0 c.c.).
Oleum Limonis, U. S. P.	1–5 min. (0.06–0.3 c.c.).
Spiritus Limonis, U. S. P.	½–1 fl. dr. (2.0–4.0 c.c.).
Acidum Citricum, U. S. P.	5–20 gr. (0.3–1.3 gm.).
Syrupus Acidi Citrici, U. S. P. (1 per cent.)	1–4 fl. dr. (4.0–15.0 c.c.).

The oil of lemon also enters into compound spirit of orange, compound spirit of ammonia, and aromatic elixir.

Therapeutics.—Lemon-juice is especially useful in *scurvy,* both as a preventive and a curative agent.   Upon what ingredient the antiscorbutic properties of the juice depend has not been determined; they do not appear to depend upon citric acid, however, as other fruits and vegetables in which this acid does not enter are almost equally efficacious.   Lemon-juice is believed by some practitioners to be of value also in *acute rheumatism.*   In the form of lemonade it makes a pleasant refrigerant drink in *febrile diseases.*

Hamm, Somers, and others have found citric acid of value in preventing the fetid odor of *atrophic rhinitis.*   A powder composed of 1 part of the acid with 3 parts of sugar-of-milk is insufflated into the nostrils two or three times daily after the parts have been thoroughly cleansed with an alkaline solution.

All the preparations of lemon are largely used for flavoring purposes.

AURANTIUM.

(Orange.)

Two varieties of the orange furnish the materials for the official preparations : Bitter orange (*Citrus vulgaris*) and sweet orange (*Citrus Aurantium*).

PREPARATIONS.	DOSE.
Aurantii Amari Cortex, U. S. P.	15–30 gr. (1.0–2.0 gm.).
Extractum Aurantii Amari Fluidum, U. S. P.	½–1 fl. dr. (2.0–4.0 c.c.).
Tinctura Aurantii Amari, U. S. P.	1–2 fl. dr. (4.0–8.0 c.c.).
Oleum Aurantii Florum, U. S. P. (Oil of Neroli : distilled from the fresh flowers of the bitter orange).	

PREPARATIONS.	DOSE.
Aqua Aurantii Flotum, U. S. P.	Indefinite.
Aqua Aurantii Florum Fortior, U. S. P.	
Syrupus Aurantii Florum, U. S. P. . . .	Indefinite.
Aurantii Dulcis Cortex, U. S. P.	15–30 gr. (1.0 2.0 gm.).
Tinctura Aurantii Dulcis, U. S. P.	1–2 fl. dr. (4.0–8.0 c.c.).
Syrupi Aurantii, U. S. P.	Indefinite.
Oleum Aurantii Corticis, U. S. P. (expressed from the fresh peel of either variety of orange)	1–5 min. (0.06–0.3 c.c.).
Spiritus Aurantii, U. S. P. (5 per cent. of the oil)	1–4 fl. dr. (4.0–15.0 c.c.).
Spiritus Aurantii Compositus, U. S. P. (oil of orange-peel, 20; oil of lemon, 5; oil of coriander, 2; oil of anise, 0.5; deodorized alcohol, to make 100)	1–4 fl. dr. (4.0–15.0 c.c.).
Elixir Aromaticum, U. S. P. (compound spirit of orange, 12; syrup, 37.5; deodorized alcohol and water, to make 100)	1–4 fl. dr. (4.0–15.0 c.c.).

Bitter orange-peel also enters into compound tincture of cinchona and compound tincture of gentian.

Therapeutics.—Bitter orange-peel is a feeble stomachic; but, like the sweet orange, it is employed almost exclusively as a flavoring agent.

ERIODICTYON, U. S. P.

(Yerba Santa.)

Eriodictyon is the leaves of *Eriodictyon glutinosum*, an evergreen shrub growing in California. It contains tannin, an acrid, bitter resin, and a trace of volatile oil.

PREPARATIONS.	DOSE.
Extractum Eriodictyi Fluidum, U. S. P. .	20–60 min. (1.2–4.0 c.c.).
Syrupus Eriodictyi Aromaticus (fluid extract, with compound tincture of cardamom, oil of sassafras, oil of lemon, oil of cloves, potassa solution, alcohol, and syrup)	1–4 fl. dr. (4.0–15.0 c.c.).

Use.—Eriodictyon has been used to some extent as a stimulant expectorant; it is chiefly valuable, however, in the form of the aromatic syrup, to obtund the bitterness of quinin. The aromatic syrup is incompatible with acids.

SARSAPARILLA, U. S. P.

Sarsaparilla is the root of *Smilax officinalis* and of other species of *Smilax*, climbing evergreens growing in swampy forests in Central and South America. It contains several glucosids of the *saponin* class.

PREPARATIONS.	DOSE.
Extractum Sarsaparillæ Fluidum, U. S. P.	½–2 fl. dr. (2.0–8.0 c.c.).
Extractum Sarsaparillæ Fluidum Compositum, U. S. P. (sarsaparilla, 75; glycyrrhiza, 12; sassafras, 10; mezereum, 3; with glycerin, 10; alcohol and water, to make 100)	½–2 fl. dr. (2.0–8.0 c.c.).
Decoctum Sarsaparillæ Compositum, U. S. P. (sarsaparilla, 10; sassafras, 2; guaiacum wood, 2; glycyrrhiza, 2; mezereum, 1; water, to make 100)	2–6 fl. oz. (60.0–180.0 c.c.).
Syrupus Sarsaparillæ Compositus, U. S. P. (fl. ext. sarsaparillæ, 20; fl. ext. glycyrrhiza, 1.5; fl. ext. senna, 1.5; sugar, 65; oil of sassafras, oil of anise, oil of gaultheria, of each, 0.1; water, to make 100)	1–4 fl. dr. (4.0–15.0 c.c.).

Uses.—Sarsaparilla has been employed for more than three hundred years as an alterative in *syphilis* and *tuberculosis*, but its beneficial effects in these diseases are very doubtful. It is chiefly useful in the form of the compound syrup as a vehicle for potassium iodid and mercury.

RUBUS IDÆUS, U. S. P.

(Raspberry.)

Raspberry is the ripe fruit of *Rubus idæus*, a shrub extensively cultivated in all temperate countries. It contains citric acid, malic acid, sugar, and a trace of volatile oil.

PREPARATION.
Syrupus Rubi Idæi, U. S. P.

Use.—Syrup of raspberry is employed only as a vehicle.

LAVANDULA.

(Lavender.)

Lavender is the flowers of *Lavandula angustifolia*, a shrub indigenous in southern Europe. It contains a volatile oil, tannin, and resin.

PREPARATIONS.	DOSE.
Oleum Lavandulæ Florum, U. S. P.	1–5 min. (0.06–0.3 c.c.).
Spiritus Lavandulæ, U. S. P.	½–1 fl. dr. (2.0–4.0 c.c.).
Tinctura Lavandulæ Composita, U. S. P. (oil of lavender, with oil of rosemary, cinnamon, cloves, nutmeg, and red saunders)	½–1 fl. dr. (2.0–4.0 c.c.).

Uses.—The preparations of lavender are used solely for their agreeable flavor.

The compound tincture is an ingredient in Fowler's solution.

ROSA.

(Rose.)

The petals of two species of the rose are official: pale rose (*Rosa centifolia*) and red rose (*Rosa gallica*). Both species are largely cultivated in western Asia and southern Europe. The chief constituents are a volatile oil, tannin, sugar, and mucilage. The official oil of rose is distilled from the fresh flowers of the Damascus rose (*Rosa damascena*).

PREPARATIONS.

Extractum Rosæ Fluidum, U. S. P. (from red rose).
Syrupus Rosæ, U. S. P. (12.5 per cent. of the fluid extract).
Mel Rosæ, U. S. P. (12 per cent. of fluid extract in clarified honey).
Confectio Rosæ, U. S. P. (red rose, 8; sugar, 64; honey, 12; stronger rose-water, 16).
Oleum Rosæ, U. S. P. (attar of rose).
Aqua Rosæ Fortior, U. S. P. (water saturated with oil of rose).
Aqua Rosæ, U. S. P. (stronger rose-water and distilled water, equal volumes).

Rose also enters into blue-mass, compound iron mixture, pills of aloes and mastic, and ointment of rose-water.

Use.—The preparations of rose are used as agreeable vehicles.

SACCHARUM, U. S. P. and SACCHARUM LACTIS, U. S. P.

(Cane-sugar or Sugar and Sugar-of-milk or Lactose.)

Cane-sugar is the refined sugar from sugar-cane (*Saccharum officinarum*), from various species of broom-corn (*Sorghum*), and from one or more varieties of sugar-beet (*Beta vulgaris*).

PREPARATION.

Syrupus, U. S. P.

Cane-sugar also enters into the various compound syrups, the various troches, several mixtures, and numerous other preparations. It is employed exclusively as a sweetening agent, a preservative, and a vehicle or excipient.

Sugar-of-milk is the sugar obtained from the whey of cow's milk. It is harder, less soluble, and less sweet than cane-sugar. It is used largely as a diluent for powders. It has some merit also as a diuretic in cardiac dropsy.

LEVULOSE.

(Fruit-sugar; Fructose.)

Levulose is a saccharine substance found in most sweet fruits, together with grape-sugar, and prepared artificially by

30

hydrolysis from cane-sugar or from inulin, a starchy body contained in inula or elecampane. It occurs as a thick, syrupy liquid or as a coarse, granular powder, and is almost as sweet as cane-sugar. It was recommended by Külz, in 1874, as a sweetening agent and as a carbohydrate food in diabetes. According to Solis-Cohen, Saundby, Tyson, and others, a moderate amount (8–12 dr.—31.0–46.0 gm. daily) can be taken by most patients with impunity.

SACCHARIN.
(Glusidum; Benzoic Sulphinid.)

Saccharin is a principle prepared from the coal-tar derivative, toluene. It is a light, white, crystalline powder, odorless, and possessing 500 times the sweetening power of cane-sugar. It is soluble in 400 parts of water, more so in alcohol and in glycerin, and readily so in alkaline solutions. With solution of sodium bicarbonate it forms a soluble sodium salt, *soluble saccharin.* Saccharin is chiefly useful as a substitute for sugar when the latter is contraindicated, as in diabetes mellitus. One grain (0.065 gm.) is sufficient to sweeten 4 or 5 ounces (120.0–150.0 c.c.) of tea or coffee.

MEL, U. S. P.
(Honey.)

Honey is a saccharine fluid prepared by the hive-bee (*Apis mellifica*). It is a mixture of dextrose (grape-sugar) and levulose (fruit-sugar), with minute quantities of wax, formic acid, volatile oil, and coloring-matter.

PREPARATIONS.
Mel Despumatum, U. S. P. (clarified honey).
Mel Rosæ, U. S. P.
Confectio Rosæ, U. S. P.

Uses.—Honey is demulcent and feebly laxative, but it is prescribed chiefly as an excipient.

Wild cherry, licorice, balsam of Tolu, and many **volatile oils** are also largely used for flavoring purposes.

REMEDIAL MEASURES OTHER THAN DRUGS.

ELECTRICITY.

The true nature of electricity is still unknown ; like the other great forces of nature, it must be studied in the light of its various manifestations. Gilbert first employed the term to designate the phenomena of attraction and repulsion which result when amber (electrum) and similar substances are briskly rubbed. Electricity produced in this simple way—by the rubbing together of certain substances—is termed **frictional** or **static electricity** or **franklinism**.

When two dissimilar metals like copper and zinc are placed in a corrosive fluid and united outside by a loop of wire, the resulting chemical action is attended with the production of an electric current, and to this manifestation of electricity the term **galvanism** is applied. The vessel, its contained fluid, and connected elements together constitute a *galvanic cell*. The union of several cells forms a *battery*. The energy which starts the electric current is termed the *electromotive force*, and it is always the same in quantity for the same materials at the same temperature, and is entirely independent of the size and shape of the plates and of the distance between them. The current is generated at the surface of the element or plate more easily corroded, which in the above illustration is the zinc, and passes through the fluid to the element or plate less affected,—that is, the copper,—and from the external end of the copper it passes through the wire loop back again to the external end of the zinc, thus completing a *closed circuit*. As the external end of the copper is constantly giving off electricity, it is termed the *positive pole*, or *anode ;* and as the external end of the zinc is constantly receiving electricity, it is termed the *negative pole*, or *cathode*.

It is evident that the electromotive force will more than represent the strength of the current, for all substances offer more or less resistance to the passage of electricity. The resistance within the cell is termed the *internal resistance*, and that without the cell the *external resistance*. The amount of internal resistance will depend upon the character of the intervening fluid, the size of the plates, and their proximity. The larger the plates and the closer they are to each other, the less becomes the internal resistance. The external resistance depends upon the length, diameter, and character of the conductor. The law of the relation of the *current-strength* to the electromotive force and to the resistance was discovered by Ohm, in 1827: *The strength of an electric current passing a section of the conductor in a unit of time is proportional to the whole electromotive force, and inversely proportional to the sum of all the resistances in the circuit.* Representing the current-strength by C, the electromotive force by E, the internal resistance by Ir, and the external resistance by Er, the law may be expressed by the following formula:

$$C = \frac{E}{Er + Ir}.$$

The unit of resistance is termed an *ohm;* the unit of electro-motive force, a *volt;* and the unit of current-strength, an *ampère.* As the last is too large a unit for medical purposes, $\frac{1}{1000}$ of an ampère, or a *milliampère*, is employed instead.

An electromotive force of one unit acting upon a conductor offering one ohm of resistance will set up in that conductor a current of one ampère.

The current-strength of a battery can be increased in three ways : (1) By diminishing the external resistance; (2) by diminishing the internal resistance; and (3) by increasing the number of cells. When the external resistance is a fixed quantity, which is often the case, the current-strength can still be increased in one of the two remaining ways; but when the external resistance is very great, as in the case of the human body, in which it is about 3000 ohms, the gain secured by enlarging the plates and lessening the distance between them is but trifling, and the current-strength must, therefore, be increased by multiplying the number of cells.

When the external resistance is slight, as in the galvano-cautery, in which the current has only to render a piece of wire incandescent, the internal resistance becomes highly important, and much is gained by reducing it to a minimum. It follows, therefore, that a battery designed for such a purpose should consist of a few cells with large plates placed very near one another. The same principles likewise apply to the construction of batteries intended for electrolytic work.

Of the accessories to a battery, the most important are the galvanometer, rheostat, and electrodes.

A galvanometer is an instrument for measuring the current-strength.

A rheostat, or resistance-coil, is an appliance for placing in the circuit a known resistance; by it the external resistance can be increased to any extent and kept uniform.

Electrodes are appliances of various sizes and shapes placed at the pole-ends so that the current can be conveniently transferred to the body.

Faradism, or Induced Electricity.—This manifestation of electric force depends upon the power which a galvanic current possesses, while passing through one conductor, of inducing momentarily a current in an overlying conductor. The induced current is instantaneous, and appears only at the making or breaking of the galvanic current. Moreover, it flows in a direction opposite to that of the galvanic current when the latter is made, and in the same direction when it is broken. By rapidly making and breaking the primary current a powerful to-and-fro current is induced in the overlying conductor. A faradic battery consists of cells for generating the galvanic current; a *primary coil*, consisting of a few turns of coarse insulated wire inclosing a coil of soft iron; a *secondary coil*, composed of many turns of very fine wire surrounding the primary coil, but not touching it; and a mechanical device for slowly or rapidly making or breaking the current in the primary circuit.

The current derived from the primary coil is termed the *primary current;* like the galvanic, it flows in one direction and affords a positive and a negative pole. It differs, however, from the galvanic current in being reinforced by the turns of the coil. The current derived from the secondary coil is termed the *secondary current*, and, since it is ever changing the direction of its flow, it cannot furnish negative and positive poles.

Physiologic Action of Electricity.—When a powerful uninterrupted galvanic current is applied to sensitive parts of the body through moistened electrodes, it produces at the points of contact severe burning pain, redness, and, finally, under the negative pole, vesication. These phenomena are the result of electrolytic changes induced by the current. When the circuit is broken and again when it is closed active muscular contraction occurs; in the interval between opening and closure, however, the muscle is quiescent.

When the negative pole, or cathode, is placed over the nerve supplying a muscle, and the positive pole, or anode, over some indifferent point, as the sternum, and a weak current is used, a contraction of the muscle occurs on closing the circuit, but none on breaking it. When the positive pole is placed over the nerve and the negative over the sternum, and a somewhat stronger current is employed, a contraction occurs when the circuit is broken. With a still stronger current a contraction occurs also when the circuit is closed. When the positive pole is over the nerve, contractions on breaking the circuit occur only with currents of very great power.

The following table represents the relative strengths of the various contractions in normal muscles, 1 being the strongest:

1. Cathodal closing contraction (CaClC).
2. Anodal opening contraction (AnOC).
3. Anodal closing contraction (AnClC).
4. Cathodal opening contraction (CaOC).

When one pole is placed over the sternum and the other over the *motor point* of the muscle,—*i. e.*, the point of entrance of its motor nerve,—contractions are obtained with weaker currents and the reactions are somewhat changed, the anodal closing contraction becoming stronger than the anodal opening contraction.

Effect of the Faradic Current.—When a strong induced current is applied to a part it causes a tingling sensation, followed by numbness. If a dry metallic electrode be used, the contact excites severe pain. When the current is passed directly through a muscle, it causes a continuous or tetanic contraction which lasts until the current ceases to flow.

The Reaction of Paralyzed Muscles to Electric Currents. —When a galvanic current is applied to a paralyzed part, the reactions may be normal, they may be altered quantitatively, *i. e.*, increased or diminished,—or they may be altered qualitatively.

A normal response of the muscles to both galvanic and faradic currents often occurs in hysteric paralysis and in paralysis of cerebral origin. An increased response to both currents without qualitative change indicates a state of irritation or hypersensitiveness of the spinal centers or peripheral nerves, and may be observed very early in apoplectic hemiplegia, very early in neuritis, and in some cases of hysteria. A diminished response to both currents without qualitative change is observed in pseudomuscular hypertrophy, in progressive muscular atrophy until a late period of disease, and sometimes in neuritis after the very early stage, and before much degenerative change has taken place in the nerve-trunk or muscles.

Reaction of Degeneration (DeR).—This consists in a qualitative change in the electric reaction, a reversal of that occurring in normal muscle. It is obtained only with the *galvanic current* when the electrode is placed over the *muscle,*—not its motor nerve or motor point,—and is observed in paralyzed muscles which are in certain stages of degeneration on account of a lesion of their supplying nerves or of that portion of the spinal cord from which those nerves have their origin; or, in other words, when the muscles are cut off from the trophic influences which emanate from the ganglionic cells in the gray matter of the spinal cord. In such cases the muscles fail to respond to the faradic current, but still respond to the galvanic current. The responses, however, instead of being immediate and short as in health, are sluggish and persistent, and, moreover, are reversed in their sequence. Thus, the anodal closing contraction may equal, or at a later period may exceed, the cathodal closing contraction, and the cathodal opening contraction may equal or exceed the anodal opening contraction. These reactions may be expressed as follows:

AnClC equals or is greater than CaClC.
CaOC equals or is greater than AnOC.

Administration.—The galvanic current may be applied in various ways. In the *stabile* or *continuous method* the electrodes are applied firmly over the desired points and are kept stationary while the current is flowing. In the *labile* method an electrode is placed over an indifferent point, while the other is slowly stroked over the affected part, but not lifted from the skin. In this way contraction of all parts of a muscle may be secured. In the *interrupted* method one electrode is kept stationary while the other is alternately applied and removed, or both electrodes are kept stationary while the current is interrupted by means of a mechanical device in the handle of one of the electrodes. Finally, the electrodes may be kept stationary while the current is repeatedly *reversed* by making pole-changes in the metallic circuit. By interrupting and reversing the current very decided stimulant effects may be secured. In all applications of the galvanic current the electrodes should be well moistened with warm water.

In applying the faradic current to muscles, one large electrode should be placed over an indifferent point and the other over the body of the muscle to be treated, or, if a more decided action is desired, over the motor point of the muscle. Both electrodes should be well moistened. If the peripheral ends of the cutaneous nerves are to be treated, the skin should be thoroughly dried and the current applied through a wire brush, the indifferent electrode only being moistened. As the dry skin has a very high resistance, it is necessary in applying the wire brush to employ a secondary coil with many turns (2000 or 3000) of fine wire.

General faradization may be practised by passing one electrode slowly over the surface of the body, the other being held stationary; or it may be practised by means of the *electric bath*. In the latter the patient's body is immersed in warm water through which a current is made to pass.

In franklinization the current is applied to the body in the form of sparks or a spray drawn from an electric machine of high potential. Such a machine should have from eight to twelve revolving plates of a diameter of not less than 20 inches.

The dosage of electricity can be estimated only very roughly. Two important factors are the current-strength and the size of the electrodes. In galvanic applications the current-strength can be accurately determined by means of the absolute galvanometer. The concentration or density of the current, however, must also be considered, and this depends largely upon the size of the electrodes. Thus, the larger the electrode,—owing to a greater opportunity for diffusion,—the less powerful the local effect. The current-density equals the current-strength divided by the surface of the electrode. We may express the dosage approximately, therefore, by a fraction, the numerator of which is the current-strength in milliampères and the denominator of which is the surface of the electrode in square centimeters. The fraction $\frac{1}{20}$, for instance, would indicate a current-strength of 1 milliampère and an exciting electrode of 20 square centimeters.

The only practical way of measuring the strength of the faradic current is by means of an arbitrary scale indicating the relative positions of the primary and secondary coils. As the thickness and length of the wire used in the coils influence decidedly the magnitude of the current, it follows that the readings of such a scale are of little value in comparing the results obtained with different instruments. Such general terms as weak, medium, and strong are usually employed to express the strength of faradic currents. As regards static electricity, there is no method of measuring the strength of the current that has even a semblance of accuracy.

Until we possess more accurate means of computing the dose of the several currents we must, perforce, be guided in their use very largely by the sensations of the patient and the activity of the muscle reactions.

The selection of the proper current cannot always be made offhand; in very many cases it must be determined empirically.

The duration of the applications in localized electrification should generally be from three to five minutes, and the treatment should be repeated two or three times a week, or even daily in acute cases. In general electrification each séance may last from ten to twenty minutes, and be repeated two or three times a week. Pain and fatigue are to be avoided.

Therapeutics.—Electricity is employed in medicine to restore functional activity to peripheral nerves and muscles when conduction has not been completely cut off by some irreparable lesion; to promote general nutrition; to arouse the nerve-centers in states of depression; to combat various sensory disturbances which are of a functional character or which depend upon minor lesions of the peripheral nerves or their centers; to effect the rapid passage of drugs through the unbroken skin (cataphoresis); to bring about the coagulation of blood in aneurysmal sacs; to favor the absorption of inflammatory or degenerative products; and in the form of the galvanocautery, to excise or destroy hypertrophied tissues or morbid growths.

Whether the good achieved by electricity, especially by franklinism, in functional nervous diseases is to be attributed to the specific action of the electricity itself, or whether it is to be regarded as being wholly the work of suggestion, is a question upon which opinion is divided. It cannot be denied, however, that in some of these cases, at least, the benefit derived is the result rather of psychologic than of physiologic effects.

In a large group of cases the efficacy of electricity depends upon its power to excite muscular contraction. In *apoplectic hemiplegia* triweekly applications of a faradic current sufficiently strong to produce good contractions of the affected muscles are often very useful. The treatment should not, however, be begun until three or four weeks after the attack, and need not last more than five or six weeks. Contractures are very rarely benefited, although occasionally slight relief is afforded by faradization of the muscles antagonistic to those which are affected.

In *paralysis from lesions of the peripheral nerves,* such as neuritis, if the damage to the nerve has not been too great, much may be hoped for from the use of electricity. It should never be employed, however, until all symptoms of irritation have subsided. The faradic current is to be used if the muscles respond to it.   In the more severe cases in which faradic excitability is lost, labile and interrupted galvanic currents should be applied with the pole which produces the best contractions.   In very severe cases with well-marked reaction of degeneration nothing is to be expected from electric treatment. In some cases of multiple neuritis the electric bath may be used with advantage.

In *infantile spinal paralysis* electric treatment persistently carried out for several months is not infrequently followed by a restoration of power in certain groups of muscles.   The applications should be begun about three weeks after the onset of the disease, and made three or four times weekly.   When faradic contractility is lost, as is usually the case, galvanism should be employed, one electrode (cathode) being placed over an indifferent point, such as the spine, while the other (anode) is slowly stroked over the affected muscles.   In spinal spastic paralysis and the progressive muscular dystrophies nothing can be hoped for from electricity.

In *locomotor ataxia*, while electricity exerts no influence on the progress of the disease, it is sometimes very useful in the form of the faradic brush in relieving lancinating pains and paresthesia.   J. K. Mitchell has found faradism especially beneficial also in rectal and vesical tenesmus, given with one pole in the rectum and in the latter altogether percutaneously.

In the so-called idiopathic forms of *neuralgia* electricity may afford great relief.   The stabile galvanic anode should be used first, and, failing with this, faradism should be tried, the current being taken from a secondary coil wound with fine wire 5000 or 6000 feet in length.   Static electricity has also been used in some cases with good results.   Temporary alleviation of pain can be secured by cataphoresis.   A satisfactory method is to cover the positive pole with several layers of tissue-paper wet with a 10 per cent. solution of cocain, and to apply it over the most painful spot.   With a current of from 5 to 10 milliampères local anesthesia can be produced in about five minutes.

In *neurasthenia* general electrization (faradism or galvanism) is often a useful adjuvant to other therapeutic measures.   It must be employed, however, with considerable circumspection, since in some cases the symptoms appear to be rather aggravated than relieved by the treatment.   Only mild currents should be used, and at first the applications should not last longer than five or ten minutes.

In *hysteria* electricity is sometimes remarkably efficacious, probably through its psychic effect, in dispelling certain local symptoms, such as anesthesia, analgesia, pain, paralysis, and contractures. Faradization with the wire brush and franklinization, to which hysteric subjects are often very tolerant, are the currents most generally useful. General faradization, also, forms a part of the Weir-Mitchell treatment, which is especially applicable to obstinate cases.

In *exophthalmic goiter* galvanic treatment systematically carried out is sometimes followed by distinct improvement, especially in the tachycardia. The best results are obtained with weak currents, the cathode being placed over the neck near the angle of the lower jaw, while the anode is applied to the opposite side next to the lower cervical vertebræ. The applications should be made daily, and each should be short, lasting not longer than from three to five minutes.

· In *chronic myalgia*—lumbago, torticollis, and pleurodynia— faradism applied by means of the wire brush and electrostatic treatment are sometimes of benefit.

In *poisoning by opium and other narcotics* a strong faradic current affords an excellent means of arousing the patient without increasing the exhaustion or causing any ill effects.

The method of treating *aneurysms of the aorta* by means of a fine wire introduced through a hollow needle and coiled within the sac was first suggested by Charles H. Moore, in 1864. Corradi, in 1879, demonstrated that the procedure could be made more effective by passing a strong galvanic current through the wire, thus producing more rapid and firm coagulation. In this country the experiments of D. D. Stewart[1] and of Guy L. Hunner[2] have done much to bring the operation into prominence. Hunner, in 1900, collected 23 cases (17 thoracic and 6 abdominal) treated by the Moore-Corradi method. Four of these (17 per cent.), 3 thoracic and 1 abdominal, were apparently cured; 9 (39 per cent.) were distinctly improved; and in 10 death was probably hastened by the treatment. The technic, according to Stewart, is briefly as follows: Asepsis, insertion into the aneurysm, where the sac-wall seems nearest to the surface, of an insulated hollow needle, through which is introduced silver wire (about No. 27). It must be well drawn, wound on a spool, and from 3 to 20 feet in length, according to the size of the cavity. The wire is then attached to the positive pole of a galvanic battery, the negative electrode, large and well moistened, being applied to the back, and a current of from 70 to 80 milliampères passed for from half an hour to an hour and a half. The wire

---

[1] *Phila. Med. Jour.*, November 12, 1898.
[2] *Johns Hopkins Hosp. Bull.*, November, 1900.

is then cut close to the skin and sunk beneath the surface. In cases of thoracic aneurysms the anesthesia may be local. The chief dangers, according to Hunner, are from sepsis, rupture of the aneurysm from increase of pressure due to coagulation in a portion of the sac only, the entering of a loop of wire into the aorta, and emboli breaking from the sac-wall. Failure is to be expected in aneurysms with fusiform sacs and when there is marked general arteriosclerosis. In view of the gravity of the disease and the inefficiency of medical treatment the results so far obtained from this method of treatment must be considered encouraging.

Galvanism has been employed for its electrolytic effects in *uterine fibroma, goitre, tonsillar hypertrophy*, etc., but the results achieved have not been sufficiently satisfactory to establish this method of treatment in the confidence of the profession.

MASSAGE.

The term massage is applied to systematic stroking, kneading, rubbing, and percussion of the body. One who practises massage is known as a *masseur*, if a male, and as a *masseuse*, if a female.

Effleurage, or stroking, consists in slowly drawing the flat of the hand or the finger-tips over the surface of the body in the direction of the venous circulation. Very slight pressure should be made in the movements. Effleurage stimulates the cutaneous nerves and quickens the lymphatic and venous circulations.

Pétrissage, or kneading, is performed by grasping the tissues between the palmar surfaces of the hands, between the fingers and thumb, or between the finger-tips only, raising them slightly, and then rolling them while alternately tightening and relaxing the hold. Kneading serves to remove waste-products from the tissues, stimulates the circulation, and promotes nutrition.

Friction, or rubbing, is done by making circular movements, under some pressure, with the palm of the hand or the finger-tips, the up-stroke being heavy and the down-stroke light. The movements should always be centripetal. Friction exerts about the same influence on the tissues as kneading.

Tapotement, or percussion, consists in delivering, with the palm of the hand, the ulnar surface of the hand, or the tips of the fingers, rapid rhythmic blows. Palmar percussion is used chiefly on the limbs; ulnar percussion on the chest and back;

and digital percussion on small uneven surfaces like the face. Percussion acts as a nerve and muscle stimulant.

General Massage.—When general massage is to be practised, the operator begins at one extremity, generally a leg, and applies to each in turn friction, kneading, palmar percussion, and palmar stroking. After the limbs, the chest receives friction, kneading, ulnar percussion, and palmar stroking. Next the abdomen is treated, deep friction and kneading being applied over the entire surface, and palmar stroking along the course of the ascending, transverse; and descending colon. Finally, vigorous stroking is applied over the spine from above downward, and the muscles of the back are subjected to friction, kneading, and ulnar percussion. In some cases various movements, such as flexion, extension, circumduction, etc., intended to secure passive exercise of the muscles, are a valuable adjuvant to general massage.

Generally, massage is best given in the forenoon, about an hour after breakfast. It should never be given on an empty stomach. At first the séances should not last more than fifteen or twenty minutes, but later they may be extended with advantage to an hour. To secure the best results the operation should be repeated daily. The patient should be recumbent and loosely wrapped, so that the hands of the operator may be applied directly to the skin. The room should be free from drafts, and the temperature of the atmosphere about 70° F. (21.1° C.). After the treatment the patient should be warmly covered and left to rest for from twenty minutes to an hour.

Unless the patient's skin is very harsh and dry or is unduly sensitive, the masseur need not use a lubricant for the hands; if, however, one is desired, cocoanut oil or cold cream is probably the best.

The effects of general massage in individuals who are required to be at rest are much the same as those of active exercise in persons who are able to be about. Thus, the immediate effects are a sense of tiredness without exhaustion and an agreeable feeling of drowsiness. The general circulation is first quickened and then slowed, the waste-products are squeezed out of the tissues, and the activity of the lymph-flow is increased. Later, massage whets the appetite, favors digestion, tends to secure restful sleep, stimulates metabolic activity, and promotes general nutrition.

Therapeutics.—In the *paralysis* following cerebral hemorrhage, acute poliomyelitis, and neuritis massage is of great value in maintaining the nutrition of the affected muscles; even when power is hopelessly lost, it tends to retard the development of contractures. Mechanical treatment, however, should not be instituted in these cases until all signs of irritation have subsided. In *neuralgia* and *sciatica* local massage is a useful adjuvant to other measures.

Writers' cramp and other *occupation spasms* are best treated by rest, massage, and galvanism.

In *sprains* rapid relief is afforded by soaking the part in hot water for half an hour, then practising massage, and subsequently applying a firm bandage. The treatment should be instituted immediately after the accident. For the first ten or fifteen minutes the massage should consist of light digital stroking applied in a centripetal direction above and below the affected joint, the latter not being touched; after this, however, gentle friction may be applied directly to the inflamed tissues. The treatment should be repeated two or three times daily.

Massage is often serviceable in hastening the absorption of inflammatory exudation and in imparting suppleness to the affected joints, in *traumatic arthritis, luxations, synovitis, chronic articular rheumatism, chronic gout,* and *rheumatoid arthritis.* In acute rheumatism and acute gout manipulations are contraindicated while the acute symptoms are present.

In *chronic constipation* the result of intestinal atony, abdominal massage systematically practised is a most effective remedy.

In *hysteria* and *neurasthenia* passive movements afford a valuable means of securing the benefits of exercise without the expenditure of strength. They are an important element in the Weir-Mitchell treatment, which includes also isolation, the administration of readily assimilable food in large quantities, and electricity. In *chronic heart-disease, diabetes, obesity,* and *severe anemia,* when the patient is unable to take active exercise, massage is a great help in stimulating metabolic processes and in improving the general nutrition.

Contraindications.—Massage should not be employed in any disease that is attended with fever. In pregnancy abdominal manipulation is contraindicated, although the treatment may be applied to other parts of the body up to the first stage of labor. Inflammatory diseases of the skin and processes attended with suppuration also forbid massage.

MOVEMENT THERAPY FOR LOCOMOTOR ATAXIA.

The treatment of locomotor ataxia by exercises intended to reëducate the muscles to perform coördinated movements, first described by Frenkel in 1890, has won for itself the support of many distinguished neurologists. Properly carried out, it may serve to keep patients on their feet for several years, when otherwise they might become bedridden, or it may serve even in getting them about again when they have been unable to walk for a long period, perhaps many months.

For exercising the upper extremities the following directions are given: Sit in front of a table, place the hand upon it, then elevate each finger as far as possible; raising the hand slightly, extend and then flex each finger and thumb as far as possible; do this with the right and then with the left hand. Touch with the end of the thumb each finger-tip separately and accurately; then touch the middle of each phalanx with the tip of the thumb. Sit at the table with a large sheet of paper and a pencil; make a dot at each corner of the paper and one in the center, and draw lines from the corner dots to the center dot, first with one hand and then with the other. Put ten coins on the paper, pick them up, and place them in a single pile, first with the right and then with the left hand.

For the body and legs, the following are sample exercises: Sit in a chair, rise slowly to erect position without the help of cane or arms of chair; then sit down slowly; stand with cane, feet together; advance left foot and return it with exactness to original position; then the same with right foot. Walk slowly ten steps forward and five back with help of canes. Stand without cane, but with the feet a little apart and the hands on the hips; in this position stoop down by flexing the knees, and rise slowly. Stand without cane with the feet separated; raise the hands from the sides above the head; carry them downward and forward, and try to touch the toes. Walk along a fixed line on the floor by the help of a cane, placing each foot in turn on the line; then repeat without using the cane.

When the patient is confined to bed, the exercises must be taken in a recumbent position. He is required to flex, extend, abduct, and adduct each leg separately and then both simultaneously. He is asked to place the heel of one foot on the big toe of the other foot, and then on various points from the ankle to the knee. He may be called upon to follow alternately, first with one foot and then with the other, lines drawn upon a small blackboard placed at the foot of the bed (Rhein).

To secure good results from Frenkel's treatment, the following points should be borne in mind: The exercises must not be applied in a routine manner; they should be carefully adapted to the particular needs of each patient. Each movement should be made slowly and deliberately, with the greatest possible exactitude, and repeated three or four times, first with the eyes open and then with the eyes closed. Owing to the benumbing of the sense of fatigue in tabetic patients, great care must be exercised to guard against exhaustion. The séances should not last longer than ten or fifteen minutes, and at first no more than two should be allowed a day. After the exercises the patient should lie upon his back until completely rested. Gymnastics of all kinds are to be forbidden during the treatment. For the first week or two, at least, the movements should be done under the direct supervision of the physician himself or a trained assistant. The duration of the treatment will depend largely upon the degree of the motor disturbances present, but in any case it should not be less than several months. Even when the exercises appear to be no longer necessary, they should be resumed at frequent intervals.

Contraindications.—According to Frenkel, the treatment is absolutely contraindicated in acute cases. As vision is essential in performing the various movements, blindness is also a contraindication. Raichline further adds that the patient should possess a certain amount of energy and intelligence, that sensation should not be absolutely lost, and that arthropathies and spontaneous fractures should not be present.

THE SCHOTT OR NAUHEIM TREATMENT.

The treatment of chronic heart-disease carried out by the brothers Schott at Nauheim, Germany, consists of bathing in natural effervescent baths and of graduated exercises. The waters of Nauheim have a temperature of from 86° to 95° F. (30°–35° C.), and are very rich in carbonic acid and in mineral ingredients, especially sodium chlorid and calcium chlorid. The baths first prescribed have a temperature of from 95° to 90° F. (35°–32° C.), contain about 1 per cent. of sodium chlorid and 0.1 per cent. of calcium chlorid, and are free from carbonic acid gas. The patient is directed to remain in the water not more than five or ten minutes. The duration of the baths and their salinity are gradually increased, while their temperature is gradually lowered. Eventually, after the lapse of a week or two, baths containing carbonic acid gas, and about 3 per cent. of sodium chlorid and 0.3 per cent. of calcium chlorid are taken, the temperature of the water being reduced as low as 80° F. (28° C.). After the bath the patient is well rubbed and allowed to rest for an hour.

Artificial baths may be prepared at home in imitation of those at Nauheim. At the beginning of the treatment the bath should contain 1 pound (454.0 gm.) of sodium chlorid and 1½ ounces (45.0 gm.) of calcium chlorid for every 10 gallons (38.0 L.) of water, and the temperature of the water should be from 95° to 90° F. (35°–32° C.).

The baths may be given every other day or every day, their duration being gradually prolonged from five or ten minutes to fifteen or twenty minutes. With the increased duration of the baths their temperature is gradually lowered until a minimum of 80° F. (28° C.) is attained, and their salinity is gradually increased until a maximum of 3 pounds (1360.0 gm.) of sodium chlorid and 4½ ounces (140.0 gm.) of calcium chlorid to every 10 gallons (38.0 L.) is reached.

The plain saline baths are to be followed by effervescent baths. The latter may be prepared artificially by dissolving in the saline baths 2 ounces (60.0 gm.) of sodium bicarbonate for every 10 gallons (38.0 L.) of water, and then adding very slowly 3 ounces (90.0 c.c.) of hydrochloric acid which has been previously well diluted. The amount of soda and acid may be gradually increased until that of the former reaches a maximum of 8 ounces (245.0 gm.), and that of the latter a maximum of 12 ounces (355.0 c.c.).

A more convenient method of generating the carbonic acid is by using prepared tablets of sodium bisulphate with an acid reaction and powders of sodium bicarbonate.

The effect of the baths is to diminish the pulse-rate, to increase the force and volume of the pulse, to lessen the frequency of the respirations, to increase the secretion of urine, and, in cases of cardiac dilatation, to diminish the area of cardiac dulness. The theory of August Schott is that the good effects of the baths are largely the result of a reflex stimulation of the heart through excitation of the cutaneous nerves.

The exercises consist of a series of simple movements of each limb and of the trunk, which are gently resisted by the hand of the attendant. Massage may also be added to the gymnastics with good results. It is highly important in carrying out the exercises to avoid fatigue. The slightest suggestion of dyspnea or of cyanosis is to be regarded as a signal for discontinuing the treatment. The effects of the exercises are very similar to those produced by the baths. It has been suggested that these effects are also brought about reflexly, a centripetal influence being exerted by the muscular contractions upon the ganglionic centers of the heart.

Broadbent, however, is of the opinion that "the baths or resisted movements give rise to a physiologic dilatation of the cutaneous capillaries or muscles respectively, so that the resistance to the onward movement of blood is lessened, and the left

31

ventricle, thus relieved, is able to complete its systole. At the same time, from the moderate and gentle character of the exercises, compression of the veins, such as occurs in severe muscular exertion, driving on the blood to the right ventricle and causing dyspnea, does not take place. There is thus a transfer of blood from the venous to the arterial system, which is the reverse of the tendency in most forms of heart disease."

Without doubt, the Schott treatment, especially when it is carried out at Nauheim, is frequently followed by brilliant results. It is not applicable, however, to every case of heart-disease; neither is it entitled to supplant the older and approved methods of treatment. At most, it should be regarded as no more than a valuable auxiliary to rest, diet, and drugs. It must be admitted, too, that home treatment with artificial baths and resistance exercises, while not wholly wanting in efficacy, is much less beneficial than a course of treatment at Nauheim. Change of air and scene, the regulation of the mode of life of the patient, rest, and quiet unquestionably contribute largely to the success which has been achieved at this celebrated German spa.

The treatment is chiefly applicable to cases of *cardiac dilatation in which there is but slight degeneration of the myocardium.* The contraindications are advanced degeneration of the myocardium, pronounced arteriosclerosis, and aneurysm.

COLD.

Cold applications may be local or general. Locally, cold may be applied by means of cloths wrung out of iced water, an ice-bag, an ice-poultice, or Leiter's tubes. Heat may also be abstracted by spraying the part with a highly volatile liquid, like ether or ethyl chlorid. An ice-bag may be made cheaply and conveniently by filling a pig's bladder with cracked ice, closing it securely, and covering it with flannel. An ice-poultice may be made by sewing up tightly in rubber cloth or a coarse towel crushed ice mixed with salt and sawdust, bran, or ground flaxseed. Leiter's tubes are flexible metal tubes arranged in the form of coils to fit the head, chest, abdomen, and other parts of the body. Two long rubber tubes are connected with the coil, one to carry ice-water from a reservoir placed a few feet above the patient's head, the other to conduct the water after its passage through the coil to a pail placed on the floor near the bed. Leiter's tubes are especially useful when the cold is to be applied continuously.

Local applications of cold abstract heat from the part, lessen the sensibility of the peripheral nerve-filaments, cause constriction of the blood-vessels traversing the tissues exposed to the cold, and even affect reflexly, as Winternitz has shown, the vascularity of parts remote from the seat of the application.

Cold may be applied to the whole or to a considerable portion of the body in several ways: by cold sponging, the cold pack, the cold douche, or the cold bath.

Cold sponging consists in freely sponging the body with water at a temperature of from 80° to 70° F. (27.5°–21° C.), care being taken not to expose more than one part at a time. The sponging may be made more effective by adding a little alcohol or vinegar to the water.

In applying the cold pack, the bedding is first protected by water-proof sheeting; the patient is then stripped and enveloped in an ordinary sheet wrung out of water at a temperature of 70°–60° F. (21°–15.5° C.). The sheet should be well tucked in at the neck and feet, and closely adapted to all parts of the body. The pack is usually continued for from ten to fifteen minutes, and during this time it is necessary to sprinkle the sheet at frequent intervals with water sufficiently cool to maintain a uniform temperature.

The cold douche may be given by placing the patient, wrapped in a sheet, in a bath-tub, and then pouring cold water on him from a pail or watering-pot, or the water may be conducted through a rubber tube attached to a faucet and directed to various parts of the body.

In the cold bath the temperature of the water may range from 90° to 60° F. (32°–15.5° C.), and the immersion may last from a minute or two to twenty or even thirty minutes, according to effect desired.

When cold is applied suddenly to a considerable portion of the body of a man in good health, the immediate effects are contraction of the cutaneous vessels, pallor of the surface, erection of the papillæ of the skin (cutis anserina), momentary catching of the breath, and a feeling of coldness, with shivering. On coming out of the bath vigorous subjects, especially if the skin be well rubbed, quickly experience a reaction. The cutaneous vessels dilate, the skin becomes flushed, and the patient feels warm and exhilarated. In delicate subjects, especially if the immersion be somewhat prolonged, this reaction may be very imperfect or altogether wanting, in which case the patient may remain chilly and depressed for several hours after the bath. When followed by a good reaction, cold bathing exerts a powerful tonic effect. It sharpens the appetite, aids digestion, promotes tissue-change, and favors the elimination of waste-products. In pyrexia, general applications of cold lower the temperature, increase the flow of urine, strengthen and slow the pulse, and invigorate the nervous system.

Therapeutics.—In the early stage of *acute inflammation*, especially that resulting from traumatism, a cold-water dressing often acts most happily. In *diphtheria, scarlet fever*, and *acute tonsillitis* the application of cold compresses or of an ice-bag to the throat generally affords more relief than poultices or fomentations. In *chilblain* or *frost-bite* the intensity of the reaction should be moderated by rubbing the part with snow or ice. In *croupous pneumonia*, when the temperature is high and the pulse strong, an ice-bag or an ice-poultice adjusted over the affected lung reduces the fever, allays the pain, and quiets the patient. In *acute dysentery* Wood has found ice-suppositories inserted for a length of time, one after another, very useful in allaying tenesmus. An ice-cap or an ice-bag may be applied to the head with advantage in *meningitis, apoplexy, cerebral congestion, headache*, and the *delirium of fevers*. In *neuralgia* dry cold sometimes affords temporary relief, but, as a rule, heat is more grateful. In *exophthalmic goiter, paroxysmal tachycardia*, and *symptomatic palpitation* an ice-bag or a Leiter's coil over the precordia often exerts a sedative influence upon the heart. An ice-bag may also temporarily relieve distress in *aortic aneurysm*, especially when the sac is superficial. Prior to the performance of acupuncture or of paracentesis or to using the actual cautery, *local anesthesia* may be secured by spraying the part with ethyl chlorid or ether, or holding over it for a few minutes a block of ice which has been sprinkled with common salt. In *internal hemorrhage* the application of an ice-bag over the affected region is a useful adjuvant to other measures.

General applications of cold are used for their antipyretic and tonic effects. In *typhoid fever* and other *continued fevers* hydrotherapy has given excellent results. There is ample evidence to show that the mortality of typhoid fever in general hospitals has been reduced at least one-half under the treatment by cold bathing, so ably advocated by Brand, of Stettin. The details of this treatment are as follows: A bath-tub half full of water at 70° F. (21° C.) is kept in readiness near the bed, and every third hour, if his temperature is above 102.2° F. (39° C.), the patient is wrapped in a sheet and carefully lifted into the water. While in the bath an ice-cap is kept upon the head or cold affusions are applied to it, and the trunk and limbs are vigorously rubbed, so as to bring new relays of blood to the surface. A stimulant is sometimes given before the bath to lessen the shock. At the end of fifteen or twenty minutes the patient is carried back to bed and covered with a dry sheet and a light blanket. After he has been thoroughly dried, the damp coverings are removed and replaced by dry ones. If the patient be delicate, it is preferable to place him in a bath of 90° F. (32° C.), and then gradually lower the temperature of the water to 70° F. (21° C.). The good effects of the bath are: reduction of temperature, increased secretion of urine, marked improvement in the pulse, and lessening of the nervous symptoms—delirium, stupor, insomnia, and sub-sultus tendinum. The only contraindications are perforation, hemorrhage, and persistent prostration after the bath. Shivering and blueness naturally follow the bath, and, unless very prolonged, are not to be considered contraindications.

When the baths are not well borne, or the temperature shows no tendency to exceed 102.2° F. (39° C.), cold sponging or the cold pack may be substituted for immersion.

The *hyperpyrexia of thermic fever* or of *acute rheumatism* should be treated by immersing the patient in water containing crushed ice, or, better, by rubbing him with blocks of ice for ten or fifteen minutes.

In *neurasthenia* and *hysteria* hydrotherapy in the form of the cold spray, cold douche, or cold bath, is often a valuable auxiliary to rest, massage, and isolation. It is very important, however, not to institute the treatment too abruptly, as otherwise it may prove harmful rather than beneficial. In *chorea of childhood* the wet-pack (80°–70° F.—26.5°–21.0° C.) is sometimes very beneficial.

HEAT.

Heat may be applied in the dry or moist form, and the application may be local or general. Locally, dry heat is applied by means of hot cloths, bran-bags, water-bags or water-bottles, or local baths of superheated dry air. Moist heat may be applied by means of fomentations, poultices, or the douche. Hot-water baths are also applicable to certain limited portions of the surface of the body. Heat, locally applied, allays irritation of the peripheral sensory nerves, dilates the cutaneous vessels, increases the secretion of sweat, and probably, like cold and counterirritation, exerts reflexly some influence on the nutrition of the superficial tissues, and even of the underlying organs. Intense dry heat (300° F.—149° C.) not only affects the tissues to which it is directly applied, but also produces general diaphoresis, raises the body-temperature two or three degrees, and increases the frequency of the pulse.

Dry heat may be applied to a considerable portion of the surface of the body by means of the hot-air bath. This is best given with the patient in bed, a spirit-lamp being used to generate the heat, which is conducted under the bed-clothes by the aid of an inverted funnel and L-shaped tube. The patient is stripped and wrapped snugly in a blanket, and the bed-clothes, over which have been spread a rubber sheet and an extra blanket, are slightly raised by means of a cradle, so that the hot air can freely enter and surround the body. The bath may last from fifteen to thirty minutes. The effect of a general application of dry heat may also be obtained in the Turkish bath. In this the patient is conducted through a series of heated apartments, the first having a temperature of 100° F. (38° C.) and the last 150° F. (66° C.) or above. After perspiration has been freely established, he is thoroughly shampooed and rubbed, and then given a cold douche or a plunge-bath.

The general application of moist heat may be obtained by means of the hot-water bath, hot-pack, or hot vapor-bath. In the hot-water bath the temperature of the water should be at first about 98° F. (37° C.), but this should be raised gradually while the patient is still in the bath to 105° F. (40.5° C.) or even to 110° F. (44° C.). The bath may last, according to circumstances, from fifteen minutes to half an hour or longer. In the warm bath the temperature may range between 98° and 85° F. (37°–30° C.).

The hot-pack is given by closely enveloping the patient in a blanket wrung out of hot water, covering him with two or more blankets and a rubber sheet, and then placing hot bottles at each side and at his feet. The pack may last from half an hour to an hour. A hot vapor-bath may be given in the same way as a hot-air bath, steam from a boiling kettle being conducted under the bed-clothes instead of hot dry air, or the steam may be generated under the bed-clothes by placing about the patient hot bricks or plates which have been wrapped in wet cloths. In the Russian bath, also, steam replaces the hot dry air of the Turkish bath.

The first effects of the general application of heat are dilatation of the cutaneous vessels and the establishment of free perspiration. The administration of cool drinks during the exposure makes the application more effective by driving the blood from the interior to the surface of the body. There is, ordinarily, no marked rise in the central temperature of the body, since the accumulation of heat is prevented by the free sweating. If, for any reason, diaphoresis does not occur, however, the central temperature may rise somewhat—even to a dangerous point if the exposure be prolonged. The chief secondary effect of the general application of heat is increased elimination of effete matters from the body. Dry heat is generally more effective than moist heat, since it induces more copious perspiration.

In sudden failure of the circulation, such as occurs in shock and collapse, the general application of heat not only serves to raise the body-temperature, but also acts as a powerful cardiac and vasomotor stimulant.

Therapeutics.—Heat, both dry and moist, makes a valuable local application in a large number of *acute inflammatory diseases.* Generally, it may be used interchangeably with cold, the chief guide in the selection of one or the other being the feelings of the patient. Moist heat in the form of fomentations or hot baths is a valuable adjuvant to morphin in relaxing the spasm of unstriped muscles which occurs in *intestinal, biliary,* and *nephritic colics,* in *spasmodic croup,* and in *retention of urine.* Hot baths are also useful in breaking up *general convulsions,* especially those occurring in children. Hot foot-baths, through their reflex effect upon the circulation, often afford relief in *cerebral congestion, acute coryza,* and *amenorrhea from exposure to cold.* In *acute and chronic parenchymatous nephritis* hot-air baths, hot-packs, and hot-water baths are of the greatest service in producing diaphoresis, especially when there is dropsy or uremia. Mention has already been made (p. 480) of the efficacy of warm effervescent baths in certain cases of *chronic heart-disease.* The continuous immersion of the patient in a warm bath (90°–100° F.—32°–38° C.) has been found useful in *extensive burns* and in some cases of *phagedena.* A course of Turkish baths is frequently beneficial in *chronic gout,* in *chronic rheumatism,* and in the early stages of *chronic Bright's disease* and *diabetes.*

In conditions attended with subnormal temperature, such as *surgical shock* and *collapse from disease* or *poisoning,* the general application of heat, especially in the form of the hot bath, is the most important factor of the treatment.

The local application of intense dry heat (250°–350° F.—121°–177° C.) by means of some such apparatus as was invented by Tallerman, in 1893, is often very advantageous in certain affections of the joints, especially in *chronic synovitis, traumatic arthritis, sprains, tendinous inflammations,* and in the after-treatment of *fractures* and *luxations.* In *subacute* and *chronic rheumatism* it is also occasionally useful, but in *rheumatoid arthritis* it is of little value.

HYPODERMOCLYSIS AND INFUSION.

Hypodermoclysis and infusion are the introduction, into the subcutaneous tissues and blood-vessels, respectively, of weak saline solutions in considerable quantities. These procedures, first brought prominently forward by Cantani during the cholera epidemic of 1892, have been found of great value in a number of conditions. They serve to restore vascular fulness when, from any cause, the volume of blood has been much reduced, to stimulate the heart, especially when cardiac failure is associated with arterial depletion, to dilute any toxic substances that may be present in the blood, and to eliminate toxic substances by promoting diuresis.

In *hemorrhage* the injection of a saline solution affords a potent means of rapidly restoring the volume of blood. Formerly, transfusion was alone practised, but this operation was unsatisfactory; it was difficult to perform, and, moreover, it exposed the patient to the danger of embolism. The blood in being transferred lost its vitality and supplied very little, if any, nutrient matter to the tissues. The injection of salt solution was found to possess all the advantages of transfusion without its attending disadvantages.

Hypodermoclysis or infusion has proved very beneficial in many severe *toxemic conditions;* thus, it has been found efficacious in *uremia, diabetic coma, puerperal eclampsia, septicemia,* and *pneumonia.* It has given good results also in *cholera, thermic fever,* and extensive *burns.* In some cases of uremia, eclampsia, pneumonia, and thermic fever it is more effective if preceded by venesection. Lastly, it is exceedingly useful in combating the profound vasomotor paresis of *shock.* The solution usually employed for these injections is a decinormal salt solution (0.6 per cent.). This solution is selected because it corresponds closely in saline strength to the blood-serum. It may be prepared by adding a heaping teaspoonful of salt to a quart or a liter of water, boiling, and filtering. Various other formulæ, some quite complicated, have been suggested, but they seem to possess no advantages over the one just mentioned. Plain water cannot be used for infusion on account of its solvent action on the blood-corpuscles.

When the circulation is fairly active and there is no great emergency, hypodermoclysis should be practised in preference to infusion, the injection being made in the cellular tissue of the iliolumbar region, chest, abdomen, or thigh. The fluid may be introduced by means of a fountain syringe to the tube of which has been attached an aspirating needle of moderate size. The apparatus, the hands of the operator, and the region selected for the puncture should be thoroughly sterilized. The bag should be suspended 2 or 3 feet above the patient, and the solution should flow from the needle while the puncture is being made. The fluid should enter the tissues at a temperature of about 105° F. (40° C.), and to insure this a temperature of 110° F. (44° C.) should be maintained in the bag. The quantity of fluid required varies from 8 ounces (236.0 c.c.) to a pint (0.5 L.) in toxemic conditions, and from a pint (0.5 L.) to a quart (1.0 L.) or more in severe hemorrhage or shock. Frequently, better results are obtained by injecting smaller quantities several times than a larger quantity at once. In children from 2 to 4 ounces (60.0–120.0 c.c.) are usually sufficient. The rapidity of absorption varies with the state of the circulation, from twenty minutes to an hour being required to introduce a pint of fluid. To prevent overdistention of the tissues, the fluid should be made to enter slowly, gentle friction being applied to aid absorption.

In cases of emergency, when the circulation is very feeble and absorption is nearly at a standstill, intravenous infusion affords a more rapid means of filling the vessels and stimulating the heart than hypodermoclysis. For infusion the apparatus may consist of a graduated glass irrigating jar, having a capacity of about 2 quarts (2.0 L.), and provided with rubber tubing. To the latter should be fitted a fine transfusion cannula having a blunt, oblique tip. In great emergency a glass funnel, a rubber tube, and an aspirating needle will suffice. The reservoir should be placed about 2 feet above the patient, and its contents should be constantly maintained at a temperature of about 110° F. (44° C.), or higher (115° F.—46° C.) in cases of profound shock. The operation is performed as follows: The arm is prepared as for venesection, a fillet being applied securely some distance above the elbow. A prominent vein, preferably the median basilic or the basilic, is then exposed by dissection, and a ligature tied around it at the distal end of the wound. A second ligature is placed around the vessel at the proximal end of the wound, but it is not tied. The vein is next incised between the ligatures, and the cannula, while fluid is escaping from it, is carefully inserted into the lumen of the vein and tied in place by the second ligature, after which the fillet is *immediately* removed. From ½ pint to 2 pints or more (236.0—1000.0 c.c.) may be infused, according to the condition of the patient. The rate of flow may be regulated by raising or lowering the reservoir or by means of a clamp attached to the tubing. At least ten minutes should be allowed for introducing each pint of fluid. If care be taken to expel all air from the tubing and cannula before inserting the latter, the danger of air-embolism need not be feared.

Shortly after infusion a rigor may occur, and the pulse and respiration may become more rapid; usually, however, these untoward symptoms gradually disappear and are followed by distinct improvement.

ENTEROCLYSIS.

By enteroclysis is meant the irrigation of the colon with large quantities of water, the latter being either plain or medicated, hot or cold, according to the effect desired. It is practised for a number of purposes: To cleanse the colon; to bring remedies in direct contact with the lower bowel in diseased conditions; to raise or lower the body-temperature; to stimulate the heart in shock or collapse; to restore water to the blood in anhydremia; to promote renal secretion; to dilute poisonous substances that may be present in the blood; to check intestinal hemorrhage; and to relieve intussusception.

In some cases of *indigestion with persistent constipation* excellent results may be obtained by thoroughly irrigating the bowel, once daily or every other day, with several quarts of water at from 90° to 70° F. (32°–21° C.). In *acute catarrhal dysentery* injections of a warm (103° F.—39.5° C.) solution of boric acid (1 dram to 1 quart—4.0 gm. to 1.0 L.), continued until the fluid returns from the bowel clear, often affords much relief. In *amebic dysentery* quinin (1 : 5000 to 1 : 2000) may be substituted for the boric acid. In *chronic dysentery* no treatment is so generally successful as intestinal irrigation with weak solutions of silver nitrate (see p. 368). In *enterocolitis* and *cholera infantum* high enteroclysis, once or twice a day, with normal salt solution, is a valuable adjuvant to other therapeutic measures. If the child's temperature is high, the water may be at 90° F. (32° C.) or lower; if, however, there is a tendency to collapse, the water may be at 105° F. (40° C.). Cantani and others have spoken highly of enteroclysis with hot tannic acid solutions (2 per cent.) as an aid to hypodermoclysis in *Asiatic cholera.*

Intestinal irrigation with hot salt solution (105°–110° F.— 40°–43.5° C.) for half an hour to an hour is a potent remedy in *shock.* It may be practised also as a preventive measure if the condition of the patient prior to a surgical operation is such that shock is to be feared. Hot saline injections are very valuable, too, in *anuria* resulting from nephritis or following operation. In *uremia* they tend not only to promote renal secretion, but also to dilute the poisonous substances circulating in the blood. Some practitioners speak favorably of very hot rectal injections (110° F.—43.5° C.) in *hemorrhage from the stomach or bowel.*

Daily irrigation with cold water—70° F. (21° C.) or lower—has been found quite effective in some cases of persistent *catarrhal jaundice.* Enemata of cold water (60°–50° F.—15.5°–10° C.) may be used successfully in connection with ice-baths to lower the temperature in *thermic fever.*

Recent cases of *intussusception* are sometimes relieved by high rectal injections, but the hydrostatic pressure must not be too great, otherwise rupture of the bowel is liable to occur. The child should be anesthetized and the colon thoroughly but very gradually distended with warm water, the elevation of the reservoir not being over 3 or 4 feet.

Enteroclysis may be practised by means of an ordinary fountain syringe to which has been adjusted a flexible colon tube, or, for children, a soft-rubber catheter (No. 20). If the latter be used, an extra eye should be cut in it near the tip. The requisite pressure is secured by elevating the water-bag 3 or 4 feet above the patient. The return flow may be collected in a bed-pan or, better, it may be conducted, by the aid of a rubber sheet, into a bucket placed on the floor. The patient having been placed in the dorsal position with the hips elevated, the tube should be well lubricated and gently inserted from 12 to 18 inches in the adult, or 10 to 14 inches in the child, and the water turned on. In the adult from 1 pint to 3 pints (0.5–1.5 L.), and in the child from ½ pint to 2 pints (0.3–1.0 L.), of water should be allowed to enter the bowel, when the tube of the fountain syringe should be detached from the rectal tube, thus permitting a return flow to occur.

While a long single tube used in the way just described gives excellent results, a double-current tube will be found far more convenient, especially for continuous irrigation. A good example of the double-current tube is the one devised by Dr. R. C. Kemp.

LAVAGE OF THE STOMACH.

There are two methods of washing out the stomach—one, by the stomach-pump; the other, by siphonage. The stomach-pump is simply a syringe with two apertures instead of one, in which are adjusted valves opening in opposite directions. By connecting the syringe alternately at each of these apertures with an incompressible tube passed into the stomach, the gastric contents may be drawn out or fluid may be introduced into the stomach.

Washing out the stomach by siphonage may be accomplished in one of two ways: In the first a large glass funnel, to which is attached a rubber tube about 2½ feet (45.7 cm.) long, is joined through the medium of a short piece of glass tubing to a previously passed stomach-tube. The requisite amount of water, usually about a pint (0.5 L.), is poured into the funnel, which is then elevated sufficiently to allow the water to flow into the stomach. Before the funnel is empty, however, it is quickly depressed, so that the water is returned from the stomach by siphonage. The escape of the fluid from the stomach may be further facilitated by the application of gentle pressure to the epigastrium. The process should be repeated until the water returns clear.

In the second method the free end of the stomach-tube is connected with a Y-shaped glass tube, to one limb of which is attached a rubber tube leading to an elevated reservoir containing water, while to the other limb is attached a rubber tube leading to a pail placed on the floor. By alternately opening and closing the two external tubes by means of clamps the water may be made to flow into or out of the stomach.

The siphoning apparatus, on account of its simplicity and safety and the ease with which it can be manipulated, even by the patient himself, has almost completely supplanted the stomach-pump. The latter may be used with advantage, however, when the stomach contains much coarse solid matter or when it is necessary to remove the gastric contents very promptly and thoroughly.

No other lubricant than water is required for the stomach-tube. The introduction of the tube is facilitated by having the patient breathe deeply and make repeated efforts at swallowing as soon as the tube reaches the posterior wall of the pharynx. Cocainization is sometimes employed to allay extreme sensitiveness of the pharynx, but it is rarely necessary. To prevent the tube from being completely swallowed, it may be secured by means of a string attached to its distal extremity.

Ordinarily, the water used for lavage may be unmedicated; when, however, there is abundant mucus in the stomach, 2 per cent. of sodium bicarbonate may be added with advantage. Potassium permanganate (1 : 5000), sodium salicylate (1 : 100), or sodium hyposulphite (1 : 100) may be used when there is excessive fermentation.

The most important indication for lavage is *retention of food with fermentation.* Thus, in organic stenosis of the pylorus thorough washing of the stomach two or three times weekly, or every day in some cases, preferably before breakfast, affords great relief. In *atonic dilatation with fermentation* lavage is also efficacious. In *acute food-poisoning* and *toxic gastritis* it is often better to resort to the tube than to employ an emetic.

In *chronic gastric catarrh,* when there is an *excessive secretion of mucus* it is often of service; in other cases, however, it does no good, and is, according to Boas, absolutely contraindicated. Practised promiscuously in cases of simple gastric catarrh, it is capable of doing considerable harm and of developing into a most pernicious habit. In *supersecretion of acid* (Reichmann's disease) washing out the stomach with a solution of sodium bicarbonate (2 per cent.) or silver nitrate (1 : 5000 to 1 : 2000) sometimes gives good results. Kussmaul and others have found lavage of the stomach very beneficial in *acute intestinal obstruction.* In this condition it serves to allay vomiting, to lessen the pressure in the bowel, and to diminish the violent peristalsis.

Lavage is contraindicated in peptic ulcer, in ulcerating carcinoma, especially when accompanied by much bleeding, and in gastric catarrh associated with atrophic cirrhosis of the liver. It should be avoided also in advanced pregnancy, in aortic aneurysm, and in aged persons with marked arteriosclerosis.

BLOODLETTING.

Bloodletting is employed both as a local and as a general remedial measure. As a local measure it may be practised by cupping or by leeching, and as a general measure by venesection.

Cupping.—There are two methods of cupping—the dry and the wet. In dry cupping the blood is not shed, but is diverted by atmospheric pressure from the deeper parts to the cutaneous surface. A small quantity, however, is actually drawn from the vessels into the cellular tissue of the skin. The special instrument used for cupping consists of a glass bell with an aperture at its summit, through which is extracted the air by means of a pump or rubber ball. In the absence of such an instrument an ordinary tumbler will serve the purpose. A small piece of blotting-paper moistened with alcohol is placed in the bottom of the tumbler and lighted. While the paper is still burning the glass is inverted and applied firmly to the skin. The flame is immediately extinguished, and the air being exhausted, the skin in the interior rises in the cup and becomes of a dark-red color. To avoid burning the skin, it is necessary to allow all excess of alcohol to run out of the tumbler before igniting the paper. From one to a dozen of these cups may be applied, and they may be allowed to remain in position for from five to ten minutes. The cup can be readily removed by making a little pressure near the rim so as to allow air to enter. The same method is pursued in wet cupping, except that numerous small incisions are made in the skin by means of a scarificator before the cup is applied. As scarification leaves permanent cicatrices, wet cups should not be applied on exposed surfaces. Dry cupping is not so efficient as wet cupping, but may be practised when the patient's strength will not warrant an absolute loss of blood.

Leeching.—Leeches accomplish the same purpose as wet cups, but on account of their size and shape they can be applied to parts to which cups cannot be adjusted, as behind the ears, about joints, over the spermatic cord, and the verge of the anus. Each leech, according to the variety employed, removes from 1 to 3 drams (4.0–11.0 c.c.) of blood. Leeches should be applied in the neighborhood of inflamed parts, not directly over them, and parts which are abundantly supplied with loose areolar tissue, like the eyelids and the scrotum, should be avoided. The eye may be depleted from the temple, and the testicle from the perineum. Before leeching, the part should be shaved and well washed, care being taken to remove all soap.

Leeches may be applied with the finger or by means of an inverted wine-glass or pill-box. A little blood or warm milk smeared over the skin will generally induce them to bite if they are not disposed to do so. They should not be detached forcibly, but allowed to drop off themselves, or made to relinquish their hold by sprinkling them with salt. If desirable, bleeding from the bites may be encouraged by applying warm fomentations. Occasionally, owing to the presence in the tissues of a large quantity of the anti-coagulant substance secreted by the leech, the hemorrhage proves very persistent, and requires for its arrest the application of firm pressure, of some styptic (ferric subsulphate or alum), or of the actual cautery. Like scarifying, leeching leaves indelible scars.

Local bloodletting may often be practised with advantage in the early stage of many acute inflammatory diseases, such as *pneumonia, nephritis, orchitis,* etc. At the onset of acute inflammation of serous membranes,—*pleuritis, pericarditis, iritis, synovitis,* and *meningitis,*—especially when the symptoms develop abruptly and are very severe, it may be of the greatest value in relieving pain and in repressing exudation. In *acute pulmonary edema with cyanosis* wet cupping is of service, although it is not generally so useful as venesection.

Venesection or Phlebotomy.—The general abstraction of blood is accomplished by incising a vein. The patient having been placed in a semirecumbent position, the arm should be constricted a few inches above the elbow by a twisted handkerchief or a few turns of a roller bandage. If this is not sufficient to render the veins prominent, the arm may be rubbed for a few minutes from below upward. A large vein (median cephalic or median basilic) having been selected, it should be fixed by the thumb of the left hand below the point of section, and then incised by a lancet or bistoury in a direction oblique to the long axis of the vessel. The pulse should be carefully observed during the operation, and when it lessens in force or becomes more compressible, the bleeding must be suspended. The loss of blood required to afford relief varies in different conditions from a few ounces to a pint or more. To arrest the flow, it is only necessary to remove the fillet from the arm and to apply a small compress and figure-of-eight bandage.

The cases in which venesection is indicated are those in which life is immediately endangered by some circulatory obstruction or by toxemia. In *overdistention of the right ventricle,* whether from mitral disease, pneumonia, or emphysema, when there is a small, weak pulse, with cyanosis and orthopnea, the venous engorgement is often most promptly and effectually relieved by the abstraction of a few ounces (10–20) of blood. In *cerebral hemorrhage,* when the pulse tension is high and there are evidences of pronounced hyperemia of the brain, venesection is advisable. In *aortic aneurysm,* when the heart-action is strong and there is severe pain, bleeding often affords great relief.

In *acute uremia* and *puerperal eclampsia* venesection affords the most rapid means available of removing toxic substances

from the body. It may be practised advantageously in connection with subcutaneous and rectal injections of normal salt
solution.

Packard and others have found bloodletting (10–20 ounces)
efficacious in *grave cases of thermic fever.* Feebleness of the
pulse is not necessarily a contraindication, as the circulation
often improves as the blood flows from the arm.

PHOTOTHERAPY AND X-RAY THERAPY.

The method devised by Finsen of treating certain localized
diseases of the skin by means of concentrated rays of either
sunlight or electric arc-light has met with considerable success. It is based upon three postulates proved experimentally by Widmark, Finsen, Bie, and Godneff: the power
of the chemical rays to penetrate the skin ; the bactericidal
properties of these rays ; and the power of the chemical rays
to excite an inflammatory reaction in the skin. Finsen's apparatus is constructed to accomplish the following objects : to
concentrate the light, to absorb or to exclude most of the
heat-rays (ultra-red, red, and yellow) without impairing the
chemical rays (ultra-violet, violet, and blue), to keep the skin
cool, and to render the tissues anemic. The last is necessary
to secure deep penetration of the chemical rays.

The apparatus for sunlight consists of a lens from 8 to 15
inches (20.0–40.0 cm.) in diameter. This lens is composed of
two glasses, one curved and the other plain, adjusted in a
brass ring and separated from each other by a weak ammoniacal solution of copper sulphate. The blue liquid does not
affect the chemical rays, but it cools the light by arresting the
heat-rays. The apparatus for sunlight consists of two quartz
lenses framed in two brass tubes, which can be moved the one
into the other, like the two pieces of a telescope. The lenses
are separated from each other by a space filled with distilled
water, and in order to prevent overheating of the latter by the
absorption of the ultra-red rays cold water is made to circu-
late through a mantle surrounding this section of the tube.
In both plans heating of the skin is further avoided by making cold water run through a hollow quartz lens, which is
applied to the affected part with sufficient pressure to make
the tissues anemic.

Excellent results have been achieved by the light treatment
in *epithelioma of the skin, lupus vulgaris,* and *lupus erythematosus.* In a few cases of *alopecia areata* in which it has been
tried the results have also been encouraging. Relapses are

said to be rare, and the applications are painless. Apart from the expense of the apparatus and the necessity of trained assistants, the main drawback to the light treatment is its duration. Daily sittings, each lasting an hour, are required for periods ranging from several weeks to a year or more.

There is considerable evidence to show that the X-rays exert a curative influence in certain diseases of the skin—*epithelioma*, *lupus*, and *hypertrichosis*—equal to, if not greater than, that exerted by the concentrated chemical rays of light. As described by Pusey, the treatment consists in repeated exposures to a weak light of definite strength. The light is produced by a secondary current generated in an induction coil of 30 centimeters' spark length, which in turn is energized by a very weak primary current. The primary current is one of 12 volts and $1\frac{1}{2}$ ampéres, interrupted from 800 to 1000 times a minute. The exposures last from five to fifteen minutes, and the distance of the tube from the skin varies from 6 to 2 inches (15.0–5.0 cm.). The adjacent cutaneous surface is protected by a leaden mask. The applications are made daily or every two or three days, and are continued for several weeks or months. The appearance of even the slightest erythema is to be regarded as an indication for interrupting the sittings. In proper hands the risk of causing X-ray burns by this treatment is slight.

LUMBAR PUNCTURE.

The procedure known as lumbar puncture consists in the removal of cerebrospinal fluid through a hollow needle introduced between the transverse processes of the lumbar vertebræ. Since 1891, when it was first resorted to by Quincke, for the relief of excessive intracranial pressure in cases of chronic hydrocephalus, it has been rather extensively practised not only as a therapeutic measure, but also for diagnostic purposes. The operation consists in introducing a needle from 5 to 8 cm. long between the third and fourth lumbar vertebræ, and from 5 to 10 mm. to one side of the median line. The needle should be directed slightly upward and toward the median line. The required depth of the puncture can be determined by the sense of touch, the disappearance of resistance indicating the entrance of the needle into the cavity. In children the fluid is usually reached at a depth of from 2 to 3 cm., and in adults at from 4 to 6 cm. It is better to allow the fluid to run from the needle than to aspirate. From 10 to 60 c.c.

32

of fluid may be removed, the amount depending largely upon the age of the patient and the nature of the disease under treatment. On the occurrence of faintness, headache, or a change in the character of the pulse, the needle should be immediately withdrawn.

The chief therapeutic indication for lumbar puncture is to relieve excessive pressure by the cerebrospinal fluid on the brain and cord. Repeated at frequent intervals, it has given very encouraging results in cases of *acute cerebrospinal meningitis.* Koplik has reported 4 recoveries in 5 cases of meningitis due to the meningococcus in which this treatment was employed, and Netter has reported 7 recoveries in 11 cases of purulent meningitis in which lumbar puncture was done from 1 to 10 times and warm (104° F.—40° C.) baths were given every three or four hours. In *tuberculous meningitis* it may be practised as a palliative measure when the pressure-symptoms are very severe. In *chronic hydrocephalus* it is of doubtful value, although in some instances it appears to have afforded temporary relief.

Jacoby has reported 2 cases of *traumatic paraplegia* in which rapid improvement followed the removal by puncture of bloody fluid. The operation is contraindicated in tumor of the brain, softening, apoplexy, and embolism. There are on record at least 5 cases in which sudden death was induced by lumbar puncture in brain-tumor.

APPLIED THERAPEUTICS.

ACUTE INFECTIOUS DISEASES.

TYPHOID FEVER.

As soon as the nature of the disease is recognized, the patient should be confined to bed. The room should be large and airy, and provided with efficient means of securing thorough ventilation. The temperature of the room should be maintained between 65° and 70° F. (18.5°–21° C.). The bed should be moderately firm. The mattress should be protected by a rubber cloth spread beneath the sheet, and the latter should be kept smooth to guard against the development of bed-sores. Even in mild cases it is advisable to have a nurse or attendant constantly at hand, for accidents resulting from sudden delirium are liable to occur. The mouth and teeth should be washed several times a day with a solution of boric acid (10 gr. to the oz.—0.6 gm. to 30.0 c.c.). The bed-pan must be used from the beginning until convalescence is well advanced. The stools and urine should be rendered innocuous before being disposed of. This may be done by treating the evacuation with twice its volume of a 1 per cent. solution of chlorinated lime or a 5 per cent. solution of carbolic acid, and allowing it to stand in a covered vessel for two hours before emptying it into the closet. Soiled clothing should be thoroughly boiled.

Diet.—The diet should be liquid or semisolid, unirritating, and easily assimilable. As a rule, milk is the best food. Most patients will be able to take from 2 to 4 pints (1.0 to 2.0 L.) in the twenty-four hours, given in portions of from 4 to 6 ounces (120.0–175.0 c.c.) every two or three hours. It is generally advisable to dilute the milk with lime-water. If curds appear in the stools, the quantity of milk should be reduced. Among other permissible articles may be mentioned buttermilk, koumiss, junket, milk-whey, albumin-water, raw eggs, oyster, mutton, or chicken broths, chicken jelly, and consommé. Cocoa, iced tea, ice-cream, wine-jelly, blanc-mange, and, unless diarrhea is present, strained fruit-juices may be given occasionally to vary the monotony of the diet. Most clinicians are of the opinion that the return to solid food should not be commenced until the temperature has been normal a week or ten days. This is a good general rule, but it is open to many exceptions. In some cases it is advisable to wait longer than the period named before permitting solid food; in others it is not only safe, but advantageous, to administer more substantial food some time before convalescence has actually begun.

Cold water or cracked ice may be given freely, and one or the other should be offered to the patient several times a day even though he may not ask for it. It is rarely advisable to awaken the patient out of a sound sleep to give him either nourishment or medicine, but when there is stupor, he should be aroused at stated intervals.

Stimulants.—Stimulants are not always required, but when the pulse becomes soft and the first sound of the heart indistinct, alcohol should be given tentatively, its effects being carefully noted.

Ordinarily, the best form of stimulant is whisky or brandy; occasionally, however, sherry, port, or champagne may be better borne. At first 1 or 2 ounces (30.0–60.0 c.c.) of whisky in the twenty-four hours may be quite sufficient; later it may be necessary to increase the amount to 10 or 12 ounces (295.0–355.0 c.c.) or more. The quantity must be determined in each case by the effect. If the pulse becomes stronger, the tongue less dry, the mind clearer, and the urine more copious under the administration of the stimulant, it is doing good; if, however, opposite effects are observed, it is doing harm, at least in the quantity in which it is being employed.

Generally, the alcohol should be given with the milk, so that it may serve as an aid to digestion, but when the symptoms are very severe, it may be given also at other times.

Special Remedies.—Various antiseptics have been recommended from time to time with the object of destroying the typhoid bacilli or of rendering them harmless. Thus calomel, corrosive sublimate, carbolic acid, iodin, chlorin water, thymol, β-naphthol, salol, and guaiacol have each had their advocates. While certain of these agents may at times be useful in checking excessive fermentation in the intestinal canal, the evidence is far from convincing that they exercise any modifying influence upon the systemic infection itself.

While antityphoid inoculations have been practised with encouraging results in preventing the occurrence of typhoid fever, the employment of antityphoid serum as a curative agent has been thus far on too limited a scale for any opinion to be formulated as to its efficacy.

Pyrexia.—High temperature is controlled best by the cold bath or the cold pack; these hydrotherapeutic measures not only reduce the temperature, but they also exert a powerful stimulant influence upon the nervous system and the circulation, and in many instances have a diuretic effect. Excellent results have been obtained by using the cold bath throughout the attack as often as the temperature in the mouth rises to 102.5° F. (39.2° C.). (See Cold.) The total mortality under hydrotherapy in a series of 1904 cases treated by J. C. Wilson in Philadelphia was 7.5 per cent.—exactly the same as that in 1902 cases treated by F. E. Hare in the Brisbane Hospital, Australia. As the death-rate under the expectant treatment is between 14 and 15 per cent., it would appear that systematic cold bathing has reduced the mortality about 50 per cent.

When the baths are inapplicable, the fever may be kept within safe limits by cold sponging or the cold pack. Occasionally it may be necessary to administer a small dose of phenacetin or of antipyrin (5 gr.—0.3 gm.), but these drugs should never be used in large doses or given repeatedly.

Heart-failure.—Cold bathing and the timely use of alcohol do much to guard against heart-failure. When the tendency to cardiac failure is pronounced, strychnin may be given in doses of $\frac{1}{40}$–$\frac{1}{20}$ gr. (0.0016–0.0032 gr.) every three or four hours. In severe cases the drug should be given hypodermically. Digitalis or strophanthus may also be tried, but in the presence of fever these remedies often prove ineffectual. If collapse is threatened, ether, alcohol, or, better still, camphor (1–2 gr.—0.06–0.13 gr.) may be given subcutaneously at frequent intervals.

Diarrhea.—When the diarrhea exceeds three or four stools a day, efforts should be made to check it. In some cases a modification of the diet—reducing the quantity of milk, peptonizing the milk, or withholding meat broths—will be all that is required. In other cases it may be necessary to use a suppository or enema of opium once or twice a day. If the diarrhea be troublesome, bismuth subnitrate or silver nitrate may be given by the mouth in combination with opium:

R Morphinæ sulphatis, gr. j (0.065 gm.);
   Orphol, ʒiss (5.8 gm.);
   Bismuthi subnitratis, ʒiij (12.0 gm.).   M.
Fiant chartulæ No. xv.
Sig.—One powder every three or four hours.

R Argenti nitratis, gr. iij (0.2 gm.);
   Pulveris opii, gr. vj (0.4 gm.).   M.
Fiant pilulæ xij.
Sig.—One pill every three or four hours.

In very obstinate cases copper sulphate with opium in pill proves efficacious.

Constipation.—Constipation lasting but two or three days requires no special treatment. Sometimes meat broths or strained oatmeal gruel given alternately with the milk will exert a mild laxative effect. When more than a modification of the diet is necessary, an enema of soapy water or glycerin should be tried before resorting to remedies by the mouth. In obstinate cases small doses of calomel, of Epsom salt, or of castor oil may be given at short intervals until the desired effect is obtained.

Tympanites.—Troublesome tympanites is sometimes entirely overcome by substituting peptonized milk for ordinary milk. In other cases it may be relieved by the external application of turpentine stupes and by the internal administration of some antiseptic remedy, such as turpentine, beta-naphthol, beta-naphthol bismuth, creasote, or salol. Enemas of turpentine or of asafœtida are also efficacious. In certain persistent cases, apparently paretic in nature, the author has successfully employed eserin salicylate in doses of $\frac{1}{60}$–$\frac{1}{40}$ gr. (0.001–0.0016 gm.) two or three times daily, as recommended by von Noorden. Finally, when the tympanites is extreme and other measures have failed to afford relief, a soft rectal tube may be introduced as far as possible into the colon.

Persistent Vomiting.—When the stomach is excessively irritable, it is better to withhold milk for a time and to give in its stead albumin-water or one of the prepared foods (pano-peptone, somatose, liquid peptonoids), in very small quantities, at frequent intervals. Occasionally it may be necessary to withhold all food by the mouth for twenty-four hours and to support the patient by rectal alimentation. The best gastric sedative in such cases is bismuth subnitrate in powder, with cerium oxalate or cocain hydrochlorate, or in water with dilute hydrocyanic acid. A mustard-plaster to the epigastrium is often efficacious.

Hemorrhage from the Bowels.—The patient must be kept at absolute rest. Cold bathing must be suspended. If the hemorrhage is severe, the bowel-movements should be received in a folded towel instead of in a bed-pan. An ice-bag may be applied with advantage to the right iliac region, and ice may be given by the mouth. It is advisable slightly to elevate the foot of the bed. For ten or twelve hours the aliment should be reduced to the smallest possible amount. The best drug is opium; it controls peristalsis and allays excitement. It may be given alone or in combination with lead acetate. Turpentine, in doses of from 10–20 min. (0.6–1.2 c.c.) every two hours, is also efficacious. Some writers speak favorably of oil of erigeron. Preparations of ergot are of very doubtful value. In cases of recurrent hemorrhage gelatin (see p. 377) subcutaneously or by the bowel should be tried. Rectal injections both of iced water and of very hot water (110°F.—43.5° C.) have been recommended. If collapse is threatened, diffusible stimulants, like ether and alcohol, should be given freely. The acute anemia following copious bleeding is best treated by subcutaneous or intravenous injections of warm saline solutions (see p. 488).

Headache.—In most cases absolute rest, quiet, and cold applications to the head suffice. If the pain is severe, it may be necessary to give sodium or potassium bromid (10–15 gr.—0.65–1.0 gm.) or small doses (5 gr.—0.3 gm.) of phenacetin.

Insomnia.—Opium is generally the best hypnotic. In some cases, however, sodium bromid or chloralamid acts better. Chloral is also very efficacious, but it should not be used when the heart is weak.

Delirium is best managed by hydrotherapy. Low, muttering delirium usually calls for stimulants. An ice-cap is useful. Camphor or musk may be tried; the former is best given hypodermically, the latter by the bowel. In active or violent delirium no drug is so generally useful as morphin. In some cases hyoscin ($\frac{1}{150}$–$\frac{1}{100}$ gr.—0.00043–0.00064 gm.) acts well.

Hypostatic Congestion of the Lungs.—Frequent change of position and the timely use of cardiac stimulants will frequently prevent serious pulmonary congestion. When severe enough to embarrass the respiration and circulation, dry cups or stupes should be applied to the chest, and strychnin or ammonium carbonate given internally. In grave cases oxygen inhalations afford relief, but their effects are usually temporary.

Perforation.—Recovery from peritonitis is so exceedingly rare under medicinal treatment that operative interference is called for in all cases which are not obviously moribund. The operation should be done at the earliest possible moment. Keen has collected 83 cases with 16 recoveries. Of 15 cases operated on within twelve hours, 4 recovered; of 20 cases operated on between the twelfth and the twenty-fourth hour, 6 recovered; of 13 cases operated on in the second twenty-four hours, only 1 recovered (Osler).

Bed-sores.—This complication can usually be prevented by absolute cleanliness, frequent change of position, the application of whisky or whisky and alum to parts most subjected to pressure, and the timely use of a water-bed or of air-cushions. Slight abrasions may be painted with flexible collodion or dusted with a powder of bismuth subnitrate and iodoform or of boric acid and starch. When there is great difficulty in keeping the parts dry, zinc ointment may be freely applied. Ulcers should be washed with weak antiseptic lotions, dusted with some antiseptic powder like iodoform or iodol, and then protected by a large piece of soap-plaster. An ointment of balsam of Peru (1 : 30) is sometimes very efficacious.

TYPHUS FEVER.

As typhus fever is a contagious disease, the patient must be isolated and every precaution must be taken to prevent the dissemination of the morbific agent, whatever it may be. Those who have had experience in typhus epidemics assert that free ventilation is often in itself sufficient to arrest the transmission of the disease. There is no specific treatment. The diet should be nutritious and easily assimilable. As in other continued fevers, liquids are better borne than solids, although special precautions are not so imperatively demanded as they are in typhoid fever. The pyrexia, nervous phenomena, heart-failure, and pulmonary congestion will require the same treatment as in typhoid fever.

RELAPSING FEVER.

As in all contagious diseases, isolation, free ventilation, and disinfection of excreta and clothing are important safeguards against the spread of the virus. The treatment is purely symptomatic. Absolute rest, good nursing, and proper diet will do much to avert complications. The chief indications are to control the fever, to quiet the stomach, to relieve the intense pains in the head, back, and limbs, and to support the heart.

On account of the short duration of the febrile paroxysms the fever itself is not so important a factor as it is in many other acute infections. Ordinarily, the application of ice to the head, with frequent cold sponging of the body, will suffice. Cold baths are not well borne on account of the severe muscular pains. Hyperpyrexia must be controlled by the cold pack and the administration of phenacetin or antipyrin or of large doses of quinin.

Vomiting is sometimes exceedingly difficult to control. Cracked ice, carbonated water, dilute hydrocyanic acid (1–2 min.—0.06–0.1 c.c.), cocain hydrochlorate ($\frac{1}{4}$ gr.—0.016 gm.), or wine of ipecac (1 min.—0.06 c.c.) should be tried. A mustard-plaster to the epigastrium is sometimes efficacious.

For the relief of the pains, restlessness, and insomnia no drug is so useful as morphin. A quarter of a grain (0.016 gm.) may be given hypodermically two or three times a day. Hot-water bags may also be ordered for the backache, and rubefacient liniments for the muscular pains. Stimulants are not generally required, but when there is any tendency to heart-failure they should be administered freely.

CEREBROSPINAL FEVER.

Cerebrospinal fever is probably not contagious, hence rigid isolation is not usually regarded as absolutely necessary. It is advisable, however, to disinfect the discharges, bed-linen, etc. The sick-room should be quiet, darkened, and well-ventilated. The diet should be liquid and supporting. Milk and animal broths may be given freely at frequent intervals. In some cases, in order to secure the ingestion of enough nourishment, it may be necessary to resort to nutrient enemata or forced feeding by means of a stomach-tube. Cardiac failure must be combated by stimulants, of which the best are whisky and brandy.

In sthenic cases, when the pain is severe and the pulse is strong, the withdrawal of several ounces of blood by wet-cups applied along the cervical vertebræ may prove very useful. General venesection is rarely indicated. Cold applied to the head and along the spine affords considerable relief. Blisters to the nape of the neck are of doubtful value, at least during the irritative stage. Morphin hypodermically is the best drug for the relief of pain, restlessness, spasms, and insomnia. According to von Ziemssen, it is the most indispensable medicine in the treatment of the disease. Chloral, bromids, phenacetin, and similar compounds, in safe doses, are not often effective.

Fever is controlled best by cold sponging or the cold pack, or, if the temperature is very high, by systematic cold bathing. Repeated lumbar punctures (see p. 497) have been found of great utility in relieving excruciating headache, delirium, somnolence, and coma. Koplik reports 4 recoveries in 5 cases treated by lumbar punctures, and Netter 7 recoveries in 11 cases of purulent meningitis treated by repeated lumbar punctures and warm baths (100°–104° F.—38°–40° C.) every three or four hours for twenty to thirty minutes at a time. Concerning laminectomy, Osler, in the *Cavendish Lecture* for 1899, says: "On the principle of a desperate remedy for a desperate disease, the operation of laminectomy seems justifiable in certain severe cases in which the spinal symptoms are very marked."

Many special remedies, like mercury, ergot, physostigma, quinin, and belladonna, have been advocated, but these are now generally believed to be of no value. Inunctions with Credé's ointment (colloidal silver), several drams daily, have recently been warmly recommended.

After the subsidence of the acute symptoms the absorption of the exudation may be attempted by the administration of potassium iodid. When symptoms indicating pressure on the brain or spinal cord persist, repeated blisters or light applications of thermocautery are sometimes useful. Tonics—iron, strychnin, cod-liver oil—are generally indicated during convalescence. Local palsies will require massage and electricity. Even after convalescence has been well established all mental and physical excitement should be avoided for a considerable period.

MALARIAL FEVER.

It has been definitely established that the hemocytozoan discovered by Laveran in 1880 is the sole cause of malarial fever, and that the mosquito plays an important rôle in transmitting the infection. Prophylactic measures include the extermination of mosquitos, the prevention of infection of mosquitos, and the prevention of infection by mosquitos (Manson). The most useful methods of suppressing mosquitos are the efficient drainage of pools and swamps and the cultivation of damp soils. Covering the surface of the water with petroleum will also free pools from larvas for from two to four weeks. To prevent the infection of mosquitos, malarial patients should be carefully screened. The chief means of preventing infection by mosquitos are avoidance of sleeping in the open air, and of exposure to the evening and early morning air, adequate protection from the insects, and the use of quinin in daily doses of from 2–5 gr. (0.13–0.3 gm.).

Quinin is the only reliable remedy for malarial fever. Methylene-blue and Warburg's tincture possess some value, but, being distinctly less efficacious than quinin, they should be employed only when the latter is not well borne. Arsenic is useful in correcting the anemia caused by malarial fever, but it probably has no parasiticidal influence. The methods of administering quinin and other antimalarial remedies has already been considered (see p. 409).

Symptomatic Treatment.—During the cold stage of the paroxysms the patient should be well covered with warm blankets and given hot drinks. Opium in the form of paregoric is sometimes useful in mitigating discomfort. It may be combined with a few minims of aromatic spirits of ammonia, chloroform, or Hoffmann's anodyne. Generally, alcoholic stimulants are better avoided, since they tend to increase the severity of the reaction. In the algid type of pernicious malarial fever, however, it may be necessary to give alcohol freely, with digitalis and strychnin, to tide the patient over the paroxysm. In the hot stage much relief is afforded by frequently sponging the body with cool water, giving cold drinks, and administering, if the symptoms are very severe, a small dose of phenacetin.

When the bowels are constipated and the tongue is thickly coated, laxative doses of calomel or of blue-mass may be given with advantage. Vomiting and purging are occasional symptoms; they are best controlled by an opiate.

In the graver congestive type, with coma and high fever, cold bathing and rectal injections of cool water are indicated.

For the anemia following malarial infection arsenic and iron should be given (see p. 304).

Malarial cachexia requires tonic and hygienic treatment. Arsenic, iron, and cod-liver oil are especially valuable. As in other manifestations of malaria, quinin is indicated so long as the blood shows parasites. According to Wood, it is much better to produce distinct cinchonism at intervals than to give the drug continuously in moderate doses. When there is constipation, mild bitter laxatives are beneficial. Change of locality is sometimes necessary to effect a cure. For the enlarged spleen, ergot has been warmly advocated, but it is of doubtful utility.

INFLUENZA.

Although we possess no certain safeguards against the spread of influenza, yet an effort should always be made to protect healthy members of the household, especially the young, old, and delicate, from infection. To this end the patient should be isolated, and the secretion from the catarrhal surfaces should be thoroughly disinfected.

No specific has yet been discovered for the disease. Hygienic measures are of the first importance; these include immediate and absolute rest in bed, a carefully selected diet, pure air without draft, and attentive nursing. Complications and relapses can generally be traced to a neglect of these rules.

In mild cases little in the way of special treatment is required. A hot foot-bath, some mild refrigerant, such as spirit of nitrous ether or solution of ammonium acetate, and at night a dose of Dover's powder (5–10 gr.—0.3–0.6 gm.), will usually suffice. If there be constipation, a few minute doses of calomel may be given with advantage; active purgation, however, should always be avoided.

In the more severe cases quinin, if well borne, may be given in small doses (2–5 gr.—0.13–0.3 gm.—thrice daily) throughout the attack. Rheumatoid pains are controlled to some extent by combinations of phenacetin with salicylates or benzoates. Only small doses, however, are admissible, and these should be carefully watched. The following is often useful:

<pre>
R   Phenacetin,
     Salophen,
     Sodii benzoatis,   aa ʒj (4.0 gm.).   M.
Fiant chartulæ No. xii.
Sig.—One every four hours.
</pre>

When the suffering is intense, there need be no hesitation in administering morphin hypodermically. Violent headache is treated best by small doses of phenacetin and the application of an ice-cap to the head. Sulphonal and chloralamid are, perhaps, the most satisfactory drugs for combating troublesome insomnia. When the stomach is persistently irritable, the diet should be restricted to liquids, and all drugs, except those of a soothing character, withheld. Antiemetics, like bismuth subnitrate, hydrocyanic acid, cerium oxalate, and cocain, are indicated. A sinapism to the epigastrium may also be recommended.

In many cases of influenza cardiac weakness is a prominent feature. It demands the free use of alcoholic stimulants and of strychnin. The latter is especially well borne, and may be given, if necessary, in doses of $\frac{1}{30}$—$\frac{1}{20}$ of a gr. (0.002–0.003 gm.) every three or four hours. If more stimulation is required, aromatic spirits of ammonia and digitalis or strophanthus may also be given.

As the course of the disease is comparatively short, the fever, even when high, will rarely call for special medication. Antipyretic drugs in large doses are always inadmissible. Delirium is another symptom that may usually be ignored.

In the respiratory type expectorants are usually needed. When the cough is hard and dry, combinations of ipecac and potassium citrate, like the following, will be found useful:

R   Potassii citratis,   ℥ij (8.0 gm.);
     Syrupi ipecacuanhæ,
     Tincturæ opii camphoratæ,   āā f℥ij (8.0 c.c.);
     Succi limonis,
     Glycerini,   āā f℥ss (15.0 c.c.);
     Aquæ, .   q. s. ad f℥iij (90.0 c.c.).   M.
     Sig.—A dessertspoonful every two or three hours.

When expectoration is established, expectorants like ammonium chlorid, squills, and terpin hydrate may be substituted for ipecac and potassium citrate. Stupes or sinapisms to the chest are also useful in relieving cough. When pneumonia is threatened, dry cupping is often efficacious.

The convalescence should be most carefully watched, and strict injunctions given as to diet, rest, and the avoidance of exposure. Tonics are usually indicated and may be continued for several weeks. Change of air will materially help to restore strength and to overcome the peculiar mental depression.

CROUPOUS PNEUMONIA.

Prophylaxis against croupous pneumonia consists largely in the avoidance of influences which tend to lower the resistance of the tissues. Undue exposure to cold and the excessive use of alcohol are important predisposing factors. Persons who have had one attack of pneumonia should be particularly attentive to their personal hygiene.

The sick-room should be well ventilated, but free from drafts. The patient's sputum and all articles liable to be soiled with the sputum should be disinfected.

The temperature of the sick-room should be between 65°–70° F. (18.5°–21° C.). The diet should be fluid or semifluid. Milk, junket, wine-whey, broths, eggs, and gruel are suitable forms of nourishment. Cool water should be given freely. In the absence of any indication for special local treatment the chest may be enveloped in a dry cotton jacket.

Specific Treatment.—Croupous pneumonia cannot be aborted. Many methods of specific treatment have been recommended, but none has proved satisfactory. From the time of Laennec down to about the middle of the last century venesection was thought to be imperative. To-day, it is rarely practised except to meet certain indications.

In robust subjects, at the very onset, when the invasion is violent and attended with high fever, a bounding pulse, marked dyspnea, and severe pleuritic pain, the abstraction of from 10–20 oz. (300.0–600.0 c.c.) of blood may afford great relief. Later in the course of the disease, if cyanosis and orthopnea develop in consequence of overdistention of the right ventricle, venesection may also prove useful.

Tartar emetic, which at one time was thought to possess the power of aborting croupous pneumonia, has been entirely abandoned. The same fate has almost overtaken veratrum viride, although some clinicians still recommend it as a substitute for venesection.

The treatment of pneumonia by large doses of digitalis,—1–2 oz. (30.0–60.0 c.c.) of the tincture daily,—originally proposed by Petresco of Bucharest, has never had many advocates in this country. In 1894 Petresco reported 1192 cases treated by this method, with a mortality of 2.66 per cent. But the author himself states that the treatment is not applicable to very grave cases, and that his observations were made in a military hospital on young and robust men. Petresco's figures lose much of their significance when we compare them with the death-rate in the German army, which, according to Osler, was only 3.6 per cent. in 40,000 cases of pneumonia among healthy picked men.

Aufrecht ardently advocates the hypodermic administration of quinin,—8 gr. (0.5 gm.) once or twice a day,—not as an antipyretic, but as a specific, remedy. The mortality under this treatment in a series of 261 patients treated in the Altstadt Hospital at Magdeburg was 8.8 per cent., but we learn from Strusberg that in the medical clinic at Bonn during the same period the mortality under symptomatic treatment was only 7.97 per cent.

Carbonate of creasote (2–3 fl. dr.—8.0–11.0 c.c. daily) has recently been highly extolled as a specific by Van Zandt, Weber, and others. In the author's experience this drug has not hastened the crisis nor exerted any appreciable influence on the general toxemia.

Of all the special remedies recommended for pneumonia, antipneumococcic serum was the most promising. Perhaps sufficient time has not yet elapsed to warrant a final opinion upon its efficacy, but certainly those who have given it a fair trial will admit that it has fallen considerably below their expectations and has yielded results in no way comparable to those obtained in diphtheria from the use of antitoxin. In a fair test made by J. C. Wilson in the German Hospital, Philadelphia, the mortality under serum treatment in one series of 18 cases was 22.2 per cent., and in another series of 18 cases, 35.3 per cent.

Although we have yet no specific treatment, much may be done in the way of symptomatic treatment to avert a fatal issue. The indications, however, are extremely variable and must be carefully considered in each case. Routine treatment is to be rigidly avoided. In a large number of cases no drugs are needed. The symptoms that most often demand special attention are the pain, cardiac weakness, cough, fever, dyspnea, insomnia, and delirium.

Pain.—Morphin hypodermically is the best drug for subduing pleuritic pain. It also serves to overcome the restlessness and anxiety which often mark the onset of the disease, and which, if allowed to continue, tend to augment the exhaustion. Hot or cold applications to the chest are also useful. When the suffering is very severe, a few wet or dry cups to the affected side, followed by poultices, will be found efficacious.

Cardiac Weakness.—Alcohol is the best stimulant. When the pulse becomes compressible and dicrotic, and the diastolic sound at the pulmonary area loses its force, it should be given freely. Often 3 or 4 ounces (90.0–118.0 c.c.) of good whisky in the twenty-four hours will suffice, but in some instances two or three times this amount may be given with advantage. The patients who need it most are the old, the debilitated, and the alcoholic. Digitalis is undoubtedly useful in some cases, but its action is uncertain and often disappointing. In the author's experience it has been almost invariably badly borne when administered for the irregular flagging pulse of the period immediately following the crisis. Its frequent failure at other times may be explained by the fact that the cardiac insufficiency in the majority of cases is not dependent on simple dilatation or relaxation of the ventricles, but on degenerative changes in the myocardium, or perhaps, in some instances, on disturbances of the motor ganglia of the heart, brought about by the infection. Mechanical distention of the right ventricle undoubtedly does occur in some cases. It may be recognized by the enlargement of the area of cardiac dulness, by the marked cyanosis, the orthopnea, and the small pulse. In this condition digitalis, by driving the blood with great force against an impassable barrier, is likely to do harm; it is better to relieve the strain on the right ventricle by judicious venesection.

As a circulatory stimulant, strychnin generally proves much more efficacious than digitalis. It should be given in ascending doses of from $\frac{1}{60}$–$\frac{1}{15}$ of a gr. (0.001–0.0043 gm.), careful watch being kept for the appearance of untoward symptoms. In order that there may be immediate absorption, large doses should always be given hypodermically. Caffein is a useful adjuvant to strychnin, but it should not be used when there is cerebral excitement or insomnia. Wood speaks well of cocain —⅙–½ gr. (0.01–0.03 gm.)—alternately with strychnin, one or the other being given in grave cases every two hours. In threatening collapse the author has found no stimulant so generally serviceable as camphor. It is especially useful in tiding the patient over the profound depression that often accompanies or follows the crisis. It is available also when pulmonary edema arises as a complication, and in this condition it may be advantageously combined with atropin. To secure the best results, camphor should be given hypodermically, dissolved in ether or sterile olive oil, in doses of from 1–2 gr. (0.06–0.13 gm.) every two or three hours.

Subcutaneous injections of normal salt solution have also been useful in overcoming adynamia.

Cough.—For the hard, dry cough, especially of the first stage, no remedy is so useful as opium in the form of Dover's powder. Expectorants are rarely needed. The inflammatory exudation in the air-vesicles is removed almost entirely by absorption, and to attempt its expulsion by means of a drug is useless. When, however, there is a good deal of bronchial catarrh, with much viscid sputum, ammonium carbonate may be given to facilitate expectoration, but it should be withdrawn immediately if it causes any disturbance of digestion.

Fever.—The fact that in a very large proportion of the gravest cases the temperature is not high indicates that the chief factor in bringing about the exhaustion is not the fever itself, but the toxemia. This being the case, and as the febrile period is generally of a comparatively short duration, antipyretic measures may be restricted in the majority of cases to frequent cold sponging and the application of an ice-bag or an ice-poultice to the affected side.

Persistent high temperature is best controlled by cool baths (80° F.—27.5° C.). Experience is strongly against the systematic employment of antipyretic drugs, like phenacetin and antipyrin.

Dyspnea.—Cardiac stimulants, since they favorably influence the pulmonary circulation, are also of service in relieving dyspnea. Naturally, the circulatory stimulants that are also respiratory stimulants, like strychnin, caffein, and cocain, are the most useful. Oxygen makes the breathing easier, lessens cyanosis, and conduces to sleep, and to this extent aids in conserving energy.

Insomnia and Delirium.—When insomnia and delirium are sufficiently severe to demand the use of drugs, morphin will generally be found the most efficacious. Of course, it should not be used when there is extreme dyspnea or when there are evidences of pulmonary edema. When the circulatory depression is not marked, chloral may be used. In other cases, chloralamid or sulphonal may be tried.

Delayed resolution is a troublesome complication. Fortunately, it is of rare occurrence. It sometimes yields to repeated blistering and the free use of tonics. Riess has recommended injections of pilocarpin, and Dieulafoy, Forbier, and Lepine the production of aseptic abscesses by means of turpentine injections.

DIPHTHERIA.

Diphtheria being a highly contagious disease, the utmost care should be taken to prevent the spread of the specific virus. General prophylactic measures include the isolation of the sick, absolute cleanliness in the sick-room, and the thorough disinfection of all articles of clothing and fomites that may have become contaminated with the secretions from the infected area. As diphtheria bacilli often remain in the throat for weeks or even months after the disappearance of the false membrane, isolation should be urged until the bacilli can no longer be detected in the throat by bacteriologic examination.

Physicians and nurses, in applying local remedies, must be exceedingly careful lest they become infected directly by shreds of false membrane expelled by the patient in coughing.

Finally, exposed persons, especially children, should receive an immunizing dose of diphtheria antitoxin (see p. 425).

The sick-room should be well ventilated, and the temperature maintained at about 70° F. (21° C.). It is desirable to have the atmosphere moist, and this may be accomplished by generating steam in an ordinary kettle or in a steam atomizer, or by slaking large quantities of quicklime in the room. Young children, especially when laryngeal symptoms are present, are best treated in a steam-moistened tent. Absolute rest must be enforced. The diet should be of the most nutritious and easily digested character. Milk, junket, ice-cream, eggs, and unseasoned broths may be allowed. After tracheotomy it may be necessary, in order to introduce sufficient nourishment, to feed through a soft-rubber nasal tube. Cool water should be given freely. The medicinal treatment of diphtheria is both local and constitutional.

Local Treatment.—Local treatment is of great importance, but often difficult to carry out. When the applications cause violent struggling and exhaust the child, it may be better to desist. Applications should be made with utmost gentleness, should be unirritating, and should be frequently repeated. Remedies may be applied by means of absorbent cotton on a swab, or by sprays, injections, or irrigations. In pharyngeal diphtheria the first method is usually the best, while in nasal diphtheria the employment of a syringe or irrigator is preferable. The nozzle of the instrument should be blunt and covered with soft rubber, and the nurse should be instructed to introduce it gently and in a horizontal direction. In both nasal and pharyngeal diphtheria, when the child cannot be raised, the fluid may be poured into the nose from a spoon, minimdropper, or, better, from a nasal cup (Jacobi). Sprays are not very efficacious. In laryngeal diphtheria local treatment is rarely feasible, except by means of medicated vapors.

Many local remedies have been recommended ; the following are, perhaps, the most generally useful: Solution of hydrogen dioxid (1 : 2 of lime-water), Löffler's solution (see p. 286), or corrosive sublimate (1 : 5000 to 1 : 1000). All applications to the nasal mucous membrane should be warm. False membrane should never be torn off, but when it is very loose, it may be carefully dislodged and removed. Emetics are sometimes useful in expelling false membrane from the larynx, but when there are marked symptoms of obstructive dyspnea, the operation of intubation or tracheotomy should not be deferred.

Externally, hot or cold applications, whichever may be more grateful to the patient, are useful in relieving pain and soreness in the throat. When there is marked swelling of the cervical glands, an ointment of mercury and belladonna is sometimes of service.

Constitutional Treatment.—Statistical reports from all parts of the world prove conclusively that the remedy of greatest value in the treatment of diphtheria is the blood-serum of animals rendered artificially immune against the disease (see article on Diphtheria Antitoxin, p. 424). Apart from antitoxin, the most important remedies are those which tend to maintain the bodily strength. Alcoholic stimulants are usually indicated, especially in the late stage of the disease. In septic cases alcohol is particularly well borne, a child of three years often being able to take several ounces of whisky a day with advantage. Next to alcohol, strychnin is the best stimulant. In profound adynamia, digitalis, caffein, camphor, and musk are also useful.

Two drugs which are believed to be of special value, but which have lost some of their reputation since the introduction of antitoxin, are the tincture of the chlorid of iron and mercury. From 3–5 min. (0.2–0.3 c.c.) of the tincture of ferric chlorid may be given in glycerin or water every two or three hours to a child of four years. It should, however, be suspended on the first appearance of digestive disturbance. Children with diphtheria are usually very tolerant of mercury, and as much as $\frac{1}{40}$ of a grain (0.0016 gm.) of corrosive sublimate may be given every three hours to a child of three or four years.

Potassium chlorate is another popular remedy. Its action, however, is a purely local one. On account of its irritant effects on the kidneys it should never be used internally. Convalescence must be managed with special care, on account of the tendency to cardiac weakness. Anemia, a common sequel, will require good hygienic surroundings, plenty of nourishing food, and iron. Diphtheric paralysis usually yields to strychnin and other tonics, combined with the use of massage and electricity.

SCARLET FEVER.

. To prevent the spread of the disease, the patient should be rigidly isolated. The sick-room should be large and airy, and, if possible, it should be at the top of the house. All unnecessary furniture should be removed from the room, and no one should be allowed to enter it except the physician and nurse. The nurse should wear a loose wrapper and cap, to be left inside the room when she is obliged to leave it. All articles that have been in contact with the patient should be thoroughly disinfected before they are permitted to be taken from the room. Finally, to prevent the dissemination of the scales, some bland ointment (cold cream, olive oil, cocoa-butter) should be applied to the patient's body at least once a day until desquamation is complete. During the period of active desquamation the inunctions should be preceded by warm alkaline baths.

The treatment of scarlet fever is purely symptomatic. The patient should be kept in bed until the end of desquamation. The diet should consist of milk, junket, koumiss, ice-cream, fruit-juices, and gruels. So as not unnecessarily to tax the kidneys, animal broths and meat should be withheld, if possible, until convalescence is well established. Water should be given freely to relieve thirst and to keep the secretions active.

In many cases no special medication is required beyond the use of a mild refrigerant mixture, such as the following:

> ℞   Spiritus ætheris nitrosi,   f℥vj (22.5 c.c.);
>   Liquor ammonii acetatis, q. s. ad f℥iij (90.0 c.c.).   M.
> Sig.—Dessertspoonful with water every three hours for a child of
>   five years.

Vomiting.—When vomiting is severe, it may be necessary to withhold food for several hours and to administer an anti-emetic, such as ice, carbonated water, bismuth subnitrate with hydrocyanic acid, or cocain ($\frac{1}{20}$–$\frac{1}{10}$ gr.—0.003–0.006 gm.).   In very obstinate vomiting opium suppositories or chloral enemata should be tried.

Fever.—Tepid sponging is very grateful throughout the febrile period.   Fever above 103° F. (39.5° C.) should be combated with cold packs or gradually cooled baths (80° F.—26.5° C.), and by cold applications to the head.   Phenacetin, antipyrin, and other antipyretic drugs should not be used except as a last resort, and then only in small doses.

Cardiac Weakness.—If the heart shows signs of failing, alcoholic stimulants will be demanded.   Many practitioners believe that the tincture of ferric chlorid (1 drop a dose for each year of the child's age) exerts a favorite influence upon the general toxemia.   Threatened collapse will require the use of such remedies as strychnin, digitalis, camphor, ammonia, and musk.

Nervous Symptoms.—Restlessness, jactitation, headache, and insomnia will generally yield to small doses of a bromid and to the application of cold to the head.   When these symptoms are severe, however, no remedy is so useful as chloral.   It may be given in doses of from 3–5 grains (0.2–0.3 gm.) every three or four hours.   Nervous symptoms the result of high fever will, of course, require vigorous antipyretic treatment.

Throat Symptoms.—Cleansing of the nose and throat with mild antiseptic sprays or washes will do much to prevent the development of septic adenitis and middle-ear disease. Weak Dobell's solution or solution of hydrogen dioxid (1 : 3) may be used for this purpose.   When tonsillitis is severe, the following application will be found useful :

> ℞   Potassii chloratis,   gr. xx (1.3 gm.);
>   Tincturæ ferri chloridi,
>   Glycerini,   aa f℥ss (15.0 c.c.);
>   Aquæ,   q. s. ad f℥ij (60.0 c.c.).   M.
> Sig.—Apply to the tonsils several times a day with a cotton swab.

Pain and swelling in the neck are best relieved by ice poultices ; in some cases, however, hot compresses are more grateful. When suppuration occurs, the abscess should be freely opened and dressed antiseptically.

In **acute otitis media** nothing affords so much relief as gently syringing the auditory canal with hot water. The application of a leech behind the ear is also useful. When the tympanic membrane bulges, indicating the presence of pent-up pus, the latter should be evacuated by puncture.

Nephritis.—A milk diet, tepid bathing during the febrile period, and the avoidance of exposure for several weeks after the subsidence of the fever are important prophylactic measures against nephritis. Should this complication develop, however, dry cupping over the loins, followed by warm fomentations, will often prove of value. Aperients, especially salines, are indicated. Warm baths, hot packs, vapor baths, or pilocarpin ($\frac{1}{16}$–$\frac{1}{10}$ gr.—0.004–0.006 gm.) should be used to promote diaphoresis. When the urine is scanty, unirritating diuretics, like potassium acetate or bitartrate and digitalis, are of service.

MEASLES AND RUBELLA.

The prophylactic measures described in connection with scarlatina are applicable in measles and rubella.

Measles is a self-limited disease and cannot be aborted. The temperature of the sick-room should be maintained at about 68° F. (20° C.). The patient should be kept in bed until all traces of the rash have disappeared, and confined to his room for a week or ten days longer. On account of the photophobia the room should be moderately darkened, and the bed so arranged that the face will be directed away from the light. When the cough is troublesome, the atmosphere should be kept moist.

Milk, junket, fruit-juices, broths, eggs, and gruels are suitable forms of nourishment. Water should be proffered at frequent intervals. Stimulants are rarely required except in malignant cases.

Daily inunctions of the body with cold cream or olive oil will serve to allay burning and itching of the skin. Spraying the nose and throat at frequent intervals with warm Dobell's solution will aid in preventing buccal and aural complications. When conjunctivitis is marked, the eyes should be protected with dark glasses and frequently cleansed with a solution of boric acid (15 gr. to the oz.—1.0 gm. to 30.0 c.c.). Hot baths and hot drinks are indicated when the rash is delayed.

Fever, unless persistently high, may be left alone. When antipyretic measures are required, tepid sponging or immersion for a few minutes in water gradually cooled from 95° F. (35° C.) to 80° F. (26.5° C.) will be found the most satisfactory. Antipyretic drugs should be used only as a last resort.

The **medicinal treatment** of measles is usually very simple. Constipation should be relieved by small doses of calomel or by glycerin suppositories. Severe vomiting may require cracked ice, lime-water, bismuth subnitrate, cocain, or suppositories of opium. Diarrhea will generally yield to a few doses of bismuth subnitrate and opium. Troublesome headache and insomnia are best treated by bromids or small doses of chloral.

Cough very frequently requires attention. The application of sinapisms to the chest, the inhalation of steam, and the administration of sedative expectorants with opium are measures calculated to afford relief. The following mixture will be found useful:

```
R   Potassii citratis,                     ℥ij (8.0 gm.);
    Vini ipecacuanhæ,                      f℥ij (8.0 c c.);
    Tincturæ opii camphoratæ,              f℥iij (11.0 c.c.);
    Syrupi tolutani,                       f℥j (30.0 c.c.);
    Aquæ cinnamomi,           q. s. ad  f℥iij (90.0 c.c.).   M.
Sig.—A teaspoonful every three or four hours.
```

A mixture of ammonium chlorid and syrup of squill will be efficacious when the cough loosens.

The treatment of **rubella** does not differ materially from that of measles.

SMALLPOX.

The preventive measures against smallpox include the complete isolation of the patient (preferably in a special hospital), the thorough disinfection of all objects that have been in contact with him, and, above all, the vaccination of all who have been or who are likely to be exposed to the contagion.

Vaccination.—Vaccination does not afford complete and permanent immunity against smallpox, but in the vast majority of cases it renders the person insusceptible to the disease for a number of years. In Prussia the death-rate from smallpox per 100,000 of population prior to the enforcement of general vaccination (1846–74) was 24.4, while after the enforcement of general vaccination (1874–86) it was only 1.5 (Abbott). In the hospitals of London, from 1876–79, there were admitted 11,412 smallpox patients who had been vaccinated in infancy, but not a single case was known to have occurred in a person who had been successfully revaccinated (Welch).

Vaccination should be performed at the third month (earlier

if smallpox be epidemic), again at puberty, and subsequently whenever smallpox is prevalent. Only bovine lymph should be used, and preference should be given to that which has been glycerinated and preserved in hermetically sealed glass tubes. The part selected should be thoroughly cleansed, first with soap and water, then with alcohol, and finally with pure water. Many methods of performing the operation are recommended. The simplest is to make, with a sterile lancet or special scarificator, a number of cross scratches, only deep enough to allow of a little oozing of pinkish serum. The withdrawal of blood is to be carefully avoided, as it tends to wash away the vaccine material. The virus is now applied and well rubbed into the exposed lymphatic spaces by additional scarification. A shield may be worn for a few hours, until the wound has become perfectly dry; after that it should be discarded, since it serves, especially if tight-fitting, to intensify the inflammatory reaction.

For a vaccination to be considered successful the vaccine lesion should pass through a papular, vesicular, and pustular stage, and leave behind a permanent cicatrix. A typical vesicle resembles "a section of a pearl on a rose leaf" (Jenner). A good scar is pale, sharply defined, and pitted.

General Management.—There is no specific for smallpox. Vaccination in the incubation period, unless delayed too long, may lessen somewhat the severity of the attack. Absolute rest in bed, light bed-clothing, a well-ventilated room of a temperature of 65° F. (18.5° C.), an easily assimilable but sustaining diet, and the free use of cool drinks are requisites of treatment. The ·severe lumbar pains will require opium and the application of hot-water or hot-bran bags. Fever above 103° F. (40° C.) is best combated by hydrotherapy—cold sponging, cold packs, or cold baths. Antipyretic drugs should be used with great caution.

Gastric irritability may be controlled by dilute hydrocyanic acid (2 min.—0.1 c.c.), subnitrate of bismuth (10 gr.—0.6 gm.), or cocain ($\frac{1}{8}$ gr.—0.008 gm.). Diarrhea should be remedied by the use of bismuth and morphin powders (see p. 372). When nervous symptoms—jactitation, delirium, and insomnia—are not relieved by hydrotherapy, opium with bromids or chloral with bromids should be tried. Alcoholic stimulants are frequently demanded, especially in confluent cases. Quinin is also useful in the adynamia of sepsis.

An attempt should be made to keep the nasopharynx clean by means of antiseptic sprays or douches. For this purpose Dobell's solution or solution of hydrogen dioxid (1 : 4) will be found useful. The eyes should also be kept clean by frequent applications of a warm boric-acid solution (15 gr. to the ounce—1.0 gm. to 30.0 c.c.).

The Eruption.—Cool antiseptic washes, such as carbolic, 1 : 200, or corrosive sublimate, 1 : 5000–1 : 10,000, are, perhaps, the most useful applications that can be made to the skin. Compresses wet with these solutions may be kept constantly on the face and hands to exclude the light, which, there is some reason to believe, acts unfavorably on the pocks. Many remedies have been recommended to prevent pitting, but it is doubtful whether any are really efficacious. An old plan was to open the vesicles and touch their base with a stick of silver nitrate. Dujardin-Beaumetz recommended an ointment of sodium salicylate (4 parts) and cold cream (100 parts). Hebra speaks favorably of continuous warm baths. Finsen, Feilberg, and others recommend the exclusion of the chemical rays of light from the sick-room, red light only being admitted.

In the stage of desiccation, warm baths followed by inunctions with cold cream or olive oil are useful in allaying itching and in hastening the removal of the crusts.

YELLOW FEVER.

In the absence of any specific remedy, the treatment of yellow fever must be mainly expectant. Absolute rest and quiet, a sick-room well ventilated but free from drafts, and careful nursing are essential. The patient should be isolated, and protected by nets and wire screens from mosquitos. All articles handled by the patient, body- and bed-clothing, and excreta must be thoroughly disinfected. Only the blandest articles of food should be allowed, and these should be given in small quantities. Peptonized milk, milk with lime-water, junket, koumiss, milk-punch, barley-water, toast-water, and iced champagne are available. Carbonated waters, like Vichy and Apollinaris, are grateful and useful. Sternberg, Touatre, and many other clinicians of wide experience in the treatment of the disease advocate the withholding of all food during the first two or three days.

If the patient be seen in the first twenty-four hours of his illness, a hot mustard foot-bath may be given with advantage. If there be a tendency toward constipation, a few small doses of calomel may be administered and followed by a Seidlitz powder, or a purgative enema may be used.

It is very questionable whether any drug has the power of modifying the course of the disease as a whole, although Sternberg is firmly convinced of the efficacy of the following combination:

> R Sodii bicarbonatis, ℥iv (15.0 gm.);
> Hydrargyri chloridi corrosivi, gr. ss (0.03 gm.);
> Aquæ destillatæ, Oj (475.0 c.c.). M.
> Sig.—In severe cases, two teaspoonfuls every hour, day and night; in mild cases, every hour by day and every two hours by night; administer ice-cold.

Obstinate vomiting should be combated by the application of an ice-bag or a sinapism to the epigastrium and by the administration of sedatives, like cracked ice, lime-water, hydrocyanic acid, or cocain. In severe cases it is best to let the stomach have complete rest and to give morphin hypodermically or by the bowel.

Systematic cold sponging keeps down the temperature, lessens nervous excitability, and favors diuresis.

High temperature should be controlled by the cold bath.

Suppression of urine will call for dry cupping over the loins, alkaline diuretics, hot-air or hot-water baths, and, perhaps, rectal or subcutaneous injections of warm saline solutions (see pp. 480–490).

If there are signs of **heart-failure,** alcoholic stimulants should be given freely : iced champagne and iced brandy with soda water are the remedies most favored. Threatened collapse will require subcutaneous injections of strychnin, ether, or camphor, hot baths, and enemata of black coffee or brandy.

Remedies have little, if any, control over **black vomit.** Guiteras recommends tincture of ferric chlorid, and Carroll, oil of turpentine.

CHOLERA.

Hygienic measures are of the utmost importance in preventing the spread of cholera. Personal prophylactic measures against the disease include removal from the infected districts; restriction of the diet to bland, easily digested food; thorough sterilization of drinking-water and milk; the protection of all food from contamination by flies and other insects; the avoidance of overwork, exposure to wet and cold, and undue excitement; and the prompt treatment of any gastrointestinal disturbance that may arise. Certain acids, especially sulphuric acid, have long been advocated as preventives of cholera. Finally, vaccination with attenuated cholera cultures, as practised by Haffkine in India, has given encouraging results.

Precautionary measures pertaining to the sick comprise isolation, absolute cleanliness, and the thorough disinfection of excreta, soiled clothing, etc.

The medicinal treatment of cholera resolves itself into that of the prodromal stage, that of the algid stage, and that of the reaction stage.

Prodromal Stage.—From the first appearance of diarrhea the patient should go to bed and remain there. Food should usually be withheld. If there be a history of indigestible food having been taken, a laxative dose of calomel or of castor-oil should be given; otherwise, aperients should be avoided. Hot stupes may be applied to the abdomen. If there be much colic, morphin may be given hypodermically. In the last European epidemic (1892) many good observers reported adversely upon the use of opium by the mouth. For the diarrhea, bismuth subnitrate is perhaps the best astringent.

Algid Stage.—Intravenous injections of warm saline solutions (see p. 488) undoubtedly afford the best means of combating the anhydremia and of restoring the failing circulation. Rectal injections of hot tannic solutions (2 per cent.), as strongly recommended by Cantani, may also be used. The body-temperature should be maintained by hot applications or hot baths. Diffusible stimulants, like ether and camphor, may be given hypodermically.

To allay thirst, ice, ice-water, or iced Seltzer water may be given at frequent intervals. The painful cramps are best treated by warm applications, hot baths, gentle friction with anodyne liniments, and, above all, by intermittent chloroform inhalations. In suppression of urine the most promising measures are dry cupping over the loins, and rectal and intravenous injections of saline solutions.

Reaction Stage.—In this stage liquid foods in small quantities are permissible. Milk with lime-water, whey, thin gruels, albumin-water, and light broths are the most appropriate. The return to ordinary food should be effected most gradually. Water should be given freely, since it tends to restore vascular fulness and favors diuresis. Any tendency to recurrent diarrhea should be met by the administration of bismuth subnitrate with an intestinal antiseptic or of silver and opium pills. Bitter tonics are often of service during convalescence. Change of air is a valuable aid to the restoration of health.

WHOOPING-COUGH.

The chief elements in the prophylaxis of whooping-cough are isolation of the patient and the thorough disinfection of all articles that have been used by him. Quarantine should last until the cough ceases, which will seldom be before the end of six or eight weeks from the onset of the disease.

As there is no specific, treatment must be directed to relieving the catarrh, allaying the irritability of the respiratory mucous membrane, and warding off complications. Fresh air, sunlight, protection from changes of weather, and a light but nutritious diet are essential. In some cases it may be desirable to keep the patient in his room, or even in bed, for the first few days, but ordinarily, if the weather be good, he need not be confined indoors. In advanced cases sea air often acts most favorably. In some instances, when the vomiting is very persistent, it may be desirable to withhold food by the mouth for a short period and to substitute nutritive enemata.

Expectorants are rarely of service except in the first stage, when the cough lacks the distinctive whoop and is hard and dry. At this period a mixture of potassium citrate, ipecac, and paregoric (see p. 518) may be useful in rendering the expectoration less viscid.

Special Remedies.—Of the numerous special remedies advocated, the chief are belladonna, quinin, bromids, antipyrin, bromoform, and alum. No one drug is of service in every case, and after failing with one, another should be tried. Belladonna has a long-established reputation ; to be effective, it must be given in ascending doses until a constitutional impression is perceptible (see p. 74). Quinin, first recommended by Letzerich, is undoubtedly of value in many cases ; it must be given, however, in large doses,—10–12 gr. (0.65–0.8 gm.) a day for a child of three or four years,—and these may disturb digestion. The bromids, especially of sodium and strontium, are often of service in lessening the frequency and severity of the paroxysms. They may often be advantageously combined with belladonna. Antipyrin has been advocated since 1886, and is, on the whole, perhaps, the most reliable remedy. The dose should be gradually increased until a child one year old receives from ½–1 gr. (0.03–0.06 gm.) every two hours. It is particularly efficacious in combination with a bromid, as in the following formula :

R   Sodii bromidi,   gr. l (3.2 gm.) ;
    Antipyrini,   gr. xv (1.0 gm.) ;
    Glycerini,   f℥ss (15.0 c.c.) ;
    Aquæ menthæ piperitæ, q. s. ad f℥iij (90.0 c.c.).   M.
Sig.—A teaspoonful every two hours for a child one year old.

Bromoform, introduced by Stepp in 1887, has been warmly recommended by Fischer, Burton-Fanning, and others. The initial dose for a child under two years of age should not exceed 1 min. (0.06 c.c.) four times a day (see p. 121).

Alum is a very old remedy ; in small doses—1 gr. (0.06 gm.) every two hours for a child of one year—it is occasionally useful when there is copious secretion.

Chloral (3 gr.—0.2 gm.—for a child of two years) or Dover's powder (1 gr.—0.06 gm.) may be given in severe cases at bedtime to secure sleep. The inhalation of chloroform or of amyl nitrite may be necessary when the paroxysms are so violent as to threaten hemorrhage or convulsions.

Local Treatment.—Mild but persistent counterirritation over the chest is sometimes useful, particularly when the catarrhal element is prominent. Medicated steam inhalations with eucalyptol, benzoin, creasote, or terebene are also very beneficial when there is excessive bronchitis or a pneumonic complication. Good effects have been claimed for oxygen and ozone inhalations. Antiseptic and sedative sprays, when feasible, sometimes afford much relief; the best are the solution of hydrogen dioxid (1 part to 6 parts of water), menthol (5–10 per cent. in liquid paraffin), resorcin (1 per cent. aqueous solution). Insufflations of drugs in the form of a fine powder—quinin, boric acid, tannin, benzoin, etc.—have been warmly advocated by many, but they are of doubtful benefit.

Convalescence.—During convalescence it is necessary to exercise considerable care, on account of the great danger of catarrhal pneumonia. Tonics, especially quinin, iron, and cod-liver oil, are very useful. When there is persistent enlargement of the bronchial glands, iodid of iron internally, with inunctions of europhen (5 per cent.), will be found efficacious. Change of air is always beneficial.

ERYSIPELAS.

As in other contagious diseases, isolation and disinfection are the most important prophylactic measures. Especially necessary is it to guard parturient and surgical patients from the contagion.

Internal Treatment.—A supporting liquid diet should be given. Alcoholic stimulants are sometimes required in considerable quantities. High fever is best controlled by cold sponging or the cold pack. Restlessness, delirium, and insomnia will call for applications of ice to the head, and perhaps the administration of morphin, chloral, or bromids.

Of the numerous special remedies recommended for erysipelas, the one which has enjoyed the most favor is the tincture of ferric chlorid (20–30 min.—1.2–2.0 c.c.—every three hours), first suggested by Bell in 1851. While apparently of some service, this drug is not a specific, and should be withdrawn if it disturbs digestion. The therapeutic status of antistreptococcus serum has not yet been definitely determined, although it has been frequently administered in severe cases with alleged good results.

Local Treatment.—Among the numerous local applications recommended for erysipelas may be mentioned mixtures of zinc oxid, boric acid, and starch on cotton-wool; lotions of lead-water and laudanum, of carbolic acid (1 : 40), of corrosive sublimate (1 : 5000), of sodium salicylate (1 : 20), and of picric acid (1 : 100); and ointments of ichthyol (30 per cent.) and of soluble silver (15 per cent.). In the author's hand, ichthyol (see p. 328) and soluble silver (see p. 370) have proved the most satisfactory local remedies.

Eschars made with the stick of silver nitrate or strong carbolic acid beyond the infected area are occasionally successful in arresting the spread of the disease. The cellular infiltration caused by these irritants probably serves to check the migration of the cocci. The injection of antiseptics, such as carbolic acid and corrosive sublimate, around the diseased area is of very doubtful utility.

Local abscesses should be incised and treated antiseptically. Extension to the nose and throat will call for antiseptic sprays or washes. Should the larynx become involved, the constant sucking of ice may serve to control the swelling; whenever, however, dyspnea becomes urgent, scarification of the larynx or tracheotomy will be demanded.

TETANUS.

The most important safeguard against the occurrence of tetanus is the thorough cleansing and disinfection of all wounds. As tetanus antitoxin is practically harmless and has been shown to possess considerable immunizing power, it is advisable to employ it as a prophylactic remedy in all suspicious wounds, especially those which have become contaminated with garden earth, street dirt, or stable refuse.

When tetanus has actually developed, the indications for treatment are to prevent the further production of toxin at the seat of injury; to neutralize the toxin that has already been absorbed and is exerting a baneful influence on the nervous system; to subdue the convulsions; and to maintain nutrition until the disease shall have run its course.

The first indication is to be met by promptly enlarging the wound, freeing it from all foreign matter, and treating it with some active antiseptic. The most hopeful means of meeting the second indication is by administering tetanus antitoxin (see p. 426). Of 290 cases of tetanus treated by the subcutaneous injection of serum, death occurred in 117 (Moschowitz).[1] While an analysis of the statistics does not justify the conclusion that antitoxin is very potent in acute cases, it does suggest that the remedy is of distinct value in subacute and chronic tetanus. As no ill effects result from the administration of the serum, and as its use does not interfere in any way with other methods of treatment, we believe that it should be used in large doses in every case, and as soon as possible after the onset of the symptoms. When the symptoms are grave it would seem to be advisable to inject the serum directly into the brain or into the subarachnoid space of the spinal cord. Forty-eight cases treated by cerebral injections presented a little over 50 per cent. of recoveries. All these cases, however, were of the severest type (Moschowitz). According to von Leyden, of 11 severe cases treated by subarachnoid injections, recovery occurred in the last 5.

Bacelli has reported excellent results (1 death in 40 cases) in the treatment of tetanus from the subcutaneous administration of carbolic acid in large doses (see p. 385), but other observers outside of Italy have not been able to repeat his success. It is not improbable that tetanus runs a milder course in Italy (Pfeiffer). As the central nervous system seems to possess some antitoxic power, the injection of brain-substance subcutaneously has been recommended.

The drugs most effective in subduing convulsions are the bromids and chloral. These should be given in large doses. Morphin, hyoscin, and eserin are useful adjuvants. Inhalations of chloroform or of amyl nitrite sometimes afford temporary relief. Of course, the patient should be kept absolutely quiet and protected from cold and from even the slightest external irritation.

The maintenance of nutrition by the administration of nutriment in liberal quantities is of the utmost importance. In severe cases, milk, raw eggs, and animal broths are suitable forms of nourishment. Alcohol is often necessary, sometimes in large quantities. When swallowing is impossible, feeding must be carried out by means of a tube passed through the nostril or through a space made by the extraction of one or more teeth.

[1] *Medical News*, October 13, 1900.

RHEUMATIC FEVER.

General Management.—Absolute rest in a comfortable bed is essential, and, with the view of preventing permanent injury to the heart, this should be maintained for at least ten days or two weeks after the temperature has become normal and all the arthritic symptoms have subsided. To guard against chilling of the body, the patient should wear a loose flannel night-dress and lie between blankets. The room should be well ventilated, but free from draft. Milk and cereals are the most suitable articles of diet. Junket, koumiss, ice-cream, milk-toast, boiled rice, and gruels may be given. During the febrile period animal food is best avoided. During convalescence, however, fish, oysters, eggs, and broths are admissible. The free use of water and of lemonade should be encouraged. Alcohol, unless the heart is seriously involved, is not often required.

Internal Treatment.—Salicylic compounds have considerable power in mitigating pain, shortening the duration of the attack, and of lessening the liability to cardiac complications. The more typical the case, the more decided are the good effects of these drugs. From 10 to 15 gr. (0.65–1.0 gm.) of ammonium or sodium salicylate should be given every two or three hours, until a decided impression is made upon the disease or the phenomena of salicylism are produced, when the interval between the doses should be lengthened to four or six hours. It is advisable to continue the drug for several days after the subsidence of the symptoms. When the ammonium or sodium salt is not well borne, strontium salicylate (see p. 394) or salophen may be substituted.

Alkalis are also remedies of great value ; they may be combined with the salicylates or they may be given alone. About 20 gr. (1.3 gm.) of potassium citrate or acetate should be given every two or three ·hours until the urine becomes decidedly alkaline ; then every four or six hours, the influence of the remedy being maintained for at least a week after all symptoms have disappeared.

Opium, in the form of Dover's powder or of morphin hypodermically, is sometimes of great value in allaying pain, subduing restlessness, and procuring sleep. Antipyrin or phenacetin, in moderate doses, is also a useful adjuvant to salicylates or alkalis when the pain is severe. When adynamia is marked, quinin (5 gr.—0.3 gm.—thrice daily), as recommended by Garrod, Duckworth, and DaCosta, is frequently beneficial. Anemic patients are usually benefited by iron. A useful combination is the ferrosalicylate mixture recommended by S. S. Cohen:

```
R   Sodii salicylatis,                      ℥iij (12.0 gm.);
    Glycerini,                              f℥vj (22.5 c.c.);
    Mucilaginis acaciæ,                     f℥iij (12.0 c.c.);
    Olei gaultheriæ,                        ℩Lxij (0.8 c.c.);
    Liquoris ammonii citratis (B. P.), f℥iss (45.0 c.c.);
Solve et adde guttatim,
    Tincturæ ferri chloridi,                f℥iij (12.0 c.c.).   M.
    Sig.—One to two teaspoonfuls in water every three hours.
```

In some cases, when the usual remedies prove unsatisfactory, syrup of iodid of iron, in large doses (1–3 fl. dr.—4.0–11.0 c.c. —daily), as recommended by J. C. Wilson, is quite effectual.

Hyperpyrexia is best controlled by the cold bath (see p. 485). The treatment of endocarditis and of pericarditis will be treated in another part of this volume. The importance of prolonged rest in cases in which the heart becomes affected cannot be overestimated. During convalescence, tonics, like iron, quinin, and arsenic, and a liberal diet are necessary.

Local Treatment.—In mild cases the application of cotton-wool to the affected joints will suffice. When the pain is severe, hot alkaline lotions containing laudanum sometimes afford relief; but, as a rule, compresses soaked in methyl-salicylate and covered with oiled paper are much more effective. Ointments containing salicylic acid (see p. 394) are also useful. In some cases small blisters applied about the inflamed joints are of great utility. According to Osler, however, they are not nearly so beneficial as light applications of the thermo-cautery.

No matter what local remedy is selected, it is highly important that the affected joints should be kept at complete rest. This may be accomplished by means of padded splints and a roller bandage, or, often better, by the use of plaster casts.

Lingering swelling will often yield to an ointment of mercury and belladonna, with firm strapping of the articulation. Blisters are also useful. When the effusion is very great and persistent, it may be necessary to aspirate the joint.

For the stiffness of the joints induced by rheumatism, massage, warm baths, and inunctions with an ointment of iodin will be found useful. The hot-air treatment (see p. 487) also does good in some cases.

MUMPS.

In the majority of cases but little treatment is required. The patient should be confined to a warm room and preferably to bed. Isolation should last about three weeks from the onset of the disease. As deglutition is painful, the nourishment must be fluid. A mild aperient may be administered at the outset. If febrile symptoms are marked, refrigerants, like aconite, spirit of nitrous ether, and solution of ammonium acetate will be found useful (for formula, see p. 61). Pain in the throat is best relieved by sucking ice.

Local Treatment.—In mild cases, covering the gland with cotton batting will be all that is necessary. When there is decided pain, hot fomentations prove soothing. Emollient applications with laudanum (1 fl. dr. to 1 oz.—4.0 c.c. to 30.0 gm.) are frequently recommended. An ointment of guaiacol (5 per cent.) has recently been warmly advocated.

Orchitis will require rest, suspension of the affected gland, and the application of lead-water and laudanum or, better still, of an ointment of guaiacol (10 per cent.). After the tenderness has subsided, an ointment of mercury and belladonna will be found useful in reducing the swelling.

CONSTITUTIONAL DISEASES.

DIABETES MELLITUS.

The treatment of diabetes mellitus is dietetic, hygienic, and medicinal.

Dietetic Treatment.—The general rule is to eliminate from the dietary, at least for a time, all starchy and saccharin foods. Some relaxation of this rigid rule, however, is often desirable and even necessary. Each case must be carefully studied from the standpoint of the urine, the body-weight, and the general health. In mild types of the disease, after the glycosuria has once been controlled by strict dietetic regimen, it is generally possible to add a certain amount of carbohydrate matter without causing the reappearance of sugar in the urine. The exact amount and character of the carbohydrates that can be tolerated should be determined in each instance by frequent quantitative estimations of the sugar excreted. Again, in severe cases it is frequently necessary, in spite of some increase of glycosuria, to permit greater latitude in diet, in order to sustain the patient's strength. "There can be no doubt that a diabetic patient whose strength is well maintained is better off than one whose urine contains only 1 per cent. of sugar, but yet is daily growing weaker" (Strümpell).

The strict diet precludes the use of the following articles: bread, cakes, and pastry made of wheat, rye, rice, or corn, sago, barley, tapioca, beets, carrots, turnips, parsnips, peas, sweet fruits, chestnuts, chocolate, syrup, preserves, sweet wines, malt liquors, cordials, and all preparations containing cane-sugar or glucose.

Potatoes are generally classed among the forbidden articles, but Mossé and Saundby have shown that they are not only permissible in many cases, but even useful. Mossé advances two hypotheses to explain the beneficial action of potatoes; the first is that they yield a sugar more easily warehoused than that from bread; and second, that they introduce a substance (potash) into the organism which favors glycolysis. The following articles may be allowed:

Animal Food.—Meats of various kinds (except liver), game, light broths and soups, fish, and eggs. Fats, when well digested, are beneficial and may be taken freely.

Vegetables.—Spinach, cabbage, Brussels sprouts, cauliflower, celery, lettuce, water-cress, string-beans, young onions, tomatoes, mushrooms, asparagus tips, cucumbers, and dandelion.

Bread.—In very many cases a small amount (2 or 3 ounces a day) of bran bread or whole-wheat bread may be permitted. Gluten bread is also used, but most so-called gluten flours contain a large percentage of starch. Many substitutes have been recommended for wheat bread, but even the best of these are rather unpalatable and heavy. The best known are bread or biscuit made of pure bran (Prout), of almond meal (Pavy), of inulin (Külz), of Soya bean (Dujardin-Beaumetz), and of aleuronat, a glutinous flour obtained as a by-product in the manufacture of starch (Ebstein, von Noorden).

Beverages.—Buttermilk, skimmed milk, carbonated waters, mineral waters (Vichy, Carlsbad, Vals, and Saratoga), tea and coffee without sugar, and alcoholic beverages containing little or no sugar (claret, Rhine wines, dry Moselle, Bordeaux, dry sherry, well diluted brandy or whisky). No restriction should be placed on the amount of water desired.

Relishes.—Caviare, olives, pickles, and cheese.

Fruits and Nuts.—Grape-fruit, lemons, sour cherries, cranberries, red currants, and nuts (except chestnuts) are allowable in small quantities.

Sweetening Agents.—Certain forms of sugar appear to be assimilable. Thus milk, although it contains a considerable quantity of lactose, is usually well borne. Levulose (see p. 465), also in moderate amounts, seems to be harmless. The chief substitutes for sugar are saccharin and glycerin.

Hygienic Treatment.—Fresh air and systematic exercise are of great value. The patient must be warned, however, against overexertion. Moderation must be observed in all things. Flannel should be worn next to the skin, and all undue exposure avoided. Massage has been strongly recommended. Hydrotherapy, as first suggested by Bouchardat, is decidedly efficacious. If the patient is vigorous, Turkish baths may prove useful. Diabetics who still possess a fair measure of health frequently derive much benefit from a visit to certain mineral springs, such as Neuenahr, Homburg, Carlsbad, and Vichy. Obese, gouty subjects often do remarkably well at Carlsbad.

Medicinal Treatment.—The medicinal treatment of diabetes is purely empiric. The drugs which have been shown to be of most value are opium, bromids, salicylic compounds, antipyrin, alkalis, and arsenic.

Opium is generally the most reliable drug, especially in the severe nervous cases. It relieves thirst, promotes sleep, allays restlessness, and diminishes the excretion of urine and sugar. Patients can often take with advantage from 5–8 gr. (0.3–0.5 gm.) a day. Upon the disappearance of the symptoms it should be gradually withdrawn. Saline aperients may be associated with it, when necessary, to prevent constipation. Codein is sometimes preferable to opium. Doses of $\frac{1}{2}$ gr. (0.03 gm.), thrice daily, may be gradually increased until the glycosuria disappears or ceases to be affected.

Bromids, either alone or as adjuvants to opium, are very useful in subduing nervous symptoms. Salicylic compounds rank next in efficacy to opium. While they yield the best results in gouty cases, their good effects are by no means limited to that class. From 40–80 gr. (2.5–5.0 gm.) of strontium or ammonium salicylate may be given in the twenty-four hours. This treatment should be discontinued upon the slightest evidence of intolerance.

Some cases do remarkably well upon antipyrin in doses of from 10–15 gr. (0.6–1.0 gm.) thrice daily. It may often be combined advantageously with an alkaline carbonate, salicylate, or bromid.

Alkaline carbonates and alkaline mineral waters have long enjoyed a reputation as remedies in diabetes. While they do not appear to have a decided influence on the glycosuria, they often relieve dyspeptic symptoms and constipation and tend to avert coma. Arsenic, especially in the form of the so-called bromid of arsenic or Clemens' solution (see p. 305), occasionally succeeds when other remedies fail.

Cod-liver oil, strychnin, iron, and alcohol are useful at times in meeting special indications.

Among other remedies which have been recommended, but which are of somewhat doubtful utility, may be mentioned: Jambul (Banatvala), uranium nitrate (West), glycolytic ferment from the pancreas (Lépine), ergot (Dougall), and calcium carbonate and phosphate (Grube).

Special Symptoms.—*Constipation* is best treated by saline cathartics or the alkaline mineral waters. Remedies which lessen the polyuria also serve to quench the *thirst*. Acidulous drinks are sometimes of service.

Pruritus of the genitals should be treated by frequent sponging, especially after micturition, with a saturated solution of boric acid. Lotions containing carbolic acid, resorcin (see p. 389), or sodium hyposulphite may also be tried.

Diabetic Coma.—When coma is threatened, prophylactic treatment consists in substituting milk with some carbohydrate for the strict proteid dietary, in the administration of sodium bicarbonate,—an ounce (31.0 gm.) or more a day,—in keeping the bowels freely opened with saline aperients, and in sustaining the circulation with strychnin and whisky. In the presence of actual coma treatment is usually unavailing. The most useful measure is the intravenous or subcutaneous injection of normal salt solution (1–2 pints—0.5–1.0 L.). The addition of sodium bicarbonate to the solution (1 dr. to 1 pint—4.0 gm. to 0.5 L.) is thought by some authorities to be an advantage. To insure rapid action of the bowels a brisk cathartic (croton oil) should be given. Hypodermic injections of ether and strychnin are useful. The inhalation of oxygen has been highly recommended.

DIABETES INSIPIDUS.

The hygienic treatment suggested for diabetes mellitus is applicable in diabetes insipidus. The patient should have a nutritious but digestible diet. No benefit is derived from restricting the quantity of water desired. Acidulated drinks, like lemonade, aid in assuaging thirst. Many remedies have been recommended; those possessing the most extended reputation are opium (4–8 gr.—0.25–0.5 gm. a day), valerian ($\frac{1}{2}$–1 fl. oz.—15.0–30.0 c.c. of ammoniated tincture daily), ergot ($\frac{1}{2}$–1 dr.—2.0–4.0 gm. of the extract or of ergotin daily), antipyrin (10 gr.—0.6 gm. three or four times a day), belladonna (5 min.—0.3 c.c. of the tincture three times a day, the dose being gradually increased), and gallic acid (1 dr.—4.0 gm. a day). Galvanism—one pole applied to the neck and the other to the loins—has also been recommended. Tonics—cod-liver oil, iron, and strychnin—are sometimes required. In syphilitic cases good results may be anticipated from a thorough trial of mercury and potassium iodid. The dryness of the skin may be relieved by warm baths, followed by inunctions with olive oil (Saundby).

GOUT.

Chronic Gout.—No absolute rule can be laid down regarding the diet. The special features in each case should receive careful study. Some patients do well upon a non-proteid diet, others do not. Simplicity and moderation are of the utmost importance. Generally speaking, a diet composed for the most part of milk, farinaceous foods, succulent vegetables, eggs, fish, and lean meats is most suitable. The foods most likely to disagree are veal, liver, sweetbreads, hashes, croquettes, concentrated soups, vegetables rich in nucleins,—peas and beans,—pastry, sweets, coffee, malt liquors, and heavy wines. Some patients are exceedingly intolerant of acid fruit. Milk taken as a beverage at meals is objectionable. Alcoholic stimulants, as a rule, should be avoided; occasionally, however, a small quantity of well-diluted whisky at meals aids digestion and is well borne. Water-drinking between meals should be encouraged.

The quantity as well as the character of the food must be regulated. No more should be eaten than is absolutely necessary to satisfy hunger. The patient should be warmly clothed and should avoid as far as possible exposure to sudden atmospheric changes. Systematic exercise in the open air is extremely beneficial. Well-nourished patients should be urged to take walking trips, to play golf or tennis, or to try horseback riding. When active exercise is not feasible, massage may be strongly recommended. All overwork of mind should be forbidden. Hydrotherapy—tepid sponge-baths and douches —is useful. Heavy robust patients often derive much benefit from the Turkish bath.

Visits to certain mineral springs—Bedford, Saratoga, Harrowgate, Carlsbad, Contrexeville, Aix-les-Bains—are sometimes of the greatest value. The good effects of the spa treatment are only in part due to the waters drunk: change of scene, fresh air, strict diet, and freedom from business and household cares are important factors. Residence during the winter months in a dry, warm, inland climate is desirable.

Remedies calculated to improve digestion are frequently indicated. In some cases a combination of an alkali with a bitter (see p. 168) before meals is of much service. Daily action of the bowels should be secured. The occasional use of calomel or blue-mass at night, with a saline in the morning, is often of value. Among the special remedies advocated for gout may be mentioned alkalis and alkaline mineral waters, colchicum, guaiac, arsenic, and iodids.

Of these, the alkalis, especially the vegetable salts of potassium, are the most useful (see pp. 220, 225, 237). Colchicum is most effective in the acute paroxysms, although small doses with alkalis may be of benefit in the interval. Guaiac (see p. 331) probably ranks next in efficacy to the alkalis. The prolonged use of arsenic in small doses has seemed to us to be of some value. Iodids are sometimes of service in relieving the concomitants and sequels of gout,—arthritis, bronchitis, arteriosclerosis,—but have little, if any, effect upon the disease itself. Salicylates relieve pain, but are distinctly inferior to colchicum. Such remedies as piperazin, lycetol, and quinic acid (see p. 238) are of doubtful utility.

Chronic affections of the joints are best treated by gentle massage, friction, and warm sulphur baths.

Acute Gout.—The most efficient remedy is colchicum; from 5–10 min. (0.3–0.6 gm.) of the wine may be given well diluted three or four times a day. It should be stopped as soon as relief is obtained. Alkalis are useful adjuvants (for formula, see p. 331). Water-drinking should be encouraged. At the outset a brisk mercurial purgative, followed by a saline, is usually advisable. Excruciating pain will demand the use of morphin. Phenacetin is sometimes effective. The diet should be light and non-stimulating.

The affected part should be elevated and wrapped in cotton-wool or covered with cloths soaked in lead-water and laudanum.

CHRONIC RHEUMATISM.

Strict attention should be paid to hygiene. The patient should be warmly clothed, wool or silk being next to the skin, and great care should be taken to guard against wet and sudden changes of temperature. The diet should be nourishing but digestible. Fatty foods, when well borne, are of much value. Hot sulphur and hot saline baths are often very useful. Massage and Swedish movements are excellent adjuvants to hydrotherapy. Vigorous patients may be benefited by Turkish baths. Residence in a dry, warm, equable climate—Southern California, Arizona, Riviera, Egypt—especially during the changeable seasons, is always desirable.

Tonics and certain alteratives, such as cod-liver oil, iron, and arsenic, are the most generally useful internal remedies. Iodides are sometimes of service, but on the whole they are inferior to arsenic. Salicylates are efficacious in relieving acute exacerbations, but they have no permanent effect. Colchicum is useless. Guaiac, sulphur, cimicifuga, ichthyol, and many other drugs have been recommended.

Local Treatment.—Massage, if employed systematically, often accomplishes much good. Superheated air-baths are occasionally useful. Electricity is of little value. Rubefacient liniments have a palliative influence in mild cases. An ointment of mercury, belladonna, and ichthyol, well rubbed into the affected part, is sometimes very efficacious. When the pain is severe and persistent, blisters or light applications of the actual cautery prove effective.

MUSCULAR RHEUMATISM.

In mild cases it will suffice to put the affected muscles at rest. In pleurodynia this is accomplished best by strapping the affected side as in fracture of the ribs, and in lumbago by applying a large piece of adhesive plaster from the floating ribs to the iliac crests. In more severe cases it will be necessary to apply rubefacient liniments, sinapisms, or, better still, hot fomentations, and to administer a salicylate, combined, perhaps, with phenacetin:

R   Salophen,   ℥ij (8.0 gm.);
    Phenacetini,   ℥j (4.0 gm.).   M.
Fiant chartulæ No. xii.
Sig.—One every three hours.

A blister is occasionally required. When the pain is intense, hypodermic injections of morphin ($\frac{1}{4}$ gr.—0.016 gm.) with atropin ($\frac{1}{120}$ gr.—0.0005 gm.) will afford great relief. In lumbago, acupuncture (see p. 430) sometimes yields excellent results. Hot packs and baths are often efficacious, but great care must be exercised to guard against exposure after their use. Persistent myalgia is often very favorably affected by massage and applications of the faradic current. In chronic cases, potassium iodid and guaiac should be tried. Gelsemium in large doses (Brunton) and ammonium chlorid (Anstie, Ringer, Roberts, DaCosta) have also been recommended.

RHEUMATOID ARTHRITIS.

The most that can be expected from treatment is amelioration of the symptoms, and, perhaps, some retardation of the progress of the disease. The hygienic treatment is as for chronic rheumatism (see p. 534). An important indication is to maintain nutrition. Tonics are frequently necessary. Cod-liver oil in such doses as the stomach will tolerate is often very beneficial. The most satisfactory special agents are arsenic and the iodids, particularly the former. Garrod has repeatedly seen good effects follow the prolonged use of the syrup of ferrous iodid. The remedies relied upon in gout are useless in rheumatoid arthritis. Salicylates are sometimes of service in acute exacerbations. Change of scene and climate, especially in the early period of the disease, frequently exerts a favorable influence. Visits to hot mineral springs (Hot Springs of Virginia or Arkansas, Bath, Baden-Baden) are recommended.

Local Treatment.—Massage is of the greatest value in preserving the mobility of the joints and in maintaining the nutrition of the muscles. Acute exacerbations should be treated by rest and the application of cold compresses. Rubefacient liniments, inunctions with mild iodin or mercurial ointments, and small blisters are sometimes of benefit. Electricity and superheated air are usually without effect.

RHACHITIS.

The treatment of rhachitis is mainly dietetic and hygienic. If the mother is deemed unfit for nursing, the child must be brought up by a wet-nurse or by hand. In most cases the latter method must be chosen. Generally speaking, the best substitute for natural food is diluted fresh cows' milk to which have been added cream and milk-sugar. The degree of dilution and the percentage of cream and sugar to be added will vary with the age of the infant and its power of assimilation. Lime-water and barley-water are the best diluents. Milk that has been partially predigested is in many cases an excellent food. Older children require undiluted milk, cream, beef-juice, veal-broth, eggs, scraped beef, pounded mutton, and fruit-juices. Starchy foods should be avoided. Very often it is necessary to correct digestive disturbances before ordering a liberal diet.

Fresh air, day and night, is essential. The child should be carefully protected, however, from the vicissitudes of the weather, and should be warmly clad in flannel. Sea-air is often very beneficial. Daily bathing, with friction and gentle massage, is to be recommended.

Dyspepsia, diarrhea, and constipation should receive appropriate treatment. When well borne, cod-liver oil is the best remedy. The dose should be increased gradually from a few minims to a teaspoonful, three or four times a day. When cod-liver oil cannot be taken by the mouth, it may be given by inunction (see p. 298). Phosphorus, which has been strongly recommended by Kassowitz and Jacobi, is of doubtful value (see p. 294). The lactophosphate and hypophosphite of calcium are frequently combined with cod-liver oil for their tonic effect. Stoeltzner has obtained good results from the use of suprarenal extract, and Mendel, from the .use of thymus extract. When there is anemia, an unirritating preparation of iron should be given. For the excessive sweating, atropin, in doses of $\frac{1}{800}$ of a grain (0.0008 gm.) two or three times a day, has been recommended (Holt).

DISEASES OF THE BLOOD AND DUCTLESS GLANDS.

CHLOROSIS.

. Fresh air, sunlight, open-air exercise, and nourishing food are valuable aids in treatment. Change of climate or a sea-voyage is often beneficial. Very severe cases sometimes require complete rest in bed and massage. If there be a good reaction, warm baths, followed by short cold douches, are efficacious. Iron is almost a specific. It is most frequently prescribed in the form of Blaud's pills (see p. 285), of which the dose is three pills, gradually increased to nine a day. In some cases other preparations of iron, such as iron and ammonium citrate, ammonioferric tartrate, albuminate, and pyrophosphate, are better borne. Laxatives, preferably mild salines, rank next in importance to iron. Arsenic is distinctly less valuable than iron. Bitters, such as nux vomica, gentian, and calumba, are often useful adjuvants, especially when there is anorexia with tardy digestion. Superacidity of the gastric juice is best treated by alkalis.

PERNICIOUS ANEMIA.

While the outlook in pernicious anemia is exceedingly grave, marked improvement and even apparent recovery are not impossible. Fresh air, rest, and a diet as liberal as the digestive power of the patient will permit, are requisite. Rest in bed, warm salt baths, and massage are valuable adjuvants to internal treatment. The teeth should receive careful attention. If there be gingivitis or pyorrhœa alveolaris, antiseptic mouth-washes should be used at frequent intervals.

Arsenic is the most valuable drug. It may be given in the form of Fowler's solution, the dose being gradually increased from 2 or 3 min. (0.1–0.2 c.c.) to 15 or 20 min. (1.0–1.2 c.c.) three times a day. Iron is rarely of service, but may be tried when arsenic fails. Bone-marrow is sometimes efficacious. Dock has often used mercuric chlorid with benefit, not only in syphilitic cases, but also in others. Inhalations of oxygen have been recommended (Shattuck). Appropriate anthelmintic remedies should be given, of course, in the cases in which intestinal parasites (Bothriocephalus latus, ankylostoma) are present. Digestive disturbances are often benefited by the administration of diluted hydrochloric acid and a bitter.

LEUKEMIA, PSEUDOLEUKEMIA, AND SPLENIC ANEMIA.

An effort should be made to maintain the general nutrition by regulating the diet and attending to hygienic measures. Rest is often advisable. Among drugs, arsenic appears to be of some service. Iron, quinin, and cod-liver oil occasionally are beneficial. Inhalations of oxygen may be tried. In pseudoleukemia Gowers and Broadbent have found phosphorus of benefit. Local treatment of the lymphatic glands or the spleen is useless. In neither leukemia nor pseudoleukemia is operative treatment of avail. In splenic anemia, however, removal of the spleen is a justifiable procedure. Of 19 cases of splenic anemia treated by splenectomy, in 14 recovery followed (Harris and Herzog).

SCURVY.

The prime requisites in the treatment of scurvy are suitable nourishment and proper hygiene. Complete rest is generally advisable. The diet should include plenty of fresh vegetables, —potatoes, lettuce, cabbage, and onions,—with 3 or 4 ounces (90–120 c.c.) of lemon-juice daily. When the digestion is feeble and the gums are tender, it may be necessary to restrict the diet to such foods as beef-juice, animal broths, milk, raw eggs, and orange- or lemon-juice. Tonics, like iron, quinin, and strychnin, are often of service. In some cases alcoholic stimulants are required. The mouth should be cleansed at frequent intervals with some antiseptic wash, such as Dobell's solution, and the spongy gums should be painted two or three times a day with a solution of silver nitrate, 20 grains (1.3 gm.) to the ounce (30.0 c.c.).

Infantile scurvy usually responds very promptly to correction of the diet. The foods most worthy of confidence are fresh cows' milk, beef-juice, and orange-juice.

PURPURA HAEMORRHAGICA.

In many cases a favorable termination follows in the course of one or two weeks without special medication, while in others the disease progresses toward a fatal issue in spite of the most careful treatment. Hemostatics, both internal and external (see p. 373), are the remedies indicated. Great efficacy has been claimed especially for oil of turpentine, gelatin, and calcium chlorid. Heart-failure should be treated by cardiac stimulants, and restlessness and insomnia by opium in small doses. Secondary anemia will require iron and arsenic.

HEMOPHILIA.

The treatment of hemophilia is largely protective and palliative. All operations liable to cause hemorrhage should be avoided. Bleeding will call for rest, the application of cold compresses and of styptics, and the administration of internal hemostatics (see p. 373). The drugs most worthy of confidence are gelatin and calcium chlorid. Combemale and Gaudier and Delace have used thyroid extract with asserted good results. Inhalations of oxygen have been strongly recommended. Wright advises inhalations of carbon dioxid when there is epistaxis. The resulting anemia is benefited by iron and arsenic.

ADDISON'S DISEASE.

As the primary lesion in Addison's disease is not amenable to therapeutic measures, little can be done beyond maintaining the patient's strength and relieving symptoms. Nutritious but readily digestible food, the avoidance of mental and physical excitement, and good hygienic surroundings are important factors in the treatment. When there is marked asthenia, absolute rest should be enjoined. Very marked improvement in the symptoms sometimes follows the administration of suprarenal extract (see p. 338). Tonics, like iron, arsenic, and strychnin, are of service. Irritability of the stomach will call for such sedatives as bismuth subnitrate, hydrocyanic acid, and cerium oxalate, and diarrhea for mild astringents and antiseptics.

MYXEDEMA.

As patients with myxedema are extremely susceptible to low temperatures, they should be warmly clad and protected from exposure to cold. Residence during the winter in a warm, sunny climate is desirable. Warm baths are often beneficial. Modern treatment consists in the administration of some preparation of thyroid gland (see p. 335). By continuing this remedy throughout life it is possible in many cases to hold the symptoms in complete abeyance.

EXOPHTHALMIC GOITER.

The general nutrition should be improved by rest, a generous, readily digestible diet, healthy hygienic surroundings, and hydrotherapy. Residence in a warm seaside locality often promotes comfort and aids recovery. In severe cases absolute rest in bed is an essential point in the treatment. Applications of cold to the precordia, by means of Leiter's tubes or ice-bags, lessen the palpitation.

Medicinal treatment is very uncertain. A drug that has been efficacious in one case may in another prove useless or harmful. Belladonna, recommended by Gowers, is undoubtedly of value in many cases. It should be given in ascending doses until some dryness of the throat is produced. When the circulation is feeble, digitalis may be found of service; on the other hand, when the heart is strong, better results may be obtained with aconite or veratrum viride. When anemia exists, iron is useful. Bromids are sometimes of service in controlling nervous symptoms. Starr has observed marked improvement from the use of sodium glycerophosphate in doses of 20 grains (1.3 gm.) three or four times a day, as originally recommended by Trachewsky. Mikulicz, Owen, Mackenzie, and others have found thymus extract (see p. 340) of some value. Wood has used glycerin extract of spleen (10–20 min.—0.6–1.2 c.c. a day hypodermically) with apparently good results. Musser has found small doses of opium successful in some cases. Among other remedies which have been recommended, but which are of more doubtful value, may be mentioned ergot, iodids, iodoform, duboisin, and bromid of gold. The consensus of opinion is decidedly adverse to the use of thyroid extract.

Galvanism (see p. 475) sometimes proves more effective in controlling the symptoms than any other remedy. As a last resort, operative interference should be considered. According to Rehn, in 177 partial excisions of the gland 57.6 per cent. of the patients were cured, 26.5 improved, 2.3 per cent. unimproved, and there were 13.6 per cent. of deaths. In 32 resections of the sympathetic 31.1 per cent. were cured, 50 per cent. were improved, 12.5 per cent. were unimproved, and death occurred in 9.5 per cent. After ligation of the thyroid arteries in 14 cases 24 per cent. were cured, 50 per cent. improved, and the remainder died. Of 59 cases operated on by Kocher and recently reported (1902), 45 were cured, 8 were decidedly improved, 2 were slightly improved, and 4 died from tetany as a result of the operation.

DISEASES OF THE DIGESTIVE TRACT.

STOMATITIS.

The source of the local irritation should be sought for and removed. Errors of hygiene should be corrected. The diet and the state of the alimentary tract should receive careful attention. The mucous membrane of the mouth should be washed at frequent intervals with cool antiseptic solutions. In mild *catarrhal stomatitis* a solution of boric acid, 5–10 gr. (0.3–0.6 gm.) to the ounce (30.0 c.c.), will suffice. In obstinate cases the mouth, after being carefully cleansed, may be lightly painted with a solution of silver nitrate, 4 gr. (0.26 gm.) to the ounce (30.0 c.c.). Ulcers may be touched with solid silver nitrate or copper sulphate.

In *aphthous stomatitis* a mild mercurial aperient is often of service. In addition to the frequent use of antiseptic washes (boric acid, potassium permanganate), the occasional application of silver nitrate—10 gr. (0.6 gm.) to the ounce (30.0 c.c.) —is desirable.

In *thrush* the mouth should be cleansed at frequent intervals, especially after every feeding, with one of the following solutions: sodium hyposulphite, 20 gr. (1.3 gm.) to the ounce (30.0 c.c.); boric acid, 15 gr. (0.9 gm.) to the ounce (30.0 c.c.); or potassium permanganate, 1 gr. (0.065 gm.) to the ounce (30.0 c.c.).

In *ulcerative stomatitis* potassium chlorate is almost a specific. It should be used both locally and internally (see p. 224). The ulcers may be painted with a solution of silver nitrate, 10 gr. (0.6 gm.) to the ounce (30.0 c.c.). Tonics, like quinin and iron, are called for in some instances.

Gangrenous stomatitis demands energetic treatment. The sloughing surface and the tissue immediately surrounding it should be promptly destroyed under anesthesia with the actual cautery or strong nitric acid. After the operation the mouth should be cleansed at frequent intervals with a solution of hydrogen dioxid (1 : 3) or of potassium permanganate (1 per cent.). Concentrated nutritious food, stimulants, and tonics are urgently indicated.

The treatment of *mercurial stomatitis* has already been considered (see p. 316).

ACUTE PHARYNGITIS.

In mild cases a gargle of potassium chlorate (see p. 356) will suffice. In severe cases the application to the throat of cloths wrung out of cold water proves grateful. The sucking of pieces of ice affords much relief. Gargles or sprays of the distillate of hamamelis (50 per cent.) are useful. A spray of menthol, 2 gr. (0.13 gm.) to the ounce (30.0 c.c.) of liquid petrolatum, is also efficacious. Lozenges containing cocain will often relieve pain and allay the tickling sensation in the throat. The following formula, recommended by Bosworth, answers the purpose admirably :

R	Cocainæ hydrochloratis,	gr. v (0.3 gm.);
	Extracti krameriæ,	gr. ij (0.13 gm.);
	Sodii bicarbonatis,	gr. xv (1.0 gm.);
	Extracti glycyrrhizæ,	℥iiss (10.0 gm.). M.
Fiant trochisci No. xxx.		

Internally, a mild aperient may be given at the outset. Sodium benzoate has a beneficial effect. From 5–10 gr. (0.3–0.6 gm.) may be given every three hours. Belladonna with aconite is also recommended. The rheumatic form usually yields promptly to a mild salicylic preparation, like salophen.

CHRONIC PHARYNGITIS.

The removal of the cause is of prime importance. All sources of local irritation, such as misuse and overuse of the voice, mouth-breathing, excessive smoking, and intemperance in eating and drinking must be avoided. Patients should be instructed to expel sounds by the aid of the diaphragm and abdominal muscles instead of the muscles of the throat. Nasal obstructions and adenoid growths must be removed. The habit of hawking and scraping to clear the throat should be rigidly interdicted. Digestive disturbances should receive careful attention. Tonics, like iron, strychnin, and cod-liver oil, are sometimes required.

Local Treatment.—The nasopharynx should be kept clean by frequent spraying with an antiseptic alkaline liquid, like Dobell's solution (see p. 165). Astringent applications are often of service; one of the following may be employed: Zinc sulphate, 5 gr. (0.3 gm.) to the ounce (30.0 c.c.); tannin, 1 dram (4.0 gm.) to the ounce (30.0 c.c.) of glycerin; silver nitrate, 10–20 gr. (0.6–1.3 gm.) to the ounce (30.0 c.c.). In the follicular variety it is advisable to destroy the enlarged follicles by means of the galvanocautery, after which the astringent applications may be made.

ACUTE TONSILLITIS.

The patient should be confined to a warm room, and, if there be much fever, to bed. A mild aperient is indicated at the outset. The diet should be light but sustaining. The sucking of ice affords relief. The most reliable internal remedies are the salicylic compounds and sodium benzoate. They should be given in full doses at frequent intervals. Guaiac is also recommended. A dram (4.0 c.c.) of the ammoniated tincture of guaiac may be given in milk every three hours. Febrile symptoms, if pronounced, may be controlled by small doses of phenacetin or by a combination of aconite and spirit of nitrous ether (see p. 61). The pain may be so intense as to require the use of opium.

Local Treatment.—Externally, cold applications aid in bringing about resolution; if, however, suppuration be inevitable, warm applications should be employed to hasten the process. Antiseptic sprays, like Dobell's solution (see p. 165) or a solution of hydrogen dioxid (1 : 4), are of decided benefit. Direct applications to the surface of the glands of the tincture of ferric chlorid—1 dram (4.0 c.c.) to the ounce (30.0 c.c.) of glycerin or of a saturated ethereal solution of iodoform—are often useful. Applications of dry sodium bicarbonate have also been highly recommended. Scarification, followed by gargling with hot water, is another measure which frequently affords relief.

Pus should be evacuated as soon as its presence can be detected. In the majority of cases it is best to make the incision not in the tonsil itself, but in the soft palate, a little above and to the outer side of the gland.

ACUTE CATARRHAL GASTRITIS.

Absolute rest is essential. If the stomach has not been completely emptied, an emetic, such as warm water or ipecac, should be employed. Locally, a mustard plaster or a turpentine stupe will aid in relieving distress. As a rule, no food should be given by the mouth until the stomach becomes retentive. Ice, however, may be allowed to quench the thirst. In delicate subjects nutrient enemata will be required. If there be constipation, a mercurial laxative may be given with advantage. Such a combination as the following usually acts favorably :

> ℞   Hydrargyri chloridi mitis,   gr. j (0.065 gm.) ;
>   Bismuthi subnitratis,   gr. xx (1.3 gm.).   M.
>   Fiant chartulæ No. vj.
>   Sig.—One on the tongue every hour, to be followed by a Seidlitz powder, if necessary.

Pain, nausea, restlessness, and insomnia are best relieved by opium suppositories or hypodermic injections of morphin.

Sedatives by the mouth are also useful; thus, bismuth subnitrate may be given alone or combined with hydrocyanic acid (see p. 69), with creasote ($\frac{1}{2}$ min.—0.03 c.c.), or with cocain ($\frac{1}{6}$ gr.—0.01 gm.). The following combination of ipecac and nux vomica is often serviceable:

R   Tincturæ nucis vomicæ,
     Vini ipecacuanhæ,   āā f℥ij (8.0 c.c.). M.
Sig.—Two drops every hour.

After the lapse of twenty-four or thirty-six hours it is generally possible to give bland nourishment by the mouth. Barley-water, champagne with soda-water, milk and lime-water, peptonized milk, and light broths may be given in small quantities at frequent intervals. The return to solid food should always be carried out very gradually.

INDIGESTION.

The first thing to be done in the treatment of all cases of indigestion is to find its cause, and to remove this, if possible. In many cases indigestion is merely a symptom of some constitutional disturbance, such as anemia, tuberculosis, Bright's disease, or diabetes, or is the result of passive hyperemia of the stomach induced by chronic cardiac, hepatic, or pulmonary disease. It is obvious that in such cases treatment must be directed very largely to the underlying condition. In the primary form of gastric indigestion the methods of eating and the character of the food are the factors of greatest importance. Regularity in the time of meals, slowness in eating, and thorough mastication of food must be insisted upon. The patient should be cautioned against overeating and the taking of large quantities of liquid, especially of iced water, during meals. Overindulgence in alcohol, tobacco, coffee, and tea should be forbidden. The resumption of mental or physical work immediately after meals should also be avoided.

Dietetic Treatment.—It is impossible to lay down any absolute rules in the matter of diet. The conditions present in each case must receive careful study, and considerable allowance must be made for individual peculiarities. Generally speaking, such articles as fried meats, salted meats, pork, veal, hashes, croquettes, coarse vegetables and fruits, fried or boiled sweet and white potatoes, fresh bread, pastry, puddings, nuts, and malt liquors must be interdicted. In many cases of *atonic dyspepsia* (*subacidity*), *nervous dyspepsia*, and *mild gastric catarrh* an ordinary mixed diet of readily digestible food is admissible. It may usually include boiled, baked, or grilled beef and mutton, chicken, sweetbread, boiled fish, oysters, soft-boiled or poached eggs, stale bread, toast, pulled bread, fresh butter, baked potato, young string-beans, small peas, spinach, hearts of celery, thoroughly cooked cereals (rice, wheat, oatmeal), baked apples, calves'-foot jelly, and junket. Tea, coffee, and cocoa may or may not be permissible.

Indigestion associated with *hyperacidity* will require a bland, unirritating diet, composed largely of albuminous foods. A moderate quantity of water or weak tea at meals is desirable. Condiments, spices, acids, oils, and alcoholic beverages are strongly contraindicated.

When the *expulsive power of the stomach is weak* and there is a tendency to *retention*, the food should be nutritious and concentrated. Foods that are bulky and are liable to ferment should be avoided. Much fluid is undesirable.

An exclusive milk diet sometimes acts exceedingly well in nervous dyspepsia and indigestion attended by hyperacidity. It is contraindicated when there is marked hypo-acidity, motor insufficiency, retention, or intestinal fermentation.

Most cases do best without alcohol, but when there is pronounced subacidity or motor insufficiency, a small amount of diluted whisky, brandy, or sherry with meals may prove beneficial. Alcohol in any form is contraindicated when there is excessive secretion of acid or when there is hyperesthesia of the gastric mucosa.

Generally, it is best to prescribe a definite bill-of-fare and to modify this from time to time as occasion demands. In determining the suitability of a given diet we must be guided chiefly by the patient's sensations and the state of his general nutrition. As a rule, no more than three meals a day should be allowed. When there is a tendency to retention, however, it may be advisable to give the food in smaller quantities and at more frequent intervals. On the other hand, in some cases of gastric catarrh with slow digestion it may be advantageous to limit the number of meals to two a day.

Hygienic Treatment.—In a large number of cases systematic exercise in the open air accomplishes much good. Anemic and neurasthenic patients, however, may require the "rest-cure." Hydrotherapy is often very serviceable. Massage is also useful, especially when there is myasthenia. Change of scene, a sunny climate, good hours, and freedom from business worry and household cares often prove more beneficial than any other measure employed.

Medicinal Treatment.—In cases of *atonic dyspepsia* and of *mild gastric catarrh with subacidity* the administration of a

35

bitter—calumba, gentian, nux vomica, cinchona—some time before meals often proves efficacious. In many cases the bitter may be combined advantageously with an alkali (see p. 168). Diluted hydrochloric acid, with or without pepsin, may also be of benefit when there is *subacidity.* It should be given after meals.

When the *stomach is unnaturally sensitive* or when there is *supersecretion,* silver nitrate will be found a valuable remedy. It may be given in pill form, with extract of hyoscyamus or of belladonna, half an hour before meals, or when this method of administration is not satisfactory, by means of the intragastric douche (see p. 368). Bismuth subnitrate in large doses before meals is also of service in such cases.

In *nervous dyspepsia* short courses of strontium bromid are sometimes efficacious. When *anemia* is marked, an unirritating preparation of iron may be given with advantage.

Systematic lavage (see p. 492) is of great value in certain cases, but it has been much abused. Its chief indications are *retention of food with fermentation, excessive secretion of mucus,* and *excessive secretion of acid.* When lavage cannot be tolerated, the stomach may be cleansed by a glass of hot alkaline water slowly sipped a half-hour or more before breakfast. *Motor insufficiency* is often benefited by strychnin. Intragastric faradism and the intragastric spray (Einhorn) may also be recommended in cases of *pronounced stagnation.*

Symptomatic Treatment.—*Deficient Secretion.*—When the digestive power is impaired owing to a deficiency of secretion, diluted hydrochloric acid with pepsin (see p. 178), taken shortly after meals, may be found useful. In many cases, however, better results are obtained from the administration of pancreatin with sodium bicarbonate.

Hyperchlorhydria.—Severe pain occurring at the height of digestion and caused by hyperacidity is best treated by the administration of an alkali (sodium bicarbonate or magnesia) in large doses. *Heartburn* and *pyrosis* are also relieved by alkalis.

Flatulence and fermentation are sometimes controlled by such antiseptic drugs as creasote, bismuth salicylate, bismuth betanaphthol (orphol), and thymol. The following combination is often of value:

R	Creasoti,	f ℥ss (2.0 c.c.);
	Orphol,	gr. c (6.5 gm.);
	Strychninæ sulphatis,	gr. ss (0.03 gm.);
	Oleoresinæ zingiberis,	ℳ v (0.3 c.c.). M.

Pone in capsulas No. xx.
Sig.—One after meals.

Vomiting usually yields to rest and proper diet. If troublesome, sedatives like bismuth subnitrate, hydrocyanic acid, and silver nitrate may be employed.

Constipation.—As far as possible constipation should be overcome by regulation of the diet, systematic exercise, and abdominal massage. Strong purgatives should be avoided. A glass of water in the morning upon rising or in the evening before retiring often suffices in mild cases. If necessary, a small quantity of sodium phosphate or of the following artificial Carlsbad salt may be added:

<pre>
R   Sodii sulphatis,   ʒx (40.0 gm.);
     Sodii bicarbonatis,   ʒiv (16.0 gm.);
     Sodii chloridi,   ʒij (8.0 gm.).   M.
     Sig.—A teaspoonful in a glass of water an hour before breakfast.
</pre>

One of the natural sulphated waters may often be used with advantage, especially when there is a tendency to excessive secretion of acid. Suppositories containing glycerin and enemata of warm water are harmless and generally effective. In some cases mild vegetable laxatives, like cascara sagrada and rhubarb, are required. The occasional use of calomel in small, frequently repeated doses sometimes serves a good purpose.

GASTRALGIA.

The hygienic and dietetic treatment described under Indigestion is applicable to gastralgia. Especially important is it to forbid the excessive use of tobacco, tea, coffee, and alcohol. When the attacks show a tendency to occur midway between meals, the administration of 3 or 4 ounces (90.0–120.0 c.c.) of milk with lime-water, two or three hours after meals, is often of service.

Treatment of the Attack.—Stupes or hot compresses should be applied over the epigastrium. Internally, such remedies as chloroform (5–10 min.—0.3–0.6 c.c.), Hoffmann's anodyne (20–30 min.—1.2–2.0 c.c.), diluted hydrocyanic acid (2–3 min.—0.1–0.2 c.c.), and antipyrin frequently afford relief The following combination will be found of value in many cases:

<pre>
R   Acidi hydrocyanici diluti,   ♏xxxv (2 0 c.c.);
     Antipyrini,   ʒij (8.0 gm.);
     Tincturæ opii camphoratæ,   f ʒj (30.0 c.c.);
     Aquæ chloroformi,   q. s. ad f ʒiij (90.0 c.c.).   M.
     Sig.—A dessertspoonful in hot water.
</pre>

If the pain be very severe, it will be necessary to resort to the hypodermic injection of morphin and atropin.

Treatment.—The two most useful remedies are silver nitrate and arsenic. The former should always be selected when there is hyperesthesia of the mucous membrane or hyperchlorhydria. It may be combined with belladonna, as in the following formula:

℞	Argenti nitratis,	gr. v (0.3 gm.);
	Extracti belladonnæ,	gr. iv (0.20 gm.);
	Mannæ,	q. s.

Fiant pilulæ No. xx.
Sig.—One pill half an hour before meals.

The best preparation of arsenic is Fowler's solution, which should be given after meals in increasing doses.

Galvanism is sometimes of service; the anode may be placed over the epigastrium, and the cathode over the dorsal spine. Very severe cases may demand a complete change of scene and air or a modified " rest-cure."

Gastralgia depending upon anemia, locomotor ataxia, or malaria is to be treated, of course, on special principles.

GASTRIC ULCER.

Rest and appropriate diet are the most important factors in the treatment of gastric ulcer. The rest should be kept up for from six to twelve weeks, and for the first two or three weeks of this period the patient should be confined to bed. If hemorrhage has recently occurred or if vomiting be urgent, it is advisable to withhold all food from the stomach for a few days and to nourish the patient by means of nutritive enemata. The following enema of Boas is usually well borne:

℞	Milk,	250 grams;
	Yolk of two eggs,	
	Salt,	2 grams;
	Claret,	15.0 cubic centimeters;
	Baked flour,	15 grams.

Use 1 to 3 in twenty-four hours.

The following preparation suggested by Ewald is also useful: A tablespoonful of baked flour (dextrinized) is boiled with one-half of a glass of a 15 per cent. solution of glucose, and a wineglassful of claret is added. Two or three eggs, which have been beaten up with a tablespoonful of water and a pinch of salt, are stirred into the mixture after it has cooled sufficiently to prevent the coagulation of the albumin. The entire quantity should not measure more than ½ pint (0.25 L.).

After the pain and vomiting have sensibly abated, feeding by the mouth should be resumed. The diet should consist of milk, buttermilk, beef-juice, animal broths, egg-white, and thin pap. As soon as the gastric symptoms have completely disappeared, which will rarely be before the lapse of three or four weeks, the patient may be allowed such articles as soft-boiled eggs, scraped beef, boiled sweetbreads, the tender part of oysters, white meat of chicken, well-made gruel, and custard pudding.

Medicinal Treatment.—The most useful drugs are alkalis, silver nitrate, and bismuth subnitrate. The alkalis are useful in overcoming the superacidity of the gastric juice. Sodium bicarbonate is one of the best; it may be combined with magnesia or chalk, according as there is constipation or diarrhea. Artificial Carlsbad salt (see p. 547) is an excellent alkaline laxative; of this a teaspoonful or more may be taken in half a pint of hot water before breakfast. Silver nitrate and bismuth subnitrate undoubtedly possess some value, although they probably exert no specific influence on the ulcer itself. The two drugs may be given alternately, each for a period of a week or ten days (see pages 368 and 371).

Vomiting usually yields to complete rest, rectal feeding, and the administration of silver or bismuth. In some cases it may be necessary to use in addition a local sedative, like hydrocyanic acid (see p. 69) or cocain. Morphin administered hypodermically is often very efficacious.

Violent pain will also demand morphin, although the danger of inducing a habit must always be borne in mind. Externally, stupes or sinapisms are sometimes useful.

In the treatment of hematemesis absolute rest is essential. No food of any kind should be given by the mouth. An ice-bag should be applied over the stomach and morphin should be given hypodermically. The application of firm bandages to the four extremities may act favorably. Ergot and local astringents like tannic acid, iron sulphate, and lead acetate are of very doubtful utility. Ewald and Minkowski have recommended rinsing the stomach with ice-water as a last resort. Collapse following hemorrhage will call for diffusible stimulants, the external application of heat, and subcutaneous and rectal injections of warm saline solutions.

Operative interference is rarely warranted unless the hemorrhage be repeated.

Surgical Treatment.—In all cases of perforation an operation should be done at the earliest possible moment. The mortality in 123 cases operated on for perforation since 1896 was 47.15 per cent. (Bidwell). In a series of 18 cases operated on within twelve hours after perforation, and reported since 1896, the mortality was only 16.66 per cent. (Keen). When life is threatened by repeated hemorrhage, operation, in the interval between attacks (gastro-enterostomy or pyloroplasty) offers the best method of relief. Again, an operation (gastro-enterostomy, pyloroplasty, or partial gastrectomy) should be considered if the disease does not yield to medical treatment and the life of the patient is endangered by malnutrition.

GASTRIC CANCER.

Medicinal Treatment.—The medicinal treatment of gastric cancer is purely palliative. The chief indications are to maintain nutrition and to relieve distressing symptoms. In the early stages of the disease, when the pylorus is still free, a mixed diet of readily digested food is often well borne. Later, when there is retention, food should be selected which leaves but little residue in the stomach. Eggs, finely divided tender meats, sweet-bread, soft part of oysters, clear soups, calf's-foot jelly, and thin gruels are admissible. A small quantity of light wine with meals is sometimes very beneficial. Milk is often badly borne. If the stomach is unretentive recourse should be had to rectal feeding.

Bitters—nux vomica, calumba, gentian—may often be employed with advantage. Condurango (see p. 170) is an excellent stomachic, wholly lacking, however, in the specific properties with which it was at one time credited. In many cases, but by no means invariably, hydrochloric acid is a useful adjuvant to the bitters.

Lavage affords the very best means of relieving the distressing symptoms resulting from retention. Vomiting that is not dependent upon retention may be treated with such remedies as carbonated water, hydrocyanic acid, bismuth subnitrate, creasote, cerium oxalate, and chloroform. In obstinate cases rectal feeding may be required for a time. Acid eructations and flatulency are sometimes relieved by antacids and antiseptics, but in the majority of cases lavage is much more effective.

The pain, if severe, will require codein or morphin. Milder attacks may be relieved by lavage, the administration of antacids or of sedatives, like carbolic acid, hydrocyanic acid, or chloroform, and by the external application of hot compresses or stupes.

Constipation may be treated by simple enemata, glycerin suppositories, or by mild vegetable laxatives.

Surgical Treatment.—Statistics of the surgical treatment of gastric cancer do not present a very favorable showing, although the mortality of the operative procedures is diminishing year by year. There are but two operations to be considered—pylorectomy and gastro-enterostomy. The former is curative in its aim, the latter purely palliative.

In 291 cases of resection of the pylorus or of a portion of the stomach recorded between 1890 and 1898, the mortality, according to Guinard, was about 35 per cent. Earlier intervention and improvement in technic will undoubtedly lessen the mortality. Unfortunately, very few patients have remained free from recurrence longer than three years after the operation. Wölfer cites three patients who lived over four years, four, over five years, one, over six years, and two, over eight years.

Gastro-enterostomy is preferable to pylorectomy in cases with extensive adhesions or extensive glandular involvement. Between 1891 and 1896 gastro-enterostomy was performed in 401 cases with a mortality of 33.91 per cent. (Chlumskij). When performed successfully this operation affords great relief and may prolong life several months, and, occasionally, even a year or two.

DILATATION OF THE STOMACH.

The treatment of dilatation of the stomach is medical and surgical.

Medical treatment is frequently curative when there is no actual stenosis of the pylorus. The diet is of the first importance. Foods should be selected which are nutritious, which are not bulky, and which do not readily ferment. Tender meats, eggs, light cereals, fresh butter, cream, and toasted bread may usually be allowed. Milk is often badly borne, and in many cases liquids of all kinds must be somewhat restricted, but it is rarely necessary to resort to the dry diet of Schroth and other writers. In severe cases predigested meat preparations are often of service. P. Cohnheim has found olive oil useful when spasm of the pylorus is a factor.

In the majority of cases it is best to limit the meals to three or four a day; occasionally, however, it will be found necessary to give the food in small quantities at frequent intervals. The time required for the stomach to empty itself should be ascertained in each case, and should be the guide in determining the number of meals a day.

To prevent retention, to control fermentation, and to cleanse the stomach no measure is so valuable as lavage (see p. 492).

In the cases due to atony, massage, exercise in the open air, and hydrotherapy are valuable aids. A carefully adjusted abdominal bandage nearly always affords comfort and gives mechanical support to the stomach. Faradization of the stomach is useful in promoting muscular contraction.

Few drugs are of value. Nux vomica may often be given with advantage. When there is hypo-acidity, diluted hydrochloric acid is indicated. Such remedies as creasote, salol, and bismuth salicylate are sometimes of benefit in checking fermentation, but the relief they afford is not to be compared to that obtained through systematic lavage. Constipation should be treated, as a rule, by simple enemata or by glycerin suppositories.

Surgical Treatment.—In the large majority of cases of non-obstructive dilatation medical treatment suffices. Occasionally, however, surgical intervention is demanded on account of persistent suffering and progressive emaciation. The operation indicated in these cases is gastroplication. In 15 cases collected by Keen (1898) all the patients recovered but one.

In cases of pyloric obstruction of a benign character an operation is indicated when it is impossible to maintain nutrition by proper medical treatment. As Loreta's digital divulsion of the pylorus has been largely abandoned, owing to its high mortality (31.1 per cent.), there may be said to be but two operations available—pyloroplasty and gastro-enterostomy. The mortality in 14 cases of pyloroplasty reported by Carle was 7 per cent., while Morison has reported 11 cases, and Mayo 8 cases without a death. The mortality after gastro-enterostomy in 27 cases reported by Carle was 7.4, and in 15 cases reported by Mayo was 6.6 per cent. The question of operative intervention in malignant stricture of the pylorus has already been considered (see p. 550).

CHRONIC CONSTIPATION.

In many cases chronic constipation can be successfully treated by attention to hygienic measures and by careful regulation of the diet without resorting to drugs. Not infrequently the normal activity of the bowels may be restored by repeated daily attempts at defecation at some special hour. Systematic exercise and cold bathing are of the greatest benefit. Abdominal massage, especially digital kneading in the direction of the colon, is often quite effectual. In persons with relaxed abdominal walls the wearing of a snugly-fitting binder will be found of service.

Food should be selected which leaves considerable undigested residue. Green vegetables, oatmeal, corn-meal, whole-wheat bread, and cooked fruits are available. Oils may often be given with advantage. Water-drinking should be encouraged. In mild cases a glass of cold water taken in the morning shortly after rising may suffice.

When hygienic treatment proves insufficient, drugs will be required. Any underlying condition of which the constipation may be but a symptom should receive careful attention. General tonics, like iron and strychnin, are often indicated. Mineral waters, like Friedrichshall, Hunyadi János, or the milder Saratoga or Bedford waters, are very useful, but possess no special advantages over the saline laxatives (sodium phosphate, or Rochelle salt), when the latter are taken in small amounts well diluted. Enemata of soapy water or of glycerin, or suppositories of gluten, soap, or glycerin, often prove highly satisfactory. Vegetable cathartics are usually necessary in obstinate cases. The mild ones should always be tried first, and even with these considerable care should be exercised lest the patient comes to rely upon drugs to the exclusion of the hygienic and dietetic measures already indicated. Of the mild laxatives, cascara sagrada is one of the best; from 10 to 30 min. (0.6–2.0 c.c.) of the fluid extract, or a corresponding dose of an agreeable elixir, may be administered at bedtime and repeated, if necessary, in the morning. In some cases a combination of several laxatives acts better than any one singly. The drugs to be preferred for conjoint use are rhubarb, aloes, podophyllin, and euonymin. As adjuvants, nux vomica or physostigma may be added to overcome atony of the bowel, and belladonna or hyoscyamus to prevent griping. The most suitable combination must be determined in each case by experience. A pill, like one of the following, will generally be found efficacious:

R	Resinæ podophylli,	gr. iv (0.26 gm.);
	Aloes purificatæ,	gr. xx–xl (1.3–2.6 gm.);
	Extracti nucis vomicæ,	
	Extracti belladonnæ,	āā gr. iv (0.26 gm.). M.
	Fiant pilulæ No. xx.	
	Sig.—One pill at bedtime.	

R	Pulveris rhei,	
	Extracti thamni purshianæ,	āā gr. xxiv (1.5 gm.);
	Extracti euonymi,	gr. xij (0.8 gm.);
	Extracti physostigmatis,	
	Extracti belladonnæ,	āā gr. iv (0.26 gm.). M.
	Fiant pilulæ No. xxiv.	
	Sig.—One pill at bedtime.	

R   Aloini,   gr. iv (0.26 gm.);
     Pulveris rhei,   gr. xxxvj (2.3 gm.);
     Pulveris ipecacuanhæ,   gr. vj (0.4 gm.) ;
     Resinæ podophylli,   gr. iij (0.2 gm.) ;
     Extracti hyoscyami,   gr. xij (0.8 gm.).   M.
Fiant pilulæ No. xxiv.
Sig.—One pill at bedtime.

ACUTE DIARRHEA.

Acute Diarrhea in Adults.—Rest in bed and the substitution of bland nourishment for the ordinary diet are all that is required in many cases. Boiled milk, milk and arrow-root, and mutton, veal, or chicken broth are suitable foods. If the patient is seen at the outset and there is reason to believe that irritant material is still present in the bowel, it is advisable to administer an unirritating purgative, such as castor oil, Epsom salts, or fractional doses of calomel. Occasionally a second dose of the purgative may be given with benefit. Externally, stupes or sinapisms are frequently efficacious.

If there be much pain, opium should be administered. Dover's powder and paregoric are deservedly popular preparations. If the diarrhea continue, mild astringents, like bismuth subnitrate and chalk, are indicated. They may be combined advantageously with an antiseptic, as in the following formulæ :

R   Bismuthi subnitratis,   ℥iij (12.0 gm.);
     Bismuthi salicylatis,   ℥j (4.0 gm.).   M.
Fiant chartulæ No. xij.
Sig.—One powder every two or three hours.

R   Salol,   ℥ss (2.0 gm.);
     Bismuthi subnitratis,
     Cretæ præparatæ,   āā ℥ij (8.0 gm.);
     Pulveris acaciæ,   q. s.
     Aquæ cinnamomi,   q. s. ad f℥iij (90.0 c.c.).   M.
Sig.—A dessertspoonful every two or three hours.

When opium is indicated, it is better to prescribe the drug separately, so that it may be discontinued more readily when no longer required. Preparations containing tannic acid are rarely needed, but when the discharges are very profuse and watery, small doses of tannalbin or of tannigen (see p. 352) may prove serviceable. Cases in which gastric indigestion is an important factor are often benefited by pancreatin in conjunction with bismuth subnitrate and an alkali. If the diarrhea shows a tendency to become chronic, mineral astringents, like silver nitrate and lead acetate, with opium, in pill form, will be found effective.

When there is reason to believe that the colon is especially involved, local treatment is of the utmost value. Copious injections of warm water both cleanse the bowel and exert a soothing effect upon the irritated mucous membrane. Small enemata (1–2 oz.—30.0–60.0 c.c.) of starch-water with laudanum (5–10 min.—0.3–0.6 c.c.) are very useful. In subacute colitis enemata containing silver nitrate (see p. 368) often prove efficacious.

Treatment of Acute Diarrhea in Infants.—Preventive treatment is of the utmost importance. Improper food and faulty methods of feeding are the two prime factors in causing the disease. Fresh cows' milk containing a minimum number of bacteria to the cubic centimeter should form the basis of all food that is intended to replace healthy breast-milk. In summer the milk should be received from the dairyman twice in the day, if possible, and should be placed immediately in a good refrigerator and kept there until such times as it is wanted. In very warm weather it may be necessary to resort to sterilization or pasteurization as a further means of preservation, but it must be remembered that neither of these processes can remove from the milk any pre-existing impurities. The strength and quantity of the food and the number of meals per day must be carefully adapted to the infant's age and its digestive powers. As an additional precaution in extremely hot weather it may be advisable to dilute the milk a little more than usual in order to lessen the percentage of fats and proteids.

Next to purity and freshness of the milk, it is important to insist upon absolute cleanliness of all the feeding apparatus. The nursing bottle and rubber nipple should be thoroughly washed after each meal. The child should have a daily morning bath, and on very warm days it is advisable to give one or two tepid sponge baths (90° F.—32° C.) in addition to the ordinary bath. In summer, if the weather be good, the child should pass most of the time in the open air, protected, of course, from the winds and the direct rays of the sun.

The first indication in the treatment of the disease is to withdraw the milk at once, and to withhold it for several days or until the stools become quite natural. Indeed, in many cases it is well to suspend all nourishment for the first twenty-four hours, allowing nothing by the mouth but barley-water or plain boiled water. Subsequently, albumin-water, fresh beef-juice, veal broth, or a liquid peptone preparation may be given in lieu of milk. Milk feeding should always be resumed very gradually. Absolute rest in the recumbent position is essential. Removal to the seashore or mountains is often of the greatest benefit.

After the withdrawal of the milk, the next most important step in the treatment is the administration of a purgative like castor oil or calomel. Of these, calomel should be given the preference when the stomach is irritable. In many cases free purgation alone suffices to cleanse the bowel, but when the stools are offensive or there is marked toxemia enteroclysis should also be practised once or twice a day, according to the severity of the attack. In cholera infantum intestinal irrigation may be employed to meet several indications: to lower temperature, to combat collapse, to relieve the anhydremia, and to stimulate renal secretion (see p. 490). In choleraic forms. of the disease lavage and hypodermoclysis are also of great value. In most cases of acute diarrhea it is necessary to follow the purge with a sedative astringent, like bismuth subnitrate or chalk. From 5–10 gr. (0.3–0.6 gm.) of one of these drugs may be given every two or three hours. Intestinal antiseptics (salol, bismuth salicylate, betanaphtol bismuth) are useful adjuvants. Some such combination as the following may be ordered with advantage:

> R Bismuthi subnitratis, ℥ij–iv (8.0–16.0 gm.);
> Salol, gr. xxiv (1.5 gm.);
> Misturæ cretæ, f℥iij (90.0 c.c.). M.
> Sig.—A teaspoonful every two hours.

A more active astringent, like tannalbin or tannigen, may be given in addition to the bismuth subnitrate or chalk when the discharges are exceedingly profuse and watery.

In ileocolitis intestinal irrigation may be followed by injections of 1 or 2 ounces (30.0–60.0 c.c.) of a solution of silver nitrate ($\frac{1}{2}$ gr. to the ounce—0.03 gm.–30.0 c.c.), of from $\frac{1}{2}$ to 1 ounce (15.0–30.0 c.c.) of a mixture of bismuth subnitrate (1 dram to the ounce—4.0 gm.–30.0 c.c.), or of from 3 to 4 ounces (90.0–120.0 c.c.) of a solution of tannin (1 per cent.). If the injection containing silver nitrate excite pain, the bowel should be flushed with salt solution.

Opium is often of the utmost value, but extreme caution must be exercised in its use. On the whole, it has undoubtedly caused more harm than good. In many cases there is absolutely no indication for it. It is called for when the diarrhea continues in spite of the thorough unloading of the bowel and the administration of mild astringents. One of the best preparations is paregoric; of this, from 3 to 5 minims (0.2–0.3 c.c.) may be given every two, three, or four hours according to circumstances. Opium should not be added to mixtures, but should be prescribed by itself. When the stomach is unretentive, laudanum in doses of from 1 to 2 minims (0.06–0.1 c.c.) may be added to starch-water ($\frac{1}{2}$ ounce—15.0 c.c.) and administered by the rectum. In cholera infantum small doses of morphin ($\frac{1}{100}$ gr.—0.0006 gm.) with atropin ($\frac{1}{800}$ gr.—0.00008 gm.) may be given hypodermically.

Stimulants are often required. Whisky and brandy are perhaps the best. From 10 to 20 minims (0.6–1.2 c.c.) of one or the other may be given in cold water every two, three, or four hours. When the stomach is unretentive, it will be necessary to administer stimulants hypodermically. Hypodermoclysis also affords a valuable means of preventing exhaustion. In threatened collapse hot packs or hot baths may prove beneficial.

Fever, according to its intensity, should be controlled by cold sponging, cold packs, or cool baths (85°–75° F.—29.5°–24° C.). Intestinal irrigation with cool water may also be used for its antipyretic effect.

CHRONIC DIARRHEA.

The first thing to do is to ascertain the cause and to remove this, if possible. The diet, clothing, habits, occupation, and mode of living of the patient should receive the most careful attention. No definite rules can be laid down in reference to the diet. Many patients do well upon an exclusive milk diet. Scraped beef, raw or underdone, is often well borne. In some cases it is an advantage to have the food peptonized. When the disease is not very severe and is confined for the most part to the colon, a selected mixed diet may be allowed. In such cases pulled bread or toast, tender beef or mutton, oysters, raw or boiled eggs, boiled rice, arrow-root, and plain custard are available. Foods that are bulky and leave much residue are always inadmissible.

Protection of the body against chilling is of vital importance. Woolens should be worn next to the skin. A snugly fitting abdominal bandage may be worn as an additional safeguard. Rest in bed is sometimes essential. The obstinate diarrhea occasionally encountered in hysterical women is often benefited by a modified rest cure. When the general nutrition is not too much impaired, a change of air and scene may prove very effectual.

The *medicinal treatment* of chronic diarrhea is very unsatisfactory. Mineral astringents, especially bismuth subnitrate (30–40 gr.—2.0–2.5 gm.), silver nitrate ($\frac{1}{4}$–$\frac{1}{2}$ gr.—0.016–0.03 gm.), copper sulphate ($\frac{1}{4}$–1 gr.—0.016–0.06 gm.), and lead acetate (1–3 gr.—0.06–0.2 gm.), enjoy the most repute. These drugs should always be given a thorough trial, as even under the most favorable conditions improvement is slow in appearing. Intestinal antiseptics—salol, thymol, bismuth salicylate, creasote, betanaphthol-bismuth, and benzonaphthol—are useful adjuvants to astringents, especially when there is considerable intestinal fermentation. Opium in small amounts is often required, particularly in acute exacerbations. Wood has found the following tar-water mixture a most effective remedy in many cases:

R Picis liquidæ, fʒiij (90.0 c.c.);
    Triturentur cum liquore calcis,
      Oviij (4.0 L.), ad saturationem,
      et percolentur per prunum vir-
        ginianam, ʒviij (250.0 gm.). M.
    Sig.—A wineglassful one to two hours after meals.

In the diarrhea of hysterical women strontium bromid is sometimes of value. Arsenic (see p. 303) has been highly recommended in that form of diarrhea characterized by an uncontrollable desire to evacuate the bowel immediately after taking food. In malarial cases a cure is generally rapidly effected with quinin. When there is decided anemia, tincture of ferric chlorid will be found serviceable. In many cases it is a good plan to resort to intestinal irrigation from time to time or to administer a purgative (castor oil, Epsom salts, or calomel) occasionally to remove from the bowel scybalous masses or viscid mucus.

When the disease is situated chiefly in the colon, local treatment is of the greatest service. Starch-water and laudanum enemata are sometimes useful. Generally, however, astringent injections are more successful. Irrigation of the bowel, two or three times a week, with a solution of silver nitrate (see p. 368) is especially to be recommended.

DYSENTERY.

Rest in bed is imperative, even in chronic cases. In acute dysentery the diet should be liquid or semiliquid. Milk with lime-water, whey, milk-toast, animal broths, and egg-white may be given. Predigested foods are often very useful. In the chronic form of the disease it may be necessary to give a somewhat more liberal diet in order to maintain nutrition. Soft-boiled eggs, pulled bread, boiled rice, oysters, and tender meats (chicken, mutton, and beef) may be allowed. The intestinal discharges should be disinfected immediately as in the case of typhoid fever.

An unirritating purgative (castor oil, Epsom salts, or calomel) is nearly always indicated at the onset. After the bowel has been thoroughly emptied, opium may be given to check peristalsis and to relieve the tormina and tenesmus. It is best given hypodermically in the form of morphin or by the bowel in the form of starch-water and laudanum injections or opium suppositories ($\frac{1}{2}$–1 gr.—0.03–0.06 gm.). Hot fomentations, turpentine stupes, or sinapisms to the abdomen also afford relief. Persistent tenesmus may sometimes be controlled by iodoform suppositories (2–5 gr.—0.13–0.3 gm.). Wood speaks favorably of ice suppositories, inserted for a length of time, one after another.

Some benefit may be derived from the administration of bismuth subnitrate (20–30 gr.—1.3–2.0 gm.) by the mouth, combined with an antiseptic like salol or betanaphthol-bismuth. Sulphur is another remedy well worthy of trial in obstinate cases. Ten grains (0.65 gm.) should be given three or four times a day, combined with a small quantity of opium. In many cases the adynamia is so pronounced that stimulants must be used freely. In cases in which the typhoid state is well developed oil of turpentine, 5–10 minims (0.3–0.6 c.c.) in emulsion, often acts favorably.

After the most acute symptoms have subsided intestinal irrigation may be employed with the greatest benefit. The patient's hips should be elevated and the fluid should be introduced very gently and slowly by means of a fountain syringe. When the rectum is very irritable, it is advisable to inject a small quantity of cocain solution (4 per cent.) before introducing the irrigator. In acute dysentery normal salt solution at a temperature of 100°–103° F. (38°–39.5° C.) may be employed for the irrigation. Weak solutions of boric acid ($\frac{1}{2}$ dram to 1 quart—2.0 gm.–1.0 L.) and of potassium permanganate (2–5 grains to 1 quart—0.13–0.3 gm. to 1.0 L.) have also been recommended. Hare has used zinc sulphocarbolate (40 grains to 1 quart—2.6 gm. to 1.0 L.) with very satisfactory results. In amebic dysentery warm injections of quinin (from 1 : 5000 to 1 : 1000) have been found very efficacious. In chronic dysentery weak solutions of silver nitrate (10–30 grains to 1 pint—0.6–2.0 gm. to 0.5 L.) are generally the most useful (see p. 368).

Three special methods of treatment merit consideration : that by ipecac, that by salines, and that by antibacterial serum.

Many practitioners who have had much experience in the treatment of dysentery in India have borne testimony to the value of ipecac. At the onset it is customary to give a large dose (30 gr.—2.0 gm.), and this usually induces vomiting. After vomiting has occurred the drug is administered in smaller doses (2–5 gr.—0.13–0.3 gm.) at intervals of two hours. To insure the retention of the latter doses opium is generally given a short time in advance of the ipecac. A successful result is indicated by the appearance, usually within twelve hours, of a copious black stool. In the sporadic dysentery of temperate zones this treatment has not proved very satisfactory.

Magnesium sulphate is another remedy which has been brought prominently forward within the last few years, especially in India and other tropical countries. In 1900, Buchanan reported a series of 453 cases of dysentery treated by this drug in which there were but 5 deaths. A purgative dose is first administered, and after the bowel has been thoroughly emptied, smaller doses (1 dr.—4.0 gm.) are given three or four times a day, preferably in combination with aromatic sulphuric acid, as in the following formula:

R   Magnesii sulphatis,   ℥iss (45.0 gm.);
     Acidi sulphurici aromatici,   f℥ij (8.0 c.c.);
     Aquæ cinnamomi,   q. s. ad f℥vj (180.0 c.c.).   M.
Sig.—Tablespoonful three or four times a day.

The treatment should be continued for several days after the stools have ceased to be dysenteric. Unlike the ipecac treatment, this plan not infrequently gives good results in the acute dysentery met with in temperate regions.

The discovery that one form of acute dysentery is caused by a special organism, the bacillus of Shiga, is sufficient ground for hope that a curative serum will be forthcoming. Indeed, Shiga claims to have reduced the mortality in a recent epidemic in Japan from 37 to 8 per cent. by means of an antibacterial serum.

APPENDICITIS.

There is a considerable diversity of opinion among physicians and surgeons as to the best method of treating appendicitis. Some advise surgical intervention in all cases as soon as the diagnosis is made; others, among whom are the greater number of representative surgeons, advocate a policy of intelligent discrimination and selection.

The author is of the opinion that operation should be urged—(1) At once in all cases in which the onset is very severe, the symptoms indicating special severity being marked right-sided tenderness and rigidity, distention, and vomiting, with or without fever; (2) in cases of moderate severity which manifest no improvement after the lapse of forty-eight hours; and (3) in cases in which the symptoms, after decided improvement, return. On the contrary, operation is rarely required, at least during the attack—(1) In cases of a mild type, in which the pain is unaccompanied by rigidity, distention, nausea, or vomiting; and (2) in cases of moderate severity in which improvement is noticeable within forty-eight hours. Operation during the quiescent stage, when the element of danger is almost entirely removed, is to be recommended—(1) When an acute attack has been followed by persistent tumefaction and tenderness, intestinal disturbances, or impairment of the general health; (2) when there have already been two attacks, even of moderate severity; and (3) when mild attacks occur with such frequency as to induce disability.

Medical Treatment.—The medical treatment of appendicitis is very simple. The patient should be kept in bed at absolute rest. The diet should be restricted to small quantities of bland liquids—milk, albumin-water, and broths. Purgatives, as a rule, should be avoided, although there is usually no objection to administering a mild saline aperient at the onset of the attack. Constipation is best relieved by enemata of warm water. Locally, cold or heat may be applied, according to the sensations of the patient. Applications should not be used which affect the integrity of the skin, as they render surgical intervention more difficult should it subsequently be required. If the pain be very severe, morphin may be administered hypodermically; only the minimum amount necessary to afford a measure of relief is to be used, however, as by obscuring the symptoms the drug prevents an accurate study of the progress of the case.

CATARRHAL HEPATITIS.

Diet is of the first importance. Fatty, starchy, and saccharine foods should be avoided. Milk, buttermilk, broths, egg-albumen, lean meats, oysters, toast, pulled bread, and well-cooked cereals — wheat, rice, and oat-meal—are admissible. Water-drinking is distinctly beneficial. Alkaline mineral waters, like Vals, Vichy, and Hathorn, are frequently of service. If constipation resists the action of these waters, daily evacuations should be secured by means of sodium phosphate, Carlsbad salts, Rochelle salts, or a sulphated water, like Hunyadi János or Friedrichshall.

Bismuth subnitrate and silver nitrate are often of value in relieving the primary gastroduodenal catarrh. Ammonium chlorid (10 gr.—0.65 gm.—thrice daily, after meals) is also useful. In obstinate cases, nitrohydrochloric acid, as originally recommended by Scott, may prove beneficial. Daily irrigation of the colon with from 1 to 2 quarts (1.0–2.0 L.) of cold water, as advocated by Krull, is sometimes followed by improvement.

Intestinal fermentation and flatulence may yield to inspissated ox-gall or to an antiseptic like salol, bismuth salicylate, or betanaphthol-bismuth. *Itching* should be treated with tepid alkaline baths and lotions of carbolic acid, boric acid, or diluted hydrocyanic acid. Internally bromids, in full doses, may be tried when the pruritus is severe.

CHOLELITHIASIS.

Medical Treatment.—Gall-stones in the gall-bladder or in the biliary passages cannot be dissolved. Turpentine and ether (Durande's mixture), chloroform, sodium succinate, sodium choleate, sodium oleate, and a host of other drugs have been recommended with a view to their solvent effects, but they have been found wholly ineffectual. Attempts to accomplish the extrusion of calculi by means of drugs have proved equally futile. Probably not more than 5 per cent. of those who have gall-stones suffer any inconvenience from their presence (Kehr). When symptoms occur, they result, in the large majority of cases, from a coincident cholecystitis. As the latter is the chief factor in exciting the violent expulsive efforts which force the stones from the gall-bladder into the ducts, it is mainly responsible for the recurrent attacks of colic. As we have no solvents for gall-stones and no remedies capable of effecting their expulsion, our efforts must be directed to keeping the stones quiescent, and this can be accomplished only by allaying catarrhal inflammation of the gall-bladder and by guarding against its recurrence.

Diet and hygiene are of the utmost importance. The food should be plain and readily digestible. Saccharine matters, fat meats, and highly seasoned dishes are undesirable. With the view of promoting a more continuous flow of bile it is often advantageous, when the gastric digestion is well maintained, to give the food in small quantities at comparatively short intervals. A light meal at bedtime has been especially recommended by Kehr. Frequent feeding may prove very harmful, however, when there is slow digestion with atony of the stomach. Water-drinking between meals should be encouraged.

Regular exercise in the open air, provided the symptoms are latent, is extremely beneficial. Constriction of the upper part of the abdomen by corsets or other articles of dress is to be rigidly avoided. The practice of manipulating the gall-bladder for the purpose of expelling calculi, which has been advocated by some, cannot be too strongly condemned.

Freedom from worry and mental effort, probably by conducing to good digestion, often exerts a marked salutary effect.

Digestive disturbances should receive appropriate treatment. Among drugs, alkalis and alkaline mineral waters are undoubtedly efficacious. Sodium bicarbonate or sodium phosphate may be taken well diluted in the morning an hour before breakfast and also between meals. If there be decided constipation, a small quantity of Rochelle salt or sodium sulphate may be added to each potation. The natural mineral waters, notably those of Carlsbad and Vichy, have acquired a high reputation. When there is a tendency to so-called bilious attacks, an occasional course of calomel in fractional doses will be found of benefit. In similar cases irrigation of the bowel, once or twice a week, with from 1 to 2 quarts (1.0–2.0 L.) of cool water is also very useful.

Finally, when there are no evidences of active inflammation in the gall-bladder and the patient's circumstances will permit, we may recommend a course of treatment at Carlsbad, Vichy, or Contrexéville, with some degree of confidence. An extended visit to one of these resorts is often followed by considerable improvement, and in rare cases by permanent relief. The benefit from a stay at one of these spas probably depends more upon change of air and scene, genial surroundings, regular hours, and freedom from worry than upon the waters themselves.

Hepatic Colic.—When the pain is violent it will be necessary to give morphin and atropin hypodermically at short intervals. As the opium habit is readily formed in these cases, it need scarcely be added that great caution should be exercised in the use of the drug. Agonizing pain often yields very promptly to a few whiffs of chloroform. In the mild but rather persistent attacks a few doses of antipyrin in hot water may suffice. The external application of heat is very useful. Hot poultices or fomentations may be applied to the region of the liver, or, when circumstances permit, the patient may be kept in a hot bath. Exceptionally, an ice-bag affords more relief than any of the hot applications. When vomiting is urgent, carbonated water, cracked ice, or small quantities of champagne may be given. In threatened collapse diffusible stimulants are needed.

Acute Obstruction of the Common Duct.—The measures best suited for promoting the advance of the stone into the bowel are rest, regulation of diet, the free use of alkaline mineral waters, the occasional exhibition of saline laxatives, and the application of heat to the hypochondriac region. Olive oil (see p. 458) has been recommended as a special remedy, but it is of doubtful efficacy. As the sequelæ of impaction of the common duct are so numerous and so grave, surgical aid should be invoked if the obstruction is not removed under medical treatment within a period of two or three weeks.

Surgical Treatment.—Surgical intervention is called for— (1) When, despite medical treatment, attacks of colic occur so frequently and are of such severity as to cause disability or make the addiction to morphin a likelihood; (2) in persistent obstruction of the common duct; (3) in hydrops of the gall-bladder due to impaction or stricture of the cystic duct; and (4) in suppurative inflammation of the gall-bladder or gall-ducts.

In the hands of experienced operators the mortality of cholecystotomy in uncomplicated cases is very low, certainly less than 1 per cent. Mayo has performed (1899) cholecystotomy 64 times for gall-stones in the gall-bladder or cystic duct, or both, with but 1 death. The mortality of choledochotomy has been from 10 to 20 per cent., although in Kehr's last 42 cases (1901) it was reduced to 4.5 per cent.

After operation medical treatment should be continued on the lines already laid down for the management of cholelithiasis that is more or less latent, with the view of preventing the further formation of stones.

CIRRHOSIS OF THE LIVER.

As cirrhosis of the liver is almost invariably associated with congestion of the stomach and gastric catarrh, these conditions are the first to demand attention. A diet of bland, readily digestible food is indicated. Alcohol should be forbidden. Tea, coffee, fatty matters, and highly spiced or seasoned dishes are inadmissible. Milk, eggs, oysters, sweetbread, tender meats, and well-cooked cereals may usually be allowed. In advanced cases an exclusive milk diet is advisable. As in simple gastric catarrh, such drugs as silver nitrate and bismuth subnitrate are sometimes of service. Lavage of the stomach should not be practised for fear of wounding the varicose veins at the lower end of the esophagus. Measures which promote the action of the skin and kidneys should not be neglected.

Portal congestion is best relieved by saline aperients and the occasional use of a mild mercurial. Mineral waters, such as those of Vichy, Carlsbad, Saratoga, Hunyadi János, and Friedrichshall, taken hot an hour before meals, often have an excellent effect.

Potassium iodid has been largely employed with the hope that it might favor the absorption of the overgrown connective tissue. While it is apparently useless in the absence of syphilis, in syphilitic cases it may prove very beneficial.

Ascites.—Effusion that is only moderate in amount can sometimes be removed by the administration of cathartics and diuretics. A concentrated solution of Epsom salts ($\frac{1}{2}$–1 oz.—15.0–30.0 gm.), taken in the morning before breakfast, is usually the most efficient purgative. Occasionally it may be desirable to substitute compound jalap powder or elaterium. The diuretics of approved value are potassium acetate or bitartrate, digitalis, and squills. The combination of blue-mass, digitalis, and squills, known as Niemeyer's pill (see p. 51), has a well-deserved reputation. Murchison and Fagge advocated resin of copaiba (15 gr.—1.0 gm., daily), but it often disturbs the stomach. The use of drugs should not be persisted in ; if the ascites does not diminish in the course of a few days, it is far better to resort to paracentesis. The operation is readily performed, and if done under antiseptic precautions, is attended with but very slight risk. It not only affords speedy relief, but, by removing the pressure from the abdominal veins, it also assists the action of diuretics. Occasionally, after one or two tappings, the fluid does not return for a long period—perhaps several months or even years; generally, however, the abdomen quickly refills, so that frequent repetition of the operation becomes necessary.

The operation of *paracentesis abdominis* is performed as follows : The bladder having been emptied, the patient is placed in a semirecumbent position, and a spot in the median line midway between the umbilicus and the symphysis pubis is anesthetized by means of a block of ice sprinkled with salt. A stout trocar is now introduced with a quick thrust into the abdominal cavity, a rubber tube is attached to the cannula for the purpose of conveying the fluid into a pail placed below the patient's bed, and the trocar is then withdrawn. While the fluid is escaping a many-tailed bandage is adjusted to the abdomen and gradually tightened. The application of such a binder should never be omitted. It gives support to the relaxed abdominal walls, and tends to prevent syncope and hematemesis. When the fluid ceases to flow, the cannula is removed, and the opening sealed with an antiseptic pad and a few strips of adhesive plaster.

Surgical Treatment.—Talma's operation (suture of the omentum to the margin of an abdominal incision, and irritation of the peritoneal surface of the liver and spleen), or one of its modifications, has proved of some benefit in a limited number of cases of liver cirrhosis with ascites. The object of the operation is to establish a compensatory circulation by making accidental adhesions and thus increasing the anastomoses between the vessels of the portal system and those of the systemic circulation. Of 105 cases collected by Greenough (1902), 31 (29.5 per cent.) died within thirty days after the operation ; 44 (42 per cent.) were more or less improved, and 29 (27.6 per cent.) showed no improvement. One case could not be traced. Thus 60 cases (57.1 per cent.) obtained no benefit from the operation. Of 44 cases reported improved, in 9 (8.5 per cent. of the whole number) the relief still continued two years after the operation. The operation is contraindicated when cardiac or renal disease coexists.

DISEASES OF THE RESPIRATORY SYSTEM.

ACUTE RHINITIS.

In mild attacks little treatment is required. If the constitutional symptoms be severe, it is advisable for the patient to remain in his room, or even in his bed. When the patient is seen at the outset and is willing to remain indoors for twenty-four hours, a hot foot-bath, with a full dose of Dover's powder, followed in the morning by a Seidlitz powder or another saline aperient, often gives excellent results. When the patient is fully able to go about, the following capsules will usually afford considerable relief:

R	Pulveris camphoræ,	gr. vj (0.4 gm.) ;
	Extracti belladonnæ,	
	Codeinæ,	āā gr. iss (0.1 gm.);
	Cinchoninæ sulphatis,	gr. xij (0.8 gm.). M.
	Pone in capsulas No. xij.	
	Sig.—One every two or three hours.	

Local treatment is very useful. In the early stage, when there is marked swelling of the mucous membrane, the application of cocain (2–4 per cent. solution) is sometimes exceedingly efficacious. If, however, no lasting effect is secured after three or four trials, its use should be discontinued. Warm Dobell's solution (see p. 165) or warm distilled extract of witch-hazel (diluted with 1 part of water), used as a spray at intervals, and followed in a few minutes by an oily application like the following, generally renders satisfactory service :

R	Menthol,	gr. iij (0.2 gm.);
	Olei pini pumilionis,	ℳ v (0.3 c.c.);
	Petrolati liquidi,	q. s. ad f ℥j (30.0 c.c.). M.

Among numerous vapor-inhalations recommended for their soothing effect the following are worthy of mention: Compound tincture of benzoin, 1 dram (4.0 c.c.), and hot water, 1 pint (0.56 L.); chloroform, 1 dram (4.0 c.c), and hot water, 1 pint (0.5 L.); iodin, 1 or 2 grains (0.06–0.13 gm.), and ether, 1 ounce (30.0 c.c.); camphor, 1 dram (4.0 gm.), and hot water, 4 ounces (120.0 c.c.).

CHRONIC RHINITIS.

The treatment of chronic rhinitis is both general and local. It is highly important to ascertain the existence of any constitutional vice and to remove this if possible. Attention to hygienic conditions is most essential. Fresh air, out-door exercise, and frequent bathing, followed by friction of the skin, are to be insisted upon. Tonics, especially strychnin and cod-liver oil, are aften required.

When the disease is not far advanced, much can be accomplished by keeping the nasal passages thoroughly clean. For this purpose detergent alkaline fluids may be applied by means of a coarse spray or the nasal douche-cup. A large number of combinations have been suggested, of which Dobell's solution (see p. 165) and the following are good examples:

℞ Sodii bicarbonatis,
   Sodii boratis,   āā gr. xxx (2.0 gm.);
   Sodii chloridi,   gr. iij (0.2 gm.);
   Thymol,
   Menthol,   āā gr. ss (0.03 gm.);
   Olei gaultheriæ,   ℥ij (0.12 c.c.);
   Glycerini,   f ℨij (8.0 c.c.);
   Alcoholis,   f ℨss (2.0 c.c.);
   Aquæ,   q. s. ad f ℥viij (240.0 c.c.). M.

In the early stage, when the redundant tissue is not very dense, local remedies of an astringent or alterative character are often very efficacious. The following applications are in common use: A mixture of iodin and glycerin containing 6 grains (0.4 gm.) of iodin, 12 grains (0.8 gm.) of potassium iodid, and 1 ounce (30.0 c.c.) each of glycerin and water; aqueous solution of ichthyol (20–40 per cent.); solution of zinc sulphocarbolate (2–5 per cent.); solution of resorcin (10 per cent.); and solution of silver nitrate (1–2 per cent.).

When the hypertrophic process has progressed so far that these simple measures prove ineffective, a more energetic treatment is demanded. The obstruction must be removed by means of caustics (chromic acid, trichloracetic acid, and acetic acid), the galvanocautery, or the snare, the proper use of which usually requires the services of a skilled specialist.

Atrophic Rhinitis (Ozena).—Improvement in the general health by the aid of fresh air, good food, and of such remedies as cod-liver oil, iron, arsenic, and strychnin is often requisite. If any syphilitic taint is suspected, potassium iodid and mercury should have a fair trial. Subcutaneous injections of antidiphtheritic serum have been used in a number of instances with alleged good results, this treatment having been tried in consequence of the assertion made by Belfanti and Della Vedova, and confirmed by other observers, that a bacillus identical with the bacillus of diphtheria is the cause of ozena.

The first requisite of local treatment is thorough cleanliness of the nasal passages, and to secure this the patient should be directed to spray the nose two or three times a day with Dobell's solution or some other mild detergent fluid. When the crusts are difficult to remove, they may be softened by pledgets of cotton soaked in a solution of hydrogen dioxid. After the nares have been cleansed, an oily solution like the following may be used to moisten and protect the parts:

R Menthol,	gr. xx (1.3 gm.);
Thymol,	gr. vj (0.4 gm.);
Eucalyptol,	ℳ xx (1.2 c.c.);
Petrolati liquidi,	f ℨvj (180.0 c.c.). M.

Among the many local remedies which have been employed for their stimulant effects ichthyol is perhaps the best (see p. 328). When there is much secretion, insufflations of zinc stearate (75 per cent. with boric acid), as recommended by C. C. Rice, are sometimes useful.

Interstitial electrolysis has been used by some observers with considerable success.   The *modus operandi* of this method of treatment is as follows : The nasal fossæ having been cleansed and anesthesia having been produced by means of a 10 per cent. solution of cocain, a copper needle attached to the positive pole is introduced beneath the mucous membrane of the middle turbinated body, and a platinum needle attached to the negative pole is introduced into the lower part of the septum or into the outer wall of the inferior meatus of the same nostril.   The intensity of the current employed has varied from 6 to 15 milliampères, and the duration of the sittings from ten to fifteen minutes.

For destroying the offensive odor a large number of drugs have been suggested, of which the following may be mentioned : Citric acid (see p. 462), formalin (1 : 1000 to 1 : 500), Labarraque's solution (1 : 30), potassium permanganate (2 grains to the ounce—0.13 gm.–30.0 c.c.), and creolin (1 : 100).

ACUTE LARYNGITIS.

In severe cases the patient should be confined to bed, and the temperature of the room kept uniformly at 70°–72° F. (21°–22° C.). The air should be rendered moist by means of steam. A hot foot-bath at the onset is often beneficial.. The use of the voice should be avoided as much as possible. The external application to the throat of cold compresses or of an ice-bag is of service. In grave forms of the disease it may be advisable to apply a few leeches to the upper part of the sternum.

Warm inhalations afford considerable relief. A dram (4.0 c.c.) of compound tincture of benzoin, half a dram (2.0 c.c.) of oil of turpentine, or half a dram (2.0 c.c.) of oil of cubebs may be added to a pint (0.5 L.) of boiling water and the fumes inhaled. For want of a special apparatus for the administration of these inhalations steam may be collected and conducted by means of an ordinary tin funnel which has been inverted over a · bowl containing the medicated hot water. In vaporizing oily substances it is advisable to mix them first with magnesium carbonate in order to aid their suspension in the water.

In the early stage direct local applications are rarely required; later in the disease, however, the application of astringents by means of the atomizer often proves useful. Alum, 5–10 gr. (0.3–0.6 gm.) to the ounce (30.0 c.c.); silver nitrate (1–2 per cent. solution) and zinc sulphate, 2–3 gr. (0.13–0.2 gm.) to the ounce (30.0 c.c.), often give satisfactory results when used in this manner.

Internal medication may do much good. At the onset of the disease it is usually advisable to administer a saline purge. If the cough be troublesome, a sedative, like Dover's powder, codein, or heroin, may be given in conjunction with a mild expectorant, such as ipecac, potassium citrate, or ammonium chlorid.

In acute *edematous laryngitis*, when the swelling does not yield promptly to local bloodletting, the external application of ice, astringent sprays, scarification of the mucous membrane, and active catharsis, tracheotomy should be performed without delay.

Spasmodic Croup.—The indications are to arrest the paroxysms and to prevent their recurrence. When the paroxysms are severe, the child may be placed in a warm bath (100° F.—38° C.) for ten or fifteen minutes, or an emetic (ipecac or alum) may be given instead. Occasionally the dyspnea is so intense as to demand the inhalation of a few drops of chloroform or amyl nitrite.   Mild attacks usually yield to the application, over the larynx, of a sponge wrung out of hot water.

The remedial measures which have been advocated for simple laryngitis are also of service in preventing attacks of pseudo-croup.   Such a combination as the following will generally be found useful:

R   Tincturæ aconiti,   ℞ ij (0.1 c.c.) ;
     Vini ipecacuanhæ,   f ʒj (4.0 c.c.) ;
     Potassii bromidi,   ʒss (2.0 gm.) ;
     Potassii citratis,   ʒj (4.0 gm.) ;
     Syrupi tolutani,   f ʒj (30.0 c.c.) ;
     Aquæ,   q. s. ad f ʒij (60.0 c.c.).   M.
Sig.—A teaspoonful every two or three hours for a child of eighteen months.

CHRONIC LARYNGITIS.

It is highly important to determine in each case the underlying cause of the local irritation, and to remove this if possible.   Injurious habits, such as the excessive use of tobacco or alcohol, must be corrected.   The diet, exercise, bathing, and clothing must be carefully regulated.   While it is essential that the body should be warmly clad, the practice of wearing thick scarfs about the neck as a means of protection from cold should be condemned.   Daily sponging the neck and chest with cool water is a far better way of guarding against recurrent attacks of inflammation.   In some cases a change of occupation is imperative.   Rest of the voice is always indicated.   Not infrequently improper or excessive use of the voice is the main etiologic factor, and when this is the case permanent relief can be secured only by the correction of the error.   As in a large group of cases a morbid process in the nasopharyngeal passages is the sole cause of the laryngeal irritation, it follows that the removal of this primary disease must often be the fundamental principle in the treatment. Any constitutional disease, such as rheumatism or gout, or any derangement of the digestion or circulation that may be present should receive careful attention.   Change of scene and climate is often of the greatest benefit, a warm, dry climate being especially favorable to the condition.

In many cases of chronic laryngitis tonics, like iron, strychnin, and cod-liver oil, are efficacious.   Even when there is no evidence of syphilis, potassium iodid in small doses is sometimes useful, especially when there is a great lack of secretion with a tendency to the formation of crusts.   Expectorants have little, if any, effect upon the chronic process, but often prove serviceable in acute exacerbations.

Local treatment is of prime importance. Remedies are best applied by means of sprays or a camel's-hair brush or cotton swab. Lozenges are also of benefit in certain conditions. Vapor inhalations and insufflations of dry powders are not so satisfactory. Thorough cleanliness, not only of the larynx itself, but also of the nose and pharynx, is always requisite. This is best secured by spraying the parts at frequent intervals with a mild alkaline solution. If there be deficient secretion, one of the following solutions may be employed as a spray after the mucous membrane has been cleansed: Sodium chlorid, 5 gr. (0.3 gm.) to the ounce (30.0 c.c.); potassium chlorate, 5 gr. (0.3 gm.) to the ounce (30.0 c.c.); ammonium chlorid, 5 gr. (0.3 gm.) to the ounce (30.0 c.c.); or potassium iodid, 5 gr. (0.3 gm.) to the ounce (30.0 c.c.). Lozenges containing 1 or 2 gr. (0.6–0.13 gm.) of ammonium chlorid may also afford relief, especially if the pharynx be much affected. When there is excessive secretion, astringent sprays are indicated, and one of the following may be selected: Alum, 3–5 gr. (0.2–0.3 gm.) to the ounce (30.0 c.c.); zinc acetate, 3–5 gr. (0.2–0.3 gm.) to the ounce (30.0 c.c.); zinc sulphocarbolate, 2–3 gr. (0.13–0.2 gm.) to the ounce (30.0 c.c.); or lead acetate, 1–3 gr. (0.06–0.2 gm.) to the ounce (30.0 c.c.). When more active astringents are required, they should be applied directly to the larynx by means of a cotton swab or camel's-hair brush. Thus, benefit may be derived from a solution of tannin, 10 gr. (0.6 gm.) to the ounce (30.0 c.c.) of glycerin, or from a solution of silver nitrate, 5–10 gr. (0.3–0.6 gm.) to the ounce (30.0 c.c.) of water. Actual abrasions may be touched with a strong solution of silver nitrate, 40–60 gr. (2.6–4.0 gm.) to the ounce (30.0 c.c.). When there is scab-formation, a 20 per cent. solution of ichthyol often proves useful. Large varicose vessels should be cauterized (chromic acid) under cocain anesthesia.

Tuberculous Laryngitis.—Hygienic, climatic, dietetic, and general medicinal treatment should be that of pulmonary tuberculosis (see p. 581). In the early stage medicated inhalations and sprays are of benefit. Such remedies as creasote, terebene, compound tincture of benzoin, and eucalyptol may be used in a respirator or inhaled from the surface of boiling water. As a rule, sprays are more satisfactory. After the mucous membrane has been thoroughly cleansed with an alkaline detergent spray, one of the following sprays may be used: Corrosive sublimate (1 : 5000 to 1 : 4000), resorcin (1–2 per cent.), formalin (1 : 1000), or menthol (1–2 per cent.).

In tuberculous ulceration the treatment may be radical or

palliative. Unfortunately, the latter is the only form of treatment that is applicable in the large majority of cases. When, however, the disease in the lung is not far advanced and the ulceration in the larynx is circumscribed and readily accessible, the thorough removal of the diseased tissue may be attempted. The most approved method of radical treatment consists in the rubbing-in of lactic acid after the ulcer has been thoroughly cureted. If the ulcer be quite superficial, the curetment may be omitted. The parts must be carefully anesthetized with cocain (20 per cent. solution), and the acid applied in the proportion of 30, 50, or 75 per cent. solution, or even stronger, according to the effects. Ichthyol (undiluted) and guaiacol (undiluted) have been used in some cases instead of lactic acid.

When the ulceration is more diffuse and is associated with considerable infiltration, better results may be obtained from one of the following applications: Carbolic acid, ½–1 dram (2.0–4.0 c.c.) to the ounce (30.0 c.c.) of glycerin; tannin, 1 dram (4.0 gm.) to the ounce (30.0 c.c.) of glycerin; formalin (1–10 per cent. in glycerin); or chromic acid (2–3 per cent.).

In the advanced stages of the disease only sedative applications to relieve pain and dysphagia are justifiable. Among the agents most generally useful as insufflations may be mentioned orthoform, iodoform, cocain, and morphin. Semon recommends iodoform (1 gr.—0.06 gm.), boric acid (1 gr.—0.06 gm.), and morphin sulphate (⅛ gr.—0.01 gm.). Fasano speaks highly of the following combination: Thiocol, 1; cocain, 2; boric acid, 10. A 5–10 per cent. solution of menthol in liquid petrolatum used as a spray often affords much relief. Intratracheal injections have also been recommended. Vacher has had excellent results from injections of 30 minims (2.0 c.c.) of the following: Guaiacol, 5 parts; eucalyptol, 2 parts; menthol, 1 part; and saturated solution of iodoform in ether, 100 parts. When dysphagia is very severe, it may be well to apply cocain to the larynx before eating, using a 2–4 per cent. solution as a spray or a 5–10 per cent. solution as a direct application. When deglutition is extremely painful, Wolfenden suggests that the patient take liquid food while lying prone, the head hanging over the edge of a couch, and the nourishment being sucked through rubber tubing from a bowl on the floor.

Syphilitic Laryngitis.—Constitutional treatment with iodids and mercurials is of the first importance. Local cleanliness should be secured by thorough spraying with some alkaline antiseptic solution. Ulcers may be touched with silver nitrate (melted on a silver probe), acid nitrate of mercury (1 to 5 parts of water), or chromic acid (1 to 8 parts of water). Insufflations of iodoform are also useful. Cicatricial stenosis may call for gradual dilatation or even tracheotomy.

ACUTE BRONCHITIS.

If the patient be weak or old, he should be confined to his room or even to bed; the atmosphere of the room should be kept warm and moist. If the patient be seen at the outset, it is useful to promote free diaphoresis, and this may be accomplished by means of a hot foot-bath, with hot drinks and a full dose of Dover's powder. Warm inhalations, such as have been suggested in acute laryngitis (see p. 569), are also of service. Counterirritation to the chest in the form of sinapisms or stupes is very beneficial. When there is much dyspnea, as there frequently is when acute bronchitis complicates emphysema, the application of dry cups, or even of a few wet cups, to the chest often has a marked salutary effect. The food should be simple and readily digestible, and the bowels should be kept regularly open by the aid of mild aperients.

Distressing dry cough is usually best controlled by one of the opium derivatives—codein or heroin—and fever by aconite with spirit of nitrous ether or by small doses of phenacetin. When for any reason opium is objectionable, hyoscyamus, spirit of chloroform, sodium bromid, or diluted hydrocyanic acid may be used instead. When the cough remains persistently dry sedative expectorants may be given to favor secretion. For this purpose potassium citrate, ipecac, apomorphin, and antimony are extensively employed. The combination of potassium citrate and ipecac given on page 265 is especially useful in this stage of the disease. In young, robust persons antimony is decidedly efficacious, and may be prescribed according to the formula suggested on page 65.

When the bronchial secretion becomes more abundant, stimulant expectorants are indicated. Ammonium chlorid is one of the most reliable members of this class; it may be prescribed in some simple vehicle, like brown mixture, or with squill, as in the formula given on page 269. Expectorants of an oily nature, such as terebene, oil of eucalyptus, oil of santal, and oil of copaiba are also very efficacious, but more liable than the ammonium salt to derange digestion. Such a combination as the following is often very serviceable when the catarrh tends to become subacute:

```
R   Terebeni,
    Olei eucalypti,              aa f ʒss (2.0 c.c.);
    Strychninæ sulphatis,        gr. ⅓ (0.02 gm.);
    Codeinæ sulphatis,           gr. ii–iij (0.13–0.2 gm).   M.
    Pone in capsulas No. xij.
    Sig.—One every four hours.
```

When the expectoration is purulent and heavy, a creasote derivative, like guaiacol carbonate, may be used with great advantage (for formula, see p. 287).

In the aged and infirm alcoholic stimulants are often required to combat general adynamia. Strychnin is a most valuable adjunct to the expectorants when there are indications that the heart is becoming strained by the violent paroxysms of cough. Should there be evidence of pronounced cardiac failure, it will be necessary to employ digitalis.

Such tonics as cod-liver oil, iodid of iron, quiniη, and arsenic are often useful during convalescence from severe and prolonged attacks. Much benefit will also be obtained from suitable change of climate.

CHRONIC BRONCHITIS.

To meet with any measure of success, treatment must be directed largely toward the prevention of recurrent attacks, and the removal, if possible, of the underlying cause. Indiscriminate routine treatment is to be rigidly avoided. Change of climate, especially in winter, is most beneficial, and should be urged if the circumstances of the patient will permit. When there is much bronchial secretion, a dry, warm climate, such as that of New Mexico or Southern California, in this country, and that of Egypt or the Riviera abroad, is generally to be recommended, whereas if there be little expectoration, a moist warm climate, such as that of Florida, the West Indies, Madeira, Pau, or Algiers, is preferable. When patients are unable to avail themselves of the benefits to be derived from a suitable climate, they should remain indoors as much as possible in bad weather, and take every precaution against exposure. Flannel should at all times be worn next to the skin, the feet should be kept perfectly dry, and the night-air should be avoided.

The diet should be simple but nutritious. In many cases alcohol in some form acts beneficially.

Underlying chronic diseases should always receive appropriate treatment. When cardiac insufficiency is present, digitalis, strychnin, or caffein may be required. When there is anemia with general malnutrition, such remedies as iron, arsenic, cod-liver oil, and hypophosphites may be given with advantage. When gout is a factor, benefit may be expected from the administration of potassium iodid and alkalis. When renal inadequacy is coexistent, the diet must be very carefully supervised, and such measures adopted as will promote the functional activity of the various emunctories.

The *direct remedies* most generally useful are the expectorants of a terebinthinate or balsamic character, such as terebene, oil of eucalyptus, myrtol, oil of santal, oil of copaiba, and oil of cubeb. Tar is another remedy of value. It may be used in substance made into pills, or in the form of tar-water or the wine of tar. When the sputum is heavy and purulent, no drug acts so well as creasote or the carbonate of guaiacol. Potassium iodid is of service in some cases, but it is difficult to state definitely the exact conditions under which it is likely to prove beneficial. It may be tried tentatively when the expectoration is very scanty and viscid, or when there is evidence of a gouty diathesis. In acute exacerbations recourse must be had to such expectorants as potassium citrate, ipecac, antimony, apomorphin, and ammonium chlorid.

When the cough is much in excess of what is required to expel the exudation, mild anodynes, like codein or heroin, may be used from time to time to keep it in subjection. Alkalis (sodium bicarbonate, aromatic spirit of ammonia), with or without a few minims of the spirit of chloroform, taken in hot water before rising, will often lessen morning cough and facilitate expectoration. Dry cough may be largely the result of habit, and if this be the case much can be accomplished by discipline.

The following formulæ will serve to illustrate the manner in which the various remedies may be combined:

R   Terebeni,
      Olei eucalypti,
      Olei santali,   aa fʒi–fʒiss (4.0–6.0 c.c.);
      Codeinæ,   gr. iij–vj (0.2–0.4 gm.).
Misce et pone in capsulas No. xxiv.
Sig.—One after each meal and at bedtime.

R   Terpini hydratis,   ʒj (4.0 gm.);
      Guaiacol carbonatis,   ʒij (8.0 gr.);
      Strychninæ sulphatis,   gr. ss (0.03 gm.);
      Codeinæ,   gr. iij (0.2 gm.).
Misce et pone in capsulas No. xxiv.
Sig.—One or two capsules three or four times a day.

R   Apomorphinæ hydrochloratis,   gr. ss (0.03 gm.);
      Syrupi pruni Virginianæ,   fʒij (60.0 c.c.);
      Syrupi picis liquidæ,   fʒiv (120.0 c.c.).   M.
Sig.—A tablespoonful thrice daily.   (Murrell.)

Inhalations are sometimes efficacious, especially when the disease appears to be situated chiefly in the trachea and larger bronchi. The volatile agent may be used in a special steam nebulizer, or preferably, if the patient must be about, in an oro-nasal respirator. The most suitable drugs for inhalation are terebene, eucalyptol, oil of Scotch fir, creasote, carbolic acid, iodin, compound tincture of benzoin, and spirit of chloroform. A number of these remedies may often be combined with advantage, as in the following formula:

<pre>
℞   Chloroformi,   f℥ss–f℥j (2.0–4.0 c.c.);
    Creosoti,
    Terebeni,
    Olei pini sylvestris,   aa f℥iss (6.0 c.c.);
    Alcoholis,   .   q. s. ad f℥j (30.0 c.c.).
Sig.—From 5 to 20 drops to be used in the inhaler several times a
    day.
</pre>

Intratracheal injections have given good results in certain instances. From 1–3 fl. dr. (4.0–11.0 c.c.) of a 1 per cent. solution of creasote or guaiacol, or of a 2 per cent. solution of menthol in olive oil may be injected between the vocal cords into the trachea once a day, a syringe with a long curved cannula being used for the purpose. The fluid should be injected in three or four separate portions, while the patient takes a slow, full inspiration.

Inhalation of compressed air by means of Waldenburg's apparatus, the pneumatic cabinet, or a pneumatic chamber like that in use at Brompton Hospital, affords considerable relief in some cases. According to Oertel, the good effects of the compressed air are to be attributed to the increased pressure within the bronchi, causing a diminution of the hyperemia, and, consequently, less exudation of serum into the bronchial walls and less pressure on the lymphatic system.

Counterirritation, preferably with iodin or small blisters, is often of great service in lessening the severity of acute exacerbations.

ASTHMA.

The cause must be sought for in every case, and removed if possible. When the source of the reflex irritation is in the nasal passages, local treatment may be followed by marked improvement, or even cure. In a certain number of cases the source of the primary irritation is in the stomach, consequently digestive disturbances should always receive careful attention. Chronic bronchitis, emphysema, and dilatation of the heart are frequent concomitants of asthma and call for special treatment.

A general toxemia, such as occurs in gout, may be responsible for the attacks, and if this be the case appropriate measures must be employed to remove or control it. Unfortunately, in the majority of cases the chief etiologic factor appears to be a peculiar hyperesthesia of the bronchial mucosa, a morbid condition which at present we have no very potent means of combating.

Although we are unable fully to meet the causal indication in many instances, nevertheless much can often be done by well-directed hygienic measures and empiric medication to lessen the frequency and severity of the paroxysms. The diet should consist of plain, readily digestible food. A light evening meal is advisable, especially when the paroxysms tend to occur at night. Vicissitudes of temperature must be carefully guarded against and flannel always worn next to the skin. Much benefit is often derived from systematic exercise in the open air and from hydrotherapy, judiciously employed.

In delicate, poorly nourished subjects the administration of iron, cod-liver oil, hypophosphites, and other tonics does much good. A change of climate, even though slight, generally proves of decided service, but the choice of locality must be determined very largely by the personal experience of the patient. Many sufferers do better in the smoky atmosphere of cities than in the country. Asthmatics with moist bronchial catarrh usually do well in a dry, warm climate, while those with dry catarrh generally derive more benefit from an atmosphere that is somewhat humid (see p. 574). Elevated regions are mostly unsuitable for old persons with emphysema, but in young adults they may prove very advantageous. The inhalation of compressed air has been highly recommended by Bertin, Sandahl, Theodore Williams, and others. If employed, it is best carried out by means of the pneumatic cabinet or the " air-bath." The inhalation of oxygen has also been advocated.

Among special remedies potassium iodid holds the first place. While it often fails entirely, it undoubtedly possesses more power in averting attacks than any other drug. To be effective, it must be given in doses of from 5–20 gr. (0.3–1.3 gm.) three times a day for long periods. Sodium or strontium iodid is sometimes better borne by the stomach than the potassium salt. Tincture of belladonna, in doses of 3–5 min. (0.2–0.3 c.c.), thrice daily, is a valuable adjunct to the iodid. Arsenic has long been esteemed as a potent remedy, and may be tried when iodids prove of no avail. Grindelia robusta (see p. 276) is sometimes useful when there is much bronchial catarrh. Strychnin is of service in cases associated with emphysema. Occasionally the prolonged administration of bromids, by allaying the nervous erethism, seems to increase the interval between the attacks. Nitroglycerin has also been highly extolled by some writers, but its action is uncertain and equivocal. VonNoorden has recently recommended the use of atropin, giving first a daily dose of $\frac{1}{120}$ gr. (0.0005 gm.) and cautiously increasing this until $\frac{1}{16}$ gr. (0.004 gm.) is given daily, and then gradually diminishing the dose. He repeats such a course of treatment every few months, using less of the drug each period.

The Attack.—The most suitable remedy for a particular case can only be determined by trial. A drug or a combination of drugs that succeeds admirably in one case may fail utterly in another. As a rule, drugs which are taken by inhalation give the most speedy relief. Some patients derive great benefit from the fumes of ignited stramonium or belladonna leaves or paper which has been impregnated with potassium nitrate. These agents may be burnt in the patient's room or smoked in a pipe or in the form of cigarets. Occasionally tobacco proves very efficacious. Marked alleviation of the paroxysms is often obtained from the inhalation of amyl nitrite (5–6 min.—0.3–0.4 c.c.) or ethyl iodid (10–20 min.—0.6–1.2 c.c.). In some cases the paroxysms yield readily to a few whiffs of chloroform, though generally the effects of the drug are only temporary.

When there is decided turgescence of the nasal mucosa, such as occurs.in the asthma of autumnal catarrh, the topical use of adrenalin solution (1 : 5000), applied on a pledget of cotton or as a spray, will be found very effective. If such measures fail to afford relief, internal remedies must be used. In some cases strong hot coffee acts most happily; in others more benefit is derived from hot whisky and water. Among the numerous special remedies which have been advocated the following appear to be the most reliable: opium, belladonna, bromids, chloral, paraldehyd, Hoffmann's anodyne, lobelia, and quebracho.

Few attacks will resist the action of morphin hypodermically with atropin, but the greatest caution must be exercised in order that the patient may not become addicted to the drug. Heroin hydrochlorid hypodermically in doses of from $\frac{1}{12}$ to $\frac{1}{10}$ gr. (0.005–0.006 gm.) may often be substituted for morphin with great advantage. When the attacks are associated with bronchial catarrh a combination like the one suggested on page 276, or the following, will be found of value:

R   Tincturæ belladonnæ,   fℨj (4.0 c.c.);
    Tincturæ lobeliæ,   fℨiij (11.0 c.c.);
    Extracti aspidospermatis fluidi, fℨss (15.0 c.c.);
    Spiritus ætheris compositi,   fℨv (18.5 c.c.);
    Strontii bromidi,   ℨijss (10.0 gm.);
    Elixiris aromatici,   q. s. ad fℨiv (120.0 c.c.).   M.
        Sig.—A dessertspoonful in water every two or three hours.

Among other measures that have been found useful in alleviating asthmatic attacks may be mentioned the application of sinapisms to the chest, the inhalation of compressed air, and the inhalation of oxygen.

EMPHYSEMA.

The treatment of emphysema is chiefly that of the accompanying disease. The various means suggested for the cure or relief of chronic bronchitis (see p. 574) may often be used here with great advantage. When asthma is the primary disease, treatment calculated to lessen the frequency and severity of the paroxysms should be instituted (see p. 576).

When fully established, emphysema is incurable, although it is susceptible to alleviation. Violent exercises and over-exertion of all kinds must be proscribed. A diet that is light but sustaining is indicated. All foods likely to induce flatulence should be avoided. Much benefit is often derived from a change of climate, the choice of locality, however, depending somewhat upon the character of the complicating bronchitis (see p. 574). As a rule, high elevations are to be avoided.

The inhalation of compressed air by means of the pneumatic cabinet or the " air-bath " is usually followed by a marked improvement in the symptoms. The inspiration of compressed air with expiration into rarefied air has been especially advocated. Strümpell speaks favorably of rhythmic compression of the lower portion of the thorax during expiration, as recommended by Gerhardt. This should be done systematically by another person two or three times a day during fifty or sixty respirations.

General tonics, like iron and cod-liver oil, are required in some cases to improve the general health. Strychnin, being both a respiratory and a cardiac stimulant as well as a general tonic, is particularly useful. Digitalis will be found a serviceable remedy when signs of cardiac insufficiency appear. Ammonium carbonate, in conjunction with strychnin and digitalis, usually has an excellent effect upon acute exacerbations of the bronchial catarrh. At such time sinapisms applied to the chest may also afford considerable relief. When sleep is disturbed by troublesome cough and oppressive dyspnea no drug is so generally useful as heroin.   When the breathing is very difficult, the face suffused and livid, and the pulmonary circulation much impeded, recourse should be had to bloodletting, either local or general, according to the urgency of the symptoms.

CATARRHAL PNEUMONIA.

Hygienic treatment is of the utmost importance.   The atmosphere of the sick-room should be kept at an even temperature of 68°–70° F. (20°–21° C.), and should be rendered moist with steam.   The diet should be liquid and nutritious.   Milk, light broths, gruels, egg-white, and junket are suitable forms of nourishment.   Alcoholic stimulants are often required.   When the circulatory depression is pronounced, whisky or brandy may be given in doses of from 10 to 30 min. (0.6–2.0 c.c.) in milk to a child of two years every two or three hours.

At the outset it is often advantageous to administer a mild purgative, preferably calomel or castor oil.   In the absence of any special indication for local treatment it will only be necessary to provide ample protection for the chest.   This may be done satisfactorily by means of the jacket of cotton-wool. When there is harsh, dry cough, mild rubefaction produced by the application of the tincture of iodin of suitable strength generally affords some relief.   In adults sinapisms or stupes may be used instead of the iodin.

Fever is best controlled by outward applications.   Compresses wrung out of cold water may be wrapped around the chest and changed for fresh ones at intervals of from twenty to thirty minutes.   This treatment not only serves to lower temperature, but it also affects very favorably the pulse and respiration, and conduces to sleep.   When the fever is high recourse should be had to cold packs or cold baths (85°–80° F.—29.5°–26.5° C.).

Expectorants are almost invariably required.   In the early stage, when the cough is harassing and the sputum viscid, potassium citrate is very serviceable.   It may often be combined advantageously with spirit of nitrous ether and solution of ammonium acetate, as in the following formula:

R   Potassii citratis,   ℥iss (6.0 gm.) ;
     Spiritus ætheris nitrosi,   fʒvj (22.5 c.c.) ;
     Liquoris ammonii acetatis,   fℨj (30.0 c.c.) ;
     Syrupi tolutani,
     Aquæ,   āā q. s. ad fℨiv (120.0 c.c.).   M.
     Sig.—Dessertspoonful every three hours for a child of three years.

Later, the ammonium salts, especially the carbonate, are more efficacious. From 1 to 2 gr. (0.06–0.13 gm.) of the latter may be given every three or four hours to a child of two years. Ammonium iodid is also useful in facilitating expectoration, and may be employed as an adjuvant, as in the following formula:

R   Ammonii carbonatis,   gr. xlviij (3.1 gm.);
     Ammonii iodidi,   gr. xxiv (1.5 gm.);
     Syrupi tolutani,
     Syrupi acaciæ,   aa q. s. ad fℨiij (90.0 c.c.).   M.
          Sig.—Teaspoonful every two or three hours for a child of three
          years.

In adults the author has found creasote carbonate to be a reliable expectorant. When the bronchial secretion is very abundant belladonna is a valuable remedy. To a child of two years a min. (0.06 c.c.) of the tincture may be given every hour or two until the desired effect is produced. When the child is unable to expel the accumulated mucus and the breathing becomes much oppressed, an emetic (ipecac, alum, or zinc sulphate) will generally prove of great service. Inhalations of oxygen sometimes lessen cyanosis and make the breathing easier. Strychnin is also of benefit at this time in combating respiratory failure. When these measures fail to afford relief alternate douches of hot and cold water should always be tried, since by provoking violent respiratory efforts they may serve to dislodge the retained secretion and to bring about a thorough inflation of the lungs.

If the symptoms of cardiac failure are especially pronounced, digitalis must be given in addition to alcohol and strychnin. Extreme restlessness and insomnia will sometimes require the use of the bromids or some other mild sedative.

Except at the onset, when their administration may be necessary to relieve pleuritic pain and to control harassing cough, opiates should not be used.

PULMONARY TUBERCULOSIS.

Prophylaxis.—Efforts intended to prevent the extension of tuberculosis should be made in two directions, that of checking the dissemination of the bacillus, and that of rendering the tissues of the individual less susceptible to infection.

The Bacillus.—As the sputum of consumptive patients is the chief source of the bacillus, the disposition of this secretion becomes of the utmost importance. Such patients should be impressed with the danger, both to themselves and others, of indiscriminate spitting, and should be taught never to expectorate except in a proper receptacle. To be suitable, the spittoon should be provided with a closely fitting cover, and should be of such construction that it can be readily cleansed and disinfected. It should contain at all times some disinfectant fluid, such as a 5 per cent. solution of carbolic acid, and should be thoroughly scalded with boiling water at least twice a day. A very good cuspidor, and one that can be burned when filled, is made of impervious cardboard. Feeble bed-ridden patients should expectorate in moist rags or paper napkins, which should be burned before they have become dry. It need scarcely be said that the sputum should never be swallowed.

Consumptive patients should always sleep alone. Their rooms should be freely accessible to the sunlight, well ventilated, and kept scrupulously clean. Stuffed furniture, carpets, and woolen hangings are undesirable. All articles used by the patient and liable to be contaminated with sputum should be disinfected by means of scalding water.

Rooms which have been vacated by consumptives should not be occupied again until they have been thoroughly disinfected (see p. 383).

Phthisical patients should refrain from kissing and other modes of salutation necessitating very intimate association. Tuberculous mothers should not suckle their children. The marriage of consumptives should be discouraged.

Through the enactment and rigid enforcement of special laws much can be done by the State to limit the dissemination of tuberculosis. Laws should be enacted providing for the systematic inspection by skilled veterinarians of all dairies and slaughter-houses with the view of declaring unmarketable the milk and meat of tuberculous animals.

Compulsory registration of phthisical patients, although it entails great hardships on individuals and their families, is desirable. Flick is probably right in recommending that consumptives in the infectious stage of the disease should be retired from occupations in which they can infect others. Spitting upon sidewalks and the floors of public buildings and conveyances should be made a penal offence. Finally, the State should provide special hospitals for the indigent suffering from tuberculosis.

The Individual.—Persons with a marked predisposition to tuberculosis, whether hereditary or acquired, can do much to increase their power of resistance by strict attention to hygiene. Fresh air and sunlight, a healthy residence, an outdoor occupation, the wearing of warm clothes, with flannel next to the skin, a diet of wholesome and nutritious food, temperate living, systematic exercise, and daily cold sponging, followed by friction of the skin, are the factors to be relied upon in attempting to overcome individual susceptibility.

Persons recovering from catarrhal pneumonia, pleurisy, measles, whooping-cough, influenza, and other diseases which predispose to tuberculosis should be treated with the utmost care. As enlarged tonsils, adenoid growths, and other obstructions in the upper air-passages, by interfering with free respiration, increase the risk of infection, they should be removed.

Finally, all local foci of tuberculosis, such as frequently appear in the cervical lymph-glands, joints, and bones, should receive immediate attention.

Sanatorium Treatment.—The value of pure air in the treatment of pulmonary tuberculosis is universally admitted. By adding to pure air an abundance of nourishment in a readily digestible form, rest alternated with graduated exercise, and constant medical supervision we supply our phthisical patient with the best means which medical science has at present to offer of securing restoration to health, or, if this be impossible, of obtaining alleviation of his symptoms. Indeed, if these measures are beyond the reach of the invalid all others are likely to prove unavailing. Nowhere are all the means just enumerated so available as in a well-managed modern sanatorium. Such an institution may be located in almost any climate, even within a few miles of a large city, the only requisites being moderate elevation, well-drained soil, abundant sunshine, and protection from cold winds.

Open Air.—In summer not less than nine or ten hours, and in winter not less than six or seven hours, are ·spent in the open air. In Falkenstein the patients remain out of doors in their chairs from seven to ten hours a day all the year round, in spite of fog, rain, wind, or snow, and even with the thermometer at 12° C. below zero, and often no sunshine (Knopf). Of course, they are sheltered from the wind and rain, and are well covered with blankets or fur robes. The bed-room windows are kept open both winter and summer. As a rule, patients soon accustom themselves to live in a low temperature without discomfort.

Nourishment.—When there is no gastric derangement the patient is given an ordinary diet of wholesome food, and encouraged to eat as heartily as his digestive capacity will permit. Milk, eggs, beef, mutton, lamb, fish, fowl, fresh vegetables, cereals, and fruits are considered suitable forms of nourishment. As a rule, the meals are given more frequently than in health. Thus, before rising the patient takes hot milk, cocoa, or gruel; at breakfast—beef-steak, chops, eggs, or fish, bread and butter, and coffee or milk; at the mid-day dinner—soups, fish, meat, vegetables, salad, fruit, and wine; at 4 o'clock—milk with bread and butter, milk with raw egg, or bouillon with toast; at supper—cold meat, bread and butter, cocoa, fresh or preserved fruit; at bed-time—a glass of hot milk or an egg-nog.

In some cases the anorexia is so pronounced that it becomes necessary to resort to forced feeding, but generally the patient can be persuaded to eat, especially if the food be presented to him in an appetizing form. Under the influence of fresh air the appetite often returns with remarkable rapidity.

Rest and Exercise.—When the disease is active the patient is kept at absolute rest. For most of the day he lies on a bamboo couch or a reclining chair in the open air, warmth being maintained by abundant covering and, if necessary, by a hot stone placed at the feet. In the incipient stage and in quiescent tuberculosis moderate exercise is generally recommended, every precaution being taken, however, to guard against fatigue.

Medical Supervision.—This is an important item of the sanatorium treatment. Everything that the patient does is reviewed by the attending physician, and the treatment is reduced to an organized system. "Success depends upon the ability of the doctor to discriminate and to shape his treatment according to the physical and mental wants of each individual case; the patient must not be made to feel that he is merely an inmate of an institution the régime and regulations of which he has rigidly to observe, but rather, that he is a well-liked guest, who, while conforming to certain rules of the house for the benefit of his health, is at liberty otherwise to enjoy life in his own fashion" (Muthu).

The results obtained under the sanatorium treatment fully justify its employment; it undoubtedly gives the patient in the early stage of the disease the best chance of complete recovery. The shortest period of treatment which will benefit a patient is rarely less than three months, and to secure a lasting improvement he should remain at least six months.

Climatic Treatment.—Since the open-air treatment has been shown to be about equally successful in nearly all countries, the selection of a special locality has become a matter of secondary importance. Still, there are some patients, with ample means at their disposal, to whom a protracted stay in a sanatorium would become exceedingly irksome and distasteful. To such patients prolonged residence in a favorable climate offers the greatest hope of cure. The requisites of a suitable climate are purity of atmosphere, equability of temperature, and abundance of sunshine. The age of the patient, the extent and type of the disease, and the condition of the other organs must be carefully considered in deciding the question of altitude, of temperature, and of humidity. Other matters which should not be overlooked in choosing a locality are wholesome food in abundance and well-prepared, good accommodations, and available medical advice.

Young subjects with considerable constitutional vigor, who have but a small area of lung involved, generally do well in high altitudes, such as are found in Colorado, Wyoming, and Montana, and in Switzerland (Davos and St. Moritz). The presence of cardiac disease, albuminuria, or emphysema is to be regarded as a contraindication to high elevations.

The moderate elevation of the Adirondack Mountains and of the mountains of North Carolina is well adapted to many patients who have but a limited area of pneumonic consolidation, and who cannot afford to remain away from home longer than a few months.

Phthisical patients who have much tendency to bronchial catarrh, and who manifest great constitutional irritability, generally do better in warm climates with little elevation, such as Southern California, Southern Arizona, and Florida in this country, and the Riviera and Egypt abroad.

In some cases an extended sea-voyage is very useful. According to Douglas Powell, it is most suitable to patients in the early stages, who have been previously healthy, who have overworked nervous systems, and in whom the disease is more or less quiescent.

Patients in the advanced stages of the disease should not be sent far from home.

Treatment at Home.—The large majority of patients are unable to avail themselves of the advantages afforded by a stay in a sanatorium or by residence in a salubrious climate. These may be consoled by the fact that not a few cases of phthisis do well at home when the conditions are not too unfavorable. Treatment at home should be made to imitate as closely as circumstances will permit that which is followed in the sanatorium. An earnest attempt should be made to secure the freest possible supply of air and of sunlight. The airiest and sunniest room in the house should be selected for the patient. *So long as he has fever absolute rest should be insisted upon.* During the day, if the weather be clement, he should rest in a reclining chair in the open air for from six to ten hours, according to the season, and at night sleep with the windows open. There is no danger, even in the presence of fever and sweating, in breathing cold air, providing the body itself is kept warm. As much nourishing food should be supplied as the digestive capacity of the patient will allow.

Artificial Aerotherapeutics.—Inhalations of *rarefied air* by means of the pneumatic cabinet have been found very beneficial by Platt, Quimby, Knopf, and others. A reduction of the atmospheric pressure about half a pound to the square inch is sufficient. According to Knopf, the séances should be given at first every day, and their duration should be gradually increased from two minutes to six or eight minutes.

C. Theodore Williams, who has had a large experience in the treatment of phthisis by means of *compressed air*, concludes that beyond lessening slightly the cough and expectoration, the baths are of no benefit. On the contrary, in several of the cases hemoptysis came on either in the bath or shortly after the treatment.

Medicinal Treatment.—Of the many attempts made to discover a specific for tuberculosis that made by Koch, which led to the introduction of the *tuberculins* (see p. 427), was the most promising. After a thorough trial, however, these preparations have been almost entirely abandoned except as an aid to diagnosis. The only agent proposed as a specific which has continued to enjoy the confidence of the profession is creasote, and even this drug is now believed to owe its efficacy to its power of modifying the expectoration rather than its influence as a germicide. As we are yet without a remedy having a direct action upon the tuberculous process, the chief indications in the medicinal treatment of phthisis are to increase the resisting power of the tissues, to reduce the local irritation, and to relieve the distressing symptoms.

In the last fifty years no remedy has been so universally esteemed in the treatment of phthisis as *cod-liver oil.* When well tolerated and digested it is undoubtedly of service in improving the general nutrition. It is to be regarded, however, as a food rather than as a medicine. If it occasion nausea, impair the appetite, or derange the digestion in any way, it should be withdrawn. The best methods of administering cod-liver oil have been considered on p. 297.

Olive oil, cream, and butter may be used as substitutes for cod-liver oil when the latter is not well borne.

Arsenic is another remedy which has many warm advocates. To be effective it must be given in small doses for several months. Jacobi speaks highly of arsenic in conjunction with digitalis in tuberculosis of childhood.

Alcohol is useful in some cases, but not in all. Each case must be carefully considered by itself. The danger of inducing the alcoholic habit must also be borne in mind. Malt liquors and wines are usually the best preparations when digestion is good, but when the digestive power is feeble whisky or brandy, well diluted, is preferable. Alcohol is best given with the food.

Various drugs which have tonic properties and tend to improve the general nutrition are of service in phthisis. Of these the most valuable are the *hypophosphites, strychnin,* and other *vegetable bitters.*

Preparations of iodin seem to be useful, especially in the early stages of the disease. A good method of administering iodin without deranging digestion is that of Flick, which is to rub into the chest twice daily an ointment (10 per cent.) of europhen, a compound containing a large percentage of loosely combined iodin.

Creasote (see p. 277).—While the view is no longer tenable that creasote exerts a specific influence on the tubercle bacilli, the testimony of numerous observers is convincing that the drug is of value in allaying cough and in lessening expectoration. It is especially useful when the sputum is very abundant and purulent. Creasote carbonate and guaiacol carbonate (see p. 279) do not have the disagreeable odor and taste of creasote itself, and appear to be equally efficacious.

Counterirritation.—This is a method of treatment less frequently employed than its merits justify. Counterirritation by iodin or, better, by small "flying" blisters, has seemed to the author to be of the greatest use, especially in chronic pulmonary tuberculosis.

Treatment of Special Symptoms.—*Digestive Disturbances.*—In a large number of phthisical cases the first indication is to correct disordered digestion. No medicines likely to irritate the stomach should be ordered. Only the most bland and readily digestible forms of nourishment should be allowed. The time of eating, as well as the character of the food, should be carefully revised. No special rules can be laid down for treatment. The various measures and drugs which are serviceable in the ordinary forms of indigestion (see p. 544) are applicable here. The most potential measures are fresh air and rest. Acute indigestion brought on by overeating is often promptly relieved by a mercurial purge followed by a saline. When anorexia and slow digestion are the chief features, an alkali with a vegetable bitter before meals (see p. 168) may have a very happy effect.

Vomiting that is dependent upon extreme irritability of the stomach is often relieved by such sedatives as bismuth subnitrate and hydrocyanic acid (see p. 69), taken before meals. Emesis excited by cough sometimes subsides when the patient is kept in bed and put on a liquid diet. Chloroform-water taken a few minutes after the meals is useful. Counterirritation over the diseased area in the lung is very often successful. Vomiting occurring shortly after a meal and preceded with a sense of suffocation, with or without coughing, is best controlled by the administration of from 5–10 minims (0.3–0.6 c.c.) of liquor potassæ (Habershon). Vomiting caused by ulceration of the epiglottis or larynx will call for local anesthetics—cocain, iodoform, orthoform, and morphin.

Cough.—In many cases of phthisis cough is indispensable, and is best treated by promoting expectoration. For this purpose creasote, guaiacol carbonate, terebene, oil of eucalyptus, myrtol, and santal oil are reliable remedies. They may often be combined advantageously with strychnin. Expectorants in the form of syrups should be avoided. Morning paroxysms of cough are often made more easy by the administration of a glass of hot water before rising, to which aromatic spirits of ammonia has been added. Intratracheal injections (see p. 576) are also useful in modifying bronchial secretion and promoting expectoration.

Hard, dry cough calls for special treatment. When such cough results from habit, as is sometimes the case, it can be suppressed by discipline. Absolute rest is a valuable aid. Inhalations of creasote, chloroform, and volatile oils (see p. 576) are often very effective. Inhalations of formalin have been highly recommended by Murrell, Muthu, Huggard, and others. The strength of the solution should at first not exceed 2 or 3 per cent. of ordinary formalin in alcohol. From 5–10 minims (0.3–0.6 c.c.) of this solution should be used from time to time in an inhaler placed over the mouth only, the entire time of the inhalation being from three to six hours daily. The strength of the solution may be gradually increased to 6 per cent., or even to 10 per cent. When the cough is very irritable chloroform may be added to the inhalation.

In many cases of chronic phthisis no measure proves so

successful in relieving cough as blistering. Periodic fits of cough at night call for warm, demulcent, nourishing liquids, such as milk or beef-tea, with a little brandy, if there be much exhaustion (J. Mitchell Bruce).

When the cough is very violent and occasions exhaustion it may be necessary to resort to internal sedatives. Of these, those least objectionable are codein, heroin, hydrocyanic acid, and spirit of chloroform. The combination mentioned on p. 69 or one like the following will be found useful:

> ℞   Codeinæ sulphatis,   gr. vi–viij (0.4–0.5 gm.);
>   Spiritus chloroformi,   f℥j (4.0 c.c.);
>   Glycerini,   f℥j (30.0 c.c.);
>   Succi limonis,   f℥ss (15.0 c.c.);
>   Aquæ,   q. s. ad   f℥iij (90.0 c.c.).—M.
>   Sig.—A teaspoonful as occasion demands.

Hemoptysis.—Absolute rest is essential. An ice-bag may be placed over the suspected seat of the hemorrhage, but it should be removed at once if it aggravate the coughing. Bits of cracked ice may be given to the patient to suck. There is no more useful internal remedy than morphin, which serves to allay excitement and to check cough. It should be given hypodermically in full doses. The application of firm ligatures to the limbs, in order to lessen the volume of blood returning to the lungs, may prove efficacious.

When the hemorrhage is protracted a saline purge is sometimes useful. It probably acts by lowering the blood-pressure. Among other remedies which seem to be of service may be mentioned oil of erigeron, fluid extract of hamamelis, and gelatin. Ergot cannot be recommended (see p. 374). Inhalations of vaporized solutions of astringent drugs, like alum and ferric subsulphate, are useless.

The assurance that pulmonary hemorrhage in phthisis is not necessarily a dangerous symptom or even a drawback to recovery has a most salutary effect on the patient. The prostration appearing after a first attack of hemoptysis is often, as Flint observed, a moral effect rather than the exhaustion caused by the loss of blood.

Night-sweats.—Sponging the body before bedtime with vinegar and water, alcohol and water, or formalin and alcohol often checks excessive sweating. Dusting the surface with a powder of tannoform (1 part) and zinc oxid (3 parts) is also effective. A glass of cold milk with brandy just before retiring sometimes affords relief. The most reliable internal remedies are atropin, picrotoxin, agaracin, sulphuric acid, and camphoric acid (see p. 256).

Pyrexia.—In many cases the pyrexia yields to absolute rest in bed or in a reclining chair, combined with life in the open air. Cold sponging is to be recommended when the temperature is high. In obstinate cases the administration of 2 or 3 grains (0.13–0.2 gm.) of phenacetin about two hours before the temperature is expected to rise may be tried. The high irregular fever resulting from secondary infection by pyogenic cocci is sometimes favorably influenced by creasote or one of its derivatives.

Pleurisy.—Mild attacks of pleuritic pain generally yield to sinapisms or the application of iodin. Strapping the affected side also affords relief. Severe pains should be treated by the application of a blister and the subcutaneous administration of morphin. The treatment of tuberculous pleurisy with effusion is much the same as that of primary pleurisy with effusion. Thoracentesis, however, is inadvisable when the effusion is only moderate and the patient is very feeble.

Diarrhea.—Diarrhea the result of indigestion usually yields promptly to restriction of the diet, rest, and the administration of a mild mercurial or saline aperient. When intestinal tuberculosis is present diarrhea becomes a most troublesome complication. Large doses of bismuth subnitrate, combined with opium in some form, may afford temporary relief. Combinations of tannigen or of tannalbin with intestinal antiseptics like betanaphtol-bismuth, salol, or thymol, sometimes prove useful. Pills of lead acetate and opium or of copper sulphate and opium occasionally succeed.

PLEURISY.

The first indications in the treatment of acute pleurisy are to relieve the pain and to check the progress of the inflammation. For the first, the application of a blister or of wet or dry cups, together with the hypodermic injection of morphin, will be found effective. Strapping the affected side from mid-spine to mid-sternum with broad strips of adhesive plaster, as originally suggested by Frederick T. Roberts, is also very useful. Cold applications are not often tolerated.

Unfortunately, we have no very potent means of checking the inflammatory process in the pleura. The patient should be kept in bed and restricted to a liquid diet. A mercurial or saline aperient may be prescribed at the onset. Acute sthenic cases, accompanied by decided fever, are often favorably influenced by the administration of salicylates, a method of treatment which has been warmly advocated by Aufrecht, Fiedler, Dock, and others. From 1–1½ drams (4.0–6.0 gm.)

of sodium or ammonium salicylate should be given in the twenty-four hours, the dose being gradually reduced as the good effects become manifest. In asthenic, protracted cases of pleurisy the salicylates are of no avail.

After the acute symptoms have subsided the indications are to accomplish the removal of the fluid, to maintain nutrition, and to secure complete expansion of the lung.

Removal of Serous Effusion.—Counterirritation by means of iodin or flying blisters appears to be useful in promoting absorption. In vigorous subjects the administration of saline purgatives according to the method suggested by Matthew Hay is sometimes followed by improvement. The object of this treatment is to deplete the blood of its serum, and thereby favor the abstraction of fluid from the lymph-spaces. The quantity of fluid consumed by the patient is restricted to a minimum, and every morning or every other morning from $\frac{1}{2}$–1 oz. (15.0–30.0 gm.) of magnesium sulphate is given in as little water as possible an hour before breakfast. This produces copious watery discharges.

Diuretics (digitalis, caffein, potassium acetate) and diaphoretics (hot-air baths, pilocarpin) are sometimes of service, but their action is much less certain than that of the purgatives. Potassium iodid has also been extensively employed as an absorbent, but it is of doubtful efficacy.

While the effusion eventually disappears in a large proportion of cases under medicinal treatment alone, much saving of time is often effected by an early resort to paracentesis. In the author's opinion this operation should not be delayed ordinarily longer than ten days or two weeks when the effusion is considerable and shows no signs of receding. The presence of fever is not a contraindication; indeed, the temperature often falls upon the removal of the fluid.

Irrespective of the period of the disease paracentesis is demanded: (1) When there is sufficient fluid to excite severe dyspnea, cyanosis, persistent cough, or failing pulse; (2) When the fluid reaches the level of the second rib, and there is marked dislocation of the neighboring organs. The suspected presence of pus is always to be regarded as sufficient ground for operative interference.

Paracentesis.—The patient should be brought to the edge of the bed, placed in a semirecumbent position with the thorax inclined slightly toward the healthy side, and supported by an assistant. The most favorable site of the puncture is usually the sixth or seventh intercostal space between the mid-axillary line and the angle of the scapula. The operation should be done under strict antiseptic precautions. The needle should be inserted with a quick thrust along the upper margin of the rib, the depth of the puncture being gauged by the forefinger. Local anesthesia is not necessary, but if desired it may be secured by means of a spray of ethyl chlorid or the application of a block of ice which has been dipped in salt. Having introduced the canula, the operator should satisfy himself that it is freely movable, should hold it in position throughout the operation, and as the evacuation proceeds should slowly raise the exposed end so as to keep the inner opening below the level of the fluid in the pleural sac. The aspiration should be effected slowly, and at intervals it should be stopped by compressing the conducting tube. Too rapid evacuation may excite engorgement of the lung and edema.

The amount of fluid which should be removed depends somewhat upon the size of the effusion and the ease with which it can be evacuated. Even with large effusions it is rarely necessary to withdraw more than a quart (1.0 L.). The removal of small quantities is in many cases followed by the rapid absorption of the remainder. Under no circumstances should extreme efforts be made to obtain the largest possible amount of fluid. The operation should be terminated at once if incessant cough, severe pain, dyspnea, palpitation, tendency to syncope, or other untoward symptoms appear.

When the requisite amount of fluid has been evacuated, the needle should be withdrawn quickly from the chest, and the puncture closed with adhesive plaster.

If the exudate reaccumulate, aspiration may be repeated after the lapse of a week or ten days. Free incision of the thoracic wall with thorough drainage has given excellent results in cases in which the fluid has reaccumulated after repeated tappings.

Occasionally attempts at aspiration are unsuccessful. The cause of failure may be plugging of the canula, great thickening of the pleura, or encapsulation of the effusion. Under these circumstances it may be necessary to make repeated trials before a flow can be established. The aspiration of pleural exudates is rarely attended by accidents of any kind. Sudden death, the result of cerebral anemia, has been reported. Such an accident is not likely to occur if the evacuation be effected slowly and arrested immediately on the first appearance of any untoward symptom. Another grave and even fatal complication of thoracocentesis, but also extremely rare, is a peculiar form of pulmonary edema, which is manifested by cough, intense dyspnea, and profuse albuminous expectoration. According to Riesman, who has collected 32 cases from the literature, the principal cause of this condition seems to be either too rapid or too great a withdrawal of fluid.

To Maintain Nutrition and Secure Complete Expansion of the Lung.—Apart from the restriction of liquids the diet during the stage of resorption should be abundant and nutritious. Tonics like iron, cod-liver oil, and arsenic are very useful. After the exudate has entirely disappeared the patient should be urged to take such exercise as will bring into play the muscles of respiration. Open-air living and change of climate are also to be recommended, especially when there is a decided phthisical tendency.

Empyema.—The treatment of empyema is surgical. The indications are to evacuate the pus and to secure thorough drainage. Simple puncture cannot be recommended, although in children repeated tappings sometimes result in cure. In the large majority of cases treatment by free incision and the introduction of a large-sized drainage-tube proves successful. Irrigation of the pleural sac should not be practised unless the exudate is putrid.

Excision of one or more ribs (Estlander's operation) or excision of the ribs and intercostal tissues (Schede's operation) may be necessary when the empyema is of long standing or occurs in elderly persons with rigid thoracic walls. Fowler recommends, in chronic empyema, decortication of the lung in preference to either Estlander's or Schede's operation, claiming that it combines the advantages of restoration of the function of the lung, so far as this is possible, with closure of the empyemic cavity.

DISEASES OF THE CIRCULATORY SYSTEM.

PERICARDITIS.

One of the chief indications in the treatment of acute pericarditis is to allay the excited action of the heart. Absolute rest of body and mind is imperative. During the first week, at least, milk is the most suitable diet. Among local applications none is so generally useful in subduing palpitation as the ice-bag. Local blood-letting is of great service when the disease occurs in a fairly robust subject and is attended with much pain. Blistering also serves to allay pain, but it does not seem to exert any influence on the progress of the disease. Internally, morphin is often required to control restlessness and to relieve pain.

Drugs which stimulate the cardiac inhibitory mechanism are of doubtful utility, although digitalis may be given when acceleration of the pulse is accompanied by decided lowering of the arterial tension. If heart failure occur such stimulants as alcohol, strychnin, caffein, and camphor must be employed.

Pericardial Effusion.—When the exudate is serofibrinous, is moderate in amount, and causes no marked subjective symptoms, an expectant plan of treatment may be pursued with the assurance that resolution will follow in due time. When, however, the effusion is extensive and causes some embarrassment of the heart's action, efforts should be made which will tend to hasten absorption. Of the value of vesicants in this stage there can be no reasonable doubt. If the patient's strength will permit, saline purges may be administered in the manner suggested by Matthew Hay (see p. 591). Diuretics, such as digitalis and caffein, may be tried. Potassium iodid is recommended, but it is of doubtful efficacy. Diaphoretics, particularly pilocarpin, should not be used.

In cases in which the symptoms become urgent, or in which absorption does not occur after a thorough trial of the remedies just mentioned, operative interference becomes necessary. In the case of serofibrinous effusions, aspiration is the operation of choice, although it yields far less favorable results than it does in pleurisy. The most suitable site for paracentesis is ordinarily in the left fifth intercostal space, about an inch or an inch and a half from the edge of the sternum. According to Osler, in large effusions the pericardium can also be readily reached without danger by thrusting the needle upward and backward close to the costal margin in the left costoxiphoid angle. In a number of instances, owing to an error in diagnosis, the needle has pierced the right ventricle. Fortunately, this accident has rarely produced serious consequences, although Broadbent recalls two cases in which it resulted fatally through hemorrhage.

In pericarditis with purulent effusion the indications are to incise the sac and to afford the freest possible outlet for the pus. Three methods of approaching the pericardium have been practised: By incision through an intercostal space, by trephining the sternum, and by epigastric incision. C. B. Porter excises the cartilage of the fifth rib, ligates the internal mammary vessels, and, after introducing a needle to corroborate the diagnosis, opens the pericardium obliquely downward and outward close to the border of the sternum. He then stitches the edges of the pericardium to the skin, irrigates, and introduces two rubber drainage-tubes. The advantages claimed for this operation are that it avoids opening the pleura and secures permanent free drainage.

The mortality of incision for pyopericardium in 47 cases was about 60 per cent.

ACUTE ENDOCARDITIS.

The treatment of acute endocarditis is mainly that of the causal condition. Too much emphasis cannot be given to the need of frequent examination of the patient's heart during the course of all acute infections, in particular of acute articular rheumatism, even of the mildest type (see page 527). The special indication is to secure complete resolution of the inflammatory process or, if this be impossible, to favor the establishment of the highest attainable degree of compensation. *Prolonged and complete rest* is the most important factor of the treatment. The patient should be confined to bed not only during the attack, but for several weeks after it has subsided, in order to allow sufficient time for the damage to be repaired or for compensatory hypertrophy to be thoroughly established. In the period of convalescence extraordinary efforts are often required to prevent those indiscretions which are so liable to cause a fresh invasion of the valves or to strain the already overburdened heart. Mental and emotional excitement must also be avoided as much as possible, since they both tend to increase the force and frequency of the pulse. The diet should be liquid and unstimulating. For a week or two milk is the most appropriate food ; later milk-toast, eggs, thin gruels, light broths, and steamed rice are admissible.

Externally, an ice-bag is often useful in allaying excitement of the heart. Internally, at the outset and from time to time throughout the illness, mild mercurial or saline aperients may be used for their depurative effect. Digitalis may be of service when the pulse is weak and irregular, but in the large majority of cases this remedy is not indicated. Heart-failure is to be combated by such stimulants as alcohol, ammonia, strychnin, and caffein. Morphin is sometimes required to relieve severe pain, allay nervous perturbation, and to promote sleep. To these measures is to be added, of course, the treatment appropriate to the disease upon which the endocarditis has supervened.

Treatment by continuous vesication and the prolonged administration of potassium iodid has seemed to most practitioners of large experience to be of doubtful utility, although it has been ardently advocated by Pepper in this country, André

Petit in France, Rosenstein in Germany, and Walsh and Caton in England.

Caton,[1] especially, has presented some very strong evidence in favor of these measures as auxiliaries to prolonged rest in bed. He reports that of 39 cases of acute endocarditis treated by this method, 29 left the hospital with normal heart-sounds and 10 with a bruit, and that of 13 treated expectantly 12 left the hospital with a bruit. Many of the cases were kept under observation subsequently for a long period, and every care was taken to make the observations accurate.

CHRONIC ENDOCARDITIS.

Period of Compensation.—Treatment during this stage must be chiefly hygienic. No specific medication should be advised unless there be a definite indication for its employment. Our aim should be to maintain the heart-muscle in the highest possible state of efficiency by carefully regulating the habits of life, preventing excessive cardiac strains, and protecting the patient against intercurrent diseases likely to induce a fresh attack of endocarditis or to weaken the myocardium. In many cases it is well to inform the patient of his valvular defect, without unnecessarily alarming him, in order to secure his thorough coöperation.

Violent exertion, mental fatigue, intense emotional excitement, intemperance in eating and drinking, and venereal excesses are to be rigidly avoided. As a protection against rheumatism and bronchitis the patient should habitually wear flannel next to the skin, and should avoid unnecessary exposure to cold and wet. In the absence of any digestive disturbances, a varied diet of plain wholesome food is suitable. The meals should be taken at regular intervals, and followed by a rest of from twenty to thirty minutes. In elderly persons and those inclined to obesity a spare diet is generally advisable. Tea, coffee, alcohol, and tobacco should be used very moderately, if at all. The bowels should be made to move daily, a mild aperient being given if necessary, and the skin should be kept active by frequent tepid baths.

Exercise judiciously employed is often of the greatest benefit in prolonging the period of compensation. It is very important, however, that the muscular efforts should be carefully graduated and adapted to the working capacity of the heart. The character and duration of the exercises must be determined experimentally in each individual case, the subjective

---

[1] *Liverpool Med.-Chirurg. Jour.,* July, 1895.

symptoms of the patient being taken as a guide. All physical effort should stop short of producing palpitation, pronounced dyspnea, or fatigue. In persons who have just recovered from acute endocarditis the exercise should be confined to massage and passive movements, as originally practised in Sweden by Ling. In cases in which compensation has been completely established, but is maintained with some difficulty, exercises consisting of movements made against slight resistance, as carried out by Schott, of Nauheim, may be recommended (see p. 480). Vigorous patients may derive some benefit from the method of treatment urged by Oertel, which consists in graduated climbing exercises taken in the open air at a moderate elevation.

It is not often necessary to recommend a change of occupation unless the patient be engaged in a pursuit requiring a great expenditure of energy. Children with chronic valvular disease in whom compensation has been perfectly established need not be debarred from outdoor games which involve only moderate exertion. Thus, baseball, cricket, golf, and lawn tennis are usually permissible. On the other hand, amusements which require sudden or prolonged effort, such as racing, swimming, football, and wrestling, had better be avoided.

General tonics, like iron, arsenic, and cod-liver oil, may be used with advantage when there is distinct anemia or impaired nutrition.

Period of Broken Compensation.—The most important element in the treatment of this stage of chronic valvular disease is *absolute rest*. Under the influence of this alone it frequently happens that the urinary secretion increases, the venous engorgement and edema subside, and compensation is gradually restored. While the recumbent posture is generally preferable, it is sometimes advisable to consult the inclinations of the patient and to allow him to sit up in an arm-chair, propped up by pillows and carefully protected from the cold.

When the digestive functions are well maintained a carefully selected mixed diet may be allowed. Eggs, oysters, sweetbread, scraped beef, lean lamb or mutton chops, white meat of chicken, boiled fish, baked potato, steamed rice, well-made gruels, toast, junket, wine jelly, and plain custard are suitable forms of nourishment. Coffee and tea should be used very sparingly, if at all. Alcohol in the form of brandy or whisky is often of great value. Full evening meals should be avoided, the chief meal being taken in the middle of the day. When there is much flatulence benefit will be derived from a considerable limitation of the carbohydrates. When the cavities of the heart are overfilled and there is decided edema, it is often very advantageous to reduce the ingestion of liquids to a minimum, although in this as in other matters pertaining to the diet we should be guided largely by observation. In severe cases it is often a good plan to restrict the diet, at least for a time, to milk, giving from 3–4 ounces (90.0–120.0 c.c.) every two hours.

Should these measures prove inadequate to restore the equilibrium of the circulation, recourse must be had to special cardiac stimulants. The best and most reliable of these is *digitalis.* The chief indications for using this drug are frequency, weakness, and irregularity of the pulse, edema of the limbs, and deficient urination. When these symptoms are present digitalis may be employed whatever the nature or seat of the lesion. From 10–20 minims (0.6–1.2 c.c.) of a good tincture may be given two or three times a day (see p. 50). The first evidences of its beneficial effects are an increase in the secretion of urine and an improvement in the rhythm and tone of the pulse. Failure to secure immediate results should not cause disappointment, as decided benefit may not be observed until thirty-six or forty-eight hours after the institution of treatment. Digitalis is of most service in mitral regurgitation. In mitral stenosis its action is less certain, but nevertheless it may be prescribed with considerable confidence when there is failure of the right ventricle with dropsy and ischuria. In disease of the aortic area it must be used with some circumspection. When the chief symptoms are slight dyspnea, precordial distress, vertigo, and tendency to syncope, it generally fails or acts unfavorably, but when venous congestion, cyanosis, and edema appear as the result of back pressure in the heart and lungs, it frequently does good. In many cases certain adjuvants must be employed before digitalis can be made effective. When the right side of the heart is embarrassed and the tissues are engorged, purgatives are essential. Without their aid digitalis may afford little or no relief Mercurial aperients are specially serviceable in lowering the venous tension. An appropriate dose of calomel or blue mass may be given in the evening, and followed in the morning by a saline. Subsequently, in some cases it is advantageous to combine the mercury with the digitalis, as in the well-known Niemeyer's pill (see p. 51). When the right ventricle is greatly overdistended and cyanosis is marked, venesection to the extent of a pint (0.5 L.) or more often affords prompt relief Indeed, in this condition digitalis usually proves wholly unavailing when used before the veins have been thus depleted.

Strychnin is one of the most valuable accessories to digitalis we possess. It is particularly useful in temporary breakdowns of compensation in which there are pulmonary complications. When the myocardium is much involved it may often be given for many months or even years with benefit. *Caffein* is another remedy that may in many cases be combined advantageously with digitalis. It is especially serviceable when dropsy is a prominent symptom. Its chief drawback is its tendency to induce restlessness and insomnia. Excellent results are frequently obtained by giving *nitroglycerin* or one of the other nitrites in conjunction with digitalis when there is high arterial tension or considerable precordial pain. In myocardial degeneration it serves a useful purpose as corrigent of digitalis in counteracting the constricting effect of the latter on the blood-vessels. In cardiorenal disease with persistent or paroxysmal dyspnea it is also useful. *Iron* and *arsenic* are valuable adjuncts to digitalis when there is anemia. The latter especially has been highly extolled by Balfour, Bramwell, and others in cases of aortic regurgitation and of fatty or fibroid heart with anginoid symptoms.

As a circulatory stimulant *strophanthus* is distinctly less reliable than digitalis, although it is somewhat more prompt in its action than the latter. It may be used in the cases in which digitalis is not well borne by the stomach or fails to give the desired result. It is also very valuable as an alternate when the administration of digitalis must be interrupted. *Convallaria*, *spartein*, and *adonidin* may be mentioned as drugs which occasionally succeed after digitalis and strophanthus have failed.

Treatment of Special Symptoms.—*Dropsy.*—In many cases absolute rest in bed, restriction of liquid ingesta, and the administration of digitalis suffice to restore compensation and, consequently, to remove the dropsy. Gentle massage of the affected parts by promoting the lymphatic and venous return is often beneficial. The application of smooth firm bandages to the swollen limbs sometimes affords much relief Purgatives are very valuable, although some caution must be exercised in their employment lest they prove exhausting to the patient. Salines in concentrated solution, compound jalap powder (30–40 gr.—2.0–2.6 gm.), and elaterin ($\frac{1}{20}$ gr.—0.003 gm.) should be tried in the order named. Diuretics are also useful. Digitalis, on account of its favorable influence on the circulation, is generally the most suitable. The best adjuvants are caffein, the vegetable salts of potassium, theobromin, spartein, and squill. The following mixture of potassium acetate and the infusion of digitalis often acts happily:

℞   Potassii acetatis,   ʒij (8.0 gm.);
      Infusi digitalis,   f℥iij (90.0 c.c.).—M.
      Sig.—A tablespoonful three or four times a day.

Another time-honored remedy is the pill containing 1 gr. (0.065 gm.) each of blue mass, digitalis, and squill, which may be given every three or four hours until diuresis, free purgation, or both, are obtained. It is generally inadvisable to employ diaphoretics like hot-air or vapor baths and pilocarpin, since they may induce profound exhaustion.

When all other measures have failed recourse may be had to paracentesis of the internal cavities and to scarification of the most swollen and most dependent parts. The insertion of fine silver canulæ with thin rubber tubes attached (Southey's tubes) often serves to drain off large amounts of fluid. Not infrequently curved incisions behind the ankles prove less annoying to the patient than either scarification or the insertion of tubes. Strict antiseptic precautions must always be taken in carrying out any of these operative measures, otherwise septic inflammation is very liable to ensue.

Dyspnea.—Chief reliance must be placed upon those measures which have already been recommended as useful in restoring the balance of the circulation. Free wet cupping or venesection is of the greatest service in relieving dyspnea the result of pulmonary engorgement. Counterirritation by means of sinapisms is much less efficacious. Oxygen inhalations are sometimes useful. Aspiration will be required in cases of hydrothorax. The nitrites are very valuable in paroxysmal dyspnea or cardiac asthma, especially when high arterial tension coexists. Dyspnea excited by flatulent distention of the stomach is usually promptly relieved by carminatives—spirit of chloroform, Hoffmann's anodyne, or spirit of mint. In many cases no remedy is so reliable as morphin. In the later stages of heart-disease it is indispensable. It should be given hypodermically in doses of from $\frac{1}{8}$–$\frac{1}{6}$ gr. (0.008 gm.–0.01 gm.) combined with atropin ($\frac{1}{200}$ gr.—0.0003 gm.). Morphin should not be used, however, when there is edema of the lungs.

Palpitation.—Cold to the precordia is a valuable sedative. When there is marked hypertrophy of the heart with a bounding pulse aconite (1–3 min.—0.06–0.25 c.c., every three hours) will be found of service. Palpitation of purely nervous origin is often benefited by the administration of bromids. When flatulence is an exciting factor carminatives usually afford relief When the attacks are obstinate and interfere with sleep morphin is to be recommended.

Pain.—Temporary precordial oppression is often relieved by warm or cold applications and the administration of Hoffmann's anodyne or brandy with aromatic spirits of ammonia. Severe continuous pain may yield to leeching or blistering. In paroxysmal pain of an anginoid character, nitroglycerin will generally be found a speedy remedy. In the anginoid pains, which frequently occur in aortic disease, potassium iodid, 10 gr. (0.65 gm.), thrice daily, is also of service. In some cases morphin is the only remedy that will afford complete relief.

Cough and Hemoptysis.—These symptoms rarely call for special treatment. Cough occurring after failure of compensation is referable to pulmonary congestion and bronchial catarrh, and can be controlled successfully only by remedies employed directly against the cardiac failure. Hemoptysis is not often excessive and is generally salutary. When it is attended with dyspnea and precordial distress it calls for venesection or leeching and aperients.

Vomiting.—Vomiting may be due to faulty diet or to digitalis or other medicines. In many cases, however, it is of serious import, being the result of stasis in the vessels of the stomach. In severe cases it is advisable to rest the stomach. No food should be given by the mouth except a little iced champagne with Apollinaris, brandy with soda-water, or milk with lime-water, whey, or liquid peptonoids. Calomel in fractional doses is useful in depleting the engorged portal system. Anti-emetics (see p. 161) are seldom efficacious. Sinapisms over the epigastric region sometimes afford relief. Cardiac stimulation is best maintained by administering digitalin and strychnin hypodermically.

Insomnia.—In mild cases Hoffmann's anodyne (f℥ss–f℥j— 2.0–4.0 c.c.) or a mixture of aromatic spirits of ammonia with whisky may suffice. Bromids are sometimes serviceable. Of the more active soporifics chloralamid, trional, and sulphonal are worthy of confidence. Bramwell has found paraldehyd very useful in cases in which there is associated bronchitis. Chloral should usually be avoided on account of the depression it causes. On the whole, no somnifacient is so generally efficacious as morphin, especially when sleeplessness is attended with pain, precordial anxiety, and restlessness. One-sixth of a grain (0.01 gm.) with $\frac{1}{150}$ gr. (0.0004 gm.) of atropin, hypodermically, is often sufficient.

Sudden heart-failure must be met by the administration of diffusible stimulants, such as ammonia, alcohol, and ether. The application of heat to the precordia is also useful.

CHRONIC MYOCARDIAL DISEASE.

Attention to the minutiæ of life is of the utmost importance. No one "rule of thumb" can be laid down for every patient. Each individual must be carefully studied as regards the state of the arteries, the character of the pulse, the digestive powers, the existence or absence of any constitutional vice, and the actual condition of the heart, so far as this can be ascertained. Exercise and diet are the points to which attention should first be directed. It is not always necessary that the patient should give up his occupation, but it is imperative that the sum of his exertions should be carefully adapted to the strength of his heart. It must be borne in mind that the fret and worry incident to many pursuits may exert as baneful an effect upon the circulation as undue physical effort. In some cases when the pathologic changes are not far advanced, and particularly if they consist in fatty overgrowth rather than degeneration of the muscular fiber, graduated exercises coupled with warm saline baths, as in the well-known Nauheim treatment (see p. 333), have a very salutary influence. When, however, palpitation, irregularity, pain, or pronounced dyspnea follow exertion, absolute rest is essential. In such cases massage is very useful, since it gives the patient the advantages of exercise without throwing any strain upon his heart.

As with exercise so with diet, no hard and fast rules can be laid down. The diet must be carefully adapted to the digestive powers and to the needs of the system. Rich foods, bulky foods, and foods which induce flatulence should be withheld. Eating between meals should be prohibited. The question of the quantity of food is also a matter of great importance. Degenerative changes in the heart usually occur at a period of life when the amount of food can often be considerably restricted without impairing the general nutrition. Chambers, Balfour, and others have dwelt upon the advisability of restricting the amount of liquids taken with meals in the case of failure of the myocardium. By this means the tension in the venous system may be lowered and the arterial resistance reduced. If thirst be complained of it is better for the patient to sip a moderate amount of hot water between meals than drink copiously during meals. Some judgment, however, must be exercised in the matter of drink, as the amount of liquid required to secure a free evacuation of waste products from the system must vary in different cases.

Tea and coffee should be used very sparingly, if at all. Alcohol also is better dispensed with. Smoking should be forbidden.   Excesses of all kinds must be rigidly avoided. Constipation must be relieved.   When it does not yield to judicious exercise and the regulation of diet, an aloetic aperient (see p. 554) should be given at night or a mild saline in the morning.   This may be supplemented from time to time by a pill of blue mass.   In many cases no single remedy affords such prompt relief from the general distress as a mercurial laxative.

If there be a gouty diathesis it should be met by appropriate treatment, and if syphilis be a factor antiluetic remedies should be employed.

As to special treatment, strychnin appears to be the most generally useful drug.   In doses of from $\frac{1}{60}$ to $\frac{1}{30}$ gr. (0.001–0.002 gm.), two or three times a day, it rarely fails to be of some benefit.   When there is anemia iron is a useful adjuvant. Arsenic in small doses, over long periods, is often of the greatest service.   Balfour considers that " arsenic is indispensable in all forms of weak heart accompanied by pain."   He recommends it in combination with iron and strychnin, twice a day, after food.   Nitrites usually prove very beneficial in cases in which high arterial tension, anginoid pains, or cardiac asthma are prominent symptoms.   Huchard and others have spoken in the highest terms of the prolonged use of the iodids, especially when angina pectoris is a feature of the cardiac failure.   They are most effective, as Osler has stated, in robust middle-aged men in whom the angina is the sole symptom.   They are contraindicated when the myocardial disease is associated with advanced arterial degeneration and chronic nephritis.

Digitalis is of service in some cases.   The less advanced the degeneration the more likely is it to be useful.   Large doses are generally badly borne, but persistently employed in ·doses of 2 or 3 minims (0.1–0.2 c.c.) of the tincture, twice a day, it often exerts a most favorable tonic effect.   Even with these small doses of digitalis it may be necessary to guard against the constricting influence of the drug on the arterioles by combining with it a vasodilator, such as nitroglycerin, erythrol tetranitrate, or sodium nitrite.   Not infrequently tincture of strophanthus, in doses of 3 or 4 minims (0.2–0.25 c.c.), twice daily, proves more efficacious than digitalis.   When neither digitalis nor strophanthus is well received, caffein, spartein, or cactus may be tried.

For the treatment of the more serious symptoms of cardiac failure the reader is referred to the article on Chronic Endocarditis (see p. 597).

ANGINA PECTORIS.

The general treatment of true angina pectoris is for the most part that of myocardial degeneration. The patient should lead a quiet, easy life, should avoid, as far as possible, all mental and physical excitement, and should abstain, as a rule, from tobacco and alcohol. Small meals of readily digested food are to be recommended. The evening meal especially should be light. Muscular exertion or mental excitement after meals is particularly injurious. The bowels should be kept regularly open, mild aperients being used for the purpose if necessary. Unless the heart is very weak and sensitive gentle exercise in the fresh air is to be recommended. Cold, however, is to be avoided. After a very severe attack it is generally necessary to enjoin absolute rest in bed for several days. Any constitutional disease that may be present, such as gout or syphilis, must receive appropriate treatment. In the anemic and debilitated, tonics, especially iron and strychnin, are very useful. The most valuable special remedies in the order of their efficacy are the nitrites, the iodids (see p. 603), and arsenic.

The Attack.—When the attacks are brought on by indigestion they can frequently be staved off by the timely use of a brisk mercurial or saline purge. Marked flatulency should be met by the prompt administration of a stimulant carminative like spirit of mint, Hoffmann's anodyne, or spirit of camphor. For the instant relief of the pain no drug is so useful as amyl nitrite. In the majority of cases a mere whiff of this speedily arrests the paroxysm. The patient should carry the remedy on his person in a small vial or, better, in the form of glass pearls containing from 3–5 min. (0.2–0.3 c.c.), which can be broken in a handkerchief and inhaled as soon as he perceives the pain. When the attacks are severe and prolonged, morphin ($\frac{1}{4}$–$\frac{1}{2}$ gr.—0.016–0.032 gm.) should be given hypodermically. In some cases, as Yeo has pointed out, there is a remarkable tolerance of this drug. When amyl nitrite and morphin fail, recourse should be had to chloroform inhalations (see p. 111). The application of a mustard plaster to the precordial region is often useful. In those cases in which the attacks occur in very rapid succession flying blisters or light touches of the actual cautery sometimes bring relief. The cardiac depression following grave seizures should be combated by such drugs as strychnin, ammonia, ether, and camphor.

DISEASES OF THE KIDNEYS.

ACUTE NEPHRITIS.

Absolute rest in bed for a period varying from four to six weeks is imperative. Even after the patient feels perfectly well he should not be permitted to leave his bed until the urine has been absolutely free from albumin for at least a week.

The patient should be well covered, woolen being worn next to the skin, and great care should be taken to guard against draughts.

The best diet for acute Bright's disease is *milk*. While it is desirable to add to the milk one-fourth or one-third part of lime-water, barley-water, carbonated water or Vichy, it is not advisable for the patient to drink excessive quantities of water. Beef-tea and broths should be interdicted. When the kidneys begin to secrete more actively, thin gruels, rice, and milk-toast may be given. During convalescence, bread and butter, baked potatoes, green vegetables, and fruits are permissible. The return to nitrogenous food should be most gradual, the effect on the urine of each addition to the diet being carefully watched.

In the absence of any direct remedies the indications are to divert the blood from the inflamed kidneys, to lessen their work as much as possible by increasing the action of the bowels and skin, and to meet the symptoms as they arise.

At the onset, if there be pain or suppression of urine, dry cupping, or, in severe cases, wet cupping, over the region of the kidneys is of value. Following the cupping warm poultices may be applied to the loins with advantage. Cantharides, turpentine, or similar drugs are not to be used.

The bowels should be kept freely opened by means of daily purges, the best in this condition being salines in concentrated solution and compound jalap powder. For children magnesia will be found sufficient. Very vigorous purgation is not called for unless there be dropsy or uremia.

Free sweating is very useful in promoting elimination by the skin. It may be effected by means of hot-water baths, hot packs, vapor baths, or hot-air baths (see p. 486.) In children hot baths and hot packs are eminently satisfactory, especially if spirit of nitrous ether or solution of ammonium acetate be administered at the same time. If the baths afford but little relief their action may be supplemented by the hypodermic administration of pilocarpin in doses of from $\frac{1}{16}-\frac{1}{10}$ gr. (0.004–0.0065 gm.).

There is much difference of opinion as to the use of diuretics. Some observers believe that all diuretics are irritating to the kidneys and therefore should be withheld at least during the stage of engorgement. Most authorities admit, however, that mild alkaline diuretics, like the vegetable salts of potassium, often serve a useful purpose in removing from the uriniferous tubules proliferated epithelium, albuminous plugs, and effete matter. From 20–30 gr. (1.3–2.0 gm.) of the citrate or acetate of potassium may be given three or four times a day, well diluted. Later in the disease, when the pulse tension falls, experience has proved that digitalis is often of great value. From 2–4 fl. dr. (8.0–15.0 c.c.) of the infusion should be given every three or four hours. It may be combined advantageously with an alkaline diuretic as in the following formula:

$$\text{R}\quad \text{Potassii acetatis,}\qquad \text{ʒij (8.0 gm.);}$$
$$\text{Infusi digitalis,}\qquad \text{fʒiij (90.0 c.c.).} \quad \text{M.}$$

Sig.—A tablespoonful, well diluted, every four hours.

Symptomatic Treatment.—The symptoms of acute nephritis which may demand special attention are excessive dropsy, uremia, nausea and vomiting, and anemia.

Excessive Dropsy.—When the dropsy is extreme and does not yield to the purgative, diaphoretic, and diuretic remedies which have already been indicated, recourse must be had to puncture of the swollen parts, to free incision on the inner or outer side of each ankle, or to the insertion beneath the skin of Southey's tubes (see p. 600).

Large accumulations in the serous sacs should be removed by paracentesis.

Uremia.—This complication calls for very prompt and energetic treatment. The chief indication is to favor the elimination of effete matter through the only available emunctories— the bowels and the skin. Two drops (0.065 c.c.) of croton oil, diluted with olive oil or glycerin, or ⅙ gr. (0.01 gm.) of elaterium should be given at once. Sweating should be promoted by hot-air or vapor baths and the hypodermic administration of pilocarpin (¼–⅓ gr.—0.015–0.02 gm.). If coma or convulsions appear, and the patient is not too feeble, venesection may be practised, the removal of from 15 to 20 ounces (450.0–600.0 c.c.) of blood sometimes exerting a very happy effect. In children a few ounces of blood may be abstracted from the loins by means of wet cups. Hypodermoclysis is another potent remedy (see p. 488). It may be resorted to whether blood-letting is deemed necessary or not, although it is more efficacious when practised after blood has been withdrawn. Enteroclysis (see p. 490) is also useful, though less effective than the subcutaneous injection of salt solution.

Convulsions may be controlled by chloroform inhalation or the administration of chloral. If the patient be unable to swallow, from 20–30 gr. (1.3–2.0 gm.) of chloral may be injected into the rectum. If these measures fail, morphin may be used cautiously (see p. 91).

Nausea and Vomiting.—When these symptoms persist food should be withheld. Pieces of ice may be given to suck. A mixture containing bismuth subnitrate and diluted hydrocyanic acid (see p. 69) often affords relief. Carbolic acid, wine of ipecac, or tincture of nux vomica, in one-drop doses, may be tried.

Anemia.—After the acute symptoms have subsided iron may be employed to combat the anemia. For adults Basham's mixture is a suitable preparation. Of this from 2–4 fl. dr. (8.0–15.0 c.c.) may be given thrice daily. For children the following mixture will be found more agreeable:

℞ Tincturæ ferri chloridi,
   Acidi phosphorici diluti,   āā fℨiss (6.0 c.c.);
   Glycerini,   fℨvj (22.5 c.c.);
   Syrupi acidi citrici,   q. s. ad fℨiij (90.0 c.c.). M.
Sig.—Teaspoonful in water three times a day.

Convalescence.—It is advisable to keep the patient under observation for many weeks after all symptoms of the disease have disappeared. He should be warmly clad and carefully guarded from exposure to wet and cold. While the diet should be liberal it should be such as will not overtax the kidneys. Overexertion must be rigidly avoided. The skin and bowels should be kept active, the former by frequent tepid baths and the latter by saline aperients if necessary. A change of air to a warm, equable climate will be found to be very beneficial.

CHRONIC NEPHRITIS.

The indications for treatment in chronic nephritis are to lessen the work of the kidneys by reducing the production of nitrogenous waste and increasing the activity of the other excretory organs, to maintain the general nutrition, and to meet the symptoms as they arise. The patient must be guarded against vicissitudes of weather by wearing flannel next to the skin in all seasons, both night and day. If he can afford it he should be urged to spend the winter months in a warm, equable climate.

In the absence of any obtrusive symptoms, moderate exercise short of fatigue should be encouraged, but all violent exertion should be prohibited. Mental strain ·and worry are also injurious, and should be avoided as far as possible. When the symptoms are well developed rest becomes an important factor in the treatment, and much time should be passed in bed. The use of tobacco and alcohol should be discontinued. Warm baths with friction are necessary to promote free action of the skin, but great care must be exercised after their use to avoid chilling. In the early stages of chronic interstitial nephritis an occasional Turkish bath is often very advantageous. The hot-air bath, vapor bath, or hot pack may be used at home once or twice a week, or oftener if there is much edema, or if uremia threaten. Cold bathing should be interdicted. The bowels should be kept active by an occasional saline aperient. A great deal depends upon diet. No hard-and-fast rules, however, can be laid down, each case being a study in itself. When the symptoms are quiescent, a simple nourishing diet comprising a moderate quantity only of nitrogenous matter, may be allowed, its effect on the urine being carefully watched. Green vegetables, potatoes, ·farinaceous food, and ripe fruits usually agree best, but a certain amount of animal food in the form of eggs, oysters, fish, and mutton, or other tender meats, is often well borne. Indeed, considerable caution must be exercised lest in our zeal to relieve the kidneys we reduce the strength of the patient by adhering too strictly to a non-nitrogenous diet. The common dogma that red meats are more harmful than white meats in chronic Bright's disease receives no support from the observations of von Noorden, Offer and Rosenquist, or Kauffmann and Mohr, who conclude that the various forms of meats are suitable in such cases purely in proportion to their digestibility. Water-drinking between meals is generally beneficial, and a mild alkaline mineral water, such as Vichy or Vals, may be taken to the extent of a pint (0.5 L.) or more a day. In many cases of chronic parenchymatous nephritis, especially when there is decided edema or a tendency to uremia, an absolute milk diet is most suitable. The patient should take at stated intervals such quantities of milk as will aggregate in the twenty-four hours from 2–4 quarts (2.0–4.0 L.). The milk should not be skimmed, but diluted with carbonated water or an alkaline mineral water.

In chronic interstitial nephritis an absolute milk diet is not always successful. Indeed, as von Noorden has pointed out, much good sometimes results from restricting in this form of nephritis the fluids to between 40 and 50 ounces (1200.0–

1500.0 c.c.), especially when there is dilatation of the heart and a tendency to cardiac asthma.

Medicinal Treatment.—We possess no remedies that are directly curative. In the interstitial form potassium iodid (1–5 gr.—0.06–0.3 gm.) or bichlorid of mercury ($\frac{1}{60}$–$\frac{1}{40}$ gr.—0.001–0.0016 gm.), two or three times a day, occasionally seems to have a beneficial effect, but in the large majority of cases these remedies are of no avail. Mild hematinics, like Basham's mixture, are often useful, especially in chronic parenchymatous nephritis, but only small doses should be used. The indiscriminate use of iron in chronic interstitial nephritis, as Tyson has shown, is often productive of harm. It is indicated only when there is evident anemia. Strychnin and the simple bitters are valuable adjuvants to iron in many cases. Late in the disease, when the urine becomes scanty and dropsy appears, hydragogue diuretics are called for, although they often prove disappointing. Those most likely to succeed are digitalis, the vegetable salts of potassium, caffein, and theobromin.

Operative Treatment.—Edebohls claims that chronic Bright's disease in its early stages is curable or susceptible to amelioration by renal decapsulation. He has performed the operation in 51 cases, including interstitial, parenchymatous, and diffuse forms of the disease. Nine patients were cured and remained cured in periods varying from two to ten years; 22 were in various stages of improvement; 2 were not improved; and 14 died, 7 within fifteen days of operation and 7 at later periods. Several patients treated by Rose, Wolff, Ferguson, and Guiteras are also reported to have done well. The operation consists in exposure of the kidneys through the loins and excision of their proper capsules. The good results are attributed to the development of numerous blood-vessels, particularly arteries, in the firm adhesions between the kidneys and the surrounding tissues, the blood stream passing from the latter to the kidneys. This, it is thought, leads to gradual absorption of the intertubular inflammatory products, thus freeing the tubules and glomeruli from compression, and permitting the re-establishment of a normal circulation, with regenerative production of new epithelium capable of carrying on the secretory function. Edebohls's operation has not been employed extensively enough as yet to warrant positive conclusions as to its therapeutic value.

Treatment of Special Symptoms.—For the *dropsy* and *anemia* the most effectual measures have already been enumerated (see p. 606).

39

High Blood-pressure.—A certain increase in the arterial tension is necessary. It compensates for the renal inadequacy. When, however, the tension in the vessels becomes excessive and gives rise to headache, vertigo, palpitation, and dyspnea, it should be reduced by saline laxatives, warm baths, and the administration of nitroglycerin ($\frac{1}{100}$ gr.—0.00065 gm.—three times a day, increased if necessary until the desired effect is accomplished).

Heart-failure with low arterial tension will require the use of such stimulants as digitalis, strychnin, caffein, and alcohol.

Asthma should be treated as a manifestation of uremia. Nitrite of amyl, nitroglycerin, or ethyl nitrite affords some relief. When there is pulmonary congestion or edema, dry or wet cupping is often useful.

Insomnia may be caused by the retention in the blood of excrementitious matter, or may be due to high arterial tension. In the first case a milk diet and the administration of eliminants are called for; in the second, vasodilators (nitroglycerin or erythrol tetranitrate) will be helpful. When the insomnia is sufficiently severe to demand the use of hypnotics, bromids, chloral, paraldelyd, and trional may be tried in the order named. Opium should be avoided.

Headache is best treated by cold to the head, and the administration of saline purges and caffein (2–3 gr.—0.13–0.2 gm.) or nitroglycerin (1–2 min.—0.06–0.12 c.c.).

Vomiting.—When vomiting is persistent it is best to withhold food entirely for a time, allowing carbonated water or cracked ice only by the mouth. Sinapisms over the epigastrium sometimes afford relief. Such gastric sedatives as diluted hydrocyanic acid, bismuth subnitrate, cocain, and oxalate of cerium are worthy of trial.

Diarrhea should not be checked too quickly. When treatment is desirable mild astringents and intestinal antiseptics may be given. Opium should be used with extreme caution, if at all, not only in this but in all other complications of chronic nephritis in which sedatives are indicated.

NEPHROLITHIASIS.

Medical treatment should be given a thorough trial in all cases of suspected renal calculus. It is doubtful whether stones once formed in the pelvis of the kidney can ever be dissolved, especially if they have already attained a considerable size, but much can be accomplished by internal remedies in preventing the formation of fresh deposits and in mechanically washing down into the bladder small concretions.

As uric-acid calculi are by far the commonest, efforts should be directed toward lessening the formation of uric acid and preventing its precipitation in the urinary passages. The patient should use only those articles of food which experience has shown to be most easily digested. The foods most likely to disagree are concentrated soups, hashes, croquettes, pastry, sweets, acid fruits, malt liquors, heavy wines, and all articles rich in nucleins, like the internal organs and young germinating plants. The quantity as well as the character of the food must be regulated. No more should be eaten than is absolutely necessary to maintain normal nutrition. Too long an interval should not be allowed to elapse between meals.

Gentle exercise in the open air and hydrotherapy are to be recommended since they favorably affect the underlying diathesis.

Much benefit accrues from abundant water-drinking between meals. This not only favors the expulsion of small concretions from the kidney, but by diluting the urine it also tends to prevent the precipitation of uric acid. Alkaline mineral waters, like Carlsbad, Vichy, Vals, and Buffalo and Londonderry lithia waters have been extensively employed, but they are no more efficacious than plain water which has been rendered alkaline by the addition of potassium citrate, $\frac{1}{2}$–1 dr. (2.0–4.0 gm.) to the quart (1.0 L.), or of potassium bicarbonate, 40 grains (2.5 gm.), and lithium citrate, 20 grains (1.3 gm.), to the quart. About a quart of such water should be consumed, as a rule, in the twenty-four hours. In mild cases a single copious draught at bedtime may suffice to prevent recurrence of colicky pains. Some care is necessary not to allow the urine to become too alkaline, since this may lead to a deposition of phosphates about the uric-acid stone.

Among special remedies brought forward as solvents of uric acid (see p. 237) may be mentioned piperazin, lycetol, lysidin, piperidin tartrate, and urosin. While some authors claim to have seen beneficial effects follow the use of these drugs, the majority of unprejudiced observers are sceptical as to their merits.

Recently Croftan has recommended calcium carbonate (15–20 gr.—1.0–1.3 gm.—thrice daily), which was first advocated by Von Noorden. The theory is that calcium, owing to its great affinity for phosphoric acid, unites with this acid in the blood to form calcium phosphate, which is subsequently eliminated in great part through the bowel, thus ridding the urine of mono-sodium phosphate, a precipitant of uric acid, while leaving in the urine di-sodium phosphate, an active solvent of uric acid.

Phosphatic Calculi.—Since phosphatic calculi can be deposited only from alkaline urine the indication is to render the urine acid, a task often difficult of accomplishment. Instead of a vegetable diet, one composed largely of readily digestible proteids should be ordered. Among special remedies recommended for the purpose of acidifying the urine may be mentioned benzoic and boric acids. From 5–15 gr. (0.3–1.0 gm.) of either of these drugs may be given in capsule, three to four times a day. Unfortunately, the continued use of these acids is prone to cause indigestion, an effect which detracts materially from their value. As in the case of uric-acid concretions, the copious ingestion of plain or distilled water is to be encouraged in order to prevent concentration of the urine.

Oxalate-of-lime Calculi.—As there is reason to believe that oxalates are formed under conditions very similar to those which give rise to an excess of uric acid, what has been said concerning the treatment of the uric-acid diathesis applies equally to that of the oxalate of lime.

Renal Colic.—The indications are to relieve the pain and to relax the spasm. This is best accomplished by hypodermic injections of morphin and atropin, coupled with hot baths or local applications—hot poultices or fomentations. If the pain is extreme it may be desirable to administer chloroform. Simple diluents should be given freely. In mild attacks full doses of phenacetin or antipyrin, with an abundant supply of hot drinks, may suffice.

Surgical Treatment.—In a certain number of cases surgical intervention affords the only hope of permanent relief Operation is urgently demanded when the attacks of renal colic occur with such frequency as to prove disabling, when there are evidences of severe pyelitis, and when there is reason to believe that the calculus is impacted in the ureter. Unless there be pyonephrosis, nephrolithotomy is the operation to be recommended, as it is attended with much less risk than that of nephrotomy or nephrectomy. The mortality in 82 cases of nephrolithotomy for non-suppurative nephrolithiasis was 2.4 per cent. (Newman and Legue). The average mortality in suppurative cases is not less than 15 per cent. The low mortality of early operation contrasted with the high mortality of late operation, when the kidney structure has been seriously damaged, is a strong argument for early interference.

DISEASES OF THE NERVOUS SYSTEM.

ACUTE CEREBRAL LEPTOMENINGITIS.

The treatment should be conducted on the same lines as that of cerebrospinal fever (see p. 505). When middle-ear disease is the exciting factor, and the symptoms can be definitely localized, operation may be justifiable. Several writers have recently spoken favorably of frequently repeated warm baths, as originally recommended by Aufrecht in 1894, combined with lumbar puncture (see p. 497).

CEREBRAL HEMORRHAGE.

Prophylaxis.—Patients predisposed to apoplexy should lead a quiet life, cne as free as possible from mental and physical excitement. The food should be nutritious and easily digestible. Moderation in eating and regularity of meals are matters of prime importance. Alcohol should be interdicted. The bowels should be made to act daily, a mild aperient being given if necessary.

The Attack.—If the patient be seen early, he should be placed in a reclining position with the head and shoulders well elevated. Bowles advocates turning the patient over on the paralyzed side to lessen the stertor. An ice-bag should be applied to the head, and warm bottles to the feet. The bowels should be promptly emptied by an enema, and later by a drastic cathartic, preferably croton oil, one or two drops of which, mixed with a little sweet oil, may be placed upon the tongue.

In sthenic cases, when the face is congested and the pulse is slow and full, venesection to the amount of $\frac{1}{2}$–1 pint (250.0–500.0 c.c.) is indicated. The abstraction of blood cannot undo damage that is already done, but by lessening the supply of blood to the brain it may serve to arrest bleeding that is still in progress or to prevent an early recurrence. If venesection does not seem justifiable, lowering of the arterial pressure may be effected by the administration of aconite or veratrum viride.

When the heart's action is feeble and the face is blanched, stimulants like alcohol, ether, and camphor may be given very cautiously, although it is unlikely that they will be of much avail.

When the respiration fails before the circulation, artificial respiration may be practised with advantage. In a case cited by H. C. Wood, in which final recovery was complete, it was necessary for nearly two days for an attendant to hold forward the jaw and tongue by raising the ramus of the jaw upward and forward, in order to prevent mechanical asphyxia.

Thorough cleanliness, bathing with alcohol, frequent change of position, and the avoidance of roughnesses in the bed are necessary in order to prevent the development of bed-sores. Since blisters are produced more readily than in health, special care should be taken in using hot applications of any kind.

Retention of urine is apt to occur, and if it does the patient must be catheterized.

After the Seizure.—Absolute rest in bed must be maintained. Even in the mildest cases the patient should not be allowed to leave his bed for two or three weeks, and during this time the diet should be light and unstimulating. The ice-cap should still be kept upon the head. Aconite in small doses with sweet spirits of nitre is often useful in subduing the fever of reaction and in decreasing arterial tension. For restlessness and wakefulness, small doses of a bromid or of chloral may be given. Potassium iodid is often prescribed with the hope that it may aid in absorbing the clot, but it is of doubtful utility, and moreover, it is very prone to disturb the digestion.

After the acute symptoms have entirely disappeared, which will rarely be earlier than ten days or two weeks after the attack, massage should be systematically practised. It often contributes much to the restoration of power, or when this is impossible, to the prevention of contractures. After the lapse of three or four weeks, triweekly applications of the faradic current may be of service (see p. 473). Strychnin is often given at this time, but it probably exerts no other influence than that of a general tonic. In some cases warm saline baths (90°–95° F.—32°–35° C.), combined with passive movements, prove useful adjuvants.

LOCOMOTOR ATAXIA.

Rest is an important factor in the treatment. The sense of fatigue being impaired in this disease, patients are very prone to overtax their strength. Erb advises that the patient should live as if he were an old man, quietly, regularly, and with no excitements. In many cases it is well to begin treatment with absolute rest in bed for several weeks. Mental fatigue should also be avoided. Sexual excesses are exceedingly injurious. The diet should be nutritious and easily digestible. Alcohol and tobacco should be used sparingly, if at all.

Patients suffering from locomotor ataxia should wear flannel next to the skin, and should avoid exposure to wet and cold, as well as sudden alterations of temperature. When practicable a change to a warm, equable climate during the winter months may be recommended with advantage.

Massage affords a valuable means of securing the benefits of exercise without the expenditure of energy.

Systematic re-education of co-ordinating movements, as originally recommended by Frenkel, has been found a most effective remedy for the ataxia (see p. 478). Even in advanced cases, in which there is marked disturbance of sensation, this method of treatment is not without benefit, and the improvement may last for years if the disease is stationary or only slowly progressive. The treatment by suspension, introduced by Motschukowski, and that by nerve-stretching, proposed by Langenbuch, have become obsolete.

Tepid baths of 80°–85° F. (26.5°–29.5° C.) are sometimes of distinct service. On the other hand, hot baths, vapor baths, and cold baths are usually harmful, and often provoke lancinating pains. Various natural springs enjoy a special reputation, the most popular being the Hot Springs of Arkansas, those of Virginia, and those of Los Vegas, New Mexico, in this country, and the thermal baths at Rehme, Nauheim, and Aix-la-Chapelle in Europe. Patients in whom the disease is far advanced should, of course, be spared the cost and discomforts attendant upon a long journey. According to Frenkel, treatment by baths should be forbidden while the exercise treatment is being used. While electricity is undoubtedly of service in relieving certain sensory disorders of ataxia, there is no ground for believing that it can affect in the slightest degree the morbid process in the spinal cord.

Mercury and iodids should be given a thorough trial in all cases in which syphilis is suspected. It must be admitted, however, that antisyphilitic treatment rarely accomplishes any good unless the infection by syphilis has been quite recent. Most of the cases of tabes that have recovered under mercury have been examples of peripheral neuritis.

Among numerous special remedies which have been in vogue at various times, but which are of very doubtful utility, may be mentioned silver nitrate, ergot, calabar bean, strychnin, belladonna, phosphorus, chlorid of gold, arsenic, corrosive sublimate, and spermin.

Treatment of Special Symptoms.—When the pains are severe the most potent remedial measure is absolute rest in bed. Light touches of the actual cautery or sinapisms over the root of the nerve supplying the affected part often afford relief. Deep massage is sometimes of service. Mitchell has found the alternate application of ice and hot water useful. Flannel bandages applied firmly from the toes up to the middle third of the thigh sometimes do much good. A snugly fitting abdominal binder may also be used to lessen girdle pain. Electricity in the form of the faradic brush, static spark, or stabile galvanic anode, is worthy of a trial.

The most generally useful anodynes are phenacetin, antipyrin, and acetanilid. Cannabis indica occasionally succeeds. According to Osler, the prolonged use of nitroglycerin, given in increasing doses until the physiologic effect is produced, is of great service in allaying pains and diminishing the frequency of crises in all cases of tabes in which there is increased arterial tension. In many cases recourse must be had to morphin, but its use should be deferred as long as possible.

Gastric Crisis may require the withholding of food by the mouth for a time, the patient being sustained by nutritive enemata. Lavage is often beneficial. The application of sinapisms over the epigastrium may do good. Morphin hypodermically may be necessary.

Laryngeal crisis may be relieved by local applications of cocain, inhalations of chloroform or amyl nitrite, or the administration of anodynes.

Numbness and *paresthesia* often yield for a time to local applications of faradism given with the wire brush.

Vesical and rectal tenesmus may be relieved by suppositories of opium with belladonna, or applications of cocain. Mitchell has found faradism of great service (see p. 474).

Obstinate priapism is best treated by tepid hip-baths or general baths and the administration of anaphrodisiacs—bromids, monobromated camphor, or hyoscin. Sexual intercourse should be avoided.

Vesical weakness should receive the most careful attention. The bladder must be thoroughly emptied, if need be by catheterization. On the first appearance of cystitis the bladder should be thoroughly washed out with weak antiseptic solutions.

MYELITIS.

Acute Myelitis.—No time should be lost in placing the patient at absolute rest. In severe cases a water-bed should be employed. Such antiphlogistic measures as cupping, blistering, and cauterizing are of doubtful utility. Cold, however, in the form of Chapman's ice-bags, may be applied to the spine. Daily warm baths (90° F.—32.2° C.) lasting from ten to fifteen minutes are useful. Drugs are of little value. Ergot is often used for the purpose of constricting the spinal vessels and thus limiting the local congestion.

Every precaution must be taken against the development of bed-sores. Frequent change of the patient's position, absolute cleanliness of the parts subjected to pressure, and bathing with alcohol and water will do much toward obviating this complication. If bed-sores are already formed they should be treated thoroughly according to surgical principles. Retention of urine must be met by systematic catheterization under the most strict antiseptic precautions. When there is constant incontinence a carefully adjusted urinal should be employed.

Any tendency to cystitis will call for daily irrigation of the bladder with a solution of boric acid or other mild antiseptic solution. If recovery with partial paralysis result, massage and electricity may aid in bringing back some of the lost power.

Chronic Myelitis.—Hygienic measures are far more important than drugs.. Prolonged rest is desirable. Residence in a warm, equable climate is often serviceable. Daily warm baths are grateful. Counterirritation in the form of light touches of the actual cautery is useful in relieving pain, but it does not seem to exert any direct influence on the progress of the disease. When there is reason to suspect the existence of syphilis, mercury and iodids should be given a thorough trial. Tonics are often indicated. Massage, passive movements, and electricity are useful in maintaining the nutrition of the affected muscles. Early measures should be taken to prevent the formation of bed-sores and the development of cystitis.

POLIOMYELITIS.

Acute Poliomyelitis.—In the acute stage absolute rest in bed is requisite. Not the slightest benefit accrues from the application of wet cups or leeches to the spine, or from any other form of counterirritation. Mild mercurial laxatives and febrifuge medicines may be used with some advantage. Ergot is often prescribed for the purpose of diminishing congestion of the ventral horns, but it is of very doubtful utility. The affected limbs should be thoroughly wrapped in cotton-wool to maintain warmth and circulation.

After the acute symptoms have subsided, which will probably be at the end of two or three weeks, an effort should be made to promote the restoration of the paralyzed muscles. General hygienic measures must not be neglected. Nutritious food and fresh air are valuable aids. Little is to be expected from internal medication. Tonics like iron and cod-liver oil are sometimes of service. Strychnin seems to do some good. Massage in the form of effleurage and kneading is the most useful measure. It should be practised for ten or fifteen minutes daily for several months, or even for a year. Local bathing with shampooing may also be employed with benefit.

Electricity, while less efficacious than massage, is worthy of a thorough trial. As faradism generally fails to elicit any response, recourse must be had to an interrupted galvanic current. One pole (cathode) may be placed over an indifferent point, such as the spine, while the other (anode) is slowly stroked over the affected muscles. The weakest current that will cause contraction should be used. The treatment should be given for ten minutes, three or four times weekly, and should be kept up, if necessary, for several months. If after six or eight weeks of electric treatment there is no demonstrable improvement the outlook is very gloomy. Still, even here a successful issue is not impossible. The treatment of the later stages of infantile paralysis is chiefly surgical, and has for its object the prevention or correction of deformities.

Chronic Poliomyelitis.—There is practically no remedy for this disease. Massage, hydrotherapy, and electricity have seemed to be of some avail, but no permanent good has been achieved by these or other measures. Gowers has spoken favorably of strychnin and arsenic. In cases with a history of syphilis, mercury and iodids should be tried.

NEURITIS.

Careful search should be made for the exciting cause—rheumatism, gout, syphilis, alcoholism, metallic poisoning, mechanical injury, or neighboring inflammation—with the view of removing it if possible. Absolute rest is imperative. When the nerves of the limbs are affected the muscles should be immobilized by means of splints and bandages, as in the treatment of fracture.

In acute cases, irrespective of the cause, counterirritation is of service. This may be accomplished by applying linear blisters or painting iodin along the course of the inflamed nerve. In mild forms of the disease hot applications of lead-water and laudanum will suffice. Occasionally cold in the form of an ice-bag proves more grateful than fomentations. Coal-tar derivatives (phenacetin, antipyrin) and salicylates are useful in relieving pain. Such a combination as the following is often efficacious:

> ℞ Phenacetini,   ʒiss (6.0 gm.);
>   Salol,   ʒj (4.0 gm.).—M.
>   Fiant chartulæ, No. xii.
>   Sig.—One every three or four hours.

When the pain is very intense subcutaneous injections of morphin must be used.

As soon as the pain and tenderness have distinctly lessened warm baths and gentle friction may be employed. In localized neuritis inunctions with mercury and belladonna ointment may also be used with great advantage in this stage. When the symptoms of irritation have completely subsided massage and electricity should be resorted to for the purpose of restoring the functions of the nerves and muscles. The faradic current should be used if the muscles respond to it. In the more severe cases in which faradic excitability is lost, labile and interrupted galvanic currents should be applied with the pole that produces the best contractions. Even with extreme atrophy these measures sometimes succeed in effecting decided improvement. Passive movements and extension will be required to overcome contractures and deformity.

Tonics, especially strychnin and arsenic, are often of service in the later stages of the disease.

Sciatica.—The first indication is to remove the cause. This may be some affection of the pelvic organs requiring surgical intervention, or it may be rheumatism, gout, syphilis, or metallic poisoning, and then general as well as local treatment must be resorted to.

In acute cases rest in bed is essential. Free evacuation of the bowels should be secured in order to deplete the pelvic veins. When there is a definite history of rheumatism or the attack has been induced by cold, salicylates should be given in full doses. When there is reason to suspect syphilis iodids should be given a fair trial. Irrespective of the cause, phenacetin or antipyrin may be useful in relieving pain. Counter-irritation often affords much relief. When the pain is very severe this is best accomplished by means of small blisters or light touches of the actual cautery applied over the points of greatest tenderness. In some cases acupuncture (see p. 430) acts very satisfactorily. In milder cases the Scottish douche—in which a stream of warm water of gradually increasing temperature is directed on the course of the nerve until the pain subsides, when it is suddenly changed for a cold jet—is an efficient remedy. Agonizing pain must be relieved by injections of cocain ($\frac{1}{4}$ gr.—0.016 gm.), chloroform (5–10 min.—0.3–0.6 c.c.), guaiacol (2–3 min.—0.1–0.2 c.c.), or morphin

($\frac{1}{8}$–$\frac{1}{4}$ gr.—0.008–0.016 gm.), made deeply and as near to the nerve as possible. Morphin should be withheld as long as possible on account of the grave danger of inducing the morphin habit. In some cases deep injections of distilled water act remarkably well. Massage is indicated only when the acute symptoms have subsided, and should then not be too energetic. Electricity is rarely needed, even if there has been much wasting. In obstinate cases of sciatica the following treatment, outlined by S. Weir Mitchell, often acts most happily: A long flannel bandage is firmly applied from the foot to the groin, and renewed twice a day. The leg is slightly bent at the knee and kept nearly straight at the thigh, and in this position secured to a light side splint extending from the axilla to the ankle by a few turns of a bandage. Care is of course taken to prevent pressure on the heel. After a few days the joint-angles are slightly changed at each dressing. Still later passive motion is carefully employed, and after about three or four weeks the splint is removed during the day and replaced at night. Finally, guided as to time by the presence or absence of pain, the splint is removed altogether, the bandage being kept on somewhat longer, and gentle massage being used between each application after pain has quite disappeared. A full diet, cod-liver oil, and iron as needed, are given during the whole period. Any points of persistent pain that may be left at the close of the treatment are treated by counterirritants, preferably by small blisters or light touches of a white-hot Paquelin button. The patient must not sit at all during the first week after getting up, but must stand or walk, and that only with crutches. He may be upright or lying on a bed, but not seated.

Among other measures which have proved successful in individual cases may be mentioned hydrotherapy in the form of warm baths or mud baths, galvanism (5–7 milliampères), the continued application of ice-bags, and nerve-stretching, either by forcibly flexing the thigh upon the body or by cutting down upon the nerve itself.

NEURALGIA.

An effort should be made to adapt the treatment to the etiology of the disease. Causes of reflex irritation—carious teeth, errors of refraction, nasal obstruction, neoplasms, and cicatrices—should be sought for, and when found removed if possible. When the disease is associated with anemia, iron and arsenic will be indicated. If there is any suspicion of syphilis, mercury and iodids should be tried. When a malarial element is present, quinin may effect a cure. When rheumatism is an etiologic factor, salicylates and alkalis may prove beneficial. In gouty subjects much may be expected from regulation of diet, systematic exercise, and the administration of alkalis. The occupation of the patient, a blue line on the gums, and lead in the urine may lead to the recognition of plumbism, and in this case the iodids will be indispensable.

All influences that tend to induce a morbid excitability of the nerves or of their centers—mental or physical fatigue, emotional excitement, sexual excess, over-indulgence in tobacco, tea, coffee or alcohol—should be removed so far as possible.

In every case of neuralgia we must endeavor to improve the general nutrition, which is almost always disturbed. The measures to be employed for this purpose include an abundance of fresh air, proper food, regular hours, adequate protection from the vicissitudes of weather, systematic exercise, frequent bathing with friction, and the use of such tonics as iron, arsenic, cod-liver oil, and hypophosphites. If the patient cannot take active exercise rest with massage should be substituted. Neuralgia in neurasthenic patients is sometimes successfully managed by a modified Weir Mitchell treatment. A change of residence to a warm dry climate has been known to effect a complete cure.

Special Treatment.—The list of agents recommended for the cure of neuralgia is bewildering. " Numerous are the means at our disposal for combating the disease, quite as numerous are the patients who, after hundreds of unsuccessful trials, give up all medicines and all physicians " (Hirt). Among the internal remedies most worthy of confidence may be mentioned the coal-tar analgesics (phenacetin, antipyrin, and acetanilid), morphin, bromids, cannabis indica, croton chloral, gelsemium, caffein, nitroglycerin, and salicylates. The coal-tar analgesics are the most generally useful anodynes for the attack. Their action, moreover, is prompt. From 5–8 gr. (0.3–0.5 gm.) of one of these drugs may be given every two or three hours, according to the intensity of the pain. The fact must be borne in mind that these agents when used continuously not only lose in potency, but may also unfavorably influence general nutrition. Morphin is undoubtedly the most certain means we possess of affording temporary relief, but on account of the danger of inducing the opium habit it should be employed only as a last resort. Under no circumstances should the patient himself be allowed to use the hypodermic syringe. In mild attacks bromids in large doses sometimes succeed. Combinations of a bromid with phenacetin or antipyrin may often be prescribed advantageously. Cannabis indica is especially valuable in migraine, in which affection it may be given in doses of from $\frac{1}{4}$–$\frac{1}{2}$ gr. (0.016–0.03 gm.). Croton chloral, in doses of from 5–10 gr. (0.3–0.65 gm.), and tincture of gelsemium, in doses of 10 min. (0.6 c.c.) or more, are occasionally serviceable in trifacial neuralgia. Caffein, in doses of from 3–5 gr. (0.2–0.3 gm.), is often efficacious. Combinations of caffein with phenacetin or with bromids in many cases do more good than single drugs. Nitroglycerin, in doses of from $\frac{1}{150}$–$\frac{1}{50}$ gr. (0.0004–0.0012 gm.), has been highly recommended in migraine-like attacks with signs of vasoconconstriction. Neuralgia brought on by exposure to cold and wet is favorably influenced by the salicylates. In such the following combination will be found of value:

R Phenacetini,
  Salophen,   aa ʒiss (6.0 gm.);
  Codeinæ,   gr. iij (0.2 gm.).—M.
 Fiant chartulæ, No. xii.
 Sig.—One powder every two or three hours.

Dana and others have spoken favorably of strychnin in heroic doses in tic douloureux (see p. 129).

Local Treatment.—Heat, dry or moist, may be applied for its soothing effect. In some cases cold in the form of an ice-bag is more grateful. Menthol and chloral camphor are useful in neuralgia of superficial nerves when the pain is slight. Congelation of the skin by ethyl chlorid occasionally has a good effect. Acupuncture and aquapuncture are effective, but are not suitable for use about the face. In obstinate cases active counterirritation by means of blisters or the thermocautery will be found a potent remedy. In trifacial neuralgia the blisters may be applied behind the ear. Even when the pain is very intense injections of cocain ($\frac{1}{8}$–$\frac{1}{4}$ gr.—0.008–0.016 gm.) upon the nerve often bring relief, but the grave danger of establishing a habit must never be forgotten. Cocain may also be introduced through the unbroken skin by cataphoric action (see p. 474). Bennett has treated 9 cases of trigeminal neuralgia successfully by the injection of osmic acid into the nerve (see p. 447). Electricity is sometimes of great service (see p. 474).

Finally, when all other measures fail, recourse may be had to surgical interference. Nerve-stretching and nerve-section are the operations usually performed. Lasting benefit is rarely obtained from either operation, although excision is more reliable than stretching. In trifacial neuralgia removal of the Gasserian ganglion usually affords permanent relief, but the mortality of this procedure is high (22.2 per cent. in 108 cases) according to Tiffany. Keen is of the opinion that the removal of the ganglion should not be attempted until the peripheral operation has been tried. Intracranial resection of the trigeminal nerve has given good results.

CHOREA.

Rest of body and mind is the most important factor in the treatment of the disease. No matter how slight the attack may be the child should be taken from school. In all except the mildest cases rest in bed in a quiet, well-ventilated room is essential. When the movements are so violent as to bring about contusions the bed should be padded or the child put in a hammock. It is almost always advisable to exclude the child's playmates, even his own brothers and sisters, from the sick-room. Excitement of all kinds should be proscribed.

Gentle massage may be of benefit, especially in poorly nourished children, but gymnastic exercises should be reserved for the period of convalescence. Daily warm baths (95° F.—35° C.) are often beneficial. In older children, and especially in neurotic cases, the wet-pack may be used with great advantage. As M. Allen Starr has pointed out, a change of air, especially a change to the seashore, is often of the utmost service in obstinate cases.

In all cases of chorea reflex irritation—ocular defects, adherent prepuce, etc.—should be carefully sought for, and when found removed if possible. On account of the tendency of chorea to recur, children who have once suffered from the disease should not be overtaxed in school, and should be guarded, as far as possible, from attacks of rheumatism.

Medicinal Treatment.—Arsenic is by far the most useful remedy. The dose should be gradually increased until physiologic effects are produced. In a child of five or six years we may give 3 min. (0.2 c.c.) of Fowler's solution three times a day, and increase the dose by a minim (0.06 c.c.) a day until physiologic effects are seen. These are puffiness about the eyes, nausea, griping pains or diarrhea. In many cases from 10–15 min. three times a day can be taken before untoward symptoms develop. When these occur the drug should be suspended for two or three days and then resumed at a dose slightly less than the child was receiving when the phenomena of salivation appeared. Arsenic should always be given after meals, and well diluted. While taking large doses of arsenic the patient should be kept under close observation, as albuminuria, conjunctivitis, inflammatory diseases of the skin, or even multiple neuritis may develop independently of the usual untoward effects.

In the cases in which arsenic fails or is not well borne, cimicifuga may succeed. A dose of 10 min. (0.6 c.c.) of good fluid extract may be given three times after meals and gradually increased to $\frac{1}{2}$ dr. (2.0 c.c.) or more. Antipyrin, in doses of from 30–60 gr. (2.0–4.0 gm.) a day, has been strongly advocated by Dujarden-Beaumetz, Legroux, Starr, Comby, and others, but when used so freely this drug may cause prostration, acute anemia, cyanosis, and hemoglobinuria. When symptoms of acute rheumatism are present salicylates are of service, otherwise they usually fail. When the movements are very violent and interfere with sleep, hyoscin hypodermically in doses of from $\frac{1}{200}-\frac{1}{100}$ of a gr. (0.0003–0.0006 gm.) twice daily, often acts most happily. If hyoscin fails recourse must be had to chloral, bromids, or morphin, but these drugs should not be used unless the symptoms are of great severity. Chloralamid and trional are reliable hypnotics.

Iron is often required to combat the marked anemia which is present in many cases.

Chorea Insaniens.—Powerful sedatives like hyoscin, chloral, and morphin are required to allay the violent excitability and jactitation. Inhalations of chloroform are sometimes useful. Stimulants are almost always required. When the patient is unable to swallow no time should be lost in resorting to forced feeding. Severe cases of chorea complicating pregnancy will call for the induction of premature labor.

EPILEPSY.

Although in the vast majority of cases epilepsy is an incurable disease, yet much can often be done to lessen the frequency of the paroxysms. Hygienic treatment is of the utmost importance. Moderate exercise, both mental and physical, is beneficial. Idleness and seclusion have a baneful effect. Children should be kept at school, but competition and overstudy should be avoided. Home training must be carried on with the greatest care, much tact and firmness being required to prevent loss of self-control. Children who are particularly irascible are often much better trained in a special institution. Adults should follow, if possible, some light and agreeable pursuit, preferably one which will permit them to spend the greater part of the day in the open air, and which will not add to the risk of physical injury should attacks come on without warning. The establishment of so-called epileptic colonies or farms, where patients can be employed in agricultural pursuits, has proved a great boon to many epileptics of the bread-winning class. In such institutions as are to be found at Biellefeld, Potsdam, and Stettin in Germany, at Zürich in Switzerland, at Chalfont St. Peter in England, and at Craig Colony in New York, large numbers of sane epileptics are able to support themselves under proper care and supervision.

The diet should be nutritious, but simple and readily digestible. As a rule a diet that is for the most part vegetable will be found to be best adapted to the patient's condition, but when the disease is associated with lowered vitality a fair amount of animal food should be permitted. Overloading of the stomach is a potent factor in precipitating attacks. The principal meal is best taken at midday, and full evening meals should be avoided. The assertion made by Richet and Toulouse that it is advantageous to reduce the sodium chlorid taken with food to a minimum—that is, to $2\frac{1}{2}$–3 gm. a day—has been confirmed by several observers.

Tea, coffee, alcohol, and tobacco should be used very sparingly, if at all. The patient must be constantly warned against excesses of every kind. The digestive functions should be brought to the highest possible state of efficiency. The bowels must be regulated by diet, and, if necessary, by mild aperients. Liberal water-drinking, frequent bathing, followed by friction of the skin, light exercise in the open air, and other measures which favor elimination are to be recommended. General tonics, like iron, arsenic, and cod-liver oil, are sometimes required to combat anemia and malnutrition.

Although very few cases of epilepsy are purely reflex, local irritation—phimosis, adherent prepuce, worms, a foreign body in the nose or ear, and painful cicatrices—should be carefully sought for, and if found removed. Epileptic patients ought not to marry.

Special Treatment.—Among palliative remedies the bromids, originally recommended by Locock in 1853, hold the first place. The ammonium and sodium salts are generally preferable to the potassium salt, being less depressing and less irritating to the stomach. In cases in which the digestive tract is especially sensitive, or in which only a mild effect is desired, the strontium salt may be selected with great advantage. A

40

combination of bromids sometimes acts more satisfactorily than a single bromid. The amount of the drug required varies with the severity of the case and the susceptibility of the individual, and must be determined experimentally in each case. The requisite daily dose usually ranges between 1 and 2 drams (4.0–8.0 gm.). The addition of one or two drops of Fowler's solution to each dose of bromid is useful in preventing the occurrence of acne. When the attacks occur at irregular intervals the bromid is best given in equal doses thrice daily after meals. When, however, some periodicity is manifested it is advisable to administer the drug in relation to the time of the attacks. Thus in nocturnal epilepsy a single large dose at bedtime may suffice. Again, in women, when the seizures occur only at the menstrual periods, active medication may be restricted to the week preceding each period. When the convulsions occur at long intervals and show no tendency to increase in frequency, it is better to dispense with special medication entirely and to rely upon hygienic and dietetic measures to lessen the excitability of the nerve-centers. In every case it is of the utmost importance to limit the dose of the bromids to the smallest possible amount that will control the seizures. Relief that comes only with saturation is generally dearly purchased.

Mention has already been made of the fact that the withdrawal of salt from the diet renders the patient more susceptible to the action of the bromids. The theory is that the tissues deprived of sodium chlorid more readily take up the bromids owing to the lack of other halogen bases in the dietary. Certain drugs, not very effective in themselves, are of service when combined with the bromids in that they lessen the amount of the latter required to control the convulsions. Antipyrin is one of the best of these auxiliary remedies. A daily dose of 10 gr. (0.6 gm.) not infrequently doubles the power of the bromids. Such a combination as that mentioned on page 140 proves valuable in many cases. Antipyrin may also be used in those cases in which the bromids, even in small doses, are not well borne. Acetanilid in daily doses of from 8–10 gr. (0.5–0.6 gm.) is sometimes substituted for antipyrin. In nocturnal epilepsy chloretone (5 gr.—0.3 gm.) or sulphonal (5–10 gr.—0.3–0.6 gm.) is often a useful adjuvant to the bromids. On account of its tendency to accumulate in the system and cause chronic poisoning, sulphonal should not be used continuously for longer periods than a week or ten days. Horse nettle (*solanum carolinense*) is another remedy which appears to increase the activity of the bromids. From ½–1 fl. dr. (2.0–

4.0 c.c.) of the fluid extract may be given thrice daily. When the circulation is weak a combination of digitalis or of adonis vernalis with the bromids, as recommended by Bechterew, will be found efficacious. In *petit mal* nitroglycerin in doses of $\frac{1}{100}$–$\frac{1}{20}$ gr. (0.0006–0.003 gm.) is sometimes of more service than the bromids.

Flechsig's treatment, which consists in giving large doses of opium for six weeks and then the bromids for two weeks (see p. 91), has been employed with some success in obstinate cases. It is not tolerated, however, by many patients and its employment necessitates a supervision that can be furnished only in fully equipped hospitals. Moreover it is not without danger.

Surgical Treatment.—Many attempts have been made to combat epilepsy by surgical treatment. Recourse has been had to castration, circumcision, clitoridectomy, tracheotomy, ligation of the carotids, ligation of the vertebral arteries, removal of the superior cervical sympathetic ganglia, and excision of certain cortical areas. Fortunately most of these measures are purely of historical interest. Trephining offers some hope of relief in certain cases of focal epilepsy, although it has to its credit less than 4 per cent. of recoveries. Clark[1] thus summarizes the present status of trephining: 1. Idiopathic epileptics with typical *grand mal* seizures should never be trephined. 2. Idiopathic epileptics in whom the seizures are of the Jacksonian type should be trephined only when infantile cerebral palsies can be excluded, and when the family and personal degeneracy is at a minimum. If operation is determined upon in such cases a very thorough removal of the epileptogenic area should be made; even then but a fraction of 1 per cent. recover from their epilepsy. Traumatic epileptics may be trephined when the injury is definitely proved and stands in direct causal relation and has existed not more than two years. The prognosis will then largely rest upon the degree of the neurotic predisposition present. The earlier trephining is resorted to after convulsions begin, the better the prognosis.

Treatment of the Attack.—When an aura is perceived it is often possible to arrest the paroxysm by the inhalation of amyl nitrite. Patients may provide themselves with this drug in the form of pearls which may be crushed in the handkerchief. When the attack is preceded by a local spasm forcible extension of the part sometimes succeeds in aborting it. If a sensory aura is felt in a limb the part may be firmly grasped or encircled with a tight ligature. The patient himself often learns by experience some method by which he can suppress

[1] *Medical Record*, Jan. 12, 1901.

seizures of which there is due warning. During the attack there is little to be done beyond protecting the patient from injuring himself. The head should be slightly raised, the clothes loosened, and a piece of cork or firm rubber pushed between the teeth. If necessary inhalations of amyl nitrite or chloroform may be used. In the *status epilepticus* the most reliable measures are inhalations of chloroform, ether, or amyl nitrite, hypodermic injections of hyoscin ($\frac{1}{100}$ gr.—0.0006 gm.) or of morphin ($\frac{1}{4}$ gr.—0.016 gm.), enemas of chloral (20–30 gr.—1.3–2.0 gm.), and the hot bath.

NEURASTHENIA.

The treatment of neurasthenia must vary with the cause of the disease and the circumstances and idiosyncrasies of the patient. In every case an earnest effort should be made to determine the exciting cause and to remove it if possible. With this in mind the family history of the patient, his occupation, habits, and amusements, and the condition of his various organs must be carefully studied.

In the milder forms of the disease, especially when overwork has been the exciting factor, a month or two of rest with change of scene will often effect a cure. In such cases quiet travel, so planned that it will interest the patient without fatiguing him, is frequently attended with excellent results. A prolonged sea-voyage is sometimes very useful. In other cases the "wilderness cure" of S. Weir Mitchell may be recommended with advantage.

In the absence of any special gastric disturbance, the diet should be simple, readily digestible, and abundant. Tea, coffee, alcohol, and tobacco are better avoided. A tepid sponge bath in the morning, provided it be followed by a good reaction, is beneficial. The wet pack, sitz-bath, spinal douche, and Scottish douche are of service in individual cases.

The question of exercise is an important one. When the physical strength is fairly well maintained exercise in the open air, systematically carried out and cautiously increased, often yields admirable results, but the fact cannot be too strongly emphasized that any exertion pushed to the point of fatigue is distinctly harmful. In many cases it is advisable to substitute for active exercise passive movements and massage.

When there is marked muscular weakness as well as nervous inefficiency rest is imperative. This may be relative or absolute. In some cases the addition of from three to five hours to the time usually spent in bed, or a rest in bed of a few hours during the day, will suffice. When, however, the symptoms are severe it will be necessary for the patient to give up all work for a period of from six to twelve weeks. In such cases much good accrues from the " rest cure " introduced by S. Weir Mitchell. This treatment includes not only rest but also full feeding, isolation, and artificial exercise. Of course, the details of the treatment must vary in each case, and only the outlines can be given here.

Rest.—In bad cases, for the first two or three weeks at least, rest must be absolute, the patient not being allowed to feed himself nor to leave the bed to pass urine or to empty the bowels. As to this point Mitchell says: " In some instances I have not permitted the patient to turn over in bed without aid, and this I have done because sometimes I think no motion desirable and because sometimes the moral influence of repose is of use." As improvement becomes manifest by a gain in weight, in muscular strength, and in nervous energy, some relaxation is permissible, and the patient may be allowed to sit up in bed to take meals and to indulge for a short time each day in reading or simple games. After four or five weeks he may be permitted to sit up in a chair for five or ten minutes a day, the time being gradually lengthened until at the end of a week or ten days he is up for from four to five hours. Active exercise is now cautiously introduced, and soon he is allowed to go out for a short walk or a drive. Finally, it is desirable that he should spend a week or two at the sea-shore or in the country before returning to his home.

Feeding.—In most cases the diet at first should be restricted to milk. From 4–5 oz. (120.0–150 0 c.c.) should be given every two hours, and this amount gradually increased until at the end of a week or ten days from 8–10 oz. (235.0–300.0 c.c.) are given every two hours. Little by little solid food may now be added until at the end of. two or three weeks the patient is getting each day three full meals with from 3–4 pints (1.5–2.0 L.) of milk in the intervals. The solid food may include stale bread and butter, soft-boiled or poached eggs, thoroughly cooked cereals, oysters, sweet breads, broiled or roasted meats, green vegetables, cooked fruits, milk-puddings, and ice cream. The evening meal should be light.

Isolation.—This is an essential element in the treatment. No one should be permitted to see the patient except the medical attendant and the nurse. Even the writing and receiving of letters are to be forbidden.

The permanent return of the patient at the close of the treatment to his family and friends should be affected very gradually. Any infringement of these rules is almost sure to mar the success of the treatment.

Artificial Exercise.—This is supplied in the form of massage and electricity. Through these measures the good effects of active exercise can be secured in a measure without the expenditure of any energy on the part of the patient. Massage should not be practised until the second or third day of the treatment, and even then it should be introduced very gradually. At first the séances should not last longer than a few minutes, but ultimately they may be increased to an hour a day. It is very important, as Dercum has urged, that the massage be performed by the nurse instead of another person with whom the patient would have to become acquainted.

Electricity, according to Mitchell, is the least necessary part of the treatment. It is, however, a useful adjuvant. Like the massage it should be introduced very gradually, otherwise it is likely to excite the patient and so prove harmful. A slowly interrupted faradic current is generally preferred. This should be applied once a day to each group of muscles in such strength as to elicit slight contractions.

Success in the rest treatment will depend quite as much upon the way in which the various measures are applied as upon the measures themselves. The fact must not be lost sight of that suggestion and discipline play a conspicuous part in the treatment; hence the importance of having all the details systematically and strictly carried out. It is always advisable to furnish a program indicating exactly what shall be done at each hour of the day. It is absolutely necessary that the nurse chosen to conduct the treatment shall be not only skilful and robust but also discrete, tactful, and agreeable to the patient. Finally, the more thoroughly the physician is able to inspire confidence in the patient and to convince him that his disease is not an incurable one, the more likely is he to effect a cure.

Medicinal Treatment.—Drugs are of little value except in meeting underlying conditions and in combating special symptoms. When there is anemia, iron and arsenic will be found useful. Small doses of strychnin are sometimes beneficial, but more often the drug is useless or actually harmful. Indigestion may be sufficiently severe to demand a modification of the dietetic treatment and the use of special remedies (see p. 544). Such drugs as asafetida, valerian, and sumbul are often helpful in allaying nervous irritability. The following pill, introduced by Goodell, may often be given with advantage:

℞ Acidi arsenosi,   gr. ss (0.03 gm.);
     Ferri sulphatis exsiccati,
     Extracti sumbul,   āā gr. xx (1.3 gm.);
     Asafœtidæ,   gr. xl (2.6 gm.).—M.
Fiant pilulæ, No. xx.
Sig.—One three or four times a day.

Every effort should be made to secure sleep by general measures—tepid baths, wet-packs, and gentle massage—before resorting to drugs. If a somnifacient becomes absolutely necessary, sodium or ammonium bromid, chloralamid, or trional may be given. Chloral and morphin should be withheld on account of the grave danger of inducing a drug habit. Severe headache may call for an occasional dose of phenacetin or of a bromid. Constipation can usually be controlled by diet and abdominal massage, but in some cases mild laxatives, like cascara sagrada, sodium phosphate, or the combination of aloin, belladonna, and strychnin will be required.

HYSTERIA.

By proper mental, moral, and physical training much can be done to prevent the occurrence of hysteria in those who through inheritance are predisposed to the disease. Prophylactic treatment includes the inculcation of absolute obedience, self-restraint, and self-denial, a judicious education, suitable outdoor exercise, hygienic surroundings, temperate living, and the avoidance of all that tends to morbid emotionalism or sentimentalism.

In developed hysteria treatment must be directed both to the mind and the body, but especially to the former. To be successful the physician must be able to inspire absolute confidence and faith in the mind of the patient. She must be impressed repeatedly with the fact that her condition is a curable one, and that with her thorough co-operation restoration to health will certainly follow. To intimate that her symptoms are feigned or are wholly within her control is an egregious error. The physician's authority must be unquestioned and his instructions must be rigidly carried out. Want of firmness and of decision is a common cause of failure. Harsh measures are occasionally needed, but they should be adopted only after the most careful consideration. In many cases no method of treatment proves successful until the patient has been removed from her customary surroundings and separated from her sympathetic relatives and friends.

Suggestion is employed consciously or unconsciously in the treatment of hysteria by every successful physician. Without it many of the remedies recognized as efficacious become wholly impotent. The cures which have been ascribed to the application of magnets and of various metals (metallotherapy) to the surface of the body seem to have been solely the result of suggestion. Complete hypnotism is by no means so generally useful as continuous suggestion. Certain symptoms—paralysis, aphonia, blindness, contracture, and anesthesia—are sometimes removed by a single hypnotic séance, but on the whole the action of hypnotism is disappointing. Moreover, in the event of failure it seems capable of still further lowering the will-power and of increasing the emotional excitability.

The physical condition of the patient must not be neglected. The general measures referred to under neurasthenia (see p. 628), such as hydrotherapy, systematic exercise in the open air, massage, and electricity, are equally applicable here. In grave cases of hysteria the treatment associated with the name of S. Weir Mitchell (see p. 629) often yields admirable results, but it is not suitable in every case, and considerable judgment must be exercised in the selection of proper subjects for it. Concerning this point Playfair writes: " The worse the case is the more easy and certain is the cure; and the only disappointments I have had have been in dubious, half-and-half cases."

Apart from their moral effect, drugs have but little influence upon hysteria. They must often be used, however, to meet underlying conditions and to combat special symptoms. Iron and arsenic are useful when there is anemia. Antispasmodics, like valerian, sumbul, asafetida, and camphor, are serviceable in allaying abnormal nervous irritability. Occasionally more powerful sedatives, like the bromids, phenacetin, and chloralamid, may be demanded, but the continuous use of such remedies is always to be condemned. Such drugs as morphin, alcohol, and chloral are distinctly dangerous.

When hysteria is complicated by local disease special treatment will be required, but no operation should ever be performed for the relief of nervous symptoms unless there exists an actual organic lesion.

Treatment of Special Symptoms.—*Convulsions.*—Isolation of the patient is imperative. Firm pressure over the ovarian region is often successful. The affusion of cold water over the face is useful. Inhalations of amyl nitrite, or even of chloroform, may be employed if necessary. Strong faradic currents applied to the spine are occasionally efficacious. *Anesthesia* is best treated by electricity, especially by the faradic wire-brush. Static electricity, owing to the profound mental impression which it produces, is also useful. *Hyperesthesia* and *pain* often yield to the continuous or interrupted galvanic current. In *paralysis* the patient should be instructed how to regain by long-continued practise the use of the affected part. This process of re-education demands the exercise of great firmness and patience. Swedish movements, massage, and faradization are useful adjuvants. In *aphonia* the faradic current, applied by means of special electrodes, is the most reliable remedy. In *contractures* the most useful measures are massage, passive movements, and faradization. In obstinate cases it may be advisable to straighten the affected joint under an anesthetic and then apply a splint to the limb.

WEIGHTS AND MEASURES.

APOTHECARIES' WEIGHT.

Troy grains.		Scruples.		Drams.		Troy ounces.		Pound
gr. 20	=	℈ 1						
60	=	3	=	ʒ 1				
480	=	24	=	8	=	℥ 1		
5760	=	288	=	96	=	12	=	℔ 1

APOTHECARIES' (WINE) MEASURE.

Minims.		Fluidrams.		Fluidounces.		Pints.		Gallon.
♏ 60	=	fʒ 1						
480	=	8	=	f℥ 1				
7680	=	128	=	16	=	O 1		
61440	=	1024	=	128	=	8	=	C. 1

APPROXIMATE FLUID MEASURES.

Teaspoonful	=	fʒj.	Tablespoonful	=	f℥iv.
Dessertspoonful	=	fʒij.	Wineglassful	=	f℥ij.
		Teacupful	=	f℥iv.	

A *drop* is usually considered equivalent to a minim, but this is the case only with a drop of water under certain conditions. The size of the drop varies with the shape of the vessel from which the liquid is being dropped and with the character of the liquid. The broader and thicker the lip of the bottle the larger is the size of the drops. Again, the lighter the liquid and the greater its viscidity the larger is the size of the drops. Chloroform and bromin, being very heavy, each drop about 250 drops to the fluidram ; tinctures, spirits, and fluid extracts drop from 130 to 150 drops to the fluidram ; oils, except castor oil, from 105 to 140 drops to the fluidram ; syrup, from 45 to 110 drops to the fluidram ; and waters and solutions, from 60 to 90 drops to the fluidram.

METRIC WEIGHTS.

1 myriagram (Mg.)	=	10,000 grams.
1 kilogram (Kg.)	=	1000 grams.
1 hectogram (Hg.)	=	100 grams.
1 decagram (Dg.)	=	10 grams.
1 gram (Gm.)	=	weight of 1 cubic centimeter of water at 4° C.
1 decigram (dg.)	=	tenth part of 1 gram, or 0.1 gram.
1 centigram (Cg.)	=	hundredth part of 1 gram, or 0.01 gram.
1 milligram (mg.)	=	thousandth part of 1 gram, or 0.001 gram.

METRIC MEASURES OF CAPACITY.

1 myrialiter (Ml. 1)	=	10 cubic meters, or the measure of 10 milliliters of water.
1 kiloliter (Kl. 1)	=	1 cubic meter, or the measure of 1 milliliter of water.
1 hectoliter (Hl. 1)	=	100 cubic decimeters, or the measure of 1 quintal of water.
1 decaliter (Dl. 1)	=	10 cubic decimeters, or the measure of 1 myriagram of water.
1 liter (L.)	=	1 cubic decimeter, or the measure of 1 kilogram of water.
1 deciliter (dl. or L. .1)	=	100 cubic centimeters, or the measure of 1 hectogram of water.
1 centiliter (cl. or L. 01)	=	10 cubic centimeters, or the measure of 1 decagram of water.
1 milliliter (ml. or L. 001)	=	1 cubic centimeter (c.c.), or the measure of 1 gram of water.

RULES FOR CONVERTING DRAMS AND GRAINS INTO THEIR METRIC EQUIVALENTS.

To convert drams into grams, *multiply* the number of drams by 3.9 grams, which is the number of grams in 1 dram. To convert grains into the corresponding metric quantity, *multiply* the number of grains by .065, which is the metric equivalent of 1 grain.

RULES FOR CONVERTING METRIC QUANTITIES INTO DRAMS AND GRAINS.

To convert grams into drams, *divide* the number of grams by 3.9. To convert grams into grains, *divide* the number of grams by .065.

EQUIVALENTS OF APOTHECARIES' IN METRIC WEIGHTS.

Grain.		Gram.	Grain.		Gram.
$\frac{1}{200}$	=	.000324	$\frac{1}{2}$	=	.0324
$\frac{1}{150}$	=	.00043	1	=	.0648
$\frac{1}{100}$	=	.00064	2	=	.1296
$\frac{1}{75}$	=	.00086	5	=	.3240
$\frac{1}{50}$	=	.00129	8	=	.5184
$\frac{1}{40}$	=	.00162	10	=	.6480
$\frac{1}{20}$	=	.00324	15	=	.9720
$\frac{1}{10}$	=	.00648	20	=	1.2960
$\frac{1}{6}$	=	.0108	30	=	1.9440
$\frac{1}{4}$	=	.0162	40	=	2.5920
$\frac{1}{3}$	=	.0216	60	=	3.888

EQUIVALENTS OF APOTHECARIES' IN METRIC MEASURES.

Minims.		Cubic centimeters.	Fluidounces.		Cubic centimeters
1	=	0.061	1	=	29.57
2	=	0.123	2	=	59 14
3	=	0.185	3	=	89.00
5	=	0.308	4	=	118.29
7	=	0.431	6	=	177.42
10	=	0.616	10	=	295.73
15	=	0.924	12	=	355.00
20	=	1.23	16	=	473.17
30	=	1.84	20	=	591.50
40	=	2.46	24	=	710 00
60	=	3.7	32	=	946.35

PERCENTAGE IN SOLUTIONS.

To estimate the quantity of a drug required to make a fluidounce or pint of a solution of a given percentage, *multiply* the weight of a fluidounce or a pint of the liquid to be used as the solvent by the percentage. Thus, in an ounce of a 10 per cent. *aqueous* solution of silver nitrate there are 455.7 grains.

455.7 grains (weight of a fluidounce of water) × 0.1 (percentage) = 45.5 grains.

In a pint of the same solution there are 729.1 grains.

7291 grains (weight of a pint of water) × 0.1 (percentage) = 729 1.

If a pint of a 1 : 1000, 1 : 100, or 1 : 40 solution be required, *divide* 7291 by 1000, 100, or 40, as the case may be, to determine the quantity of the drug. Thus, in a pint of corrosive sublimate solution (1 : 1000) there are 7.29 grains of the drug.

7291 grains ÷ 1000 = 7.29 grains.

To determine the weight of a fluidounce or pint of a liquid other than water, *multiply* 455.7 or 7291 by the specific gravity of the liquid, and the product will be the weight desired.

INDEX.

[To obviate the inconveniences attending the use of two separate indexes—an Index of Diseases and an Index of Remedies—the author has employed but one general index. The name of each disease is printed in **bold-faced type**, and under it are grouped the remedies employed in the disease.]

Abortion, threatened
opium, 90
viburnum prunifolium, 124
Abrin, 440
Abrus, 440
Abscesses
carbolic acid, 385
formalin, 399
hydrogen dioxid, 403
iodoform, 313
Acacia, 453
Aceta, 19
Acetanilidum, 343
Acetonemia. See also *Diabetes mellitus.*
hypodermoclysis, 488
infusion, 488
sodium bicarbonate, 165
Acetum, 444
Acid, acetic, 444
agaric, 258
arsenous, 299; as an antimalarial, 409; as a caustic, 450; as a poison, 300
benzoic, 397
boracic, 405
camphoric, 258
carbazotic, 389
carbolic, 383; as a local anesthetic, 151; as an anti-emetic, 163; as a poison, 384
cathartinic, 201
chromic, 445
cinnamic, 183, 271
citric, 462
gallic, 353
hydriodic, 310
hydrobromic, 142
hydrochloric, 175
lactic, 446
hydrocyanic, 68
muriatic, 175
nitric, 443
nitrohydrochloric, 443
osmic, 447

Acid, oxalic, 243
phosphoric, 294
picric, 389
pyrogallic, 446
quinic, 238
salicylic, 391
sozoiodolic, 387
sulphuric, 441
sulphurous, 402
tannic, 350
thymic, 388
trichloracetic, 444
Acidum aceticum, 444
arsenosum, 299: as a caustic, 450; as a poison, 300; as an antimalarial, 409
benzoicum, 397
boricum, 405
camphoricum, 258
carbolicum, 383; as a local anesthetic, 151; as an anti-emetic, 163; as a poison, 384
chromicum, 445
citricum, 462
gallicum, 353
hydriodicum, 310
hydrobromicum dilutum, 142
hydrochloricum, 175
hydrocyanicum, 68; as an anti-emetic, 163; as a circulatory depressant, 66; as a poison, 68
lacticum, 446
nitricum, 443
nitrohydrochloricum, 443
osmicum, 447
oxalicum, 243
phosphoricum, 294
picricum, 389
salicylicum, 391
sulphuricum, 441; as an antihydrotic, 258; as a poison, 442
sulphurosum, 402
tannicum, 350
trichloraceticum, 444

Made in the USA
Monee, IL
07 July 2026

56551413R00385